Solvent Extraction: Classical and Novel Approaches

Solvent Extraction: Classical and Novel Approaches

Vladimir S. Kislik
Casali Institute of Applied Chemistry
The Hebrew University of Jerusalem
Campus Givat Ram, Jerusalem 91904, Israel

ELSEVIER

AMSTERDAM • BOSTON • HEIDELBERG • LONDON • NEW YORK • OXFORD
PARIS • SAN DIEGO • SAN FRANCISCO • SINGAPORE • SYDNEY • TOKYO

Elsevier
The Boulevard, Langford Lane, Kidlington, Oxford OX5 1GB, UK
Radarweg 29, PO Box 211, 1000 AE Amsterdam, The Netherlands

First edition 2012

British Library Cataloguing in Publication Data
A catalogue record for this book is available from the British Library

Library of Congress Cataloging-in-Publication Data
A catalog record for this book is available from the Library of Congress

ISBN: 978-0-444-53778-2

For information on all Elsevier publications
visit our web site at elsevierdirect.com

Printed and bound in Great Britain

12 13 14 10 9 8 7 6 5 4 3 2 1

Contents

Preface

Solvent extraction (SX) or liquid-liquid distribution[1] techniques have a broad field of applications in inorganic and organic chemistry and large-scale industrial separations, in analytical chemistry, in pharmaceutical and biochemical industries, and in waste treatment. In addition, SX is a good instrument for studying fundamental understanding of equilibrium and kinetics of complex formation processes. Extraction methods have now become a routine procedure in separation technologies.

The literature volumes of publications on the subject, and the number of newly discovered extraction methods, increase every year. Many methods based on chelate extraction have been considered as routine practice. The hitherto unexplored potentialities are also very considerable. Extraction is a major tool in the study of the processes of complex formation and of the state of compounds in solution.

This monograph differs in scope and approach from the recent books by Rydberg et al. and Ritcey (Rydberg J, Cox M, Musikas C, Choppin GR, editors. *Solvent extraction: Principles and practice*. New York: CRC; 2004 and Ritcey GM. *Solvent extraction principles and applications to process metallurgy.* Elsevier; 2006, Amsterdam, The Netherlands) which give a detailed modern treatment of the extraction processes and is therefore an excellent reference works.

The book consists of three parts. The first presents the fundamental[2] principles of solvent extraction processes and practice during the last decade. In the second part novel principles of extraction processes interpretation are presented. Examples of carboxylic acids and metals solvent extraction interpreted using theses principles are presented and analyzed in comparison with conventional techniques. A more detailed treatment has been reserved for the subjects that have been studied by the author and his colleagues over the last decade. Some unpublished data have also been included. Special stress has been laid on the general theory of extraction, on extraction of coordination-unsaturated and charged complexes, and on the methods of determination of the composition of the extracted compounds. In studying the theoretical aspects of extraction, the chemist often has to deal with a number of interconnected problems. He has to consider the original form in which the species

1. IUPAC designation, but in this book the more commonly term "solvent extraction" is used.

2. In the book we use more often the nomination "classical" taking in mind modern fundamental principles of extraction.

exists in the aqueous solution, all the stages of the interaction with the extractants and the solvents, and the states of the extractant and the complex compound in aqueous and organic phases.

The Competitive complexation/solvation theory is unified one, developed for interpretation and quantification of interactions in solvent extraction systems and may be used at R&D of many new separation systems in different Industries. The theory overcomes some limitations of the classic ion-exchange models, permits to make some standardization of parameters and requires less experimentation at R&D.

In the third part of the book the modern and future trends in Solvent Extracion theoretical principles, new solvents and techniques development are discussed. Topics concerned with extraction, calculation, and analytical techniques are mentioned only at the description of novel theory. Specific subjects treated include the techniques of increasing the selectivity of the extraction, ways of effective combination of the extraction with subsequent determination methods, characteristic features of the extraction of this class of compounds, and radiochemical methods.

Introduction

The most common approach in explanation of extraction mechanisms is to perform equilibrium distribution studies, treated by Nernst's distribution law and supplemented by sets of mass action law equilibria.[1–4] As a rule, three categories of extraction are considered: non-specific solvation, specific solvation, and ionic interactions. The first two categories are based on the solution theories and on the strength of the solvation bonds. In many cases, the bonds are too weak for a specific number of solvating molecules per solute molecule to be identified. The significantly more strong donor (or acceptor) properties render the specific solvation and the number of solvating molecules per solute molecule may be experimentally determined. The third category of the extraction process is characterized by reversible ion-exchange reactions, by proton/electron transfer, or by ion pair structures formation. The readers can find detailed description in Refs. 1–11.

The complicated behavior of extraction systems conflicts, in some cases, with the mass action law models. Because of quite complex associative interactions in extraction systems, the theories developed for processes taking place in dilute solutions can not be extended on systems of high solute concentrations. A generalization that can be drawn from these studies is that the stoichiometry (and maybe the nature) of the solute–extractant interactions are changing with increasing solute concentration in organic phase. Most of the researchers modeled the process as a simple cation- or anion-exchange reaction with additional (over-stoichiometric) complexation at high solute concentrations. Some of the authors described over-stoichiometry through hydrogen bonding,[2–5] aggregation,[7,8,12] and even polymerization.[13] The readers can find various aspects of the stoichiometric ion-exchange theories in Refs. 1–20. In this book these theories are called as conventional theories in common and are presented in Chapters 1–5 of Part I.

The difference in the extraction trends must be attributed to solvation (coordination) effects. In order to analyze solvation effects, identification and, most important, quantification of solute–solvent interactions must be accomplished. Complexity of solvent structures makes this task extremely difficult.

The methodology of coordination chemistry in solution has been applied extensively to the determination of the equilibrium constants for solvent replacement in the coordination spheres of ions. Coordination models often result in good values for metal solvent extraction systems but within limited ranges of conditions. They meet difficulties when modeling anion exchangers, acid solvent extraction and, especially, at the attempts of quantification of the models.

The competitive complexation/solvation model, presented in Part II, is developed as an attempt to quantify solvent extraction systems in a broad range of solute concentrations. The theory is based on the competitive preferential solvation (COPS) theory[21] applicable also to the coordination models[22] and modified for extraction systems,[23,24] the concepts of Lewis acid–base, of hydrogen bonding and proton transfer,[2] of chemical quantum mechanics,[25–28] of amphoteric properties of solvents,[23] and of different aggregation structures formation at increasing solute concentrations in organic phase.[24]

In this book, the general statements of the competitive complexation/solvation theory, CCST, and the applications of the theory to solvent extraction of acids by amines and amine-based mixed extractants, of metal ions by acidic, basic, neutral, and mixed extractants are presented in Chapters 6–11 of Part II. Influence of active solvents as synergistic agents and of temperature is analyzed. Experimental arrangements for the novel techniques are introduced. Suitable techniques for experimental verification of the model are presented. The data available from the literature[3,5,7,29] and our experiments[30–34] are used for interpretation.

Modern developments (during the last two decades) in theoretical science and practice of solvent extraction–based techniques are presented in Chapters 12–14 of Part III. Potential of these techniques and future trends of the direction in common are also suggested.

REFERENCES

1. Lo TC, Hanson C, Baird MHI, editors. *Handbook of solvent extraction*. New York: Wiley; 1983.
2. Szafran M. Recent aspects of the proton transfer reaction in H-bonded complexes. *J Mol Struct* 1996;**381**:39–64.
3. Tamada J, King C. *Extraction of Carboxylic Acids by Amine Extractants. Ph.D. Thesis*. Lawrence Berkeley Laboratory, University of California, Applied Science Division; 1989.
4. Tamada J, Kertes A, King C. Extraction of carboxylic acids with amine extractant. 1. Equilibria and law of mass action modeling. *Ind Eng Chem Res* 1990;**29**:1319–26.
5. Eyal AM, Canari R. pH Dependence of carboxylic and mineral acid extraction by amine-based extractants: effects of pKa, amine basicity, and diluent properties. *Ind Eng Chem Res* 1995;**34**(5):1789–98.
6. Rydberg J, Cox M, Musikas C, Choppin GR, editors. *Solvent Extraction Principles and Practice*. 2nd ed. New York: CRC Press; 2004.
7. Neuman RD, Nai-Fu-Zhou, Wu J, Jons MA, Gaonkar AG. General model for aggregation of metal-extractant complexes in acidic organophosphorus solvent extraction systems. *Sep Sci Technol* 1990;**25**(13–5):1655–74.
8. Hogfeldt E, Fredlund F, Odberg L, Merenge G. Extraction of water, nitric and perchloric acids by nitrobenzene. *Acta Chem Scand* 1973;**18**(5):1781–94.
9. Yang S, White S, Hsu S. Extraction of carboxylic acids with tertiary and quaternary amines: effect of pH. *Eng Chem Res* 1991;**30**:1335–48.
10. Grinstead RR. Base strengths of amine in liquid-liquid extraction systems. *Proceedings of ISEC'67* 1968;**1**:427–36.

11. Marcus Y, Kertes AS. *Ion exchange and solvent extraction of metal complexes*. London: Wiley-Interscience; 1969.
12. Komasawa I, Otake T, Hattori I. The effects of diluent in the liquid-liquid extraction of copper and nickel using 2-hydroxy-5-nonylbenzophenone oxime. *J Chem Eng Japan* 1983;**16**(5):377–88.
13. Kolarik Z, Grimm R. Acidic organophosphorus extractants. XXIII. Distribution of mono-2-ethylhexyl phosphoric acid in two liquid phase systems. *J Inorg Nucl Chem* 1976;**38**(5):1761–3.
14. Sekine T, Hasegawa Y. *Solvent extraction chemistry*. New York: Marcel Dekker Inc; 1977.
15. Tamada J, Kertes A, King C. Extraction of carboxylic acids with amine extractant. 2. Chemical interaction and interpretation of data. *Ind Eng Chem Res* 1990;**29**:1327–32.
16. Tamada J, Kertes A, King C. Extraction of carboxylic acids with amine extractant. 3. Effect of temperature, water coextraction and process considerations. *Ind Eng Chem Res* 1990;**29**:1333–41.
17. Canari R, Hazan B, Eyal A. Carboxylic acids extraction by basic extractants: the effects of pH, ion concentration and acid-base property of the extractant and extracted acid. *Proceedings of ASEC'96* 1996;**2**:1517–22.
18. Vieux AS. Extraction of some dicarboxylic acids by tri-isooctylamine. *Anal Chim Acta* 1974;**68**:415–24.
19. Wardel JM, King CJ. Solvent equilibria for extraction of carboxylic acids from water. *J Chem Eng Data* 1979;**23**(2):144–8.
20. Wen Mei, Yang Yiyan, Dai Youyuan. Studies on the mechanism of extraction of carboxylic acids with tri-N-octylamine by chemical complexation. *Proceedings ISEC'96* 1996;**1**: 195–200.
21. Nagy OB, Muanda H, Nagy JB. Competitive preferential solvation theory of weak molecular interactions. *J Chem Soc Faraday Trans 1* 1978;**74**:2210–29.
22. Szpakowska M, Nagy OB. Application of the preferential solvation theory to coordinative solute-solvent interactions. *J Chem Soc Faraday Trans 1* 1989;**85**(9):2891–9.
23. Kislik V. Competitive complexation/solvation theory of solvent extraction. II. Solvent extraction of metals by acidic extractants. *Sep Sci Technol* 2002;**37**(11):2623–57.
24. Kislik V. Application of preferential solvation concept for interpretation of the mechanisms of acids solvent extraction by amines. *Sep Sci Technol* 2001;**36**(13):2899–926.
25. Pearson RG, Vogelsong DC. Developments in crystal field theory. *J Am Chem Soc* 1958;**80**:1038.
26. Pearson RG. Hard and soft acids and bases. *J Am Chem Soc* 1963;**85**:3533.
27. Luder WF, Zuffanti S. *The electronic theory of acids and bases*. New York: Wiley; 1946.
28. Douglas B, et al. *Concepts and models of inorganic chemistry*. 3rd ed. New York: John Wiley & Sons; 1994. p. 309–49.
29. Neuman RD, Jons MA, Zhou NF. Photo-correlation spectroscopy applied to hydrometallurgical solvent-extraction systems. *Colloids Surf* 1990;**46**(1):45–61.
30. Kislik V, Eyal A. Acidity dependence of Ti(IV) extraction: a critical analysis. *Solvent Extr Ion Exch* 1993;**11**(2):258.
31. Kislik V, Eyal A. Extraction of titanium(IV) by mixtures of mono- and d-(2-ethylhexyl) phosphoric acids esters. *Solvent Extr Ion Exch* 1993;**11**(2):285–310.
32. Kislik V, Eyal A. Mechanisms of titanium (IV) and some other transition metals extraction by organophosphorus acids. *Proceedings of ISEC'93* 1993;**2**:1353–60.
33. Kislik V, Eyal A. Hybrid liquid membrane (HLM) and supported liquid membrane (SLM) based transport of titanium (IV). *J Membr Sci* 1996;**111**:273–81.
34. Kislik V, Eyal A. *Unpublished experimental data*; 1998.

Part I

Conventional (Classical) Principles and Practice of Solvent Extraction

Chapter 1

Modern (Classical) Fundamental Principles of Solvent Extraction

Chapter Outline

Solvent Extraction. DOI: 10.1016/B978-0-4445-3778-2.10001-7

1. INTRODUCTION

Solvent extraction (SX) is one of the favored separation techniques because of its simplicity, speed, and wide scope. By utilizing relatively simple equipment and requiring a few time to perform, extraction procedures offer much to the chemists and engineers. Frequently it appears to be the ideal method of separating trace constituents from large amounts of other substances.

Solution diffusion (with or without chemical reactions) is a commonly accepted mechanism for SX. The process is often very selective and the isolation of the solute of interest can usually be made as complete as desired by several repetitions of the extraction procedure.

Even at extraction without chemical reactions at partition of solute neutral molecules between two immiscible phases, there is a chemical change of the solute in its solvation environment. More drastic chemical changes of the solute species take place with the presence of extractant, when different chemical interactions (reversible or irreversible), formation of new coordination compound, dissociation or association, and aggregation are possible. This is facilitated, enhanced SX.

Kinetics of the SX is a function of the rates of chemical changes that occur in the system and the rates of diffusion of the various species that control the chemical reactions. At least one of the chemical or diffusion steps is slow enough to control the overall rate of the process. So, analysis of mechanisms and kinetics of the chemical and diffusion steps of the overall process are needed.

In chemical technology, the SX plays an important role in the purification of chemical reagents and semiconductor materials. This method is also widely used in nuclear chemistry and technology for the separation of various radioisotopes and for the reprocessing of nuclear fuels.

Using SX, important theoretical problems concerning the composition and stability of soluble as well as insoluble complexes can be solved. This quality is very useful especially for the case when it is not possible to obtain reliable results by other methods.

In this chapter, modern theoretical considerations in SX techniques are presented and are reviewed with a classification and grouping. Better understanding of the principles of SX allows the optimization of solute separations by SX technique. Factors that influence the effectiveness and selectivity of separation are analyzed.

2. SOLVENT EXTRACTION BY SOLVATION

2.1. Solubility of Liquids

SX is the distribution of a solute or solutes between two immiscible liquids or phases. SX commonly takes place with aqueous and organic solutions. SX is generally carried out at ambient pressures and temperatures. Sometimes high pressures are used and liquid state can extend up to critical temperature. Organic phase usually consists of extractant dissolved in diluent. Normally, the organic phase contains a diluent, an extractant, but in some cases, a modifier, and a synergistic agent. The modifier is added to improve the physical properties of the system (e.g., to cause phase disengagement after mixing the two phases); synergistic agent is used for improvement and enhancement of extractant. The aqueous phase, or feed phase, consists of solute or mixture of solutes, contamination compounds, and acids dissolved in water. As an example, solubilities of water in some organic solvents and these solvents in water are presented in Table 1.1.

Liquids have some properties that are important in SX techniques. These are viscosity, surface tension, vapor pressure, density, polarity (electronegativity, dipole moment), and polarization ability.

Many liquids used in SX are polar. Polarity is determined by electric dipole in molecule, since their atoms have different electronegativities. The value of difference is characterized by dipole moment, δ_d, e.g., dipole moment of nitrobenzene, $\delta_d = 4.0$ D (1 Debye unit $= 3.34 * 10^{-30}$ C $*$ m) and water, $\delta_d = 1.83$ D.

Liquid compounds that have not permanent dipole moment are called nonpolar. These are many of hydrocarbons. If this liquid is placed in electric field, the electrons of atoms respond to the external field by atomic polarization which is called relative permittivity and determined by dielectric constant, ε. The forces between charges, such as ions, of opposite signs are inversely proportional to the relative permittivity of the liquid. So the dissociation of molecules into ions strongly depends on the relative permittivity of the solvent used for electrolyte.

Forces that keep the molecules of liquids together and are responsible for the above-mentioned properties are called as cohesive forces: dispersion (or London) forces, Van der Waals forces, dipole forces, and hydrogen bonding. Parameter that is called square root of the cohesive energy density, δ^2, connects the molar heat of vaporization, ΔH^V, with the liquid volume, V, and temperature, T:

$$\delta^2 = \frac{\Delta H^V - RT}{V} \tag{1.1}$$

The δ parameter is nominated as solubility parameter of the liquid.[1]

2.2. Mixtures of Solutions

The general behavior of the nature, the phase rule of Gibbs says:

$$p + f = n + 2 \tag{1.2}$$

where p is the number of phases, f is the number of degrees of freedom, and n is the number of components.

TABLE 1.1 Solubility of Water in Organic Solvents and These Solvents in Water at 25–30 °C

Solvent	Solvent in water (% wt)	Water in solvent (% wt)	Solvent	Solvent in water (% wt)	Water in solvent (% wt)
n-Hexane	0.00123	0.0111[a]	Benzene	0.179	0.0334
n-Octane	6.6×10^{-7}	0.0095[a]	Toluene	0.0515	0.043
n-Decane	5.2×10^{-8}	7.2×10^{-5}	*p*-Xylene	0.0152	0.0456
n-Dodecane	3.7×10^{-9}	6.5×10^{-5}	Dichloromethane	1.30	0.093
Chloroform	0.815[a]	0.0135[b]	Trichloroethylene	0.137	0.0327
Carbon tetrachloride	0.077	0.096	Chlorobenzene	0.0488[b]	0.309
1,1-Dichloroethane	5.03[a]	0.187	1,2-Dichlorobenzene	0.0156	0.0142
1,2-Dichloroethane	0.81[a]	0.32	1-Hexanol	0.706	7.42
l-Butanol	7.45	16.9	1-Octanol	0.0538	
Isobutyl alcohol	10	7.46	2-Ethyl-1-hexanol	0.07	2.6
1-Pentanol	2.19	9.19	Phenol	8.66	28.72
Isoamyl alcohol	2.97	7.42	Methyl ethyl ketone	24.0	10.0
Diethyl ether	6.04	1.47	Methyl isobutyl ketone	1.7	1.9
Diisopropyl ether	1.2	0.57	Propylene carbonate	17.5	8.3
Bis(2-chloroethyl ether)	1.02	0.1	Tri-*n*-butyl phosphate	0.039	4.67
Benzonitrile	0.2	1	Nitrobenzene	0.19	0.24
Cyclohexanone	2.3	8.0	Acetylacetone	16.6	4.5
Butyl acetate	0.68	1.2	Di(2-ethyl hexyl) ether		

[a]*At 25 °C.*
[b]*At 30 °C.*

A system consisting of two immiscible solvents and one solute distributed between them has one degree of freedom at constant temperature and pressure. The solubility of each component at equilibrium or miscibility of the solutions is the main question for two-component, two-phase system in the extraction practice.

2.2.1. *Ideal Mixtures*

Two liquids very similar in properties such as water and ethanol or benzene and toluene are mixed practically completely. It means that energy or heat of self-interactions between molecules of every liquid is similar to that of interaction between molecules of A and B liquids. In this case the entropy of mixing is maximal and the mixture of such liquids is called ideal mixture. Chemical potential of A in the ideal mixture is

$$\mu_A = \mu_A^o + RT \ln \chi_A \tag{1.3}$$

where μ_A^o is the molar chemical potential of pure A and χ_A is the molar fraction of A in the mixture. A similar equation holds for the B in the mixture. The ideal mixtures obey Raoult's and Henry's laws.

2.2.2. *Real Mixtures*

Real (not very diluted) mixture systems have deviations from Raoult's and Henry's laws. Chemical potentials of such mixtures related to activities (thermodynamical concentrations) of the substance in the mixture but not their concentrations:

$$\mu_A = \mu_A^o + RT \ln \gamma_A \tag{1.4}$$

where γ_A is the activity of the substance A in the mixture.

Activity coefficients may be smaller or larger than unity; therefore, deviations from the Raoult's and Henry's laws may be negative or positive.

Distribution ratio remains constant only if the ratio of activity coefficients is independent of the total concentration of A (or B) in the mixture. This holds (approximately) in dilute solutions. At the moderate concentrations, which is particularly important in industrial SX, the deviation from the Nernst's distribution law leads to the considerable errors in preliminary predictions at new SX processes development.

2.3. Solute–Solvent Interactions (Extractability)

Distribution of a solute between phases takes place according to its solubility in every phase. The system comes to mass transfer equilibrium. The distribution coefficient (or constant at equilibrium), K_D, of a solute is approximately equal to the ratio of its solubilities in the organic and aqueous phases. Thus only such compounds can be usefully extracted as are only slightly soluble in water but are readily soluble in organic solvents.

2.3.1. The Solubility Parameters

Solubility parameter of the liquid,[1] δ, connects cohesive forces with physico-chemical potentials ΔG, ΔH of the solutions (see Eqn (1.1)). This relation holds for ideal and regular solutions that show ideal entropy effects in mixing solute and solvent, and no interactions occur besides the cohesive forces between the solute and solvent molecules. Regular solutions exhibit heat changes when mixed, while for ideal solutions, the heat of mixing is equal to zero. There is no change of state in association or in orientation.

The works correlating various types of cohesive forces with the Hansen solubility parameters[2,3] are summarized by relation:

$$\delta^2_{\text{total}} = \delta^2_{\text{dispersion}} + \delta^2_{\text{polar}} + \delta^2_{\text{hydrogen}} \tag{1.5}$$

Here, interactions between the solute and the solvent are described by contributions from the various types of cohesive forces. In general, the dispersion (or London) forces dominate. In Fig. 1.1 the measured distribution ratios of an americium complex between an aqueous nitrate solution and various organic solvent combinations are compared with distribution ratios calculated from tabulated Hansen parameter values.[4]

The distribution constant for the extractant, E, can be expressed as

$$RT \ln K_{\text{dE}} = V_{\text{E}}\left[(\delta_{\text{aq}} - \delta_{\text{E}})^2 - (\delta_{\text{org}} - \delta_{\text{aq}})^2\right] + RT\, V_{\text{E}}\left(\frac{1}{V_{\text{org}}} - \frac{1}{V_{\text{aq}}}\right) \tag{1.6}$$

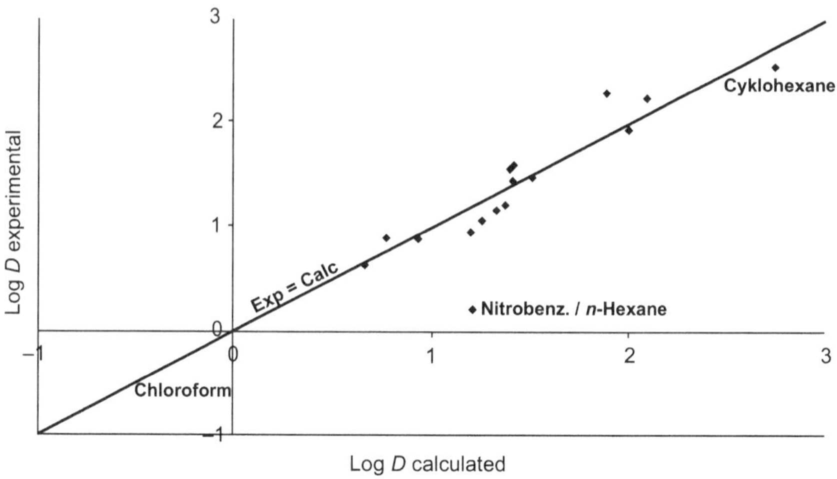

FIGURE 1.1 Comparison of measured and calculated distribution ratios D_{Am} of americium(III)–terpyridine–decanoic acid complexes between 0.05 M HNO_3 and various organic solvent combinations. The calculated values are obtained with the Hansen partial solubility parameters. *(Source: From Ref. 4 with permission)*

where K_{dE} is distribution constant of extractant given in terms of mole fractions; molar volumes V_E and solubility parameters have to be empirically estimated in each set of experiments, assuming constant values for the aqueous systems in the two different two-phase systems.

The overall enthalpy and entropy changes for the distribution reaction (i.e., transfer of the complex from the aqueous to the organic phase) can be obtained from the temperature dependence of K_{dE} according to

$$RT \ln K_{dE} = -\Delta H + T\Delta S \tag{1.7}$$

From Eqn (1.6) an enthalpy and entropy formally can be obtained as

$$\Delta H = -V_E\left[(\delta_{aq} - \delta_E)^2 - (\delta_{org} - \delta_{aq})^2\right] \tag{1.8}$$

and

$$\Delta S = RV_E\left(\frac{1}{V_{org}} - \frac{1}{V_{aq}}\right) \tag{1.9}$$

If the distribution constant for a standard system is used as a reference, changes in enthalpy and entropy relative to this standard system can be assessed by replacing δ_{aq} and V_{aq} with the corresponding data for the organic diluent in the reference system.

Rearrangement of Eqn (1.6) shows that $\log K_{DE}/(\delta_{aq} - \delta_{org})$ vs. δ_E, where

$$\delta_E = \delta_{org}\frac{RT(V_{org}^{-1} - V_{aq}^{-1})}{(\delta_{aq} - \delta_{org})} \tag{1.10}$$

should yield a straight line with slope $V_E/RT \ln 10$. This relation demonstrates a satisfactory correlation between distribution constants and solvent parameters, indicating the usefulness of the solubility parameter concept.

2.3.2. Thermodynamic Relations

If concentration of the solute in one phase is constant, the concentration of the solute in the other phase is also fixed. The relationship between concentration of solute in each of the phases led to Nernst's formulation of the distribution law.[5] Distribution constant, K_d (sometimes designated as partition constant, P):

$$K_d = \frac{\text{Solute species : conc. org. phase}}{\text{Solute species : conc. aq. phase}} \tag{1.11}$$

Distribution constant has to be distinguished from the distribution ratio, D. When solute S is present in various differently complexed forms in the aqueous phase and in the organic phase, $[S]_{tot}$ refers to the sum of the concentrations of all S species in a given phase. It is important to distinguish between the distribution constant, K_d, which is valid only for a single specified species, and the distribution ratio, D_S, which may involve sums of species and thus is not constant:

$$D_S = \frac{\text{Conc. of all species of } S \text{ in org. phase}}{\text{Conc. of all species of } S \text{ in aq. phase}} = \frac{[S]_{\text{tot.org}}}{[S]_{\text{tot.aq}}} \tag{1.12}$$

Often in practice, a distribution ratio is less informative than one of percentage extraction S, $\%E_S$:

$$\%E_S = \frac{100D_S}{(1 + D_S)} \tag{1.13}$$

Extraction percentage curves are particularly useful for designing separation schemes. Often these curves are not linear but S shaped. Figure 1.2 illustrates S-shaped curves where distribution (or extraction) depends on some of variables (pH, free ligand concentration, pK_a, etc.). Sometimes, the authors use the log variable value of 50% extraction, e.g., $\log[Cl^-]_{50}$. The pH_{50} value indicates $-\log[H^+]$ for 50% extraction.

In many extraction systems, some or all of the solvated water molecules must be removed to obtain a species extractable by organic solvents. These species must, however, be uncharged because of the low dielectric constants of the organic solvents that are generally used in extraction procedures.

Organic reagents having one anionic group (OH^-), (SH^-), etc. and one uncharged basic group can easily replace water molecules forming neutral, essentially covalent chelate compounds.

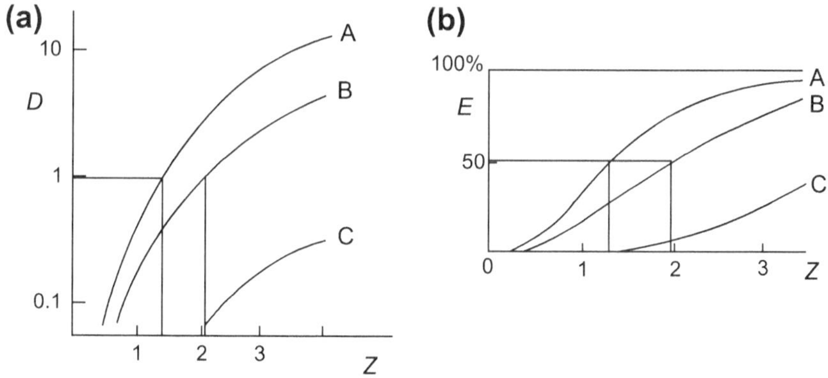

FIGURE 1.2 Solvent extraction plots. (a) The distribution ratio D_S for three different solutes A, B, and C plotted against the variable Z of the aqueous phase. Z may represent pH, concentration of extractant ($[HL]_{org}$) in organic phase, free ligand ion concentration in the aqueous phase ($[A_{aq}]$), salt concentration in aqueous phase, etc. (b) Same systems showing percentage extraction $\%E$ as a function of Z. D_S and Z are usually plotted on logarithmic scale.

3. SOLVENT EXTRACTION WITH CHEMICAL REACTIONS (BY COMPLEXATION)

Species separated by SX may be organic as well as inorganic. Organic compounds are usually easily dissolved in organic solvents (lipophilic and hydrophobic). Inorganic substances are opposite, particularly metal salts (lipophobic and hydrophilic). So, for extraction of inorganic substances, they must be changed to lipophilic by reaction with organic compound extractants (also extracting agents, reagents). This process is described as reactive extraction or extraction with complexation.

3.1. Thermodynamics of Extraction Systems

Extraction from aqueous into organic solutions can be achieved by solvation and through different chemical reactions. As a rule, they (chemical reactions) are very complicated and are usually presented through a number of simplified steps. The subdivision of an extraction reaction into simple steps is useful for understanding the dependence of distribution ratio values on components of the extraction system variations. These models allow to determine some of thermodynamic equilibrium constants.

Solutes may be both inorganic and organic nonelectrolytes and electrolytes (e.g., metal–organic complexes, metal ions soluble in organic solvents through reactions with organic components). Reactions are considered in the aqueous phase, in the organic phase, or at the interface between the phases. The system comes to equilibrium state. Let us illustrate some subprocesses of the extraction process with an example of titanium(IV) extraction from the strong acidic (>7 mol/kg HCl) aqueous solution by di(2-ethylhexyl)phosphoric acid (DEHPA) in benzene.[6] Overall extraction reaction is

$$[TiCl_4]_{aq} + 4HOPO(OR)_{2\,org} \rightarrow TiO_4[PO(OR)_2]_4 * 2HCl_{org} + 2HCl_{aq} \qquad (1.14)$$

This extraction process may be described in four steps:

First step—complexation of titanium ion by HCl in aqueous phase

$$Ti^{4+} + 6HCl \leftrightarrow TiCl_4 * 2HCl + 4H^+ \qquad (1.15)$$

It is known that titanium central atom may be surrounded by hydrated water molecules (or HCl), but they are usually omitted for simplification, although water of hydration (or HCl of solvation) plays a significant role (see Part II).

Second step—interaction between titanium chloride complex and DEHPA on the phase interface with formation of Ti–DEHP organic complex

$$TiCl_4 * 2HCl_{aq} + 4HOPO(OR)_{2\,org} \leftrightarrow TiO_4[PO(OR)_2]_4 * 2HCl_{org} + 4HCl_{aq} \qquad (1.16)$$

Third step—distribution (mass transfer) from the interface of the organic complex to the bulk organic solution and hydrochloric acid to the bulk aqueous solution

It is possible to make more intermediate step reactions, but this is of little significance since the overall free energy change, ΔG^0_{ex}, depends only on the starting and final states of the system. If ΔG^0_i of the steps are known, the relation between ΔG^0_{ex} and K_{ex} may be given as

$$\Delta G^0_{ex} = \sum_{i=1}^{n} \Delta G^0_i = -\mathrm{RT} \log K_{ex} \tag{1.17}$$

K_{ex} may be determined experimentally and shows the efficiency of an extraction process. It depends on the "internal chemical parameters" of the system, i.e., the chemical reactions and the concentration of reactants of both phases. The latter determine the numerical value of the distribution ratio of the solute.

3.2. Solute–Solvent Interactions in Aqueous Phase

3.2.1. Dissociation

Acids (or salts) dissociate in the aqueous phase with a dissociation constant K_a:

$$\mathrm{HL} = \mathrm{H}^+ + \mathrm{L}^- \quad K_a = \frac{[\mathrm{H}^+][\mathrm{L}^-]}{[\mathrm{HL}]} \tag{1.18}$$

The distribution ratio incorporates the K_a for extraction of acids, HL, as:

$$D = \frac{[\mathrm{HL}]_{org}}{[\mathrm{H}^+][\mathrm{L}^-]} = K_{d\mathrm{HL}}\left(1 + \frac{K_a}{[\mathrm{H}^+]}\right) \tag{1.19}$$

Solubility in organic solvents is not a characteristic usually attributed to simple metal salts. As can be expected from their highly ionic nature, most metal salts are strong electrolytes: molecules dissociated to ions completely or partly. The principle of electroneutrality stands for all solutions.

$$\sum a * z^+ + \sum b * z^- = 0 \tag{1.20}$$

Ionic strength I is defined as summation over all ions present in the solution:

$$I = \frac{1}{2}\sum a_i * z_i^2 \tag{1.21}$$

The ions in the solution are likely to be solvated by the solvent or hydrated in the aqueous solutions. As a rule, electrolytes have large solubility in aqueous media.

Chemical potential may be formally considered as made by summation of individual ions.

$$\mu_{ab} = \mu^o_\pm + \mathrm{RT}\ \ln[S_\pm] + \mathrm{RT}\ \ln \gamma_\pm \tag{1.22}$$

Ionic activities and activity coefficients have a subscript $\pm$ for ions with different charges. The calculation of ionic activity coefficients of an electrolyte in solution is required for the estimation of SX systems. The osmotic coefficient of water activity is also important, related to ability of solutions to dissolve electrolytes and nonelectrolytes.

3.3. Solute–Solvent Interactions in Organic Phase

In organic solution, if its relative permittivity is $\varepsilon > 40$, electrolytes are more or less completely dissociate into ions. If $\varepsilon < 10$, dissociation is only slight or particularly does not take place.

At the transfer of electrolyte ab from aqueous to organic solution, the Gibbs free energy of transfer, ΔG_{abtr}, may be determined:

$$\Delta G^{\mathrm{o}}_{\mathrm{abtr}} = \mu^{\mathrm{o}}_{\mathrm{ab}}(\text{organic}) - \mu^{\mathrm{o}}_{\mathrm{ab}}(\text{aqueous}) \tag{1.23}$$

Activity coefficient can be defined as

$$\ln \gamma^{\mathrm{o}}_{\mathrm{abtr}} = \frac{\Delta G^{\mathrm{o}}_{\mathrm{abtr}}}{RT} \tag{1.24}$$

These equations hold at the standard state of infinite dilution when only solute–solvent interactions take place. At visible concentrations of solute, ion–ion interactions take place and the above dependences need strong corrections.

In weak polar or nonpolar organic solutions, at $\varepsilon < 40$, ion pairing with the solute is observed. It close but should be distinguished from the ionic complex formation. Contrary to metal ion–anionic ligand complexes that have coordinative bonds, the ion pair cannot be presented as geometrically oriented association, although it is presented as reaction with an equilibrium constant K_{ass}:

$$K_{\mathrm{ass}} = \frac{(1-\alpha)\gamma_n}{C_{\mathrm{ab}}\alpha^2\gamma_{\pm}^2} \tag{1.25}$$

where α is dissociated electrolyte fraction, $(1-\alpha)$ is associated fraction, γ_n is activity coefficient of the undissociated part, and $\gamma_{\pm}$ is the ionic activity coefficient of the dissociated part of electrolyte.

At relative permittivity of organic solvent less than 10, ionic dissociation can be completely neglected and solutes behave as if they are nonelectrolytes. As a rule these solutions have negligible electrical conductivity. Interactions between solute and solvent or solute–solute are taking place through solvation. These are donor–acceptor, hydrogen bond, dipole–dipole, and other cohesive forces interactions.

Dipole–dipole and hydrogen bond interactions can lead to solute aggregation. Linear (head to head or head to tail), chainlike, or cyclic aggregates can be formed.

One more phenomenon, formation of the second organic phase, often happens in SX techniques. When organic solution comes to saturation by solute, a new phase appears. It may be a liquid or a solid. One of the two organic phases is saturated by solute phase and the second is solvent-rich dilute solution of the solute. According to phase rule, there are no degrees of freedom in these compositions.

4. DRIVING FORCES OF SOLVENT EXTRACTION

Solutes have different strength of interactions with different solvents or different solubilities in different solvents. So, in the system of two immiscible or partly miscible solvents, different solvents become differently distributed between these two phases. This is a base of solvent extraction technique.

At the simple solute distribution (without any chemical interactions), the transport of the solute to and from the phase interface is proportional to the chemical potentials or to the concentration gradients of the solute in the phases and to the partition coefficient K_{dSF}: the constant giving the concentration ratio of the solute between both phases at equilibrium. It is called the (stoichiometric) distribution coefficient (constant) and it can be expressed by the following equation:

$$[C_{SF}]_{aq} \underset{k_{-1}}{\overset{k_1}{\longleftrightarrow}} [C_{SE}]_{org} \tag{1.26}$$

$$K_{dSF} = \frac{[C_{SE}]_{org}}{[C_{SF}]_{aq}} \tag{1.27}$$

Where $[C_{SE}]_{org}$ and $[C_{SF}]_{aq}$ denote the concentration of the distributed compound in the organic and aqueous phases, respectively, at equilibrium.

It follows from Eqn (1.25) that the distribution coefficient K_{dSF} does not depend on the total solute concentration or on the actual phase volumes. The distribution law is only valid if the solute is in both phases in the identical chemical form.

Equation (1.25) can also be derived thermodynamically from the equality of the chemical potentials of the solute in organic and aqueous phases at equilibrium:

$$\mu_{Sorg} = \mu_{Saq} \tag{1.28}$$

Substituting suitable expressions for μ_{Sorg} and μ_{Saq}, the following relation can be obtained:

$$\mu^0_{Sorg} + RT \ln C_{org} + RT \ln \gamma_{org} = \mu^0_{aq} + RT \ln C_{aq} \ln \gamma_{aq} \tag{1.29}$$

where $\mu^0_{S\ org}$ and $\mu^0_{S\ aq}$ represent the standard chemical potentials of the solute in the organic and aqueous phases and γ_{org} and γ_{aq} represent the activity coefficients in the organic and aqueous phases respectively.

From the above relation, we may obtain the expression for the distribution coefficient at equilibrium:

$$K_{dSF} = \frac{C_{org}}{C_{aq}} = \frac{\gamma_{aq}}{\gamma_{org}} \exp\left(-\left[\gamma^0_{org} - \gamma^0_{aq}\right]/RT\right) \tag{1.30}$$

When the solute concentration is low, the activity coefficients γ_{org} and γ_{aq} approach unity and the stoichiometric distribution coefficient K_{dSF} becomes constant because γ^0_{org} and γ^0_{aq} are also constant. Here, the extraction depends on the partition (distribution) coefficient K_{dSF} only.

When the organic phase contains an extractant that is able to form a complex with the solute:

$$S_{aq} \overset{k_1}{\longleftrightarrow} S_{org} + E \overset{k_2}{\longleftrightarrow} SE_{org} \overset{k_3}{\longleftrightarrow} (SE)_n \tag{1.31}$$

Considering here only first two steps (chemical reaction of complexation):

$$K_{FE} = \frac{[S_{SE}E]}{[S_{Sorg}][E]} \tag{1.32}$$

the forward extraction step becomes

$$K_{SFex} = K_{dSF}K_{FE} \tag{1.33}$$

where driving factor of extractant at complex formation D_{SFex}:

$$D_{SFex} = \frac{K_{SFex}[E]_0}{(1 + K_{SFex}[C_{SF}])(1 + K_{SFex}[C_{SE}])} \tag{1.34}$$

Now, let us consider the coupling effect, using the specific extraction reactions of metal ion, e.g., Cu^{2+} with acidic extractant HL_{org}:

$$Cu^{2+}_{aq} + 2HL_{org} \leftrightarrow CuL_{2\,org} + 2H^+ \tag{1.35}$$

and total driving force of extraction process:

$$K_{tot} = \frac{[CuL_2]_{org} * [H]^2_{aq}}{[Cu]_{aq} * [HL]^2_{org}} = K_{SFex} * \left(\frac{[H]_{aq}}{[HL]_{org}}\right)^2 = \frac{K_{SFex}}{K_{Scoup}} \tag{1.36}$$

where K_{Scoup} is the coupling coefficient (metal ion–proton) at forward extraction.

The process of solute extraction is driven by solute concentration gradient, partition (distribution) coefficient, and Donnan equilibrium coupling.

Driving force coefficients K_{SFex} and K_{bex} may be determined by membrane-based extraction experiments.[7]

Basic features of extractant-mediated transport are as follows: (1) the effect of extraction is proportional to the extractant concentration and (2) the extraction shows typical saturation behavior at low solute concentrations, and is close to be independent of $[C_S]$ at high solute concentrations. A very strong complexing extractant becomes fully loaded with solute even at low concentrations. This leads to the conclusion that the best extractant is not necessarily the strongest complexing one. Apart from this thermodynamic reason, there may also be a kinetic reason: strong complexing agents frequently show a slow rate.

5. INFLUENCE OF KINETICS FACTORS

Extraction kinetics is a function of both the kinetics of the various chemical reactions occurring in the system and the diffusion rate of the various species that control the chemistry.

Products of any extraction process are usually in a chemical state different from initial species. Even at simple partition of neutral molecules between two immiscible phases, there is a chemical change in solvation environment. Using extractants that chemically react with the solute, drastic chemical changes take place. They can be described by subsequent partitioning, interaction (complexation), and diffusion of the initial species from the bulk phase to the reaction interface and product (solute–extractant complexes) from the interface to the bulk phase. At first two processes, the solvated water molecules can be removed from the solute ion; the carrier molecule can undergo an acid dissociation reaction; a new compound, soluble in the organic phase, may be formed with chelating group of the extractant; extractant–solute complex can undergo changes in aggregation; and so on. Usually these interactions are going stepwise through several adducts, including dissociation, aggregation, etc. Every step has its own rate. At least one of the chemical steps of the overall reaction mechanisms or diffusion rate steps may be slow enough, and overall transport kinetics would depend on this slow rate. So, we have to consider both chemical kinetics and diffusion rate steps to determine rate-controlling step for the overall process.

Very few tools have been developed up-to-date to investigate chemical changes occurring at liquid–liquid interfaces, and our knowledge is still limited and is based on indirect experimental evidence and speculations.

5.1. Diffusion Transport Regime

Most extraction systems involve stirring of the phases to minimize the time for diffusion of dissolved species toward and away from the interface. It follows that transport of species from the bulk phase to a region very close to interface can be considered instantaneous and the diffusion in the bulk phase can be neglected. But even the most vigorously stirred systems possess two thin films at the aqueous–organic interface which are essentially stagnant. These films, often referred to as diffusion films, Nernst films, and diffusion layers, vary from

50 to 500 μm thick[8,9] and can be crossed only by diffusion processes. The two-film model is used here to describe the diffusional transport (see Fig. 1.3). The thickness of the diffusion films never goes down to zero. The time required for the diffusional crossing of the films by the solute may be longer or comparable with the time required for chemical reactions. So diffusion across the boundary layers may control the overall kinetics.

The basic idea of the chemical kinetics–diffusion method is that the solute presented in the different phases is considered as different chemical species obeying the laws of chemical kinetics. The general assumptions of the transport may be formulated as follows:

1. Steady-state conditions of the solute transport through the phase interfaces: all fluxes are necessarily the same. At thin diffusion films only one dimension can be considered.

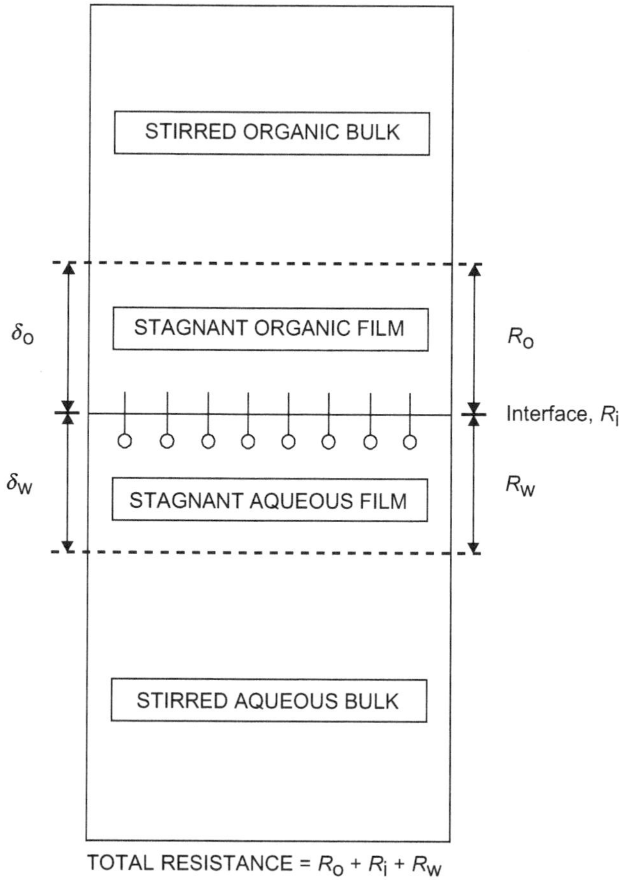

FIGURE 1.3 Interfacial diffusion films. δ_o and δ_w are the thicknesses of the organic and aqueous films, respectively. Adsorbed layer of extractant molecules at the interface is shown.

2. The overall mass transfer rate of species can be controlled by any of the chemical reaction–diffusion resistances.
3. The overall mass transfer resistance at steady state is the sum of individual mass transfer resistances at diffusional regime through the diffusion film and chemical reaction resistances at the phase interface.
4. Thermodynamic local equilibrium at the interface.

Experience has shown that the film diffusion is very often predominating rate-controlling factor in the SX practice. The barriers to transport imposed by the need for diffusion across diffusion layers can be minimized by decreasing film thickness or by increasing the mobility of the diffusible species. The thickness of the diffusional films depend on stirring rate of the phases and type of the stirrers used, on the geometry of the extraction equipment, and on the viscosity and density of both phases. The thickness, up to a certain limiting value, is inversely related to mechanical energy supplied (e.g., by stirring).[10] Viscosity and density of the liquids used as well as equipment geometry also affect film thickness but interfacial films apparently cannot be completely eliminated.

5.1.1. Mathematical description of the diffusion transport

The diffusion flux J_S (M, g/cm^2/s) of the specie is defined as the amount of matter passing perpendicularly through the unit area during the unit time. Different solutes have different solubilities and diffusion coefficients, D_S. The flux of a species S through a film thickness, h, is related to the concentration gradient through Fick's first law:

$$J_S = -D_S \frac{\partial C_S}{\partial h} \tag{1.37}$$

High fluxes can be obtained when a large chemical potential (concentration gradient) is maintained over a thin film in which the diffusivity, D_S, of the species is high (as a rule, 10^{-5}–10^{-6} cm^2/s). This diffusion is expressed by Fick's second law:

$$\frac{\partial C}{\partial t} = D_S \frac{\partial^2 C}{\partial h^2} \tag{1.38}$$

For steady-state diffusion occurring across thin films, only one dimension can be considered and solute flux, J_S, can be presented as (see Fig. 1.3)

$$J_S = -k_S\Big([C_{S2}] - [C_{S1}]\big) = k_S\big([C_{S1}] - [C_{S2}]\Big) \tag{1.39}$$

where

$$k_S = \frac{D_S}{h} \tag{1.40}$$

is individual mass transfer coefficient dependent on the thickness of the diffusion film, h, which is constant (as the diffusion coefficient) at the process parameters used.

At one-dimensional diffusion and assuming a linear solute (or extractant, or solute–extractant complex) concentration profile, $C_i - C_j$ within the film thickness, h, the flux relation becomes

$$J = -D_J \frac{C_i - C_j}{h} \tag{1.41}$$

Following are descriptions of the diffusion steps of the system:

1. Diffusion of species through the aqueous-phase diffusion film:

$$J_f = k_f([C_{SF}] - [C_{Sf}]) \tag{1.42}$$

2. Diffusion of species through the organic-phase diffusion film:

$$J_{fO} = k_{fO}([C_{SO}] - [C_{Sf}]) \tag{1.43}$$

Assuming the volumes of the organic and aqueous phases are equal, $V_F = V_O = V$, the amount of diffusion matter inside the films is negligible to that in the bulk phases; we obtain simple correlation between concentrations C_{SF} and C_{SO} and time:

$$-\frac{dC_{SF}}{dt}\frac{V_F}{H} = +\frac{dC_{SO}}{dt}\frac{V_O}{H} = J \tag{1.44}$$

where V_F and V_O are volumes of aqueous and organic phases, respectively, and H is the area of the interface.

At initial conditions: $t = 0$, $C_O = 0$, and $C_F = C_F^0$, and integration we obtain

$$\ln\left(\frac{2C_F - C_F^0}{C_F^0}\right) = -\frac{2HD_S}{Vh}t \tag{1.45}$$

Some conclusions can be formulated from these considerations.

1. At equilibrium concentrations of the solute in organic and aqueous phases are equal.
2. At approaching equilibrium the flux rate is proportional to the interface area and diffusion coefficient and inverse to the phase volume and the thickness of diffusion film.
3. Diffusion-controlled process can be experimentally identified by measuring kinetic parameters (forward and reverse rates) at film thicknesses h_F and h_O variations.

This simple mathematical treatment with many assumptions helps better understanding of the fundamental role of diffusion films in the rate-controlling

SX. Of course, diffusion through two liquid films (organic and aqueous) on the opposite sides of the interface has to be considered.

Decreased film thicknesses are not the only way to increase diffusion rate. Structural features of phases can themselves alter diffusion. Thus manipulation of extractant structural features offers minimal benefit for increasing transport rates.

There is another, more general description for time dependency of the solute fluxes. Using postulates of nonequilibrium thermodynamics,[11] the general equation that relates the flux, J_S, of the solute to its concentration C and its derivative is

$$J_S = UC - D_S\frac{\mathrm{d}C}{\mathrm{d}h} \tag{1.46}$$

where U is the phase flow or stirring rate.

Referring to equation continuity, as h approaches zero, the steady-state layers are formed next to the phase interface and separation occurs by differential displacement permeation. According Giddings' analysis[12] of such a system:

$$\ln\left[\frac{\left(C - \frac{J_0}{U}\right)}{\left(C_0 - \frac{J_0}{U}\right)}\right] = U\frac{h}{D} \tag{1.47}$$

where C_0 is initial solute concentration and J_0 is initial flux.

Using Eqn (1.45) we obtain individual mass transfer coefficient k_i for every layer at sampling time t_i:

$$k_i = U\left\{\ln\frac{\left[\left(C_i - \frac{J_{ss}}{U}\right)\right]}{\left[\left(C_{i-1} - \frac{J_{ss}}{U}\right)\right]}\right\}^{-1} \tag{1.48}$$

where C_i and C_{i-1} are the concentrations of the solute in the bulk phase at time t_i and time of previous sampling, t_{i-1}, respectively, and J_{ss} is the flux at steady state.

5.1.2. Determination of diffusion coefficients

Different methods can be applied to determine diffusion coefficients independently from transport experiments, e.g., determination of the lag-time,[13] pulsed-field gradient nuclear magnetic resonance (NMR),[14] and permeability measurements.[15]

The bulk diffusion coefficient D_b is derived by Stokes–Einstein relationship.[16] Simplest relationship is

$$D_b = \frac{KT}{6\pi\eta r} \tag{1.49}$$

where k is the solute mass transfer coefficient based on concentration, T is temperature, η is the solvent viscosity, and r is the solute molecular radius.

And the Wilke–Chung relation[17] is

$$D_b = 7.4 * 10^{-8}\left(\frac{[S]^{0.5}T}{\eta V^{0.6}}\right) \tag{1.50}$$

This relationship is accurate for neutral molecules. For ionic species much more complex models are required, taking into account factors such as ionic charge, ionic strength, and the presence of electric fields. In either case, the range of values for D_b is quite narrow, usually 10^{-5}–10^{-6} cm^2/s.

Diffusion coefficient D_b can be obtained from lag-time experiments.[18] A lag time is defined as the time required for the complex to diffuse through the films from the feed phase to the organic phase, assuming dilute conditions.

$$t_{lag} = \frac{h_S^2}{6D_{lag}} \tag{1.51}$$

Lag times can be obtained from lag-time experiments (see Fig. 1.4). The resulting diffusion coefficient D_{lag} has to be corrected according to the interface properties to obtain the bulk diffusion coefficient ($D_b \approx D_{lag}$).

The diffusion coefficient may be determined by permeability experiments in the absence of extractant. The solute diffuses through the interface and the increase of concentration in the organic phase is monitored by UV/Vis spectroscopy (A_m) as a function of time:

$$\ln\left(\frac{C_F^0 - C_{Ft}}{C_F^0}\right) = -2\frac{HD_b}{Vh}t \tag{1.52}$$

Pulse field gradient (PFG) NMR spectroscopy[14] can be applied to investigate self-diffusion of molecules in solution. PFG NMR is a direct method to measure the mean square distance $<r^2(t)>$ which is traveled by a tracer during a time period Δt. Under the conditions of free isotropic diffusion in three

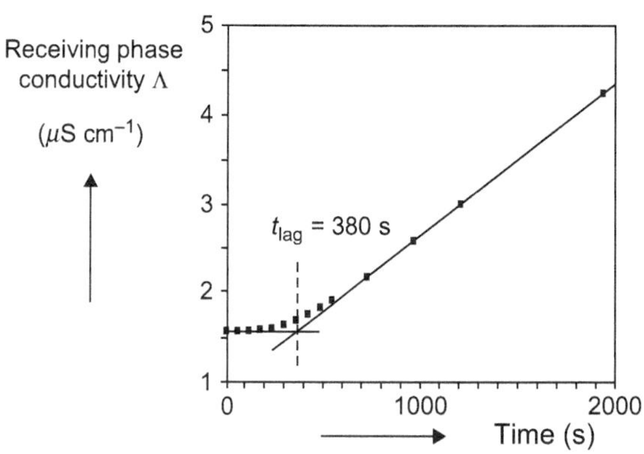

FIGURE 1.4 Lag-time experiment for the transport of $NaClO_4$ mediated by carrier. *(Source: From Ref. 15 with permission)*

dimensions of dilute solutions), the replacement is related to the self-diffusion coefficient D_{sd} by equation:

$$<r^2(t)> = 6D_{sd}\Delta t \quad (1.53)$$

With PFG NMR the molecular displacements by self-diffusion can be measured to give the microscopic diffusion coefficient D_{sd} in the range of 10^{-6}–10^{-14} m^2/s with an optimum between 10^{-9} and 10^{-13} m^2/s. The radial and axial diffusion coefficients were measured by changing the orientation of the magnetic field.

5.2. Chemical Reaction Kinetics Regime Transport

For the last time more and more investigations report that chemical reaction kinetics play sometimes critical role for overall extraction process kinetics.[19] When one or more of the chemical reactions are sufficiently slow in comparison with the rate of diffusion to and away from the interfaces, diffusion can be considered "instantaneous," and the solute transport kinetics occur in a kinetic regime. Kinetic studies of chemical reactions between solute and reagent (extractant) seek to elucidate the mechanisms of such reactions. Information on the mechanisms that control solvent exchange and complex formation is reported briefly below.

5.2.1. Complexation–Decomplexation

Two series of chemical reactions mechanisms and their kinetics have to be analyzed:

1. Solute partition and chemical interactions with extractant (or solvents) and formation of solute–extractant complex (forward extraction or complexation).
2. Solute release from organic phase with chemical interactions and destruction of the complex (decomplexation or back extraction) and partition of the solute between organic and aqueous strip phases.

In the solvent exchange, the composition of the coordination sphere often changes, either because of the formation of complexes between the solutes and a complexing reagent, preferentially soluble in an organic phase, or because of the replacement of a ligand in the aqueous phase solute complex with another more lipophilic one in the organic phase. Solvent exchange and complex formation are special cases of nucleophilic substitution reactions. The basic classification of nucleophilic substitution is found on the consideration that when a new complex is formed through the breaking of a coordination bond with the first ligand (or water) and the formation of a new coordination bond with the organic ligand, the rupture and formation of the bonds can occur with different rates and through the formation of transition intermediates (adducts). The rate at which solvent molecules are exchanged between the primary solvation shell of an ion and the bulk solvent is of primary importance in the kinetics of complex formation from ions in aqueous solutions. In both water exchange and complex formation, a solvent molecule in the solvated ion is replaced with a new molecule (another water molecule or a ligand). Therefore, strong correlations exist between the kinetics and the mechanisms of the two types of reactions. Observations[20] showed that the rates and activation parameters for complex formation are similar to those for water exchange, with the complex formation rate constants usually about a factor of 10 lower than those for water exchange with little dependence on the identity of the ligand. This means that, at least as a first approximation, the complex formation mechanism can be described by the rapid equilibrium formation.

For ligand displacement reactions, very few generalizations can be made, since the reaction mechanisms tend to be specific to each chemical system. However, it has been experimentally observed, at least for aqueous-phase reactions, that the variation of the rates with the identity of the ligand correlates well with the variation in the thermodynamic stability of the complex. Thermodynamically very stable complexes can be expected to react slowly with the extractant. For example, the rate of trivalent lanthanide or actinide cation extraction from aqueous solutions with weakly complexing ligands (Cl^- or NO_3^-) by DEHPA is very fast. On the other hand, in the presence of polyaminocarboxylic acids, such as Ethylene diamine tetraacetic acid (EDTA) (powerful complexing agent), the extraction reaction proceeds only slowly.[20] Ligand substitution reactions with planar tetracoordinated complexes are often slow in comparison with the rate of diffusion through the interfacial diffusional films.

5.2.2. *Liquid–Liquid Interface*

Contact, interaction, and transfer of chemical species on the liquid–liquid interface of two immiscible phases have to be mentioned in the general consideration of chemical kinetics. Little direct information is available on physicochemical properties (interfacial tension, dielectric constant, viscosity, density, charge distribution, etc.) of the interface. The physical depth of the interfacial region can be estimated in the distance in which molecular and ionic forces have their influence. On the aqueous side (monolayers of charged or polar groups), this is several nanometers; on the organic side is the influence of Van der Waals forces. These interfacial zone interactions may slower exchange and complex formation reactions, but as a rule enough fast to be not rate controlling for the most chemical interactions on the interface. The reader may obtain more detailed information about interfacial phenomena in Refs. 21,22.

All considerations, presented for the kinetics of forward extraction reactions, may be applied in the reverse mode to the chemical interactions at strip interface with the complex destruction (decomplexation or back extraction), the solute release, and partition between organic and aqueous strip phases. It has to be taken into consideration two basic differences: quite different thermodynamic conditions on the organic–strip phase interface that may lead to different interaction mechanisms and kinetics and different, more slow kinetics of complex destruction (comparing to the complexation), especially at destruction of aggregates or oligomers that can be formed in the organic phase at high initial concentration of the solute[23,24]

Spectrometric NMR studies showed that the complexes can be kinetically stable[25,26] and, as a consequence, decomplexation rates can be very slow. Recently, at cation interaction experiments with different calix crown ether derivatives,[27] it was proven that the rate of decomplexation can be rate controlling in the extraction–back extraction techniques.

5.2.3. Mathematical Description of Kinetic Regime

The first goal of any kinetic study is to devise experiments that establish the algebraic form of the rate law and to evaluate the rate constants. Rate laws can be derived by measuring concentration variations as function of time or the initial rates as function of the initial concentrations. Unfortunately, there is no general method for finding the rate law and the reaction order. Usually, a trial-and-error procedure is used based upon intelligent guesses.

Experimental kinetic data derived by variables (concentration, temperature, nature of the solvent, presence of other solutes, structural variations of the reactants, etc.) refer to a reaction rate. Reaction mechanism is always only indirectly derived from primary data. Stoichiometry of the reaction, even when this is a simple one, cannot be directly related with its mechanism, and when the reaction occurs through a series of elementary steps, the possibility that the experimental rate law may be interpreted in terms of alternative mechanism increases. Therefore, to resolve ambiguities as much as possible, one must use all the physicochemical information available on the system. Particularly useful here is information on the structural relations between the reactants, the intermediate, and the reaction products.

Mathematical descriptions of simple rate laws used, as a rule, by investigators are presented below.

1. Irreversible first-order reactions depend only on concentration of a solute C_S because the rate of the reverse reaction is always close to or equal to zero. Reaction and rate expressions are

$$C_S + E \xrightarrow{k} C_SE \tag{1.54}$$

$$-\frac{\mathrm{d}[C_S]}{\mathrm{d}t} = k[C_S] \tag{1.55}$$

$$C_S = C_S^0 e^{-kt} \tag{1.56}$$

where E is the extractant concentration, C_S^0 is the solute initial concentration, k is the reaction rate constant, and t is time.

2. Reversible first-order reactions are

$$C_S + E \underset{k_2}{\overset{k_1}{\longleftrightarrow}} C_SE \tag{1.57}$$

$$-\frac{d[C_S]}{dt} = k_1[C_S] - k_2[C_SE] \tag{1.58}$$

At $t = 0$, $[C_S] = C_S^0$, and $[C_SE] = 0$:

$$[C_S] = \frac{[C_S^0]}{k_1 + k_2}\left(k_2 + k_1 e^{-(k_1+k_2)t}\right) \tag{1.59}$$

At $t = \infty$ (at equilibrium, eq), $k_1[C_S]_{eq} = k_2[C_SE]_{eq}$ and equilibrium constant K_{eq} is

$$K_{eq} = \frac{k_1}{k_2} = \frac{[C_SE]_{eq}}{[C_S]_{eq}} \tag{1.60}$$

and

$$\ln\left(\frac{[C_S] - [C_S]_{eq}}{[C_S^0] - [C_S]_{eq}}\right) = -(k_1 + k_2)t \tag{1.61}$$

The individual rate constants of the reaction can be evaluated from the slope of a plot, providing the equilibrium constant is available. Many distribution processes between immiscible liquid phases of noncharged species, as well as distribution of solute ions (e.g., metal ions) performed at very low solute concentrations, can be treated as first-order reversible reactions when the value of the equilibrium (partition) constant is not very high.

3. Series first-order reactions may be referred to the cases when the mechanism goes through an intermediate C_{Sorg}, e.g., at an interfacial adsorption of the solute:

$$C_{Saq} \xrightarrow{k_1} C_{Sorg} + E_{org} \xrightarrow{k_2} C_SE_{org} \tag{1.62}$$

$$\frac{d[C_S]_{org}}{dt} = k_1[C_S^0]e^{-k_1 t} - k_2[C_S]_{org} \tag{1.63}$$

At steady-state approximation $d[C_S]_{org}/dt = 0$

$$[C_S]_{org} = [C_S^0]\frac{k_1}{k_2}e^{-k_1 t} \tag{1.64}$$

and

$$[C_S E] = [C_S^0]\left[1 - \left(1 + \frac{k_1}{k_2}\right)e^{-k_1 t}\right] \tag{1.65}$$

$k_2 >> k_1$ means that reactive intermediate (complex) is formed at low solute concentration which is then interacted with increasing concentration to form the stable final complex aggregate $(C_S E)_n$:[28,29]

$$C_{Saq} \xrightarrow{k_1} C_{Sorg} + E \xrightarrow{k_2} C_S E \xrightarrow{k_3} (C_S E)_n \tag{1.66}$$

5.2.4. Determination of Kinetic Parameters

Determination of α (Dimensionless Parameter Which Relates Diffusion-Limited Species Transfer to Kinetically Limited)

In general for the layer:

$$\alpha = \frac{D}{kh} \tag{1.67}$$

where h is layer (film) thickness and k is the rate of reaction.

When the transport is purely limited by diffusion ($\alpha \sim 0$), parameters can be obtained from measurements of the flux as a function of concentration. In principle, all parameters can be derived by measuring the flux with varying concentrations and thickness of layers. In the case when the extractant at the feed phase interface is fully loaded by solute, the flux reaches its maximum value ($J_{S\max}$):[9]

$$J_{S\max} = \left(\frac{D_S}{h_S}E_0\right)\left(\frac{1}{1+\alpha}\right) \tag{1.68}$$

where E_0 is activation energy of the chemical reaction.

The first term in Eqn (1.68) describes the diffusion-limited flux, while the second term $l/(l + \alpha)$ is a correction factor for slow reaction kinetics. The value of D_S can be determined independently. Direct determination of k and D_S from flux measurements as a function of the layer thickness may be obtained by varying the stirring rates. As a result α is also obtained. If $\alpha < 1$, the transport is mainly limited by diffusion of the species, while the transport is primarily controlled by the reaction rate in the case that $\alpha > 1$.

By combining Eqn (1.66) with relationship $D_S = D_{\text{lag}}$, obtained by lag-time experiments, α may be obtained from D_{lag} and $J_{S\max}$:

$$\alpha = \frac{D_{\text{lag}} E_0}{h_S J_{S\max}} - 1 \tag{1.69}$$

Separation of the diffusional and kinetic terms is achieved by expressing the flux as the ratio of driving force and flux:

$$\frac{E_0}{J_{S\max}} = \frac{h_S}{D_S} + \frac{1}{k_S} \tag{1.70}$$

Consequently, a plot of $E_0/J_{S\max}$ vs. h_S gives a straight line with a slope $1/D_S$ and an intercept of $1/k$.

The value of α can be determined directly by calculation:

$$\alpha = \frac{1/k}{h_S/D_S} \tag{1.71}$$

Determination of Activation Energy

Activation energy of transport gives information about the rate-limiting step in the transport process.[10,30] The activation energy for self-diffusion of a solvent often correlates well with the activation energy for diffusion of a solute species, since on a molecular level diffusion of a solute can be considered as a process in which either a solute or a solvent molecule jumps from solvent cavity to cavity (see Fig. 1.5). Since the activation energy for self-diffusion varies with the solvent used, it is important to determine the activation energy E_a at the transport process design.

The influence of the temperature on the transport rate is related by Arrhenius equation:

$$J = J_0 \exp\left(\frac{-E_a}{RT}\right) \tag{1.72}$$

The authors[32] varied the temperature in order to determine the activation energy E_a from an Eyring plot (slope of a curve ln J as a function of $1/T$, see Fig. 1.6). They indicated that E_a values below 20 kJ/mol are generally accepted as indicative of pure diffusion-limited transport. Generally, at activation energies are above 40 kJ/mol and chemical reactions do play a role in the transport.[33] The authors[15] obtained activation energies for the transport of $KClO_4$ with two Calix[4]-crown-5 extractants, measuring the flux J_{max} as a function of operating temperature. Eyring plot of ln J_{max} as a function of $1/T$ showed values of the apparent activation energies: $E_a = 32 \pm 2$ kJ/mol and $E_a = 59 \pm 7$ kJ/mol. Consequently, they concluded that the transport of $KClO_4$

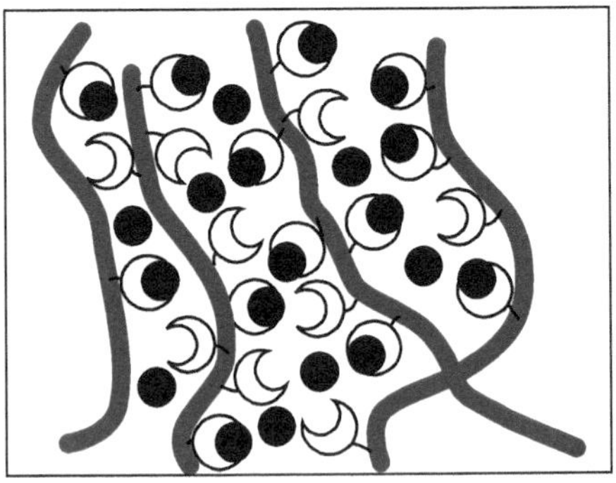

FIGURE 1.5 Facilitated transport through a complexing polymer matrix. *(Source: Adapted from Ref. 31 with permission)*

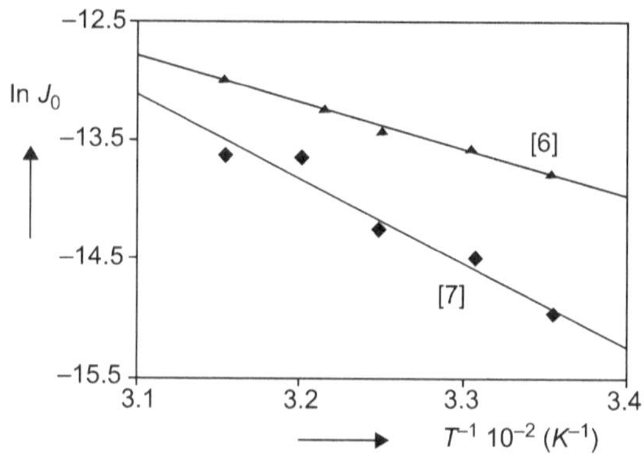

FIGURE 1.6 Eyring plots for $KClO_4$ transport by two [6] and [7] calix[4]-crown-5 carriers across an NPOE–Accurel membranes. *(Source: Adapted from Ref. 15)*

by first carrier extractant is diffusion limited and that of the second is determined by the slow kinetics release.

The temperature dependency of the viscosity of organic solvents η has an Arrhenius-type behavior:

$$\eta(T) = \eta_0 \exp\left(\frac{E_a}{RT}\right) \tag{1.73}$$

Activation energy for self-diffusion in viscous flow may be calculated from Eqn (1.71) by measuring the kinematic viscosity as a function of temperature.

5.3. Mixed Diffusional–Kinetic Transport Regime

When both chemical reactions and film diffusion processes occur at rates that are comparable, the SX kinetics takes place in a mixed diffusional–kinetic regime, which, in engineering, is often referred to as "mass transfer with slow chemical reactions." This is the most complicated case, since the rate of extraction must be described in terms of both diffusional processes and chemical reactions, and a complete mathematical description can be obtained only by simultaneously solving the differential equations of diffusion and those of chemical kinetics. The unambiguous identification of the extraction rate regime (diffusional, kinetic, or mixed) is difficult from both the experimental and theoretical viewpoints.[34,35] Experimental difficulties exist because a large set of different experimental information. Very broad range of several chemical and physical variables are needed. Unless simplifying assumptions can be used, frequently the differential equations have no analytical solutions, and boundary conditions have to be determined by specific experiments.

5.3.1. Identification of the Rate-Controlling Transport Regimes

The experimental identification of the regime that controls the process kinetics is, in general, a problem that cannot be solved by reference to only one set of measurements. In some systems, a definite situation cannot be obtained even when the process rate is studied as a function of both hydrodynamic parameters (viscosity and density of the liquids, geometry of the module, stirring or flow rates) and concentrations of the chemical species involved. It is because the rates may sometimes show the same dependence on hydrodynamic and concentration parameters, even though the processes responsible for the rate are quite different. For a correct hypothesis on the type of regime that controls the transport kinetics, it is necessary to supplement kinetic investigations with other information concerning the biphasic system. This may be the interfacial tension, the solubility of the extractant in the aqueous phase, the composition of the solutes in solution, and so on. The criteria that are often used to distinguish between a diffusional regime and a chemical kinetic regime are as follows:

1. Comparison of the heat transfer and the mass transfer coefficients.
   If the same dependence of the heat transfer coefficient and the mass transfer coefficient on the stirring or flowing rate of the phases is observed, the conclusion can be reached that the transport occurs in a diffusional regime.
2. The reference substance method.
   This method is based on the addition of another inert component with known diffusion rate. By following the simultaneous transfer of the species of interest and of the reference component as function of the hydrodynamic conditions, a diffusional regime will be indicated by a similar functional dependence, whereas a kinetic regime is indicated by a sharply different one. Criteria 1 and 2 are complicated and may be used only occasionally to evaluate the transport regime.
3. Evaluation of parameter α: relation between diffusion and reaction kinetics regimes (see subsection 'Determination of α (Dimensionless Parameter Which Relates Diffusion-Limited Species Transfer to Kinetically Limited)').
4. Evaluation of the activation energy of chemical reactions (see subsection Determination of Activation Energy').
   This criterion is not always very meaningful, since many chemical reactions occurring in separation processes exhibit activation energies of only a few kilocalories per mole, i.e. having the same order of magnitude as those of diffusional processes.
5. Dependency of the transport rate on the rate of stirring or flowing of the phases.
   This criterion is simplest, as proved by its widespread use and will be discussed here in more detail.

A typical curve of transport rate vs. stirring or flowing rate is shown in Fig. 1.7. In general, a process occurring under the influence of diffusional contributions is characterized by an increase of the transport rate as long as the stirring or flowing rate of the phases is increased (Fig. 1.7, zone A). On the other hand, when the transport rate is close to be independent of the stirring rate, it is

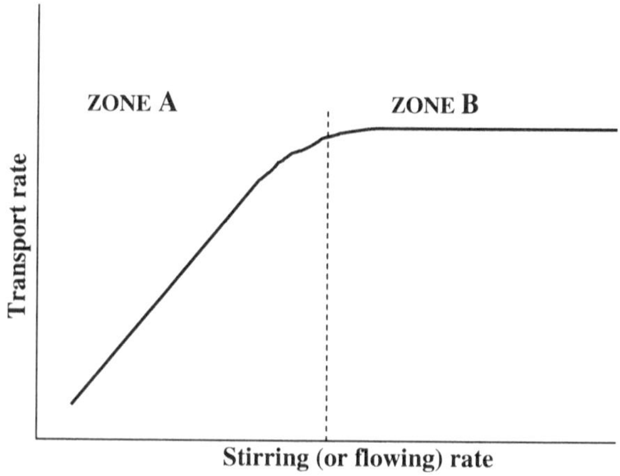

FIGURE 1.7 Typical curve of transport rate vs. phase stirring (flowing) rate at constant interfacial area.

sometimes possible to assume that the process occurs in a kinetic regime (Fig. 1.7, zone B). An increase in stirring or flowing rate produces a decrease in thickness of the diffusion films: the relationship is approximately linear. The rate of transport will increase with the rate of stirring or flowing, as long as a process is totally or partially diffusion controlled. When the thickness of the diffusion films is reduced to minimum, chemical reactions can be rate controlling, and the rate of transport becomes close to independent of the stirring rate.

Unfortunately, this kind of reasoning can lead to erroneous conclusions. Although zone A is certainly an indication that the process is controlled by diffusional processes, the opposite sometimes is not true for zone B: in spite of the increased stirring rate, it may happen that the thickness of the diffusion films never decreases below a sufficiently low value to make diffusion so fast that it can be completely neglected relative to the rate of the chemical reactions. This effect, sometimes called "slip effect," depends on the specific hydrodynamic conditions.

Sometimes an increase in the transport rate can take place with the increase in stirring rate when the system is in a kinetic regime. For example, the increase in zone A may indicate here an increase in the number of droplets of the dispersed phase (proportional to the overall interfacial area) and not a decrease in the thickness of the diffusion film. Moreover, the plateau region of zone B does not necessarily prove that the extraction occurs in a kinetic regime: at high stirring rates the number of droplets of the dispersed phase eventually becomes constant, since the rate of drop formation equals to the rate of drop coalescence. Lack of internal circulation and poor mixing can occur inside the droplets of the dispersed phase. This is particularly true in the systems with the presence of surfactants and small droplets. Therefore, here also, a plateau region may simulate a diffusion-controlled regime. It is then apparent that criterion 5 can also lead to erroneous conclusions.

Finally, it has to be emphasized that both the hydrodynamic parameters and the concentrations of the species involved simultaneously determine whether the transport regime is of kinetic, diffusional, or mixed diffusional–kinetic type. Therefore, it is not surprising that different investigators, who studied the same SX systems in different hydrodynamic and concentration conditions, may have interpreted their results in terms of completely different extraction regimes.

5.3.2. Basic Parameters of Transport Regime

The basic parameters of an extraction regime are related to the properties of the solutes, extractants, solvents, and diluents. These are individual and overall mass transfer coefficients (in diffusional and chemical reactions kinetics regime), distribution constants, extraction and coupling coefficients of forward extraction—K_{fd} and K_{fex}, respectively, diffusion coefficients of complexes in all phases, and selectivity. Influence of these parameters on the solute transport in different configurations will be generally analyzed in the next sections and the respective chapters in detail.

5.3.3. Determination of Extraction Kinetic Parameters

Let us consider an extraction process. Individual mass transfer coefficients of solute species in the aqueous feed and organic extractant solutions through the interfacial stagnant layers are determined experimentally by feed and extractant phases with stirring or flow rate variations using Eqn (1.46).

For feed stagnant layer:

$$k_{(SF)_i} = U_{F_j}\left\{-\ln\left[\frac{\left([C_{SF_i}] - \dfrac{J_{F_{SS}}}{U_{F_i}}\right)}{\left(\left[C_{SF_{(i-1)}}\right] - \dfrac{J_{F_{SS}}}{U_{F_i}}\right)}\right]\right\}^{-1} \tag{1.74}$$

$$k_{(SE)_i} = U_{E_j}\left\{\pm\ln\left[\frac{\left([C_{SE_i}] - \dfrac{J_{E_{SS}}}{U_{E_j}}\right)}{\left([C_{SE_{i-1}}] - \dfrac{J_{E_{SS}}}{U_{E_j}}\right)}\right]\right\}^{-1} \tag{1.75}$$

where U_{F_j} and U_{E_j} are j-flow velocities; $J_{F_{SS}}$ and $J_{E_{SS}}$ are fluxes at steady state; and $[C_{SF_i}]$ and $[C_{SE_i}]$ are concentrations of solute species, sampled at time i, in the feed and extractant solutions, respectively. In Eqn (1.75), "plus" (positive) is at increasing concentrations vs. time, and "minus" (negative) is at decreasing ones.

D_S-"effective" diffusion coefficients of solute species in the feed and extractant solutions are evaluated by extrapolating the plots of $t_i = f(U)$ at $U \to 0$. The magnitude of D_S is far from the real diffusion coefficient of solute complexes in liquids, because of some assumptions mentioned above. Equation (1.23) at $U \to 0$ becomes undefined; however, for calculating mass-transfer coefficients of solutes at visible flow rates, this parameter (coefficient) is quite applicable.

Correlation factor between $[C_S]_i$ and J_{ss}/U_i in Eqns (1.74) and (1.75) has been checked in experiments with several metal ions transport[36–38] in the range of $U = 0.1–1.5\ cm^3/s$. Results showed that $[C_S] > J_{ss}/U$ in 1.5–4.0 orders of magnitude. It means that the J_{ss}/U ratio may be excluded from Eqns (1.74) and (1.75). Therefore, we obtain

$$k_{(SF)_i} = U_{F_i}\left\{-\ \ln\frac{[C_{SF_i}]}{[C_{SF_{i-1}}]}\right\}^{-1} \tag{1.76}$$

or

$$k_{SF_i} = U_{F_i}\left\{\ln\frac{[C_{SF_{i-1}}]}{[C_{SF_i}]}\right\}^{-1} \tag{1.77}$$

$$k_{SE_i} = U_{E_j}\left\{\pm\ \ln\left[\frac{[C_{SE_i}]}{[C_{SE_{i-1}}]}\right]\right\}^{-1} \tag{1.78}$$

Equations (1.74)–(1.76) are similar to those used by many researchers.[18,19,25–28,32–35] Referenced authors obtained these equations by considering the basic Stokes–Einstein equation. Here the same equations were obtained as particular case from Eqn (1.47), based on kinetics of irreversible processes (nonequilibrium thermodynamics).

According to assumptions and linear concentration profiles, solute being extracted can be described by the solute flux from the bulk feed aqueous phase to the bulk organic phase in terms of individual mass transfer coefficients at steady state and additivity of a one-dimensional series of diffusion resistances, $R_i = 1/k_i$. Overall mass transfer coefficient, $K_{F/E}$, is

$$\frac{1}{K_{SF/E}} = \frac{1}{k_{SF}} + \frac{1}{K_{Fex}k_{com}} + \frac{1}{K_{Fex}k_{SE}} \tag{1.79}$$

where K_{SFex} is solute distribution (extraction) coefficient; k_{SF} and k_{SE} are solute individual mass transfer coefficients in stagnant films; and k_{com} is individual coefficient of chemical reaction (complexation) rate at the interface.

First and third components of the sum in Eqn (1.79) present the diffusional step resistances and second component is the chemical kinetics regime at forward extraction. Analogical equation may be presented for back extraction technique.

If the system is polar organic–nonpolar organic, then the relation (1.79) for aqueous–organic systems are valid with the polar organic phase used instead of the aqueous phase.

6. SELECTIVITY

Selectivity is a main parameter of the SX technique. Effective separation of individual solutes of interest from the mixtures or of groups of close by properties of solutes, purification of solutes of interest—this task is held by technologists. The simplest and most obvious separation technique is to use a reagent, extractant, which readily reacts with solute of interest to form complex, is soluble in organic phase, but does not react with others. Utilization of different reactivities of extractants with different solutes, their products with different solubilities in aqueous and organic phases, and the differences in extraction constants are the basic tasks of technologists. In this section we will try to make some generalizations for above extraction parameters to make some descriptions of the selectivity in the extraction theory.

Let us consider the selectivity parameter on the example of metal ion separation. According to the extraction model equations, the selectivity of two solutes, e.g., two metal species, $S_{\mathrm{M}_1/\mathrm{M}_2}$ is determined by relation:

$$S_{\mathrm{M}_1/\mathrm{M}_2} = \frac{K_{\mathrm{M}_1\,\mathrm{tot}}[C^0_{\mathrm{SF}}]_{\mathrm{M}_1}}{K_{\mathrm{M}_2\,\mathrm{tot}}[C^0_{\mathrm{SF}}]_{\mathrm{M}_2}} \tag{1.80}$$

where subscripts M_1 and M_2 refer to the two metal species; $K_{\mathrm{M}_1\,\mathrm{tot}}$ and $K_{\mathrm{M}_2\,\mathrm{tot}}$ are the overall extraction (distribution) constants); and $[C^0_{\mathrm{SF}}]_{\mathrm{M}_1}$ and $[C^0_{\mathrm{SF}}]_{\mathrm{M}_2}$ are the initial concentrations of two metal ions in the treated feed solution.

Introducing a separation factor, A, defined as a ratio of the overall extraction coefficients of the metal species:

$$A_{\mathrm{M}_1/\mathrm{M}_2} = \frac{K_{\mathrm{M}_1}}{K_{\mathrm{M}_2}} \tag{1.81}$$

an equation for the system selectivity is obtained

$$S_{\mathrm{M}_1/\mathrm{M}_2} = A_{\mathrm{M}_1/\mathrm{M}_2}\frac{[C^0_{\mathrm{SF}}]_{\mathrm{M}_1}}{[C^0_{\mathrm{SF}}]_{\mathrm{M}_2}} \tag{1.82}$$

Based on the principle of resistance additivity, the overall extraction coefficient, $K_{\mathrm{M}_{\mathrm{tot}}}$, of every solute passing through the separation system (forward and back extraction) is related to the overall extraction coefficients as follows:

$$K_{\mathrm{M}_{\mathrm{tot}}} = \frac{\left(\dfrac{K_{\mathrm{SFE}_{\mathrm{tot}}}}{K_{\mathrm{SER}_{\mathrm{tot}}}}\right)_{\mathrm{M}}}{(K_{\mathrm{SFE}_{\mathrm{tot}}} + K_{\mathrm{SER}_{\mathrm{tot}}})_{\mathrm{M}}} \tag{1.83}$$

where K_{SFE} and $K_{\mathrm{SER}_{\mathrm{tot}}}$ are overall forward and back extraction coefficients of metal species.

Thus, the separation factor of the two metal species is

$$A_{M_1/M_2} = \frac{(K_{SFE_{tot}} * K_{SER_{tot}})_{M_1} * (K_{SFE_{tot}} + K_{SER_{tot}})_{M_2}}{(K_{SFE_{tot}} * K_{SER_{tot}})_{M_2} * (K_{SFE_{tot}} + K_{SER_{tot}})_{M_1}} \tag{1.84}$$

For evaluation of the selectivity of two metal species separation, we can assume that in the same solution environment (water or organic) the diffusion coefficients of these metal ions with the same charge have similar values and the diffusion coefficients of the metal–extractant complexes have similar values. Thus, we can represent separation factor as dependent only on the extraction coefficients at forward and back extraction, determined experimentally (see Section 4).

$$A_{M_1/M_2} \approx \frac{\left(\frac{K_{SFex}}{K_{SBex}}\right)_{M_1}}{\left(\frac{K_{SFex}}{K_{SBex}}\right)_{M_2}} \tag{1.85}$$

where K_{SFex} and K_{SBex} are extraction coefficients of forward and back extraction, respectively.

The extraction coefficient may be expressed as a function of the metal association (stability) constants in the organic phase, the association constants of metal ions with solvent environment (water) in the feed and in the strip solutions, and partition coefficients of the extractant and metal ion. In this case, the separation factor can be determined by stability constants of the metal complexes, formed with functional groups of an extractant, if we assume that the metal ions are predominantly present (a) as free ions in the aqueous solution, so that complex concentrations can be disregarded, and (b) as complexes in the organic solutions, so that free ion concentrations can be disregarded.

$$A_{M_1/M_2} \approx \frac{\left(\frac{\beta_{SFE}}{\beta_{SER}}\right)_{M_1}}{\left(\frac{\beta_{SFE}}{\beta_{SER}}\right)_{M_2}} \tag{1.86}$$

where β_{SFE} and β_{SER} are stability constants of the metal–extractant complexes in equilibrium with the feed and strip solvent environment (as a rule acids for metal ions), respectively. Therefore, preliminary selectivity data of metal species separation may be evaluated without experimentation, if stability constant data are available in the literature.

From Eqns (1.85) and (1.86) it follows that high separation factors are favored when approaching conditions $(K_{SFex}/K_{SBex})_{M_1} >> (K_{SFex}/K_{SBex})_{M_2}$ or $(\beta_{SFE}/\beta_{SER})_{M_1} >> (\beta_{SFE}/\beta_{SER})_{M_2}$. On the other hand, the system loses its selectivity when extraction parameters of both metal species are either extremely high or extremely low.[39,40] Selectivity can be increased by choosing a selective extractant with intermediate distribution data values and adjusting its concentration, its volume, and the acidity of the feed and strip solutions in such a way as to approach the above conditions. Another way to improve selectivity parameters is to choose some mixtures of the strong and relatively weak extractants. In every case it has to be checked by experimentation.

7. FACTORS AFFECTING EXTRACTION PROCESS

7.1. Extractant Properties

Numerous reagents are used as extractants, solvents, and diluents in SX. They can be divided into different classes:[10]

1. Species capable of forming three-dimensional networks of strong hydrogen bonds (e.g., water, poly- and amino alcohols, hydroxy acids).
2. Species that do not form three-dimensional networks (e.g., primary alcohols, acids, primary and secondary amines, ammonia, nitro compounds with α-positioned hydrogen).
3. Species composed of molecules containing donor atoms, but no active hydrogen atoms (e.g., ethers, ketones, aldehydes, esters, tertiary amines, phosphoryls, nitro compounds without α-hydrogen).
4. Species composed of molecules containing active hydrogen atoms, but no donor atoms (e.g., $CHCl_3$ and other aliphatic halides).
5. Species with no hydrogen bond-forming capability and no donor atoms (e.g., pure hydrocarbons, CS_2, CCl_4).

This diversity in solvent properties results in large differences in the distribution ratios of extracted solutes. Some solvents, particularly those of class 3, easily react directly (owing to strong donor properties) with inorganic compounds and extract them without need for any additional extractant, whereas others (classes 4 and 5) do not dissolve salts without the aid of other extractants. As a rule these reagents are used as diluents. The class 1 reagents are very soluble in water and are useless for extractants of metal species although they find use in separation of biochemicals.

Solvents of low dipole moment and low permittivity, which contain no oxygen atoms, have the lowest solubility and dissolve less water than other types of solvents. Solvents of class 2 may completely dissolve in water, class 3 to about a several percents, most solvents of classes 4 and 5 dissolve very little. Solubility in water of the solvents as the solubility of water in these solvents usually decreases with increasing molecular weight.

Extractant properties play a basic role in selective separation and concentration of solutes from dilute solutions. The reader can find detailed descriptions in the solvent extraction handbooks (e.g., Ref. 39). Many synthetic extractants have been developed during the past few decades. As an example, crown ether macrocycles, which selectively bind alkali and some other metal ions, have been synthesized and used as extractants for the selective recognition of neutral, charged, or zwitter-ionic species.[41,42] The reader can find descriptions of many extractants in the solvent extraction handbooks (e.g., Ref. 43).

Extraction performance is strongly related to the characteristics of an extractant. The main parameters of extractant are as follows:

1. High selectivity to species has to be separated.
2. High capacity of the species has to be extracted.
3. High ability of an extractant to complexate a solute of interest from an aqueous feed phase into organic phase—high extraction, or distribution, or partition constant, $E_{F/E}$)
4. High ability of an extractant–solute complex to be decomplexed and stripped from loaded organic to an aqueous strip phase—high decomplexation, or stripping constant, $E_{E/R}$).
5. Rapid kinetics of formation (complexation) and destruction (decomplexation) of the complex.
6. Rapid kinetics of diffusion of the extractant–solute complex through the interface organic layer (a high diffusion coefficient, D_{LM}).
7. Stability of the extractant.
8. No side reactions.
9. No irreversible or degradation reactions.
10. Low solubility of the extractant in the aqueous phase.
11. Low coextraction of water.
12. Extractant should be easily regenerated.
13. Extractant should have suitable physical properties, such as density, viscosity, and surface tension.
14. Low toxicity for biological systems and low corrosivity.
15. Reasonable price at industrial applications.

A great variety of extractants may be divided to cation and anion exchangers and neutral ligands. First group are the big number of organic acids and their derivatives and related proton donors. For example, some commercially available extractants: DEHPA, bis(2,2,4-trimethylpentyl) phosphinic acid (Cyanex 272), some hydroxyoximes (LIX or Acorga series), and oligoamide compounds containing 8-hydroxyquinolyl groups (Kelex or LIX-26 series).[10,28–31,39,40,44,45] Crown ethers and related macrocyclic multidentate ligands have a pronounced selectivity for cations (metal ions).[10,41–43,46] Their ability to selectively and reversibly bind metal ions may enable to perform difficult separations. Wide variety of macrocyclic carriers exist and are developing in the past few years.[40–43] The complex stability, donor abilities of functional atoms (O, N, S, P, etc.), and/or groups of macrocyclic carriers are intensively studied to develop some rules for predicting their properties.

Second group are water-immiscible primary, secondary, and tertiary amines and their derivatives, quaternary amine salts, and other proton acceptors (see Chapters 2, 3, 5). Considerable effort has been devoted to the development of anion agents, playing an important role in the biological and biochemical processes. Several amines, their derivatives, surfactants, lipophilic metal complexes, and macrocycles with positively charged subunits are known and developing as anion extractants.[8–10,47–49]

The third group are water-immiscible organic species with electron donor or acceptor properties or solvating extractants. They include carbon–oxygen compounds (amides, ethers, ketones); phosphorus–oxygen compounds (tri-*n*-butylphosphate (TBP), dibutyl-phosphate (DBP), or dibutyl-phosphonate (DBBP); phosphine oxides (tri-*n*-octylphosphine oxide (TOPO); phosphine sulfides (Cyanex 471); alkyl sulfides (dihexyl, diheptyl sulfides); nitrogen-containing compounds (CLX 50), etc.[41–43,52,53] All of them are known as selective extractants.

7.1.1. Mixtures of Extractants (Ionic Additives)

Cotransport of oppositely charged ions is the most obvious way to maintain electroneutrality. In recent years, many studies have been conducted to examine the use of anionic additives for maintenance of electroneutrality at cation extraction.[25,40,51–53] The anionic additives are typically lipophilic carboxylic, phosphoric, or sulfonic acids. Cation or neutral macrocyclic carriers coupled with anionic additives result in a synergistic extraction of cations which exceeds that accomplished by each component individually. Example of such synergism is demonstrated in Fig. 1.8. The authors of Ref. 54 observed a 10–100-fold enhancement of copper extraction while adding the anionic group to the cation coordination macrocycle.

Parthasarathy and Buffle[55] have systematically varied the chain length of a series of lipophilic carboxylic acids with 1,10-didecyldiaza-18-crown-6 as an extractant. Chain lengths ranged from 10 to 18 carbons. Optimal Cu^{2+} separation was achieved with additives from 12 to 14 carbons in length, and lauric acid ($n = 12$) yielded the best results due to its decreased tendency to form precipitates.

As mentioned above, hydrogen ions can be solvated by strong donor molecules like TBP, sometimes leading to the extraction of a solvated hydrogen salt. For example, TBP extracts nitric acid, HNO_3, in the form of the complex TBP–HNO_3, which could be considered as the complex $H(TBP+)^+NO_3^-$ ion pair. The amines form stronger adducts with hydrogen ions; in fact, they are so strong that they remain protonated while exchanging the anion. A classical reaction is the formation of the ammonium ion NH_4^+ when NH_3 is dissolved in water.

The organic amines have a nitrogen atom N attached to a large organic molecule R usually containing aliphatic or aromatic carbon atoms. They are highly soluble in organic solvents (diluents) and almost insoluble in water. In contact with an aqueous phase containing acid HL, the amine base RN reacts with the acid HL to form $(RNH)^+ + L^-$, but extracts with an excess amount of acid HL (over the 1:1 HL/RN ratio) into the organic phase, and also with

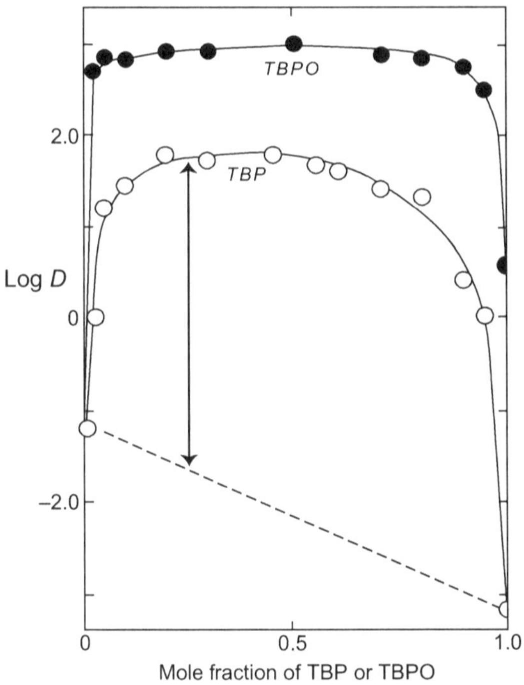

FIGURE 1.8 Synergistic extraction: distribution of U(VI) between 0.01 M HNO3 and mixtures of thenoyltrifluoroacetone (TTA) and TBP, or TBPO, at constant total molarity $[TTA]_{org}$ plus $[TBP]_{org}$ or $[TBPO]_{org}$. *(Source: From Ref. 10 with permission)*

additional water. The practical concentration of amine in the organic phase is usually less than 20%; at higher concentrations the amine salt solutions become rather viscous. The amine salts dissociate in highly polar solvents, in more inert diluents they easily form aggregates, micelles and third phase (second organic phase).

It appears that the ratio of anionic additive is the most critical measure.[55–57] While the optimal amount of additive varies from system to system, the potential benefit of the additives is well established.

In order to achieve high selectivity, a substrate-specific receptor must be added. In biological systems this task is fulfilled by ionophores such as valinomycin (l); in extraction systems we can use synthetic macrocyclic receptors developed during the past two decades.[51]

7.1.2. Acidity–Basicity of the Extractant Influencing Extraction

Distribution of the solute (e.g., metal), D_M, depends on the concentrations of H^+ and HL_{org} to the power z of the metal ion charge and on the concentration of extractant to the power of b and number of extractant molecules in the extracted complex:

$$K_{ex} = D_S[H^+]^z[HL]_{org}^{-z}[E]_{org}^{-b} = K_a^z b_z K_{dE}^{-b} K_{dS} \tag{1.87}$$

According to Eqn (1.87), K_{ex} increases to the z power of the dissociation constant, K_a, for the extractant acid HL. Thus it appears desirable to use stronger acids as extractants, i.e., those with large K_a (or small pK_a). Thus factors that affect the dissociation constant of an acid are of importance for extraction behavior. The acidity of organic acids increases (i.e., pK_a decreases) by electrophilic substitution in the carbon chain. For example, replacing an H atom in the CH_3 group of acetic acid ($pK_a = 4.75$) by I, Br, Cl, and F atoms yields monohalide acetic acids with pK_a of 3.0, 2.9, 2.8, and 2.7, respectively. The mono-, di-, and trichloroactic acids have pK_a equal to 2.8, 1.48, and 1.1, respectively. Other electrophilic groups that can increase acidity are HO^-, NC^-, NO_2^-, etc.

Conversely, nucleophilic substitution by aliphatic or aromatic groups usually has little effect on the pK_a, though it may affect the distribution constant K_{DR}: e.g., the addition of a CH_3- or C_6H_5- group in acetic acid changes the pK_a to 4.9 (propionic acid) and 4.3 (phenyl acetic acid). The further substitution is from the carboxylic group, the less the effect: e.g., while pK_a for benzoic acid is 4.19, a Cl in orthoposition yields pK_a of 2.92 and in paraposition, 3.98.

These general rules hold rather well for the acidic organic extractants and can be used to extrapolate from related compounds to new ones as well as to develop new extractants. The effect of various substituents on pK_a is extensively discussed in textbooks on organic chemistry.[58]

7.1.3. Influence of Dissociation Constant and Distribution Coefficient of the Extractant

K_{ex} also increases with increasing formation constant of the uncharged metal complex, β_z. Thermodynamic factors and geometrical aspects that influence the size of β_z are discussed in Chapter 3.

To achieve the highest possible D value, the concentration of the extracted uncharged complex (ML_z, ML_zE_b, or E_bML_{z+b}) must be maximized. A large value for the formation constant of the neutral complex favors this goal. In Chapter 3, it is pointed out that for hard acids the complex formation constant increases with the charge density of the metal ion, provided there is no steric hindrance. This is seen for the metal fluoride complexes with no steric hindrance. For the lanthanide acetylacetonates, the increase in β_z from ^{57}La to ^{64}Gd is due to a reduction in the ionic radius and, accordingly, to the increased charge density. It should be noted that the diminishing lanthanide size leads to a successive diminution of the coordination number (from 9 to 8 for H_2O) for La → Gd. For Gd → Lu, the coordination number is probably constant.[59]

Because the H^{++} ion acts as a hard metal ion, one should expect a close correlation between the formation constants for HL and MA_z. Since K_a is defined as the *dissociation* constant of HL, β_z is the *formation* constant of ML_z; therefore, a correlation is expected between pK_a ($-\log K_a$) and $\log \beta_z$: large $\log \beta_z$ values are usually observed for organic acids with large $\log K_a$. Since K_{ex}

is directly proportional to both β_Z and K_a^Z (see Eqn (1.87)), a large β_Z value is likely to be accompanied by a small K_a value (large pKa); in Eqn (87) the two parameters counteract each other. Thus in order to obtain high free ligand concentrations, which favors the formation of the ML_Z complex, $[H^{++}]$ must be low, i.e., pH must be high. However, at high pH the metal may be hydrolyzed. A careful balance must be struck between β_Z, K_a, and pH to achieve an optimum maximum concentration of the uncharged complex.

7.2. Aggregate Formation

Some extractants tested meet serious problems: they show low rate of decomplexation which becomes a rate-limiting step of total extraction–back extraction process. In order to understand the phenomenon, mechanisms of Cu, Cd, and Zn extraction[45] with DEHPA were investigated. Depending on metal concentration in the organic phase, three different interaction stages were suggested. At low Cu concentration in the organic phase, cation exchange extraction takes place with formation of tetrametric complexes as $CuL_2 * 2HL$ (stage I). With increasing Cu concentration, these nuclei grow in size by coordination bonding (extraction by solvation) with undissociated copper salt molecules, forming linear (or planar) aggregates (stage II). Upon reaching a critical size, a structural reorganization occurs: polyhedral aggregates, or "clusters"[70,71] with polydentate, asymmetric bonds are forming (stage III). Some spectroscopic and chemical techniques showed results, which indirectly confirmed mechanisms, taking place at stages II and III. Kinetics of these aggregate decomposition and consequently back extraction of Cu to the strip phase is very slow. Of course, proposed aggregation mechanisms have to be proved, using direct techniques, e.g., NMR, light scattering, zeta potential, etc. The knowledge of the extraction mechanisms is important for designing the extraction technology, processing feed solutions with significant metal ions concentrations. The reader can find more details in Part II.

7.3. Solvent (Diluent) Properties Influencing Extraction

Nearly every extraction system devised to date involves the solvation of extraxtant in an organic solvent (diluent) through similar interactions ranging from that of cavity formation for very inert diluents like hexane and through dipole–dipole interactions, π-electron interaction, and hydrogen bonding for the more reactive solvents. The expected degree of interaction can be judged from values of permittivity and solubility parameters. So extraction rate and selectivity depend heavily upon the properties of the diluent–solvent. Authors of the review[60] divided the free energy of extraction process into four components in a thermodynamic cycle: free energy of desolvation of the cation, free energy of desolvation of the anionic ligand, free energy of the gas-phase interaction between cation and ligand, and free energy of solvation of the solute–extractant complex. Three of the four components are intimately related to the nature of the organic solvent. Solvent characteristics influence the thickness of the stagnant film at the phase interface, equilibrium constants for solute–extractant interaction, partition coefficients, and diffusivities of the species in the system[61] Not only is the influence of the solvent type rather large, but also it is quite complex. These cases involve the aqueous phase, so some phenomena may be attributed to the interaction with water molecules.

Many groups have investigated the suitability of various solvents for use in LM systems and have attempted to describe the relationship between solvent characteristics and transport properties.[62–65] Of all solvent properties, dielectric constant seems to be most predictable in its effect on extraction.[60] For solvents, such as the halocarbons, extraction usually decreases with increasing dielectric constants.[61] Figure 1.9 shows this trend for alkali metal binding by dicyclohexano-18-crown-6 in a number of alcohols. This trend holds true for many simple systems, but it breaks down under more complex conditions. Solvent donor number, molecule size, solvent viscosity, extractant solubility in the solvent, permanent and induced dipole moments, and heats of vaporization are important.[66]

Solvent characteristics that influence the diffusion and extraction are found to be viscosity (η) and polarity (ε). For spherical solutes, the diffusion coefficient depends on the solvent according to the Stokes–Einstein relation (Eqn (1.64)). From this, it follows that the diffusion coefficient linearly increases with T/η. Hence, the extraction increases linearly with the reciprocal viscosity.[62]

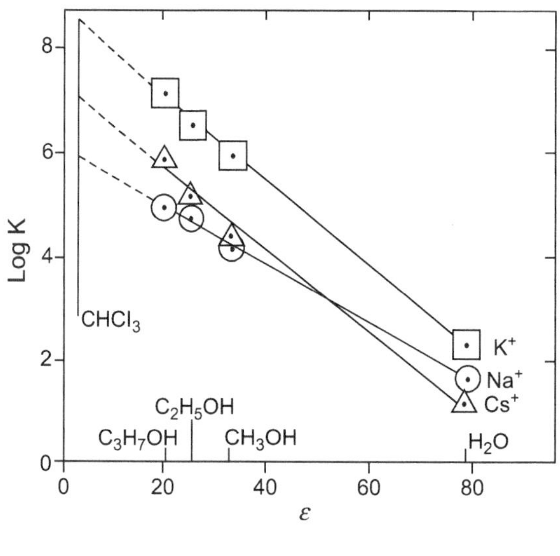

FIGURE 1.9 Relationship between extractant (macrocycle)–cation complex stability (extraction constant) and solvent dielectric constant ε for several solvents. *(Source: Adapted from Ref. 64 with permission)*

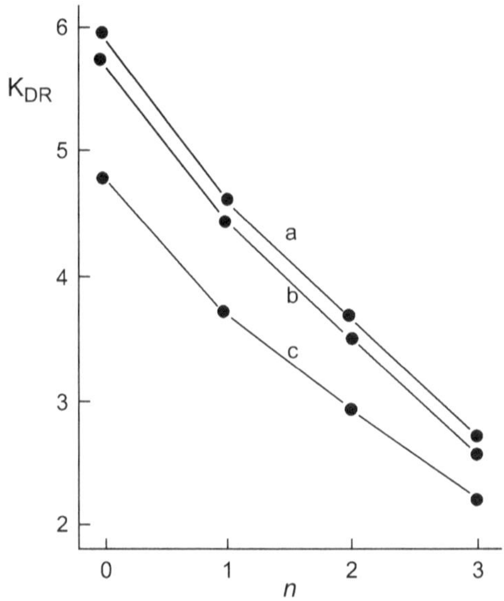

FIGURE 1.10 Distribution constants K_{DR} of acetylacetone between methyl-substituted benzenes at different aqueous ionic strengths: (a) 0.001 M, (b) 0.1 M, and (c) 1.0 M $NaClO_4$ at 25 °C. The number of methyl groups, *n*, changes the aromatic character at $n=0$ to more aliphatic at $n=3$. *(Source: From Ref. 67)*

The solvent effect on extraction constants is a combination of the influence on salt partition and association. Both processes are influenced by the polarity of the solvent.[63] The Kirkwood function describes the relation between polarity, ε, and extraction constant: a more polar solvent promotes extraction. In general, polar solvents favor salt partition, but the tendency toward complexation diminishes. Since the overall effect of solvent polarity on the extraction is positive, the polarity appears to affect the partition coefficient to a higher degree.

Figure 1.10 shows[67] a decrease in K_d for acetylacetone with increasing number of methyl groups in the substituted benzene solvent. Since acetylacetone is more soluble in aromatics than in aliphatics, as illustrated by the K_{dR} sequence: hexane 0.95, cyclohexane 1.03, ethylbenzene 3.31, and benzene 5.93 (aqueous phase 1 M $NaClO_4$), the decrease observed in Fig. 1.10 may simply be an effect of the increasing aliphatic character of the solvents.

The use of mixed solvents introduces some significant advance in the SX process.[60,62,65] Figure 1.11 shows the synergistic effect of binary mixtures of chloroform and nitrobenzene: maximum Na^+ transport occurs with an equimolar solution of the two solvents. Parthasarathy and Buffle[55] in a similar study report optimal extraction of Cu^{2+} with an equimolar solution of phenylhexane and toluene.

Many equations have been developed for consideration of solvent properties predicting separation rates,[57] but their predictions still suffer from considerable inaccuracy. Moreover, the issue of solvent effect is a complex one and considerable room remains for further study of this subject.

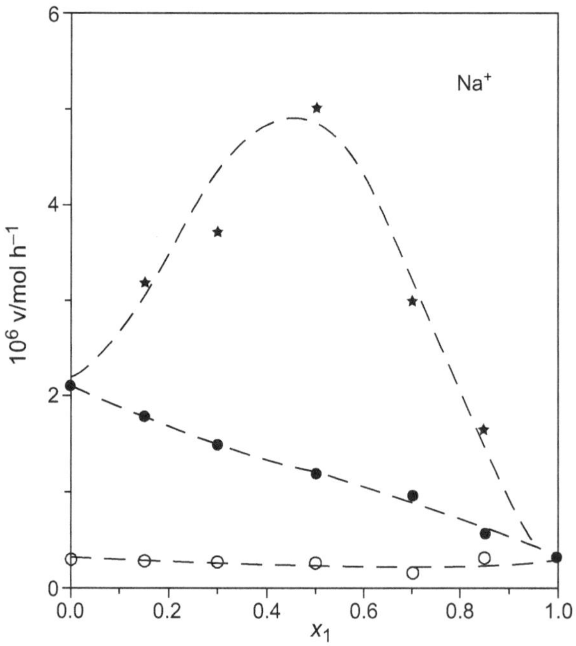

FIGURE 1.11 Synergistic effect of mixed solvents on Na^+ transport by dibenzo-18-crown-6. The feed phase is 1 M NaCl. Three mixed solvent systems were tested: (*) chloroform(1)–nitrobenzene(2); (●) dichloroethane(1)–nitrobenzene(2); and (○) chloroform(1)–dichloroethane(2). *(Source: Reproduced with permission from Ref. 68)*

7.4. Coupling Ions: Anion Type

In order to maintain electroneutrality and solute uphill pumping, many extraction systems require a coupling ion to be counter- or cotransported along with the solute ion. Because the coupling ion must also cross the organic phase, it is bound to influence extraction efficiency.

Proton or sometimes alkali metal cations are used for countertransport of cationic or cotransport of anionic solutes because of their good transport properties. It is not the case with the coupling anions. Many studies of the anion effect on extraction efficiency have been conducted.[64,65] The effects of anion hydration free energy, anion lipophilicity, and anion interactions with solvents have been mentioned, although anion hydration free energy seems to be the major determinant. For example, extraction of K^+ by dibenzo-18-crown-6 decreased with anions in the order: picrate > PF_6^- > ClO_4^- > IO_4^- > BF_4^- > I^- > SCN^- > NO_3^- > Br^- > BrO_3^- > Cl^- > OH^- > F^- > acetate > SO_4^-. This order is almost identical to that for increasing anion hydration free energy. This example demonstrates the strong correlation between anion hydration and transport efficiency: larger anions are more easily dehydrated and thus more readily facilitate process.

While nearly all investigations of anion effects have focused on kinetics, a few recent works suggest a correlation between anion type and selectivity. At extraction of alkali metals into chloroform by dicyclohexano-18-crown-6,[69] selectivity for K^+ over both Rb^+ and Cs^+ decreases dramatically depending upon anion type in the order: $NO_3^- > SCN > ClO_4^-$ $I^- > Br^-$. Selectivity for K^+/Cs^+ decreases from 16.0 for nitrate to 3.5 for bromide. The authors were unable to tie this trend to any particular parameter, though they discounted the possibility that it is correlated to anion radius, hydration enthalpy, or anion softness.

In research[57] the authors demonstrated that anions are capable not only of altering selectivities but indeed of reversing them. Typically, 18-crown-6 analogs show a strong preference for Hg^{2+} over Cd^{2+}. However, by using SCN^- as a counter-anion, completely reverse selectivity was obtained, resulting in highly selective extraction of Cd^{2+} over Hg^{2+}. This result is due to the fact that the SCN ion forms coordination complexes with these cations, which lead to reverse selectivity. A similar result is present when Br^- is used as an anion for altering selectivity between Cd^{2+} and Zn^{2+}. These results suggest that careful consideration of coupling anion is crucial when designing an extraction system or when comparing results listed in the literature.

7.5. Influence of Temperature

In general, the distribution into organic phase of a metal complex increases with increasing temperature for complexes with significant hydrophobic character. The introduction of a complex into organic phase involves a number of processes that can be associated with large changes in enthalpy (solvation processes) and in entropy (solvent orientation and restructuring), leading to considerable temperature effects. The metal acetylacetonate systems in Fig. 1.12 illustrate this effect. In these particular systems, the free energy of distribution between the organic and aqueous phases is dominated by the enthalpy term.

The extraction coefficient K_{ex} decreases with increasing temperature while the diffusion coefficient increases.

7.6. Influence of Complex Structure

The effect of various structures on the bond strengths in organic complexes has been extensively treated in publications.[71–75]

7.6.1. Chelate Ring Size

Formation of metal–organic chelate complexes results in stronger complexation (i.e., larger β_n values) compared to interaction with monodentate ligands. The chemistry of these complexes has been discussed in the literature.[72–75] Chapter 3 presents the most important factors in the formation of such complexes: (1) the type of binding atom; (2) the chelate ring size (or "bite"); and (3) the number of chelate rings formed (mono- or polydentate).

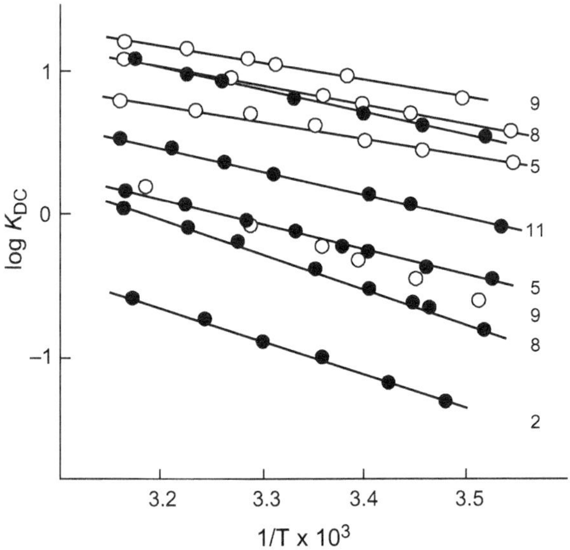

FIGURE 1.12 Distribution constant K_{DC} for metal diacetylacetonates, MA2, between organic solvents and 1 M $NaClO_4$ as a function of temperature: Cu(II) (open circles) and Zn(II) (solid circles) with various solvents. *(Source: From Ref. 70 with permission)*

Table 1.2 shows that extraction (i.e., largest K_{ex} value) is favored by six-membered rings. The extractants TBOA, TBMA, and TBSA are very similar, but their structural differences (see formulas in Table 1.2) allow the formation of only one type of metal chelate complex: five-, six-, and seven-membered rings, respectively. Similarly, the reactants DMDOMA and DMDOSA form only six- and seven-membered chelates.

TABLE 1.2 Extraction of Actinide Amides from Nitrate Solutions into an Organic Solvent Showing the Effect of Chelate Ring Size

Metal ion	Extractant[a]	K_{ex}	Ring size
UO_2^{2+}	TBOA	0.005	5
UO_2^{2+}	TBMA	10.6	6
UO_2^{2+}	TBSA	9.3	7
Am^{3+}	DMDOMA	0.47	6
Am^{3+}	DMDOSA	<0.000 2	7

[a] *TBOA* $(C_4H_9)_2NCO_2$; *TBMA* $[(C_4H_9)_2NCO]_2CH_2$; *TBSA* $[(C_4H_9)_2NCO]_2(CH_2)_2$; *DMDOMA* $(CH_3C_8H_{17}NCO)_2CH_2$; *DMDOSA* $(CH_3C_8H_{17}NCO)_2(CH_2)_2$.

The electron shells of the M^{3+}+ elements with unfilled or partially filled 3d, 4d, and 5f orbitals contract as these shells are being filled with electrons, increasing the charge density (z/r) of the cation. This leads to increased stability constants for the ML_3 complexes with increasing atomic number. For bidentate ligands, the *f*-electron (lanthanide) ML_3 complexes are coordinatively unsaturated, i.e., only six of eight available coordination sites are filled. Therefore, the two remaining sites are occupied by H_2O or HL or some other donor molecules *E*. The electronic structure of the central atom is also important for synergism, as illustrated by the easy complex formation of metal ions with unfilled or partially filled d or f orbitals.

7.6.2. Steric Hindrance Effect

The closer the reactive ligand atoms get to the central metal ion, the stronger the bond, and, consequently, the greater the formation constant. This can be studied by making substitutions in the organic ligand, which structurally interferes with the formation of the complex. For example, Dyrssen[76] substituted oxines (8-hydroxyquinolinols), in various positions, and measured their distribution constants, K_d. Although the pK_a and K_{dHL} did not vary much with the substitution, the K_d for the ThL_4 complex with unsubstituted oxine was 425, and for the 5-methyl substituted complex around 1000, while the 2-methyl-substituted complex was not extractable. This can be attributed to the 2-methyl group blocking the complex formation.

Branching has complex effect on steric hindrance. Isoforms of TBP (TiBP) and DOBA (DOiBA) are more branched. But the branching has little effect on K_{ex} when TBP is replaced by TiBP due to the free rotation of the substituents around the phosphorus atom. On the contrary, at using the lower group, the amide DOBA, and its branched isomer DOiBA, the branched substituent strongly depresses the extraction constant and forms less stable complex due to the restricted rotation of substituents around the amide moiety.

Steric hindrance can be predicted by modeling the molecular structures and chemical reactions. But advanced programs require mainframe computers.

7.6.3. Donor Ligand Effects

Table 1.3 shows extraction constants for some metal ions with three alkyl phosphates substituted by 0, 1, and 2 sulfur atoms, but with almost identical aliphatic branching. Previously (see Section 2.3) we discussed hard and soft acids and bases (HSAB theory) due to the extraction processes. According to this theory, hard acids form strong complexes with hard bases, while weak acids form strong complexes with weak bases. In Table 1.3, the metals are ordered in increasing hardness, the subgroup II.b being rather soft. Presumably, the K_{ex} values reflect this pattern, as they are proportional to K_a z and β_z. In Table 1.3, the acidity of the acids increases (i.e., K_a increases) in order R–PSSH,

TABLE 1.3 Comparison of the Extraction Constant K_{ex} for the IIb-Subgroup Divalent Ions, and IIIb Lanthanide Ions, with Oxygen (R–POOH) or Sulfur (R–POSH, R–PSSH) Dialkyl Phosphoric Acids[a]

Metal ion	Hg^{2+}	Cd^{2+}	Zn^{2+}	$_{57}La^{3+}$	$_{63}Eu^{3+}$	$_{71}Lu^{3+}$
Charge density Z^2/r	3.6	4.1	5.4	7.83	8.74	9.68
Log K_{ex} for R–POOH	−2.20	−1.80	−1.20	−2.52	−0.44	2.9
Log K_{ex} for $(C_4H_9O)_2\cdot POSH$	5.40	3.70	0.70	−4.78	−4.23	0.34
Log K_{ex} for $(C_4H_9O)_2\cdot PSSH$	4.40	3.49	2.40		−8.28	

[a]*R is* $(C_4H_9CH(C_2H_5)CH_2O)_2$.

R–POSH, and R–POOH, as sulfur is less basic than oxygen. In general, the K_{ex} increases for the dialkyl phosporic acid (hard donor ligand) with increasing metal charge density within each group as is predicted in Chapter 3. For the soft metals, K_{ex} also increases with increasing softness of the ligand, while the opposite effect is seen for the hard Ln–metal ions. The divalent subgroup II.b metals prefer to bind to sulfur rather than to oxygen because they have a rather soft acceptor character, while the hard metals III.b prefer to bind to the hard O atom of the ligand.

7.6.4. Inner and Outer Sphere Coordination Effect

The extractabilities of metal–organic complexes depend on whether inner or outer sphere complexes are formed. Extraction of uranyl nitrate by TBP is a good example. The free uranyl ion is surrounded by water molecules forming $UO_2(H_2O)_6^{2+}$ Thus, in solution (as well as in the solid salt), the uranyl ion is surrounded by six molecules of H_2O in an *inner coordination sphere*. With nitrate anions it forms $[UO_2(H_2O)_6]^{2+}(NO_3)_2^{2-}$, though it is commonly written $UO_2(NO_3)_2(H_2O)_6$. In the solid nitrate salt, the distance $d_{U\text{–}O}$(nitrate) between the closest oxygen atoms of the nitrate anions, $(O)_2NO$, and the U–atom is longer than the corresponding distance, $d_{U\text{–}O}$(water), to the water molecules. Thus the nitrate anions are in an *outer coordination sphere*.

When the extractant TBP is added, the $OP(OC_4H_9)_3$ groups are closer to the U atoms than the OH_2's, $d_{U\text{–}O}(TBP) < d_{U\text{–}O}(water)$. Two waters are expelled, although the OP bond only occupies one coordination site. The charge distribution around the UO_2 group allows the nitrate group to enter the inner coordination sphere of the uranyl complex. Thus, the extraction reaction can more correctly be written:

$$UO_2(H_2O)_6^{2+} + 2NO_3^- + 2TBP_{(org)} \leftrightarrow UO_2TBP_2(NO_3)_{2(org)} + 6H_2O_{(aq)} \tag{1.88}$$

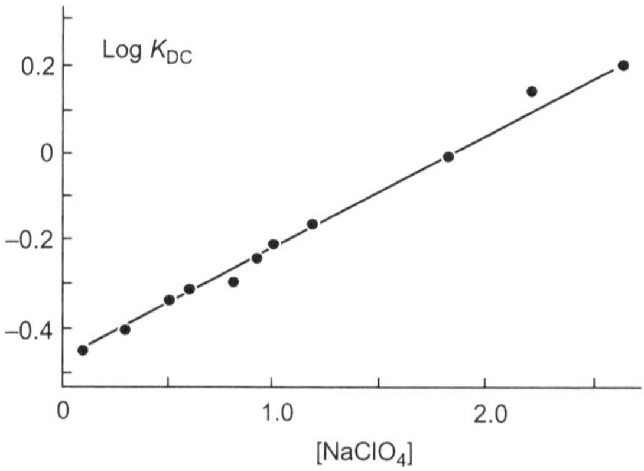

FIGURE 1.13 The distribution constant K_{DC} for zinc diacetylacetonate, ZnL_2, between benzene and various concentrations of $NaClO_4$ in the aqueous phase at 25 °C. *(Source: From Ref. 70)*

In many systems, it is difficult to distinguish between inner and outer sphere complexation. Knowledge about the thermodynamic parameters ΔH and ΔS may be of help. For inner sphere complexes, the hydration is disrupted more extensively and the net enthalpy and entropy changes are positive. In outer sphere complexes, the dehydration is less disrupted, and the net enthalpies and entropies are negative owing to the complexation with its decrease in randomness without a compensatory disruption of the hydration spheres.

Migration experiments have shown that the hydrated cations carry with them not only the water in the inner coordination sphere, but also one or more shells of additional water molecules, for typical total values of 10–15. When the metal ion leaves the aqueous phase in the SX step, this ordered coordinated water returns to the bulk water structure, contributing an additional factor to consider in evaluating the thermodynamics of extraction.

7.7. Effects of the Aqueous Medium

The effect of the water activity of the aqueous phase, discussed in the Section 2, is determined by the total concentration and nature of the salts. Generally, the distribution constant for a neutral complex would increase with increasing ionic strength. This *salting-out effect* (see Fig. 1.13) is often ascribed to a reduction in free water available for hydration. On the other hand, the salt also breaks down the water structure, which could reduce the energy to form a hole in the phase.

8. MODULE DESIGN CONSIDERATIONS

Solvent extraction results are mostly presented in the form of diagrams. This is schematically illustrated in Fig. 1.2 for three hypothetical substances,

A, B, and C. The range of distribution ratio, D_S, is usually from about 0.1 to 10, although ranges from about 10^{-5} to 10^4 can be measured. The distribution ratio is measured as a function of the variables of the extraction system: concentration of some reactant Z, pH, concentration of extractant in the organic phase, [HL], the extractant anion concentration in the aqueous phase, salt concentration in the aqueous phase, etc.

In many cases, more informative is extraction percentage, %*E*. Such a plot is shown in Fig. 1.2b for the same system as shown in Fig. 1.2a. Percentage extraction curves are particularly useful for designing separation schemes.

A convenient way to characterize the S-shaped curves in Fig. 1.2b is by the pH value of 50% extraction (the pH_{50} value). This is indicated in the figure for distribuents A and B, assuming that *Z* is the pH value.

Very efficient separation is needed in industry, and a single extraction stage may be insufficient. In the design of separation using multistage extraction, other extraction diagrams are preferred. Only single-stage extraction is discussed in this chapter and multistage extraction is discussed in Chapter 4.

8.1. Calculation of Equilibrium Constants

Previous sections presented chemical models for the extraction of acids and metals into organic solvents, and showed that these models, expressed mathematically, agree with experimental data at trace metal concentrations and at constant activity coefficients. These models provide a rationale for understanding the chemical principles of SX.

From plots of the distribution ratio against the variables of the system: [M], pH, $[HA]_{org}$, [B], etc.—an indication of the species involved in the SX process can be obtained from a comparison with the extraction curves presented in this chapter. Sometimes this may not be sufficient, and some additional methods are required for identifying the species in SX. These and a summary of various methods for calculating equilibrium constants from the experimental data, using graphical as well as numerical techniques, is discussed below. Calculation of equilibrium constants from SX is described in several monographs.[77–81]

8.2. Identification of Species

An initial step in data analysis is to develop an equation that represents the experimental data reasonably. Although previous sections dealt with this issue, the approach assumed that certain species are formed. Two alternatives to this procedure are discussed here, both yielding the approximate stoichiometry of the complexes formed in the system. The most elementary is referred to as *Job's method*, while the *ligand number method*, developed by J. Bjerrum, is slightly more advanced.

8.2.1. Job's Method

When a metal M with a ligand L forms an extractable complex of the chemical form M_mL_a, the extraction is at maximum when the molar ratio of these two in the system is m:a. Similarly, if the extracted species is ML_aE_b, a plot of the molar ratio of A and E yields a maximum at the ratio a:b. The curves are almost symmetrical around the mole fractions 0.5:0.5, indicating that the extracted species has a 1:1 ratio.

This method is useful when only one species is extracted, but it has little value for the study of SX systems that contain several complex species.

8.2.2. Ligand Number Method

This method[82–84] is useful for identifying the average composition of the metal species in the system. Consider the extraction of ML_z, and assume that only one species, ML_n, exists in the aqueous phase. Taking the derivative of the logarithm of distribution ratio vs. anion concentration yields

$$\left(\frac{\mathrm{d}\log D_S}{\mathrm{d}\log[\mathrm{L}]}\right) = z - n \quad (1.89)$$

The concept of average ligand number, $\bar{n}$, defined as the number of ligands per central atom:

$$\bar{n} = \frac{\sum n[\mathrm{ML}_n]}{[\mathrm{M}]_t} \quad (1.90)$$

which can be rewritten

$$\bar{n} = \frac{z - \mathrm{d}\log D_S}{\mathrm{d}\log[\mathrm{L}]} \quad (1.91)$$

For example, in Fig. 1.14 is shown a smoothed curve of measurements of the distribution of Th(IV) from 0.1 M $NaClO_4$ into chloroform containing the extractant acetylacetone (HL).[83] Taking the derivative of this curve according to Eqn (1.10), the average ligand number is derived as shown in the lower insert. Th(IV) is successively complexed by L^- forming ThL^{3++}, ThL_2^{2+}, ThL_3^+, and uncharged ThL_4, which is extracted. At pL > 8.5, the n value is zero, i.e., Th is uncomplexed, while at pL < 3.5 the average ligand number of 4 is reached, i.e., Th is fully complexed as ThL_4. The average ligand number can be used to obtain approximate equilibrium constants, as described by Bjerrum,[82] assuming that at half integer $\bar{n}$ values the two adjacent complexes dominate: e.g., at $\bar{n} = 0.5$ the species M^{z+} and ML^{z-1} dominate, while at $n = 1.5$ ML^{z-1} and ML_2^{z-2} dominate, etc. The following expression for the stepwise formation constant is approximately valid at

$$\bar{n} = n - 0.5, \quad \log K_n \approx -\log[\mathrm{L}]_n \quad (1.92)$$

This method of obtaining an estimate of the formation constants was done as a first step in the Th(IV)–acetylacetone system in Fig 1.14, where in the lower figure, n is plotted against $-\log[L^-]$, yielding the preliminary log K_n values.

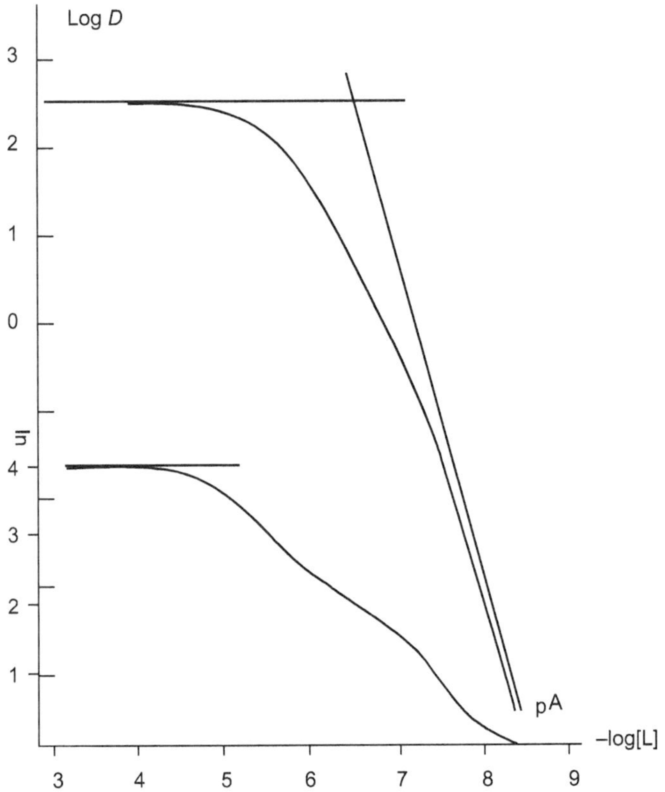

FIGURE 1.14 (Upper curve) The distribution of Th(IV) between benzene and 0.1 M $NaClO_4$ as a function of aqueous acetylacetonate ion concentration pL = −log[L] the asymptotes have slopes 0 and −4. (Lower curve) The average number of ligands per central atom, *n*, in the same system, as obtained from a derivation of the log *D*(pL) curve. Using the ligand number method, the following equilibrium constants were estimated (with values from graphical slope analysis within parenthesis): log K_1 8.0 (7.85), log K_2 7.6 (7.7), log K_3 6.4 (6.3), and log K_4 5.1 (5.0). Log *K*D4 2.50 is obtained from the horizontal asymptote. *(Source: From Ref. 83)*

8.2.3. Graphic Slope Analysis

For a rational application of slope analysis, it is important to measure the variation of the distribution ratio *D* with one component *x* at a time while the others, *C*, are kept constant. The SX equation can then be expressed in the form of a simple polynomial of type

$$y = a_0 + a_1 x + a_2 x^2 + \dots \qquad (1.93)$$

where y is a function of the distribution ratio $D_{S\,\mathrm{const}}$ and x a function of the variable (e.g., pH, the free ligand ion concentration $[L^-]$, the concentration of free extractant HL or adduct former E in the organic phase, etc.).

8.2.4. Linear Plots

When the distribution equation can be expressed in the form $y = a_0 + a_1x$, from a plot of y vs. x the intercept on the y axis yields the a_0 parameter and the slope the a_1 parameter. This treatment is referred to as the *limiting value method.*

At extraction of Th(IV) by acetylacetone (see subsection 8.2.2 and Fig. 1.14) it was concluded that the aqueous phase contained all ThL_n complexes with $0 < n < 4$, and the organic phase contained only the uncharged ThL_4 complex. We can therefore write the distribution of Th(IV) between chloroform and water:

$$D_{Th} = \frac{[ThL_4]_{org}}{\sum[ThL_n]_{aq}} = \frac{K_{ThL_4}[L]^4\beta_4}{\sum\beta_n[L]^n} \tag{1.94}$$

This is rearranged to yield

$$\frac{1}{D_{Th}} = Q_0 = \left(1 + \frac{a_1}{x} + \frac{a_2}{x^2} + \frac{a_3}{x^3} + \frac{a_4}{x^4}\right) \Big/ K_{ThL_4} \tag{1.95}$$

where K_{ThL_4} is the distribution constant of ThL_4, x equals [L], and $a_{4-n} = \beta_n/\beta_4$. In a plot of F_0 against x^{-1}, the intercept becomes $\frac{1}{K_{ThL_4}}$ and the slope is $\frac{a_1}{K_{ThL_4}}$.

In the next step, the function Q_1 is calculated:

$$Q_1 = \left(\frac{K_{ThL_4}}{D_{Th}} - 1\right)x = a_1 + \frac{a_2}{x^1} + \frac{a_3}{x^2} + \frac{a_4}{x^3} \tag{1.96}$$

A plot of Q_1 vs. x^{-1} yields a_1 at the intercept and a_2 at the slope. In third step, $Q_2 = (Q_1 - a_1)x$ is calculated, yielding a_2 and a_3, etc. Using this technique, all four a_n values are obtained, from which one can deduce the β_n values, K_n values, and the K_{ThL_4} values. This principle can be useful for estimations when there is insufficient reliable experimental data.

8.2.5. Nonlinear Plots

It is not possible to obtain simple linear relations between D_S and the variables when both the aqueous and organic phases contain several solute (metal) species. The series of polynomials such as

$$D_S = \frac{b_0 + b_1y + b_2y^2 + b_3y^3 + \ldots}{a_0 + a_1x + a_2x^2 + a_3x^3 + \ldots} = \frac{\sum b_iy^i}{\sum a_nx^n} \tag{1.97}$$

is obtained which has no simple solution.

Here different complexes can be expected. Stepwise complexes may be formed with a ratio of Eqn (97) type. Even in this case, a graphic extrapolation technique was useful by determining a set of intermediary constants for each constant $[HL]_{org}$ from $\log D_S$ vs. pL. Plotting these intermediate constants vs. $[HL]_{org}$, a new set of constants were obtained, from which both the stepwise formation constants $\beta_{n,p}$ and the adduct formation constants K_{ad1} and K_{ad2} may be obtained. However, with this model and the constants calculated, $\log D_S$ could be predicted.

8.2.6. Numerical Methods

The most complicated equations presented in this chapter are of the types shown by Eqn (1.97), where a series of complexes are formed in the organic phase and in the aqueous phase. More complicated equations with mixed complexes and/or polynuclear complexes are not discussed here.

The SX Eqn (1.97) where x and y are independent variables and a_n and b_n are unknown independent parameters. The polynomial in the denominator refers to the formation of aqueous phase complexes; in this case the metal forms several series of complexes with different ligands and contains several polynomials. The polynomial in the numerator always refers to the formation of organic phase complexes.

There is no exact numerical solution to Eqn (1.97) when both x and $y > 0$. In practice, therefore, one variable must be kept constant (or zero), while the value of the other changes. If x is kept constant, the double polynomial is reduced to a simple one:

$$y = C_x \sum \beta_i y^i \left(\frac{1}{C_x} = a_0 + a_1 x + a_2 x^2 + \ldots \right) \tag{1.98}$$

where C_x is a constant at constant x. The problem is reduced to the determination of the parameters of a simple polynomial, to which many numerical methods have been applied with varying degrees of success. The main requirement is that all computed parameters should be positive numbers (or zero). In order to solve an algebraic system of n parameters, only n equations are needed.

When the SX reaction can be described by Eqn (1.98), there are as many equations as experimental points. As a rule, the experimental points are much more than the number of unknown parameters. It is important to use the complete suite of experimental data that give greater significance to the a_n or b_n values.

There are several ways to solve a large number of polynomials like Eqn (1.98), e.g., by minimizing the sum of the residuals, the least square (LSQ) method being the standard procedure. It is an objective regression analysis method, which yields the same results for a given chemical system, independent of method of investigation. Because the method may be applied in somewhat different ways, it may sometimes give slightly different results. Therefore, in judging a set of equilibrium constants, it is useful to learn how an author has applied the minimization technique. x is usually the free ligand concentration $[L^-]$, or concentration [E] of the extractant, while y is a simple function of the distribution ratio D. The (LSQ) method requires that S (the *weighted squared residuals*) in Eqn (1.99) is minimized:

$$S = \sum_{i=1}^{F} w_i \left(\sum_{n=0}^{N} (a_n x_i^n) - y_i \right)^2 \tag{1.99}$$

where $N + 1$ is the number of parameters and F is the number of experimentals points; $N + F - 1$ is referred as the number of degrees of freedom of the system; each point has a value x_i/y_i.

Because experiments are carried out over a large range of D, $[L^-]$, and/or [E] values, the points carry different algebraic weight (e.g., the value 1000 obscures a value of 0.001). Therefore, in order to use the LSQ technique properly, each point must be correctly weighted, w_i. This can be done in several ways, the most common being to weight it by y_i, or by σ_i^{-2}, or by a percentage value of y_i. σ_i is the standard deviation in the measurement of y_i. The difference $a_i x_i^n - y_i$ (the *residual*) is not zero, because it has to be taken between a measured value, y_i (meas.), and the corresponding calculated value, y_i (calc.), by the $a_n x_i^n$ function (i.e., y_i(calc.) − y_i(meas.)), using the actual a_n values at the time of the operation.

The principle of the LSQ technique is to compute the set of positive a_n values that give the smallest sum of the residuals; Eqn (1.99) reaches the S_{min} value. If the residuals equal zero (which rarely occurs in practice), S_{min} would be zero, and there would be a perfect fit between the experimental points and the calculated curve.

There are several mathematically different ways to conduct the minimization of S.[85–87] Many programs yield errors of internal consistency (i.e., the standard deviations in the calculated parameters are due to the deviations of the measured points from the calculated function) and do not consider external errors (i.e., the uncertainty of the measured points). The latter can be accommodated by weighting the points by this uncertainty.

More advanced statistical tests for systems of this kind have been described by Ekberg.[88]

9. EXPERIMENTAL DETERMINATION OF DISTRIBUTION RATIOS

After identifying the reacting species in an SX process and developing a quantitative model for their interactions, the fundamental parameter, the distribution ratio, from which extraction curves are derived, has to be determined. SX research work can be carried out with simple batchwise (or point-by-point) technique or continuous on-line measurements which give faster and more accurate results. Some experimental techniques are described below.

9.1. Stirred Cell Semicontinuous Techniques

Each point-by-point experiment requires a complete set of mixing, separation, sampling, and analysis. This usually leads to scattered results, though it may not be critical, if the D values cover a limited range from 0.1 to 10. However, the more the D values deviate from 1, the more accurate must be the measurements; also the number of points required for a reliable extraction curve usually increases.

To reduce the uncertainty and labor involved with the batch technique, the stirred cell technique has become popular. Figure 1.15 shows a typical thermostated stirred cell. The liquid volumes are commonly $50 + 50$ mL. The stirrer may consist of a single paddle at the interface or a double paddle, one in the center of each phase. The cell may contain baffles to improve the mixing. Stirring may be violent, completely destroying the interface and producing very small ($\ll$1 mm) droplets, or slow in order not to destroy the interface. The stirring rate is optimized to the time for reaching equilibrium and complete

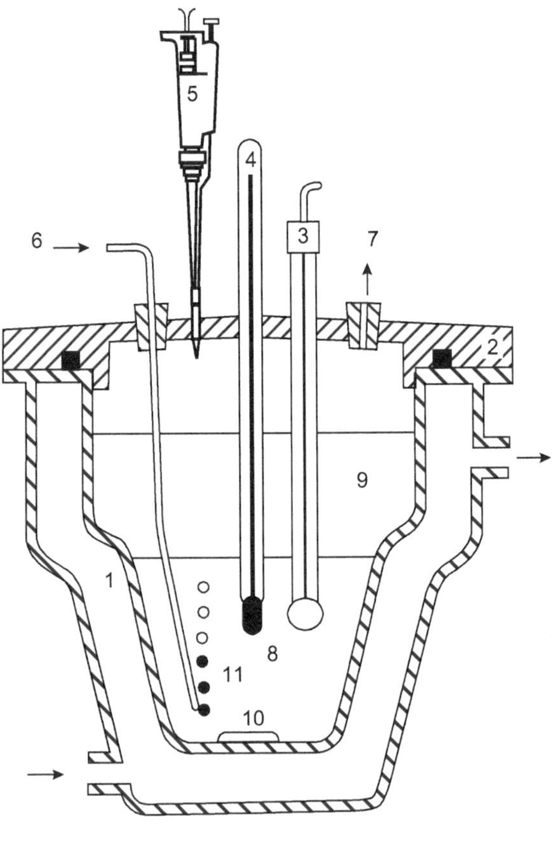

FIGURE 1.15 A thermostated double jacket (1) cell for solvent extraction studies Cover of cell (2), (heavier phase 8, lighter phase 9) under nonoxidizing conditions, using a hydrogen gas inlet (6) to a Pd black catalyst (11), pH glass electrode (3), magnetic stirrer (10), connections for additions (5, 7, thermometer (4)). Alternative constructions contain rotating paddles and fixed pipings connected to the two phases for frequent sample withdrawals.

phase separation. The experiments are carried out either with intermittent violent stirring, in which case samples are withdrawn after each complete phase separation, or with mild stirring during which it is possible to continuously withraw samples. Equal volumes are sampled each time, commonly <1 mL. The simple stirred cell has been improved by introducing phase-discriminating membranes in the sampling outlet.[89] This is particularly advantageous for kinetic experiments. The sampling of the stirred cell can be automated, so that at regular intervals pH and temperature are recorded, and samples are withdrawn for automatic analysis of concentration of interesting species in a more or less standard fashion. It is also possible to use ion-sensitive probes in one of the phases instead of sampling.

9.2. Centrifugal Extraction–Separation Systems

A different approach to SX experimentation is AKUFVE principle.[90] The AKUFVE is titanium made, or PEEK (polyether ether ketone) to allow measurements under very corrosive conditions. The separated phases pass AMX gadgets for on-line detection (radiometric, spectrophotometric, etc.) or phase sampling for external measurements (atomic absorption, spectrometric, etc.), depending on the system studied. The aqueous phase is also provided with cells for pH measurement, redox control (e.g., by reduction cells using platinum black and hydrogen, metal ion determination, etc.), and temperature control (thermocouples). The AKUFVE technique allows a large number of points (50–100) to be determined in a one-day experiment over a D-range of better than 10^3–10^{-3}, not counting time of preparation (Fig. 1.16).

The centrifugal separator of the AKUFVE system is also used for phase separation in the SISAK technique.[91] SISAK is a multistage SX system that is used for studies of properties of short-lived radionuclides, e.g., the chemical properties of the heaviest elements, and SX behavior of compounds with exotic chemical states. In a typical SISAK experiment, radionuclides are continuously transported from a production site in an accelerator or a research reactor to the SISAK equipment via a gas jet transport system. The nuclides are dissolved in an aqueous phase that is fed into a centrifuge battery comprising 1–4 SX steps. The product solution leaving the last step is pumped, e.g., to a nuclear radiation detection system. The transport time from the target site to the detection system depends on the centrifuge size, number of centrifuge steps, and flow rate. For a one-step chemistry, i.e., dissolution step and a single centrifuge extraction, and maximum flow rates, the overall transport time is around 2.5 s. This fast transport has allowed detailed γ-spectroscopic studies of radionuclides with half-lives around 1 s. Recently, the SISAK equipment was successfully applied to studies of the heaviest elements, and SX data were obtained for element 104.[92] Fast chemical separation systems have been developed for quite a large number of elements.[93]

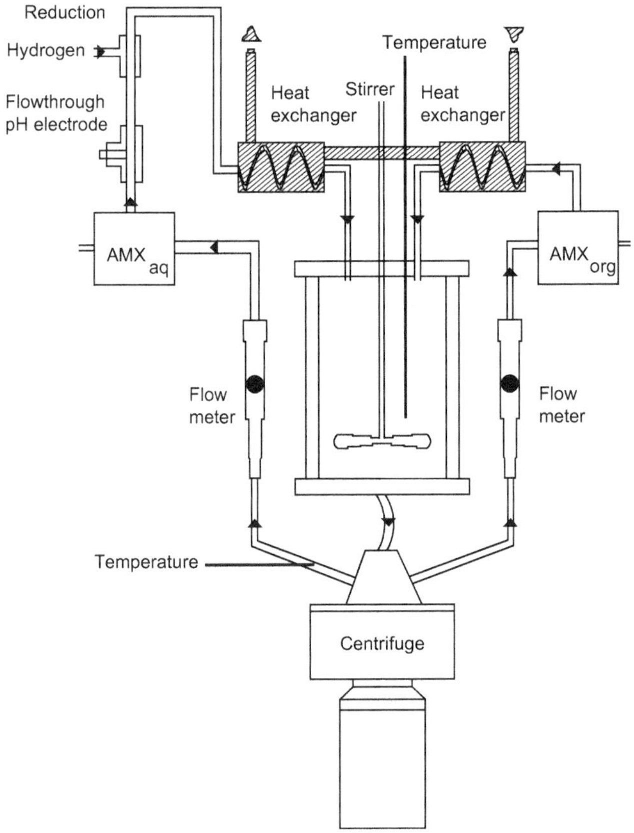

FIGURE 1.16 The AKUFVE solvent extraction apparatus: efficient mixing is achieved in the separate mixing vessel, from which the mixture flows down into the continuous liquid flow centrifugal separator (the H-centrifuge, hold-up time <1 s) (*Source: From Ref. 90*). The outflow from the centrifuge consists of two pure phases, which pass on-line detectors, AMXs, for on-line detectors or continuous sampling. *(Source: From Refs. 80, 83)*

Centrifugal extractors have been designed for a number of industrial uses (see Chapter 4). In some cases, they have been scaled down to laboratory size but mainly been used for developing industrial multistage processes.

10. SUMMARIZING REMARKS

The SX process is usually described by a single net reaction, defined by the extraction constant K_{ex}. Variations in K_{ex} caused by modifications of the solvent system, such as changes in the temperature or aqueous ionic strength, by replacing one solvent by another, or by making substitutions in the extractant molecule may be explained by careful consideration of the parameters of the system. However, such studies are difficult and not always sufficient for predicting new systems. A better foundation for understanding the extraction process is to consider the steps in the process contributing to the net extraction reaction, particularly when these steps are governed by regularities. Knowledge of these regularities helps in interpreting systems as well as in predicting new ones. Extension of these distribution data to thermodynamic constants is likely to give benefits in increased chemical knowledge of the behavior of solutes in different solvent systems. Advanced quantum chemistry calculations and computer modeling of extraction processes can help us in designing new, selective SX systems, as well as in interpreting extraction phenomena.

Many organic substances as well as metal complexes are less extracted from aqueous solutions into organic solvents than expected from simple considerations such as the amount of organic matter in the solute or their solubility in organic solvents. Such substances are hydrated. More basic donor molecules can replace such water, forming adducts. For the most common oxygen-containing adduct molecules, the efficiency of the replacement depends on the charge density, also referred as basicity, of the oxygen atoms. The basicity is presented in the form of *donor number*. The larger the difference between the donor number of water and the adduct former, the larger the adduct formation constant. Often the donor property has to be rather strong, which is the case for many phosphoryl compounds (such as TBP, tributylphosphineoxide (TBPO), and TOPO), because the concentration of H_2O in the aqueous phase is very large (often >50 M), even though H_2O is only a moderately strong donor.

Table 1.4 gives a survey of the most common extraction processes. In general, Type I extraction refers to the distribution of nonelectrolytes, without (A) or with adduct former (B). Type II refers to extraction of (mainly organic) acids, Type III to the extraction of metal complexes, and Type IV to the special (but common) use of SX for evaluation of formation constants for hydrophilic complexes. An arrow within parentheses suggests a reaction of secondary importance. Three examples are all of Type III-C, but also contain elements of Types I-A (the distribution of TBP) and II-A (the distribution of the weak acid Thenoyltrifluoroacetic acid (HTTA)), though the presence of undissociated acid (HTTA) or the acid adduct (TBP-HTTA) is not discussed. In the evaluation of experiments, all molecular species are present and all equilibria must be taken into account, as demonstrated subsequently for a number of cases.

Solutes containing metals can further be classified according to the type of ligand; *N* refers to the maximum coordination number of the metal relative to the ligand:

Class A: Type MX_N. (Note: It is generally assumed that the ligand is monovalent.) A small number of almost purely covalent inorganic compounds that are extracted by nonsolvating organic solvents. These complexes are nonelectrolytes and almost as inert as the solutes of Type I-A.

TABLE 1.4 Survey of Fundamental Liquid–Liquid Distribution Processes[a]

Type I-A	*Nonelectrolyte extraction.* Solute A extracted into organic phase[b] (equilibrium governed by the Nernst distribution law). Solute is the nonelectrolyte A in water	A $\downarrow\uparrow$ A
Type I-B	*Nonelectrolyte adduct formation and extraction.*[c] Adduct AB in organic phase (plus eventually B). Solute A and adduct former (or extractant) B	$B \quad AB$ $(\downarrow\uparrow) \quad \downarrow\uparrow$ $A + B \leftrightarrow AB$
Type II-A	*Extraction of nonadduct organic acids.* Acid and dimer (and possible polymers) in organic phase. Acid dissociation in aqueous phase	$HA \leftrightarrow H1/2H_2A_2 + \ldots$ $\downarrow\uparrow$ $HA \leftrightarrow H^+ + A^-$
Type II-B	*Extraction of acid as adduct.* Acid adduct (and acid and adduct former) in organic phase. Acid dissociation in aqueous phase	$HAB(\leftrightarrow)B\ (+)\ HA$ $\downarrow\uparrow \quad (\downarrow\uparrow)\ (\downarrow\uparrow)$ $HAB \leftrightarrow B + HA \leftrightarrow H^+ + A^-$
Type III-B[d]	*Extraction of saturated metal complex.* Neutral, coordinatively saturated metal complex in organic phase. Metal ion Mz^+ is complexed by zA^- ligands	MA_z $\downarrow\uparrow$ $M^{z+} + zA^- \leftrightarrow MA_z$

Type III-C	*Adduct extraction of unsaturated metal complex.* Formation of coordinatively saturated metal complex trough adduct former B[e]	$\begin{array}{c} B \quad MA_zB_x \\ (\downarrow\uparrow) \quad \downarrow\uparrow \\ M^{z+} + zA^- + xB \leftrightarrow MA_zB_x \end{array}$
Type III-D	*Liquid anion exchange extraction.* Organic phase with metal complex, anion exchanger L^- and organic amine	$\begin{array}{c} RNH^+ + L^- \; (RNH^+)_p(ML^-{}_x)_p \\ \downarrow\uparrow\downarrow\uparrow \\ M^{z+} + xL^- + pRNH^+ + L^- \leftrightarrow (RNH^+)_x(ML_z^-)_p \end{array}$
Type III-E	*Extraction of ion pairs and other unusual complexes.* Ion pair $C_1^+A_2^-$ (and counter species) in organic phase. Aqueous cation C_1^+ and anion A_2^- associated into ion pair $C_1^+A_2^-$	$\begin{array}{c} C_1^+A_2^- \\ \downarrow\uparrow \\ C_1^+A_1^- + C_2^+A_2^- \leftrightarrow C_1^+A_2^- + C_2^+A_1^- \end{array}$
Type IV	*Hydrophilic complex formation and solvent extraction.* Coordinatively saturated metal complex in organic phase. Formation of extractable and nonextractable complexes	$\begin{array}{c} MA_z \\ \downarrow\uparrow \\ M^{z+} + zA^- + nX^- \leftrightarrow MA_z + MX_n^{z-n} \end{array}$

[a]*The organic phase (solvent, diluent) is assumed to be "inert" (shaded area). The aqueous phase (nonshaded area) is unspecified, but may contain various salting agents, not considered here.*

[b]*A nonelectrolyte solute is denoted A; an electrolyte solute is assumed to be the cation Mz^+ and anion A−, L−, or X−.*

[c]*The extractant (or reactant) is denoted A− (from acid HA), or ligand L−, and by B (for adduct).*

[d]*Type III-A (denoted Class A in first edition of this book) is closely related to and covered by Type I-A.*

[e]*If B is undissociated HA, the self-adduct $MA_z(HA)x$ may be formed forming bifunctional chelates, z = N/2. They belong to the extraction Type III-B.*

Class B: Type MA_z. Neutral coordinatively saturated complexes formed between the metal ion and a lipophilic organic acid. This class contains the large group of metal–organic chelate compounds.

Class C: Type MA_zB_b or ML_zB_b. These are Type III-C complexes. They are neutral complexes formed between the metal ion and ligands A^- or L^-, where the neutral complex MA_z or ML_z is coordinatively unsaturated ($N > z$ or $2z$) and acts as an acceptor for uncharged organic compounds (adducts B) containing lipophilic donor groups. If the system does not contain any donor molecules B, the water of hydration may be replaced by undissociated HA (assuming the ligand A^- to be a dissociation product of HA), at least at high HA concentrations; the $MA_z(HA)_x$ complexes are refered to as *self-adducts*.

Class D: Ion pairs, consisting of the metal bound in an anionic complex (e.g., ML_n^{z-n}, where $n > z$) and one or more large organic (usually monovalent) cations (symbolized by RNH^+); the extracted complex is written $(RNH)_{n-z}ML_n$.

Class E: Metal complexes that do not fit into these categories; e.g., other types of ion pairs and chlathrate compounds. All metal ions in water are hydrated, and at higher pH most of them also hydrolyze. It can be difficult to distinguish between the hydrolyzed and complexed species, as well as their self-adducts. For such systems, plots of D_M against $[A^-]$ at various pH and total concentrations of [HA] show three types of curves: (a) for the simple chelate MA_n, (b) for the self-adduct $MA_n(HA)_b$, and (c) for the (mixed) hydroxide $MA_n(OH)_p$ (see Fig. 1.17). It should be noted that the mixed $MA_n(OH)_p$ complexes include the $MA_n + M(OH)_p$ complexes. As mixed complexes are more difficult to determine, they are less often described. However, it is important to realize that if metal hydroxy- complexes are formed and not corrected for, the result of the investigation can be misleading. A test of the system according to Fig. 1.17 rapidly establishes the type of metal complexation.

Because metals differ in size, charge, and electronic structure, no two metals behave exactly the same in the same SX system, not even for the same

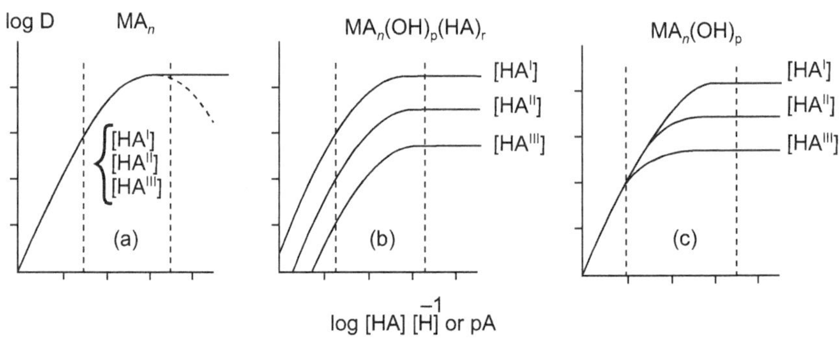

FIGURE 1.17 Extraction curves for various types of metal chelate complexes.

class of solutes. Nevertheless, there are systematic trends in the formation and extraction of these complexes, as described in Chapter 3. Here, the emphasis is on models that give a quantitative description of the extraction within each type or class.

In the subsequent discussion, the following simplifications are made:

1. The systems behave "ideally," i.e., the activity factors are assumed to be unity, unless specifically discussed.
2. The metal extracted is in trace concentration: [M]t_ [M] $\ll$ [Extractant] and this simplifies the equations.
3. The reactants are at very low concentrations in both phases.

These are great simplifications in comparison with the industrial SX systems described in later chapters. Nevertheless, the same basic reactions occur also in the industrial systems, although activity factors must be introduced or other adjustments are made to fit the data, and the calculation of free ligand concentrations are more complex. Some of these simplifications are not used in later chapters.

REFERENCES

1. Hildebrand JH, Scott RC. *Solubilities of nonelectrolytes*. New York, USA; London, UK: Dover Publishers; 1964.
2. Barton AFM. *CRC handbook of solubility parameters and other cohesion parameters*. New York, USA; London, UK: CRC Press; 1991.
3. Hansen C. *Hansen solubility parameters: a users handbook*. New York, USA; London, UK: CRC Press; 2000.
4. Nilsson M. *Influence of organic phase composition on the extraction of trivalent actinides. Diploma work, Department of Nuclear Chemistry*. Gothenburg: Chalmers University of Technology; 2000.
5. Nernst WZ. *Phys Chem* 1891;**8**:110.
6. Kislik V, Eyal A. *Mechanisms of titanium (IV) and some other transition metals extraction by organophosphorus acids*. Proceedings of ISEC'93, 1993;**2**:1353–60.
7. Kislik VS, editor. *Liquid membranes. Principles and applications in chemical separations and wastewater treatment*. Elsevier; 2010.
8. Bird RB, Stewart WE, Lightfoot EN. *Transport phenomena*. New York: John Wiley & Sons; 1960.
9. Cussler EL. *Diffusion, mass transfer in fluid systems*. Cambridge: Cambridge University Press; 1984.
10. Ridberg J, Cox M, Musikas C, Choppin GR, editors. *Principles and practices of solvent extraction*. New York: CRC; 2004.
11. Giddings JC. *Unified separation science*. New York: John Wiley & Sons; 1991 17–48.
12. Giddings JC. *Unified separation science*. New York: John Wiley & Sons; 1991 112–4.
13. Bromberg L, Levin G, Kedem O. Transport of metals through gelled supported liquid membranes containing carrier. *J Membr Sci* 1992;**71**:41.
14. Zhu XX, MacDonald PM. Pulsed-gradient spin-echo NMR measurements of the diffusion coefficients of ketones in poly(methyl methacrylate). *Macromolecules* 1992;**25**:4345.

15. Chrisstoffels LAJ, de Jong F, Reinhoudt DN. Mechanistic studies of carrier-mediated transport through supported liquid membranes. In: Bartsch RA, Way JD, editors. *Chemical separations with liquid membranes. ACS Symposium Series 642*. Washington, DC: American Chemical Society; 1996 18–56. Ch 3.
16. Inoue Y. In: Araki T, Tsukube H, editors. *Liquid membranes: chemical applications*. Boca Raton, FL: CRC Press; 1990. 77–102.
17. Noble RD, Koval CA, Pellegrino L. *J Chem Eng Prog*; 1989 March:58–70.
18. Kiani A, Bhave RR, Sirkar KK. Further studies on solvent extraction with immobilized interfaces in a microporous hydrophobic membrane. *J Membr Sci* 1986;**26**:79.
19. LiY Gokel G. Hernandez 1, Echegoyen, L. Cation-exchange kinetics of sodium complexes with amide- and ester-substituted crown ethers in homogeneous solution. *J Am Chem Soc* 1994;**116**:3087–96.
20. Eigen M, Wilkins RW. Mechanisms of inorganic reactions, *Adv Chem Ser.* Washington, D.C: American Chemical Society; 1965: vol. 76, 49–68.
21. Adamson AW. *Physical chemistry of surfaces*. New York: John Wiley & Sons; 1982.
22. Davies T, Rideal EK. *Interfacial phenomena*. New York: Academic Press; 1961.
23. Neuman RD, Nai-Fu-Zhou, Wu J, Jons MA, Gaonkar AG. General model for aggregationof metal-extractant complexes in acidic organophosphorus solvent extraction systems. *Sep Sci Technol* 1990;**25**(13–15):1655–74.
24. Kislik V, Eyal A. Hybrid liquid membrane (HLM) and supported liquid membrane (SLM) based transport of titanium(IV). *J Mem Sci* 1996;**111**:273–81.
25. Ghidini E, Ugozzoli F, Ungaro R, Harkema S, Abu EI-Fadl A, Reinhoudt DN. Complexation of alkali metal cations by conformationally rigid, stereoisomeric calix[4]arene crown ethers: a quantitative evaluation of preorganization. *J Am Chem Soc* 1990;**112**:6979.
26. Reinhoudt DN, Dijkstra PJ, in 't Veld PJA, Bugge KE, Harkema S, Ungaro R, Ghidini E. Kinetically stable complexes of alkali cations with rigidified calix[4]arenes. X-ray structure of a calixspherand sodium picrate complex. *J Am Chem Soc* 1987;**109**:4761.
27. Casnati A, Pochini A, Ungaro R, Bocchi C, Ugozzoli F, Egberink RJM, Struijk W, Lugtenberg RJM, de Jong F, Reinhoudt DN. 1,3-Alternate calix[4]arenecrown-5 conformers: new synthetic ionophores with better K+/Na+ selectivity than valinomycin. *Chem Eur J* 1996;**2**(4):436–45.
28. Neuman RD, Nai-Fu-Zhou, Wu J, Jons MA, Gaonkar AG. General model for aggregationof metal-extractant complexes in acidic organophosphorus solvent extraction systems. *Sep Sci Technol* 1990;**25**(13–15):1655–74.
29 Eyal AM, Kislik VS. Hybrid liquid membrane (HLM) process for metal recovery. Avicenne joint research project report: Membrane recovery of metal pollutants from waste waters of the fertilizers industry (MERMEP), 1996, p. 25–9.
30. Danesi PR, Horwitz EP, Vandergrift GF, Chiarizia R. Mass transfer rate through liquid membranes: interfacial chemical reactions and diffusion as simultaneous permeability controlling factors. *Sep Sci Technol* 1981;**16**:201–10.
31. Michaels AS. Membranes, membrane processes, and their applications: needs, unsolved problems, and challenges of the 1990's. *Desalination* 1990;**77**:5–34.
32. Szpakowska M, Nagy OB. Chemical kinetic approach to the mechanism of coupled transport of cu(ii) ions through bulk liquid membranes. *J Phys Chem A* 1999;**103**:1553–9.
33. O'Hara PA, Bohrer MP. Supported liquid membranes for copper transport. *J Membr Sci* 1989;**44**:273–81.
34. Schlosser S. Pertraction through liquid and polymeric membranes. In: Belafi-Bako K, Gubicza L, Mulder M, editors. *Integration of membrane processes into bioconversions.*

Proceedings of the16th European Membrane Society Annual Summer School. Veszprem, Hungary, **2000**. New York: Kluwer Academic/Plenum; August 1999. 73–100.
35. Wodzki R, Szczepanska G, Szczepanski P. Unsteady state pertraction and separation of cations in a liquid membrane system: Simple network and numerical model of competitive M^{2+} /H^{+} counter-transport. *Sep Pur Tech* 2004;**36**(1):1–16.
36. Kislik V, Eyal A. Hybrid liquid membrane (HLM) system in separation technologies. *J Mem Sci* 1996;**111**:259–72.
37. Kislik V, Eyal A. Hybrid liquid membrane (HLM) and supported liquid membrane (SLM) based transport of titanium(IV). *J Mem Sci* 1996;**111**:273–81.
38. Eyal AM, Kislik VS. Hybrid liquid membrane (HLM) process for metal recovery. Avicenne joint research project report: membrane recovery of metal pollutants from waste waters of the fertilizers industry (MERMEP), 1996, p. 25–9.
39. Lo TH, Baird MHI, Hanson C, editors. *Handbook of solvent extraction*. John Wiley; 1983.
40. Ensor DD, McDonald GR, Pippin CG. Extraction of trivalent lanthanides by a mixture of didodecylnaphthalenesulfonic acid and a crown ether. *Anal Chem* 1986;**58**:1814–6.
41. Nijenhuis WF, Buitnuis EG, et al. Calixcrowns as selective potassium cation carriers in supported liquid membranes. *J Am Chem Soc* 1991;**113**:7963–8.
42. Cooper SR. *Crown compounds, toward future applications*. New York: VCH; 1992.
43. Lo TH, Baird MHI, Hanson C, editors. *Handbook of solvent extraction*. John Wiley; 1983.
44. Davies T, Rideal EK. *Interfacial phenomena*. New York: Academic Press; 1961.
45. Neuman RD, Nai-Fu-Zhou, Wu J, Jons MA, Gaonkar AG. General model for aggregationof metal-extractant complexes in acidic organophosphorus solvent extraction systems. *Sep Sci Technol* 1990;**25**(13–15):1655–74.
46. Ghidini E, Ugozzoli F, Ungaro R, Harkema S, Abu El-Fadl A, Reinhoudt DN. Complexation of alkali metal cations by conformationally rigid, stereoisomeric calix[4]arene crown ethers: a quantitative evaluation of preorganization. *J Am Chem Soc* 1990;**112**:6979.
47. Marcus Y, Kertes S. *Ion exchange and solvent extraction of metals*. Chichester: Wiley-Interscience; 1969.
48. Van Ooyen J, Dyrssen D, Liljenzin JO, Rydberg J, editors. *Solvent extraction chemistry (ISEC 1966)*. Amsterdam: North-Holland; 1967.
49. Sekine T, Sakairi M. *Bull Chem Soc Japan* 1967;**40**:261.
50. Liljenzin JO, Vadasdi K, Rydberg J. Coordination chemistry in solution. Högfeldt, E. Swedish Nat Res Council; Trans Roy Inst Techn., 280, Stockholm, 1972.
51. Cooper SR. *Crown compounds, toward future applications*. New York: VCH; 1992.
52. Ansari SA, Mohapatra PK, Prabhu DR, Manchanda VK. Evaluation of N, N, N′, N′-tetraoctyl-3-oxapentane-diamide (TODGA) as a mobile carrier in remediation of nuclear waste using supported liquid membrane. *J Mem Sci* 2007;**298**(1–2):169–74.
53. Shu-Lan Ma, Xiao-Zhu Wang, Cui-Hong Fan, Qian-Ling Guo, Wen-Xiang Zhu, Jing Zhang. The syntheses of phenol-containing diazamacrocycles and liquid membrane transports of alkali cations. *J Mol Struct* 2007;**2**:46–52.
54. Moyer BA, Case GN, Alexandratos SD, Kriger AA. Extraction of copper(II) from sulfuric acid by macrocycle-synergized cation exchange: Comparing a novel impregnated resin with its solvent-extraction analog. *Anal Chem* 1993;**65**:3389–95.
55. Parthasarathy N, Buffle J. *Anal Chim Acta* 1994;**284**:649–59.
56. Schaller U, Bakker E, Spichiger UE, Pretsch E. Ionic additives for ion-selective electrodes based on electrically charged carriers. *Anal Chem* 1994;**66**:391–8.

57. Bakker E, Malinowska E, Schiller RD, Meyerhoff ME. Anion-selective membrane electrodes based on metalloporphyrins: the influence of lipophilic anionic and cationic sites on potentiometric selectivity. *Talanta* 1994;**41**:881–90.
58. Carey FA, Sundberg RJ. *Advanced organic chemistry.* 3rd ed. New York: Plenum Press; 1990.
59. Fuks L, Majdan M. *Min Pro Ext Met Rev* 2000;**21**:25.
60. Lamb JD, Izatt RM, Christensen JJ. In: Izatt RM, Christensen JJ, editors. *Progress in macrocyclic chemistry,* Vol. 2. New York: Wiley; 1981. 41–90.
61. Izatt RM, McBride DW, Brown PR, Lamb JD, Christensen JJ. The influence of halocarbon solvent on macrocycle-mediated cation transport through liquid membranes. *J Membr Sci* 1986;**28**:69–76.
62. Dernini S, Paimas S, Poicara AM, Maronglu B. Extraction and transport of sodium ion and potassium ion in a liquid membrane containing crown ethers: effect of the mixed solvent. *J Chem Eng Data* 1992;**37**:281–4.
63. Christensen LL, Lamb LD, Izatt SR, Starr SE, Weed GC, Astin MS, Stitt BD, Izatt RM. Effect of anion type on rate of facilitated transport of cations across liquid membranes via neutral macrocyclic carriers. *J Am Chem Soc* 1978;**100**:3219–20.
64. Olsher U, Hankins MG, Kim YD, Bartsch RA. Anion effect on selectivity in crown ether extraction of alkali metal cations. *J Am Chem Soc* 1993;**115**:3370–1.
65. Christensen JJ, Christensen SP, Biehl MP, Lowe SA, LambLD Izatt RM. Effect of receiving phase anion on macrocycle-mediated cation transport rates and selectivities in water-toluene-water emulsion membranes. *Sep Sci Technol* 1983;**18**:363–73.
66. Szpakowska M, Nagy OB. Application of the competitive preferential solvation theory to facilitated ion transport through binary liquid membranes. *J Phys Chem* 1989;**93**:3851–4.
67. Johansson H, Rydberg J. *Acta Chem Scand* 1969;**23**:2797.
68. Izatt RM, Bruening RL, Bruening ML, Lamb JD. Macrocycle-mediated cation transport through liquid membranes. *Isr J Chem* 1990;**30**:239–45.
69. Brown PR, Hallman JL, Whaley LW, Desai DH, Pugia MJ, Bartsch RA. Competitive, proton-coupled, alkali metal cation transport across polymer-supported liquid membranes containing sym-(decyl)-dibenzo-16-crown-5-oxyacetic acid: variation of the alkyl 2-nitrophenyl ether membrane solvent. *J Membr Sci* 1991;**56**:195–206.
70. Allard B. *The coordination of tetravalent actnide chelate complexes with β - diketones.* Gothenburg: Chalmers University of Technology; 1975. Dissertation.
71. Johnson BFG, Rodgers A. Polyhedral rearrangements and fragmentation reactions in cluster complexes. In: Shriver DF, et al., editors. *The chemistry of metal cluster complexes. VCH*; 1990. 303.
72. Echt O. van der Waals clusters. In: Benedek et al., editors. Metastable decay, coulomb explosion, and electron attachment, in elemental and molecular clusters. In: Proceedings of the 13th International School, Erice, Italy, July 1987, p. 263.
73. Hancock RD, Martell A. *Chem Rev* 1989;**89**:1875–913.
74. Hancock RD. *Prog Inorg Chem* 1989;**36**:187–290.
75. Martell AE, Calvin M. *Chemistry of the metal chelate compounds.* Englewood Cliffs: Prentice-Hall; 1952.
76. Dyrssen D. *Svensk Kem Tidskr* 1956;**68**:212.
77. Rossotti FJC, Rossotti H. *Determination of stability constants.* New York: McGraw Hill; 1961.
78. Froneus S. Metal complexes in Aqueous Solution. In: Jonassen HB, Weissberger A, editors. *Technique of inorganic chemistry.* Chichester: Wiley-Interscience; 1963.
79. Beck MT. *Chemistry of complex equilibria.* New York: Van Nostrand Reinhold; 1970.

80. Hartley FR, Burgess C, Alcock R. *Solution equilibria*. New York: Ellis Hoorwood, John Wiley & Sons; 1980.
81. Meloun M, Havel J, Högfeldt E. *Computation of solution equilibria*. New York: John Wiley Sons; 1988.
82. Bjerrum J. *Metal ammine formation in aqueous solution: theory of the reversible step reactions*. Copenhagen, Denmark: P. Haase and Son; 1941. p. 296–325.
83. Rydberg J. *Arkiv Kemi* 1953;**5**:413.
84. Irving H, Williams RJP, Rossotti FJC. *J Chem Soc* 1955;**1955**:1906.
85. Fisher RA. *Statistical methods for research workers*. New York: Hafner Publishing; 1958.
86. Dixon LCW. *Nonlinear optimization*. London, UK: English University Press; 1972.
87. Albinsson Y, Mahmood A, Majdan M, Rydberg J. *Radiochim Acta* 1989;**48**:49.
88. Ekberg CH. *Uncertainties in actinide solubility calculations illustrated using the Th– OH–PO4 System*. Gothenburg: Chalmers University of Technology; 1999. Dissertation.
89. Watari H, Cunningham L, Freiser H. *Anal Chem* 1982;**54**:2390.
90. Reinhardt H, Rydberg. *J Chem Ind* 1970: 488.
91. Persson H, Skarnemark G, Skalberg M, Alstad J, Liljenzin JO, Bauer G, Heberberger F, Kaffrell N, Rogowski J, Trautmann N. *Radiochim Acta* 1989;**48**:177.
92. Skarnemark G, et al. *J Nucl Radiochem Sci* 2002;**3**:1.
93. Skarnemark G. *J Radioanal Nucl Chem* 2000;**243**(1):219.

Chapter 2

Principles of Solvent Extraction of Organic and Mineral Acids

Chapter Outline

1. INTRODUCTION

Recovery and concentration of organic acids, as well as separation of acid mixtures, have attracted a great interest among researchers, especially in connection with their recovery from fermentation broths, reaction mixtures, and waste solutions. Liquid extraction has been proposed as an alternative to

Solvent Extraction. DOI: 10.1016/B978-0-444-53778-2.10002-0

the classical precipitation process, and there have been several recent efforts to develop process technologies based on extraction.[1–8] The existing information on liquid extraction recovery of the acids is still rather limited in scope.

Extraction of acetic, propanoic, lactic, tartaric, succinic, and citric acids has received more attention because of industrial importance. Most of them are freely soluble in water, some in alcohols, and some in polar solvents.

Extraction chemistry is concerned solely with the state of equilibrium in multicomponent heterogeneous systems. Experience has shown that they are best treated by Nernst's distribution law supplemented by sets of mass action law equilibria (for details see Chapter 1). The diversity of the extraction processes stems from the type of reaction governing the transfer.

The extraction of carboxylic acids may be presented according to three extraction categories:

1. Extraction by solvation with carbon-bonded oxygen-bearing extractants. The partition of the acids between their aqueous solutions and inert aliphatic and aromatic hydrocarbons and some of their substituted homologs are also included here because of the similarity of the partition chemistry involved.
2. Acid extraction by solvation with phosphorus-bonded oxygen-bearing extractants.
3. Extraction by proton transfer or by ion-pair formation, with extractants being high-molecular-weight aliphatic amines.

The first two categories involve the solvation of the acid by donor bonds of some kind which are to be distinguished from strong covalent bonds and from ionic interactions. The latter is the reaction involved in the third extraction category.

Conventional extraction systems using water-immiscible alcohols, ketones, or ethers are relatively inefficient for acid recovery from the dilute aqueous acid solutions found in most fermentation streams. Organophosphorus oxygen-bearing extractants and aliphatic amines are opening new rooms for the recovery of organic acids from a wide variety of aqueous solutions, including fermentation broths, waste waters, etc.

The fundamental chemistry involving all three categories of extractants is known and defined. The basic mass action law presentations of the equilibria involved in the extraction process are given separately for the three categories.

The distinction between the first two categories is based on the strength of the solvation bonds and the specificity of solvation. Solvation with alcohols, ethers, and ketones, the typical carbon-bonded oxygen donor extractants, is not being specific. The coordinate bonds between the acid hydrogen and the oxygen donor are weak for a specific number of solvating molecules per acid to be clearly identified.

On the other hand, the significantly more basic donor properties of the phosphorus-bonded oxygen move the solvation process to specific, and the number of solvating molecules per extracted acid is experimentally accessible.

The specific behavior of the various classes of phosphorus and amine extractants, their compatibility with common diluents, and the basic chemistry involved in the process of acid extraction are aspects that have been extensively investigated, reviewed, presented in the chemical literature,[9–13] and discussed in this chapter.

2. EXTRACTION OF ACIDS BY CARBON-BONDED OXYGEN-DONOR EXTRACTANTS AND SUBSTITUTED HYDROCARBON SOLVENTS

Extraction of carboxylic acids into inert noninteracting hydrocarbons and substituted hydrocarbons is considered with oxygen-bearing organic solvents. In spite of the substantial difference in the extractive capacity between these two types of solvents, the same mass action law description of the process applies to both.

When acids do not react with other components and are not protonated, dissociated, polymerized, and hydrated, the distribution ratio of the acid D_{HL} is constant. The distribution constant $K_{D,HL}$ equals the measured distribution ratio. When HL is used for the extraction of a metal, $K_{D,HL}$ is abbreviated K_{DE} as the distribution constant of extractant.

Figure 2.1a shows the effect of aqueous salt concentrations on the D_{HL} value of acetylacetone at constant total HL concentration and pH. The salt has two effects: (a) it ties up H_2O molecules in the aqueous phase (forming hydrated ions) so that less free water is available for solvation of HL; and (b) it breaks down the hydrogen bond structure of the water, making it easier for HL to dissolve in the aqueous phase. Figure 2.1 shows that the former effect dominates for NH_4Cl, while for $NaClO_4$ the latter dominates. Increase of the distribution ratio with increasing aqueous salt concentration is described as a *salting-out* effect, and the reverse as a *salting-in* effect. Figure 2.1b shows D_{HL} for the extraction of acetylacetone into $CHCl_3$ and C_6H_6 for two constant aqueous $NaClO_4$ concentrations at pH 3, but with varying concentrations of HL. Acetylacetone is infinitely soluble in both $CHCl_3$ and C_6H_6; at $[HL]_{org} = 9$ M, about 90% of the organic phase is acetylacetone ($M_w = 100$), so the figure depicts a case for a changing organic phase. Polar $CHCl_3$ interacts with HL, making it more soluble in the organic phase; it is also understandable why the distribution of HL decreases with decreasing concentration (mole fraction) of $CHCl_3$. C_6H_6 and aromatic solvents do not behave as do most aliphatic solvents: in some cases the aromatics seem to be inert or even antagonistic to the extracted organic species, while in other cases their pi-electrons interact in a favorable way with the solute. For acetylacetone, the interaction seems to be very weak. The salting-in effect is seen in both Figs. 2.1a and 1b.

The distribution of an acid between water and a nonpolar hydrocarbon is nonreactive and can be regarded as physical distribution according to the

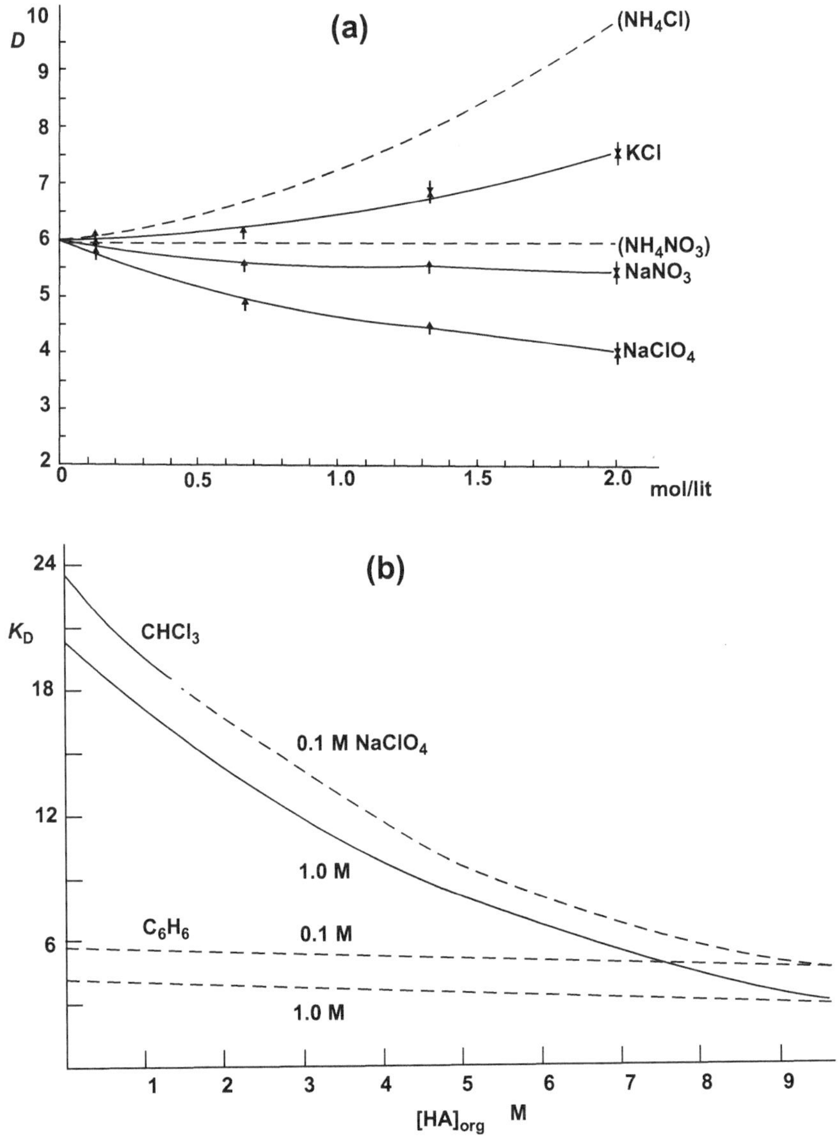

FIGURE 2.1 Distribution ratio D_{HA} of undissociated acetylacetone. (a) Distribution between benzene and aqueous phase containing different inorganic salts at 25 °C. (b) Distribution between $CHCl_3$ (upper curves) or C_6H_6 (lower curves) and aqueous phase 0.1 and 1.0 M $NaClO_4$ as a function of $[HA]_{org}$. The uncertainty at the lowest D values is ±1 for $CHCl_3$ and ±0.2 for C_6H_6.

Nernst's distribution law. But account has to be made for partial dissociation of the acid in the aqueous phase, its usual protonation, and its extensive dimerization (aggregation, polymerization) in the organic phase. On the other hand, acids extracted by carbon-bonded oxygen donor solvents are strongly hydrated by varying numbers of water molecules. The exact solvation number of the acid molecules in the organic phase is usually undetermined, but it is known that a large number of solvent molecules are needed for an efficient competition with water molecules that hydrate the acid at the interface.

2.1. Thermodynamic (Mass Action Law) Extraction Equilibria

The set of mass action equations describing the transfer of a weak (monobasic) organic acid from water into an organic phase with which the acid does not interact chemically must be taken into account.

2.1.1. Dissociation of the Acid

One of the properties have to be taken into account is the ionization (dissociation) of the acid in the aqueous solution. When the acid dissociates in the aqueous phase, the distribution ratio becomes:

$$\mathrm{HL} \leftrightarrow \mathrm{H}^+ + \mathrm{L}^-; \qquad K_a = \frac{[\mathrm{H}^+][\mathrm{L}^-]}{[\mathrm{HL}]\overrightarrow{\mathrm{AB}}} \tag{2.1}$$

$$D_{\mathrm{HL}} = \frac{[\mathrm{HL}]_o}{[\mathrm{HL}]_w [\mathrm{L}^-]_w} \tag{2.2}$$

where the aqueous (w) and organic (o) phases in the denominator are concentrations of all acids—as anions and as undissociated in the aqueous phase.

Equation (2.2) can be given as

$$D_{\mathrm{HL}} = \frac{K_{\mathrm{D,HL}}}{1 + K_a[\mathrm{H}^+]_{(w)}} \tag{2.3}$$

If acid is extracted as a solute in its undissociated form HL distribution constant is $K_{\mathrm{D,HL}}$. If the acid is used as extractant $K_{\mathrm{D,HL}}$ is identical to distribution constant of the extractant, K_{DE}.

At low proton concentration (high pH), distribution ratio becomes inversely proportional to proton concentration and dissociated form of acid increases. The concentration of the free anion, $[\mathrm{L}^-]$, is an important parameter in the formation of complexes at extraction of metals by acidic extractants, referred to as the *free ligand concentration* and can be calculated by

$$\log[\mathrm{L}^-] = \log K_a - \log[\mathrm{H}^+] + \log C^0_{\mathrm{HL}} - \log\left[K_{\mathrm{D,HL}} + \frac{V_w}{V_o}(1 + K_a[\mathrm{H}^+])\right] \tag{2.4}$$

where C^0_{HL} is the original concentration of HL in the organic phase at the beginning of the experiment. At $V_o = V_w$ and pH << pK_a, Eqn (2.4) will be

$$\log[L^-] = \log K_a - \log[H^+] + \log C^0_{HL} - \log(K_{D,HL} + 1) \qquad (2.5)$$

From Eqns (2.4) and (2.5), it can be deduced that $[L^-]$ increases with increasing pH, but tends to become constant as the pH value approaches the pK_a value. As it was mentioned above at the extraction of metal complexes where HL is extractant, $K_{D,HL} \equiv K_{DE}$.

2.1.2. Protonation

At low pH, some organic acids accept an extra proton to form the H_2L^+ complex:

$$HL + H^+ \leftrightarrow H_2L^+ \qquad K_{aH} = \frac{[HL][H^+]}{[H_2L^+]} \qquad (2.6)$$

Because H_2L^+ is ionic, it is not extracted into the organic phase, and thus the distribution ratio becomes

$$D_a = \frac{[HL]_o}{[H_2L^+] + [HL] + [L^-]} = \frac{K_{D,HL}}{\left(\frac{[H^+]}{K_{aH}}\right) + 1 + \left(\frac{K_a}{[H^+]}\right)} \qquad (2.7)$$

2.1.3. Dimerization

The dimerization of the acid in the organic phase:

$$2HL_{(o)} \leftrightarrow (HL)_{2(o)}; \qquad K_D = \frac{[(HL)_2]_{(o)}}{[HL]^2_{(o)}} \qquad (2.8)$$

The experimentally accessible distribution ratio D_{HL} on the molar concentration scale and expressed in terms of the total (analytical) concentrations of the acid in all its possible forms in the aqueous phase, $C_{HL(w)}$, and the organic phase, $C_{HL(o)}$, is formulated accordingly:

$$D_{HL} = \frac{C_{HL(o)}}{C_{HL(w)}} = \frac{[HL]_{(o)} + 2[(HL)_2]_{(o)}}{[HL]_{(w)} + [L^-]_{(w)}} = \frac{[HL]_{(w)}P + 2K_D[HL]^2_{(o)}}{[HL]_{(w)} + \left(\frac{K_{HL}[HL]_{(w)}}{[H^+]_{(w)}}\right)}$$

$$= \frac{[HL]_{(w)}P + 2K_D P^2[HL]^2_{(w)}}{[HL]_{(w)}\left(1 + \left(\frac{K_{HL}}{[H^+]_{(w)}}\right)\right)} = \frac{P + 2P^2 K_D[HL]_{(w)}}{1 + \left(\frac{K_{HL}}{[H^+]_{(w)}}\right)} \qquad (2.9)$$

The square brackets denote concentrations in moles per liter. If for dilute solutions the term $K_{HL}/[H^+]_{(w)}$ is negligibly small, Eqn (2.9) allows P and K_D to be determined from a plot of D_{HL} versus $[HL]_{(w)}$ which in turn is obtained from $C_{HL,(w)}$ via the known dissociation constant K_{HL}. If there are sufficient data points at moderate enough $C_{HL,(w)}$ values (absolute concentrations depend on the nature of both the acid and the solvent but are usually below 0.5 mol/L), any aberrant D value will become apparent, and the concentration beyond which the linear relationship called for by Eqn (2.9) no longer holds becomes obvious.

A strict thermodynamic expression of the partition process represented by Eqn (2.9) requires quantities that measure the activity of the distributed: the activity coefficient of the species in both phases and P on the mole fraction scale. The activity coefficients correct for the acid that might dimerize in the aqueous phase at higher concentrations and for the acid that might aggregate beyond the dimer in the organic phase in the case of polybasic acids. Although none of these reactions is likely to be significant, the activity coefficient may still vary, mostly because the acid is hydrated in the organic phase to an unknown and varying degree depending on its concentration.

For similar reasons P and also K_D, as derived via Eqn (2.9), may be somewhat misleading. Namely, the values so obtained refer to an equilibrium between two mutually saturated phases rather than to pure water and solvent. The values of P and K_D, thus, need not be equal to those derived by other means that do not involve mutual saturation. For example, K_D values are usually larger[14] in anhydrous than in wet (water-saturated) solvents, probably because of the competing H-bonding between the water and the acid in the organic phase. However, if the acid concentration is kept low and there is low mutual solubility of the phases, the P and K_D values should not be much at variance from the values in pure media.

With equally good approximation, Eqn (2.9) can usually be applied to evaluate the P and K_D functions of di- and tricarboxylic acids by neglecting the effect of the second and third dissociation steps, which are usually smaller than the first by 2 or 3 orders of magnitude (see Table 2.1).

Most of the distribution data in the literature are either too fragmentary in terms of the concentration range covered or too imprecise because of presentation in graphical form for the P and K_D values to be derived via Eqn (2.9).[15–22] Distribution data in the dilute concentration range obey Eqn (2.9) reasonably well, yielding self-consistent and constant partition and dimerization constants in a large variety of solvents. Presented in Table 2.2, information was critically evaluated for P and K_D values.

2.2. Mutual Solubilities and Phase Miscibilities

Mutual solubility between an aqueous solution and a solvent at a fixed temperature is affected by the nature of the acid distributed and its total concentration in the system. At extraction of weak organic acids mutual solubilities cause dramatic volume changes. The trend is that the volume of

TABLE 2.1 Dissociation Constants pK_a of Some Carboxylic Acids

Acids	Structure	pK_{a1}	pK_{a2}	pK_{a3}
Acetic acid		4.76		
Propanoic acid	CH_3CH_2COOH	4.85		
Lactic acid	$CH_3CHOHCOOH$	3.86		
Pyruvic acid	$CH_3COCOOH$	2.44		
Succinic acid	$HOOCCH_2CH_2COOH$	4.20	5.64	
Fumaric acid	HOOCCHHCCOOH	3.02	4.38	
Maleic acid	HOOCCHCHCOOH	1.93	6.14	
Malic acid	$HOOCCH_2CHOHCOOH$	3.22	4.70	
Tartaric acid	HOOCCHOHCHOHCOOH	3.01	4.38	
Isocitric acid	$HOOCCH_2(CHCOOH)CH(OH)COOH$	3.29	4.71	6.40
Citric acid	$HOOCCH2(C(OH)COOH)CH_2COOH$	3.13	4.76	6.40

organic phase increases as the total concentration of the acid in the system increases. For example, at 25 °C, a 3 mol/L aqueous solution of propanoic acid is completely miscible with isobutyl alcohol, or a 4 mol/L solution with isoamyl alcohol, while chloroform or carbon tetrachloride requires a 7 mol/L toluene or xylene and an 8 mol/L aqueous propanoic acid solution to form a single homogenous phase.[23] These figures indicate that there is clear distinction to be made between the behavior of hydrocarbon solvents and those with some functional group. The trend is that a water–solvent pair with a considerable mutual solubility, such as isobutyl alcohol + water, e.g., will "tolerate" much less carboxylic acid before a complete miscibility sets in than a solvent–water pair with only traces of mutual solubility, such as heptane–water. The latter pairs behave close to ideality in terms of volume changes when the carboxylic acid, at low concentrations, partitions between them. Conversely, solvent pairs that exhibit mutual solubilities at higher than trace levels will also exhibit measurable deviation from a quasi-ideal behavior.

The extent of the volume change is, of course, related to the coextraction of water along with that of the acid. The numerical information on the water–propanoic acid–rnethylisobutyl carbinol[24] and water–propanoic acid–*n*-octane[25] systems, e.g., demonstrates the effect. It should be noted that the relevant binary systems, water + propanoic acid and *n*-octane or methylisobutyl carbinol with propanoic acid, are completely miscible at 25 °C, and that the *n*-octane-containing ternary system,[25] as well as similar ones with *n*-heptane[26] rather than

TABLE 2.2 Partition P and Dimerization K_D Constants of Carboxylic Acids at 25 °C

Acid/solvent	Partition, P	Dimerization constant, K_D
Propanoic acid		
n-Hexane	0.005	9000
Cyclohexane	0.006	6500
Benzene	0.043	190
Toluene	0.034	230
Xylene	0.030	310
Carbon tetrachloride	0.015	940
Chloroform	0.11	30
Nitrobenzene	0.16	11
Diethyl ether	1.75	0.1
Diisopropyl ether	0.80	0.5
Methylisobutyl ketone	2.15	
Cyclohexanone	3.30	
n-Butanol	3.20	
n-Pentanol	2.95	
Lactic acid		
Diethyl ether	0.10	
Diisopropyl ether	0.04	
Methylisobutyl ketone	0.14	
n-Butanol	0.73	
Isobutanol	0.66	
n-Pentanol	0.40	
n-Hexanol	0.37	
n-Octanol	0.32	
Pyruvic acid		
Diethyl ether	0.16	

(Continued)

TABLE 2.2 Partition P and Dimerization K_D Constants of Carboxylic Acids at 25 °C—cont'd

Acid/solvent	Partition, P	Dimerization constant, K_D
Succinic acid		
Diethyl ether	0.15	
Methylisobutyl ketone	0.19	
n-Butanol	1.20	
Isobutanol	0.96	
n-Pentanol	0.66	
n-Octanol	0.26	
Fumaric acid		
Diethyl ether	1.50	
Methylisobutyl ketone	1.40	
n-Butanol	3.30	
Isobutanol	4.60	
Maleic acid		
Diethyl ether	0.15	
Methylisobutyl ketone	0.21	
Isobutanol	0.92	
Malic acid		
Diethyl ether	0.02	
Methylisobutyl ketone	0.04	
Isobutanol	0.36	
Itaconic acid		
Diethyl ether	0.35	
Methylisobutyl ketone	0.55	
Isobutanol	1.80	
Tartaric acid		
Diethyl ether	0.003	
Methylisobutyl ketone	0.02	

TABLE 2.2 Partition *P* and Dimerization K_D Constants of Carboxylic Acids at 25 °C—cont'd

Acid/solvent	Partition, *P*	Dimerization constant, K_D
n-Butanol	0.16	
Citric acid		
Diethyl ether	0.009	
Methylisobutyl ketone	0.09	
n-Butanol	0.29	
Isobutanol	0.30	

n-octane, forms one homogeneous phase at an acid content of about 65 mol% regardless of the water–alkane ratio in the mixture. In this context, one should note perhaps that at high acid content, which is a water-deficient situation, the solvation sheath around the carboxylic acid head must be composed of both water and solvent molecules. This process, which increases the lipophilic character of the aqueous phase, is the one that enhances the complete miscibility of the layers by decreasing the interfacial tension between the two liquid phases. At the same time, the energy required to transfer the polar head groups of the acid decreases. This consideration implies that the weak acids do not have a specific hydration number in either of the phases. Rather, there is a mixture of different hydrate solvents where the water–organic solvent ratio varies from one solvent to another and, more importantly, with the overall solute concentration. For some of these ternary systems available data permit construction of phase diagrams such as the one reproduced[26] in Fig. 2.2 for the water–propanoic acid–*n*-heptane system at 25 °C. This triangular solubility curve, on a mole percent scale, is exceptionally symmetrical. The lack of symmetry in most of such plots,

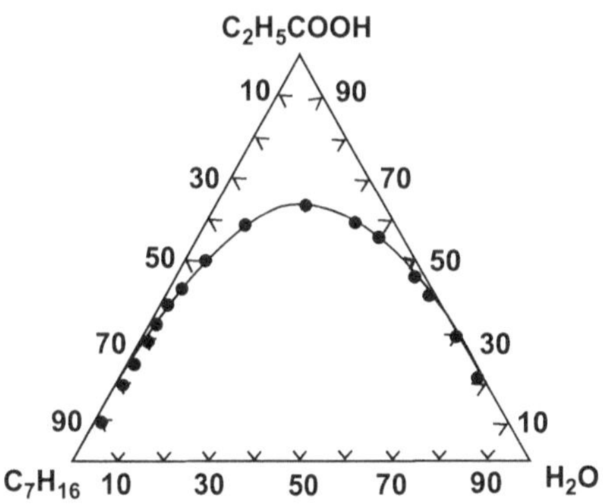

FIGURE 2.2 Mutual solubility in the water + propanoic acid + heptane system at 25 °C (mole percent scale). *(Source: From Ref. 26 with permission)*

on the mole or weight percent scale, is due to the preferential solubility of the water in the organic phase.

For lactic acid[27–29] systems, the area of heterogeneity increases in the order butyl lactate, cyclohexanone, and diethyl carbinol, at both 27 and 90 °C. With alcohols the two phase regions are the most limited. Temperature has been influenced mainly on the mutual solubility between water and solvent in the absence of acid; otherwise, mutual solubilities increase only slightly with increasing temperature.[27,29]

2.3. Parameters of Aqueous Phase

The most important factors associated with the characteristics of the acids which affect their extractability should be identified with the number of carboxylic groups and their acid strength, the nature and number of additional functional groups (keto acids, hydroxo acids) on the molecules, and the size and hydration of the anion. Since only the undissociated acid is extractable, the experimentally determined *D* values depend much on the acid strength. In the case of polybasic acids, it is essentially the first dissociation constant that determines the acid strength, the contribution of others being rather marginal. Nevertheless, monocarboxylic acids are more extractable than di- or polybasic acids with an equal number of carbon atoms due to an increased affinity for the aqueous solution of acids with two or more functional groups. Hydrophilicity of the acid radical is increased by hydroxo or keto functional groups, the differences in *D* values often being considerable.

In solvent extraction, the organic phase is always saturated with water, and the organic extractant may become hydrated. For example, in the extraction of benzoic acid, HBz,[14] it was found that the organic phase contained four different species: the monomer HBz, the monomer hydrate HBz H_2O, the dimer H_2Bz_2, and the dimer hydrate $H_2Bz_2(H_2O)_2$. Only by considering all these species it is possible to explain the extraction of some metal complexes with this extractant. The extent of hydration of the acid and the energy of the bond to the water molecules are obviously the factors affecting extractability. The distribution ratio of the weak acids into inert hydrocarbons is only occasionally higher than 0.01 up to 6 or 7 molar initial acid concentration. At such a high acid content, not all of the solution is fully hydrated. As mentioned earlier, the solvation shell around the functional group in such a water-deficient situation may consist of a mixture of water and solvent molecules, thus making the solute species more organic-solvent-like. This in turn accounts for the relatively abrupt increase in *D* values at high acid concentration in initial solution. It should be noted in this context that the distribution ratios remain essentially independent of acid concentration when alcohols and ketones are used as solvents.[21,30–32]

As one would expect, the distribution ratio of an acid is greater when the aqueous solution contains an essentially unextractable electrolyte. The few systematic investigations of the effect of the supporting electrolyte on the

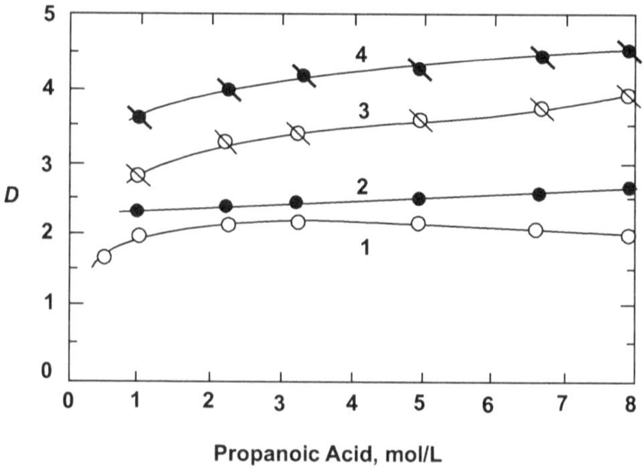

FIGURE 2.3 Effect of sulfuric acid concentration in the initial aqueous solution on distribution ratio of propanoic acid into methylisobutyl ketone at 25 °C: (1) No H_2SO_4 in aqueous phase; (2) 0.28 mol/L; (3) 1.14 mol/L; (4) 1.77 mol/L. *(Source: From Ref. 21 with permission)*

extraction of succinic[33] and propanoic[21,34] acids show the effect using carbon tetrachloride,[34] isoamyl alcohol,[33] and methylisobutyl ketone[21] as extractants. The effect for the water–propanoic acid–sulfuric acid–methylisobutyl ketone system at 25 °C is shown in Fig. 2.3. As long as the supporting electrolyte is not extracted along with the organic acid, the nature of the electrolyte for a given ionic strength is only marginally important.[33] The effect of strong mineral acids on suppressing the dissociation of the weak carboxylic acid, thus increasing its extractability, although necessarily present, is not easily discernable.

2.4. Organic Phase Parameters

The numerical values of the constants P and K_D listed in Table 2.2 indicate that there is a close relationship between these functions and the chemical nature of the extractant. It will be shown later that such a relationship is by no means restricted to these functions. The chemical nature of the extractant is characterized in particular by its donor–acceptor capability, specifically featuring the interaction with the carboxylic group.

The aliphatic and aromatic solvents, including substituted hydrocarbons, are characterized by a low degree of acid extractability. The partition coefficients are higher by 2, or sometimes 3, orders of magnitude for extractants having a carbon-bound oxygen donor than for inert hydrocarbons. Here it is the tendency of the solvent to participate in hydrogen bonding with the carboxylic group, giving the primary effect on the experimentally determined distribution ratio. Alcohols, capable of being both acceptors and donors, as are the acids, give the highest partition coefficients; followed by the purely basic solvents such as ethers and ketones.

As a consequence of the strong donor–acceptor interaction, the extracted acid molecules will dimerize to only a small extent, or not at all. As the solute–solvent hydrogen bond is apparently stronger than the solute–solute one that leads to the formation of the dimer. Thus systems with low P values exhibit high K_D values and vice versa. For a given acid, a plot of log K_D versus log P yields a fairly good conforming straight line with a slope of −2, indicating that P^2K_D is constant for different solvents. Furthermore, this implies that for systems in which this relationship holds, the activity coefficient of the organic phase remains relatively constant.

Although attempts have been made to correlate the experimentally determined P and K_D values with the physical properties of the organic solvent, not much in the way of useful generalization has been made, except for the obvious qualitative observations that the dimerization constant decreases with increasing dielectric constant and dipole moment of the solvent. The partition coefficient changes in the opposite direction, as does the mutual solubility of solvent–water pair. This in turn reduces the differences in dielectric constant and dipole moment between the two phases.

The attempt to correlate the partition coefficient of propanoic acid between water and several organic solvents resulted in a seemingly self-consistent set of solubility parameters values. The conventional relationship between partition coefficient P^* on a mole fraction scale and the solubility parameter of the components, water (δ_w), solvent (δ_o), and propanoic acid (δ_{HL})

$$RT \ln P^* = \nu_{HL}[(\delta_{HL} - \delta_w)^2 - (\delta_{HL} - \delta_o)^2] \tag{2.10}$$

where the ν_{HL} is the molar volume of the acid.

This value for the solubility parameter of propanoic acid yields 16.3. Because of the large number of adjustable parameters needed for the least-squares analysis of the experimental data, and because of the inadequacy of the theory for polar liquids, the utility of correlation (10) is rather limited.

Finally, the interfacial tension of the solvent–water pair has been correlated with the partition coefficient and the distribution ratio of propanoic acid. Authors[35] have found that log D of the acid in the heptane and dichloroethane systems is linear function of the interfacial tension σ at 20 °C which decreases with increasing total acid concentration. The difference between the interfacial tension in the presence and in the absence of the acid is a function of log D only and is independent of the solute concentration in the system. Furthermore, a plot of log P versus the interfacial tension σ at 20 °C gives a straight line for solvents with σ values higher than about 9 dynes/cm.[36] The relationship appears to be valid for n-heptane ($\sigma = 51$), carbon tetrachloride ($\sigma = 45$), benzene ($\sigma = 35$), chloroform ($\sigma = 32$), nitrobenzene ($\sigma = 26$), isopropyl ether ($\sigma = 18.5$), methylisobutyl ketone ($\sigma = 10.8$), and methyl cyclohexanone ($\sigma = 9.1$). This order of decreasing interfacial tension of the solvent–water pair parallels the increasing solubility of the solvent in water for which values are (mol/L): 0.0005, 0.005, 0.0073, 0.068, etc. The relation is not surprising in view of lipophilic–hydrophobic balance in the character of propanoic acid.

2.5. Thermodynamic Functions of Transfer

There is only a very slight effect, if any, of temperature in the 20–90 °C range on the distribution ratio of propanoic,[19] lactic,[27,37] and succinic[20,38] acids into alcohols, ketones, diethyl carbinol, and ethers. On the other hand, the partition coefficient of propanoic acid[34,39] increases with increasing temperature at extraction into an aliphatic hydrocarbon benzene, chloroform, carbon tetrachloride, or nitrobenzene, frequently by as much as a factor of 2 in the 20–50 °C range.

The most detailed set of thermodynamic data on the extraction of propanoic acid has been reported in the Refs. 19,40. (see Table 2.3). The data have been derived via the relationship $\Delta H_P^0 = RT^2(\mathrm{d}\ln P/\mathrm{d}T)$ (see Chapter 1) on the usual assumption that the standard molar enthalpy change is not temperature dependent in the range studied. The probably low precision of the enthalpy and entropy values in Table 2.3, as well as the differences in the values reported,[19,36,39] is due, at least to some extent, to the subtle changes in the mutual solubilities of the phases with temperature. These changes are not necessarily reflected in the P values derived from the experiments.

In a similar way, the enthalpies and entropies of acid dimerization, as derived via the K_D values determined at various temperatures, are compiled in Table 2.4. The negative sign of the functions reflects, of course, that K_D decreases with increasing temperature, indicating that the driving force of the dimerization process is enthalpy governed. This is in contrast to the partition coefficient which increases with temperature, thus resulting in positive ΔH_P^0 values. For these solvents, partitioning must be an entropy-governed transfer process in which the main driving force for extraction is the dimerization of the acid. It should be remembered that there is practically no acid dimerization in solvents with an oxygen donor functional group (see Table 2.2) and that in these solvents P is only slightly affected by temperature. The acid solvates formed in these extractants are of considerable stability, stable enough to prevent measurable dimerization of the acid.

Rationalizing the thermodynamic functions (X) of partition (P) of several members of the homologous series of n-carboxylic acids into a variety of aliphatic and aromatic solvents, ethers, and alcohols, the authors[19,40] calculated that the transfer functions are additively composed of two separate terms—one corresponding to the transfer of the carboxylic group (COOH) of the acid and the other to its hydrocarbon tail (R) $\Delta X_P^0 = \Delta X_{P,\mathrm{COOH}}^0 + \Delta X_{P,\mathrm{R}}^0$. The numerical values of the terms of COOH and R depend much on the nature of the extractant. In inert hydrocarbon solvents with no functional groups, the free energy change associated with the transfer of the methylene group is close to its enthalpy of transfer, and practically independent of the acid–solvent pair, with

TABLE 2.3 Thermodynamic Functions of Partition of Propanoic Acid Between Water and Various Organic Solvents at 20 °C

Solvent	Free Energy ΔG_P^0	Enthalpy ΔH_P^0	Entropy ΔS_P^0
n-Hexane	11.2	–	–
n-Heptane	12.0	–	–
n-Octane	12.7	–	–
Chloroform	5.4	19.0	47
Benzene	7.5	20.1	43
Toluene	8.2	17.8	81
p-Xylene	8.5	–	–
Nitrobenzene	5.1	17.5	42
Dibutyl ether	2.4	2.1	−1
Dipentyl ether	3.1	–	–
Diheptyl ether	4.5	–	–
n-Pentanol	−2.7	–	–
n-Hexanol	−2.3	–	–
n-Heptanol	−2.0	0	7
n-Nonanol	−1.1	–	–
Acetophenone	−1.1	0	4

Note: ΔG_P^0 – Free Energy gradient; ΔH_P^0 – Enthalpy gradient; ΔS_P^0 – Entropy gradient.
Source: From Ref. 19

TABLE 2.4 Thermodynamic Functions of Dimerization of Propanoic Acid in Water-Saturated Solvents at 20 °C

Solvent	K_D (L/mol)	$-\Delta G_P^0$ (kJ/mol)	$-\Delta H_P^0$ (kJ/mol)	$-\Delta S_P^0$ (j/mol deg)
Benzene	154	12.3	41	98
Toluene	196	12.8	38	152
Chloroform	73	10.5	41	105
Dibutyl ether	0.8	−0.5	9	28
Nitrobenzene	13.2	6.3	37	103

Source: From Refs. 48,64

an average value of $\Delta H^0_{P,\mathrm{CH_2}} = -3.4$ kJ/mol. The corresponding entropy term, $\Delta S^0_{P,\mathrm{CH_2}}$, is close to zero. On the other hand, both enthalpy and entropy of transfer of the carboxylic group are positive.

For solvents with an oxygen-containing functional group, $\Delta S^0_{P,\mathrm{CH_2}}$ has a value of about 10 J/mol deg. The $\Delta S^0_{P,\mathrm{COOH}}$ term is negative, and its absolute value depends on the solvent. This implies that the difference in the extractive capacity of a solvent is governed essentially by the difference $\Delta S^0_{P,\mathrm{R}} - \Delta S^0_{P,\mathrm{COOH}}$, where the first term is positive and the second negative, with $\Delta H^0_P \approx 0$.

3. PHOSPHORUS-BONDED OXYGEN DONOR EXTRACTANTS

Generally speaking, weak organic acids are extracted by organophosphorus compounds with a significantly higher distribution ratio than by carbon-bonded oxygen donor extractants under comparable experimental conditions. Furthermore, their extractability is higher than that of mineral acids.[10] Distribution ratios for weak carboxylic acids extracted by undiluted tributyl phosphate (TBP) are relatively high: malic (1.3), lactic (1.4), citric (2.0), maleic (3.6), succinic (5.7), and propanoic (8.4) at a 0.1 mol/L initial acid concentration at 25 °C, similar to that behavior in extraction systems of TBP with alcohols, ethers, and ketones. Monobasic acids are more extractable than dibasic; hydroxyl groups strongly depress extractability. So, this group of extractants has strong extractive power. During the past few decades, several separation and recovery technologies for the organic acid recovery were described[1,41,42] and patented.[6,43]

3.1. Mass Action Law Equilibria

In describing the equilibria involved in the extraction of weak monocarboxylic acids by strong solvating extractants, E, such as the organophosphorus compounds[10] and sulfoxides,[22] the pertinent steps of the heterogeneous process are the dissociation of the acid in the aqueous solution (see Eqn (2.1)) and the formation of the acid solvates in the organic phase:

$$\mathrm{HL_{(w)}} + p\mathrm{E_o} \leftrightarrow \mathrm{HL} * \mathrm{E}_{p(\mathrm{o})} \quad K_{\mathrm{DE}} = [\mathrm{HL} * \mathrm{E}_p]_{(\mathrm{o})}/[\mathrm{HL}]_{(\mathrm{w})}[\mathrm{E}]^p_{(\mathrm{o})} \tag{2.11}$$

The experimentally accessible distribution ratio D on the molar concentration scale and expressed in terms of the total concentration of the acid in all possible forms in the aqueous, $C_{\mathrm{HL(w)}}$, and the organic, $C_{\mathrm{HL(o)}}$, phases is given by

$$D = \frac{C_{\mathrm{HL(o)}}}{C_{\mathrm{HL(w)}}} = \frac{[\mathrm{HL} * \mathrm{E}_p]_{(\mathrm{o})}}{[\mathrm{HL}]_{(\mathrm{w})} + [\mathrm{L}^-]_{(\mathrm{w})}} = \frac{K_{\mathrm{DE}}[\mathrm{HL}]_{(\mathrm{w})}[\mathrm{E}]^p_{(\mathrm{o})}}{[\mathrm{HL}]_{(\mathrm{w})} + \dfrac{K_{\mathrm{HL}}[\mathrm{HL}]_{(\mathrm{w})}}{[\mathrm{H}^+]_{(\mathrm{w})}}} = \frac{K_{\mathrm{DE}}[\mathrm{E}]^p_{(\mathrm{o})}}{1 + \dfrac{K_{\mathrm{HL}}}{[\mathrm{H}^+]_{(\mathrm{w})}}} \tag{2.12}$$

The concentration of free, unbound extractant at equilibrium is

$$[\mathrm{E}]_{(\mathrm{o})} = C_{\mathrm{E(o)}} - p[\mathrm{HL}]_{(\mathrm{o})} \tag{2.13}$$

If the ratio $K_{HL}/[H^+]_{(w)}$ is constant or has a negligibly small value, $[H^+] << K_{HL}$, the solvation number p can be obtained by partial differentiation of D with respect to $[E]_{(o)}$, a plot of

$$\log D = \log K_{\mathrm{DE}} + p \log[\mathrm{E}]_{(\mathrm{o})} \tag{2.14}$$

yielding the slope p and the intercept K_{DE}. Both of these quantities are known to depend on both the solute and extractant concentrations; however, they usually remain reasonably constant over some range. At high solute concentration, as an appreciable fraction of the extractant molecules becomes bound to the acid, lower values of p will predominate and the plot of Eqn (2.14) will deviate, sometimes significantly, from the straight-line requirement.

The two-equation set, Eqns (2.1) and (2.9), is a simplified, though realistic, version of a more general and thermodynamically strict one that includes activity coefficient terms and/or allows for the simultaneous formation of two, or perhaps three, different solvates, depending on the organic phase loading. It is also simplified by neglecting the partition equilibrium of the molecular acid, Eqn (2.3), and that of its dimerization in the organic phase, Eqn (2.4), on the grounds that in the presence of strong solvating agents there are no unsolvated acid species $HL_{(o)}$, and no acid dimers $(HL)_{2(o)}$, can be formed since the extracting agent E is a much stronger acceptor than the carboxyl oxygen of the acid's carboxylic group. Under certain experimental conditions, however, this simplification is unjustified. Thus, e.g., when there is a stoichiometric deficiency of the solvating agent E in the system ($[E]_{(o)}$ of Eqn (2.13) approaching zero), the distribution ratio (Eqn (2.12)) requires the addition of appropriate terms to account for the acid extraction by the diluent alone and its dimerization.

3.2. Extraction Characteristics

Few systems have been investigated extensively enough to permit the evaluation of the association equilibrium constant K_{DE} using Eqn (2.9). For those which were, the constants are shown in Table 2.5. The numerical data are in line with what is known for the extractive capacity of various organophosphorus extractants;[10] the extractive power increases markedly as the number of direct C–P linkage increases. Thus, the acid adduct with trioctylphosphine oxide (TOPO) is 2 orders of magnitude more stable than that with TBP.

The solvation numbers of the acids in Table 2.5 appear to be the same as the number of COOH groups on the acid molecule. This indicates a stoichiometric association between individual phosphoryl group and individual carboxylic group and displays the strong effect of the extractant concentration on the experimentally determined distribution ratios. Figure 2.4 shows the effect of

TABLE 2.5 Equilibrium Constants K_s for the Formation of Solvates Between Carboxylic Acid and Organophosphorus Extractants [Eqn (2.6)]

Acid	Extractant	Diluent	Solvation number	Extractant constant, K_{DE}	Temperature (°C)
Propanoic	TOPO	*n*-Hexane	1	59.0	25
Succinic	TOPO	*n*-Hexane	2	152.0	25
Malic	TBP	*n*-Dodecane	2	0.11	20
Tartaric	TBP	*n*-Dodecane	2	0.04	20
Lactic	TBP	*n*-Dodecane	–	0.26	20
Lactic	DEHPA	Diethylbenzene	1	38.0	–
Citric	TBP	Carbon tetrachloride	3	0.04	25

Note: TOPO, trioctylphosphine oxide; TBP tributylphoshate; DEHPA di-2-ethylhexylphosphorlc acid.

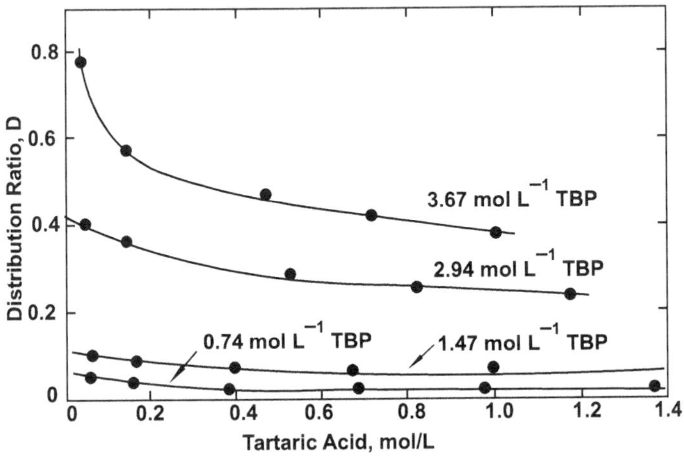

FIGURE 2.4 Distribution ratio of tartaric acid between water and TBP solutions in *n*-dodecane as a function of the acid concentration in the equilibrium aqueous phase at 20 °C.

acid concentration on the extraction of tartaric acid by TBP at different concentrations in dodecane.[45] In the TBP systems, D values are usually below 0.2 when the TBP concentration is below 2 mol/L (or 50% v/v). With a 0.7 mol/L (or 20% v/v) TBP solution, D is less than 0.05. As a rule, D decreases, initially rather sharply, with increasing acid concentration in the initial aqueous solution at any TBP level, but more so at high TBP concentration[45–47].

The exact solvation number p of lactic acid in the TBP system[46] could not be determined unequivocally; the slope of Eqn (2.13) has the value of around 1.4. This and the similar citric acid system[47] deviate from the quasi-ideal behavior apparently more than other systems probably because of the significant mutual solubility of two phases. The volume changes occurring on equilibration depend on the acid concentration in the initial aqueous solution and are so dramatic that a third phase is formed on the equilibrium of a 6 mol/L aqueous lactic acid solution with any TBP–dodecane mixture. The ejected phase is almost pure diluent. Or, in the citric acid system, the extracted acid, which is hydrated with two water molecules,[47] causes the aqueous solution to decrease in volume. The changes are significant: an initial phase ratio of unity between undiluted TBP and a 0.02 mol/L aqueous acid solution at 20 °C increases to 1.13 at equilibrium and to 1.40 when the acid concentration is around 1 mol/L.

For the acids under consideration, there is now enough comparable evidence to show that the effect of diluent on the extractive properties of the alkylphosphates is much less marked than one would assume on the basis of the solvent effect discussed under the previous heading. When cithic acid is extracted[47] from a 0.26 mol/L aqueous solution at 18 °C by an organic phase containing 50% v/v TBP, its distribution ratio is 0.56 with disopropyl ether as the diluent, 0.50 with methylisobutyl ketone, 0.33 with Shellsol H (an aliphatic hydrocarbon mixture), and 0.31 with Shellsol A (an aromatic hydrocarbon mixture). The most extreme phenomenon, which is usually linked to the nature of the diluent, is the splitting of the organic phase to form a third phase, a second organic one, as described above for the lactic acid system. There appears to be a stronger diluent effect when TOPO is the extractant rather than TBP, at least in the -extraction of acetic acid.[48]

Although the high distribution ratios of the acids are due to the successful competition of the solvating phosphoryl group against water molecules at the interface, the organic phase adducts with alkylphosphates are hydrated with one or more water molecules.[44,46] It is rather remarkable that the hydration number appears to remain constant over wide concentration ranges of both the acid and the extractant.

The very significant effect of temperature on the distribution ratio may be linked, at least in part,[10] to a change in the hydration number of extracted species. In TBP systems, D for citric acid[47,49] decreases by a factor of up to 4 in the 0–80 °C range and that for tartaric acid[49] by at least 2 in the 0–35 °C range.

Finally, mention should be made of alkylsulfoxides as extractants for the weak carboxylic acids under discussion.[6,22,50] An unspecified mixture of alkylsulfoxides of an average molecular weight of 270 has been shown to extract acids, among them propanoic, lactic, and pyruvic, with a distribution ratio higher than that of TBP, under comparable experimental conditions.[22] The extracted species is a monosolvate of propanoic and pyruvic acids and a disolvate of lactic acid, as determined from plots of Eqn (2.13).

4. ALIPHATIC AMINE EXTRACTANTS

Extraction of proton-bearing organic and inorganic compounds from aqueous media by long-chain aliphatic amines dissolved in water-immiscible organic solvents is one of the novel developments in separation technology.[3,4,8,28,51–62] Amine-based extractants are highly favorable for carboxylic acid extraction (tridodecylamine is used in the citric acid process). In addition to high efficiency and selectivity, they provide for product concentration through extraction at about ambient temperature followed by back-extraction at an elevated temperature.[5] Production of lactic acid (pK_a of 3.86) fermented at a pH > 4.5. Lactic acid can be extracted by solvating extractants from the fermentation liquor into which a strong acid was added.[4] Such processes are, however, less attractive as they consume a base and an acid and form a salt as a byproduct. Alternatively, extraction from aqueous solutions of pH > pK_a was proposed[56,57] Mineral and carboxylic acid distribution between aqueous solutions of relatively high pH and amine-based extractants are of interest also in salt decomposition.[5,8] Effect of equilibrium pH on extraction of lactic, acetic, propionic, and butyric acids by a tertiary amine (Alamine 336) and by a quaternary amine chloride (Aliquat 336) was studied.[58] The amines were tested as single component extractants and in kerosene or in 2-octanol. A theory was developed.

The extractability of acids in contrast to that of anionic acidic metal complexes depends more on the composition of the organic phase, the amine, and the diluent than on the aqueous phase conditions. The requirements for practical extraction applications are rather general for both strong and weak acids and are fairly well described.

4.1. Mass Action Law Equilibria

At the extraction of acids, the fundamental difference between oxygen- and nitrogen-bearing basic extractants is the behavior of the acid proton during the transfer from an aqueous into an organic solution: In the systems with oxygen-bearing solvents, whether carbon, phosphorus, or sulfur bound, the acid strength in the aqueous solution and that of the hydrogen bond in the organic solution are the measures of extractability. On the other hand, the acid extracted into an amine-containing organic phase is no longer regarded as an acid but as an ammonium salt. It is thus the extent of ion-pair association between the alkylammonium cation and the acid radical that is the measure of extractability, or more precisely, the stability of the organic phase species. Thus, the extraction process is based on an acid–base-type reaction between the alkylamine, E, and the acid HL:

$$HL_{(w)} + E_{(o)} = EHL_{(o)} K_{DE} = \frac{[EHL]_{(o)}}{[HL]_{(w)}[E]_{(o)}} \tag{2.15}$$

where $[HL]_{(w)}$ is calculated as before, from $C_{HL,(w)}$ and dissociation constant of acid. According to the theory,[58] carboxylic acids distribute either as undisocciated acids with a distribution coefficient of

$$K_1 = \frac{[HL]_{(o)}}{[HL]_{(w)}} \tag{2.16}$$

or as anions with a distribution coefficient of

$$K_2 = \frac{[L^-]_{(o)}}{[L^-]_{(w)}} \tag{2.17}$$

Using these coefficients and the acid dissociation constant, Yang et al. derived an equation for the total distribution coefficient:

$$K_{D,HL} = \frac{[HL]_{(o)} + [L^-]_{(o)}}{[HL]_{(w)} + [L^-]_{(w)}} = \frac{K_1 + \dfrac{K_2 K_a}{[H^+]}}{1 + \dfrac{K_a}{[H^+]}} \tag{2.18}$$

The authors have determined the constants, K_1, K_2, and K_a by fitting the experimental data to this equation using computer nonlinear regression. Based on these constants, they calculated the dependence of $K_{D,HL}$ on $[H^+]$ and found good correlations with experimental data. Equation (2.18) does not comprise the coefficient for the contribution of amine basicity. In a more recent publications of this group,[51–57] the authors explain their results by extraction of undissociated acid

$$E_{(o)} + nHL_{(w)} \leftrightarrow E(HL)_{n(o)} \tag{2.19}$$

$$K_{exE,n} = \frac{\left[E(HL)_{n(o)}\right]}{[E]_{(o)}[HL]''_{(o)}} \tag{2.20}$$

where HL, E, and K_{ex} are the undissociated acid, the base (amine), and the equilibrium constant, respectively. The authors also explain the strong drop in extraction on pH elevation by the effect of pH on the concentration of the undissociated acid. According to Ref. [57] strong amines show efficient extraction even at high pH (low concentration of undissociated acid) since for such amines K_{ex} magnitude is high. But high efficiency of mineral acids extraction, which are about completely dissociated, cannot be explained with this theory.

At investigations of carboxylic and strong mineral acids extraction with amines by Eyal's group[29,59–62] improved model was presented. Four major mechanisms determine acid extraction by amine-based extractants: anion exchange, ion-pair formation, H-bond formation, and solvation. Anion exchange may take place on equilibrating an aqueous solution of an acid or its salt with an extractant comprising an amine salt (not necessarily quaternary). HL represents both dissociated and undissociated acids. Ion exchange depends on the organic phase hydrophilicity, on the pK_a and pH of HL in aqueous phase, and on the organic phase composition.

The term ion-pair formation mechanism is used for extraction by amines that are basic enough to bind a proton to form the ammonium cation $R_nNH^+_{4-n}$. This cation further binds an anion to maintain neutrality.

Several kinds of H-bonds may form. With an amine that is not basic enough to dissociate, the extracted acid may form an H-bond. Ion pairs formed on extraction by strong amines are in many cases basic enough to bind additional acid molecules through H-bonds (the two anions may be identical or different). This has been shown to be the case for some of the monocarboxylic but not dicarboxylic, acids under consideration.[63–65] Though the exact nature of the chemistry involved in the uptake of extra acid is not known, and in spite of the obvious nonideality of the organic phase under these conditions, distribution data have been interpreted in terms of simple mass action equations:

$$EHL_{(o)} + nHL_{(w)} \leftrightarrow EHL(HL)_{n(o)} \tag{2.21}$$

$$K_{DE,n} = \frac{[EHL(HL)_{n(o)}]}{[EHL]_{(o)}[HL]''_{(w)}} \tag{2.22}$$

H-bonds are also formed between ion pairs, between ion pairs and basic or acidic diluents and coextracted water.

The term solvation is used here for extraction due to replacement of water molecules in the solvation sphere by extractant molecules with only *no specific* interaction.

One should keep in mind that, except for extreme cases (a very strong acid in dilute aqueous phase or a very weak acid), extraction is determined by more than one mechanism. Hence extraction of carboxylic acids has in many cases an intermediate nature between ion pair and H-bond. The distinction between mechanisms is not always clear. Thus, the above-stoichiometric extraction may result from H-bonding to the ion pair as well as from solvation in the organic phase, the polarity of which increases on ion-pair formation.

Phase neutrality is maintained, in the case of anion exchange, through replacement of one anion by another. Explaining neutrality in all other mechanisms by transport of an undissociated acid[3,4,8,51–58] may, however, be misleading. It does not distinguish between ion-pair formation (acid–base neutralization and formation of charged species in the organic phase) and binding of an undissociated acid through H-bonds (lower energies and formation of less polar species). In addition, it leads to the conclusion that extraction is mainly determined by the concentration of the undissociated acid. Such a conclusion contradicts much of the experimental data. Understanding is achieved by analyzing the extraction in terms of acid–base interaction. Energies involved with neutralizing a strong aqueous base by a strong acid are in most cases much higher than those of H-bond formation. This also applies for neutralizing a base in the organic phase by an acid from an aqueous phase. As a result, in many cases, the dominating extraction mechanism is ion-pair formation.

Authors[59–62] added to dissociation constant of acid the association constant of extractant with proton, or protonation constant. Inversive to protonation constant is the dissociation constant of the protonated amine, $K_{a,E}$:

$$E_{(o)} + H^+_{(w)} = EH^+_{(o)} \qquad K_{a,E} = \frac{[H^+]_{(w)}[E]_{(o)}}{[EH^+]_{(o)}} \tag{2.23}$$

For each acid at a determined pH, the formation of the ammonium cation, therefore, the extraction through ion-pair formation is dependent on the relative basicity of the amine and the acid anion, as well as on the concentration of the undissociated acid:

$$\begin{aligned} \log\frac{[EH^+]_{(o)}}{[E]_{(o)}} &= \log\frac{[HL]_{(w)}}{[L^-]_{(w)}} - \log K_{a,E} + \log K_{aHL} \\ &= \log\frac{[HL]_{(w)}}{[L^-]_{(w)}} + pK_{a,E} - pK_{a,HL} \end{aligned} \tag{2.24}$$

According to Eqn (2.24), in systems with $pK_{a,E} >> pK_{a,HL}$ ion-pair formation has an important contribution to the extraction mechanism.

For cases where extraction through other mechanisms is relatively small, pH dependence can be calculated according to

$$\log\frac{[EH^+]_{(o)}}{[E]_{(o)}} = \log[H^+] - \log K_{a,E} = pK_{a,E} - pH \tag{2.25}$$

Similarly, anion exchange at low pH is determined, in addition to the parameters listed above, by acid–base interaction, if the anions strongly differ in their basicity. If HL is a much weaker acid than HX, the anion exchange is driven forward by anion reaction with the proton. Thus, if HX is a strong acid, anion exchange dependence on pH is calculated according to

$$\log\frac{[R_4NX]_{(o)}}{[R_4NL]_{(o)}} = \log K_{a,exch} + \log[X^-]_{(w)} - \log[HL]_{(w)} + pK_{a,HL} - pH \tag{2.26}$$

where $K_{a,exch}$ is the anion exchange mass action constant:

$$K_{a,exch} = \frac{[R_4N^+X^-]_{(o)}[L^-]_{(w)}}{[R_4N^+L^-]_{(o)}[X^-]_{(w)}} \tag{2.27}$$

If only one anion is present in the system (no anion exchange) and if $pK_{a,HL} >> pK_{a,E}$ (the contribution of ion-pair formation is small), extraction by amine is mainly affected through H-bonding of undissociated acid molecules. Extraction dependence on pH is calculated according to

$$\log\frac{[EHL]_{(o)}}{[E]_{(o)}} = \log K_H + \log[L^-]_{(w)} + pK_{a,HL} - pH \tag{2.28}$$

where K_H is the H-bond formation constant:

$$K_H = \frac{[E:HL]_{(o)}}{[E]_{(o)}[HL]_{(w)}} \tag{2.29}$$

Equation (2.28) holds for cases where the basicity of L^- (the anion of the acid extracted) is strong compared to that of the amine. It is, therefore, also suitable for other H-bonding extractions in the system such as acid binding to ion pairs. It holds for nonanionic bases also, e.g., alkanols, esters, ketones, etc., and also for solvation. To explain mechanisms and make predictions, some authors used the Gibbs–Donnan model[60] in acid extraction:

$$pK_{a,E} - pH = \log\frac{[L^-]_{(o)}}{[L^-]_{(w)}} + \log\frac{\gamma_{R_3NH^+}}{\gamma_{R_3N}} + \log\frac{\gamma_{HL,(o)}}{\gamma_{HL,(w)}} \tag{2.30}$$

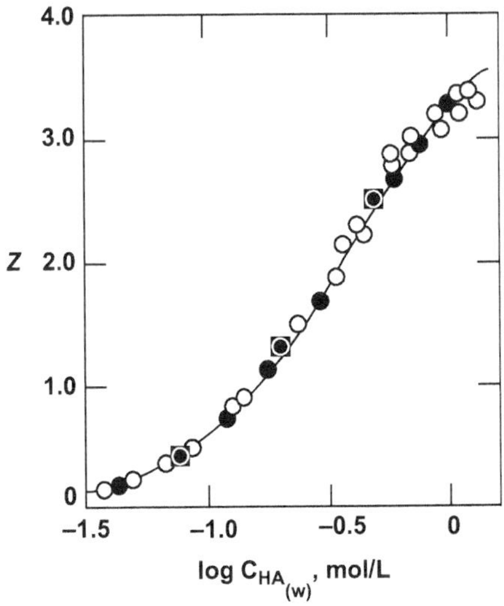

FIGURE 2.5 Plot of Z versus log $C_{HA_{(w)}}$ for the system water + propanoic acid + Amberlite LA-2 + n-hexane for different concentrations of the amine at 30 °C: (●)– 0.01 mol/L; (○)– 0.02 mol/L, and (◙)– 0.05 mol/L.

where γ are activity coefficients of species in two phases.

The extent to which the organic phase (amine + diluent) can be loaded with acid is expressed as the loading ratio Z:

$$Z = \frac{C_{HL(o)}}{C_{E(o)}} \tag{2.31}$$

When the diluent used has an acid-interacting functional group (alcohols, ethers, ketones) or a solvent that dissolves the acid to a considerable extent, $C_{HL(o)}$ should be corrected for the acid extracted into the diluent alone (blank). The value of Z depends on the extractability of the acid (strength of the acid–base interaction) and its aqueous concentration and is independent of the amine content in the inert diluent. The extraction of propanoic acid by Amberlite LA-2[62] in an inert diluent is shown in Fig. 2.5. If the organic phase is not highly loaded, $Z < 0.5$, the constant K_{DE} of Eqn (2.17) can be expressed via the experimentally accessible loading ratio

$$\frac{Z}{(1-Z)} = K_{DE}[HL]_{(w)} \tag{2.32}$$

and the quasi-ideal behavior of the system can be demonstrated by a linear plot of $Z/(1-Z)$ against $[HL]_{(w)}$, the slope yielding K_{DE}. An example is shown in Fig. 2.6 for the succinic acid–tridodecylamine system.[63]

It should be strongly emphasized that Eqn (2.32) is valid only for very dilute, slightly loaded organic solutions when Z has a value of 0.5 or less. The reason for the commonly strong deviation of the plot from the requirements of

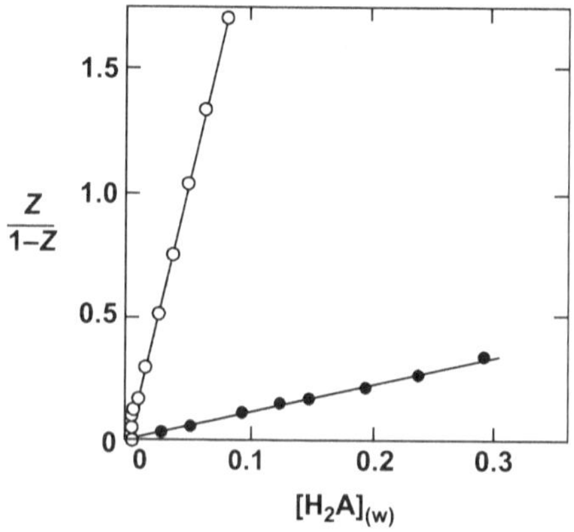

FIGURE 2.6 Plot of $Z/(1-Z)$ versus $[H_2A]$ for the extraction of succinic acid by 0.05 mol/L solution of tri-*n*-dodecylamine in chloroform (○) and benzene (●) at 25 °C.

Eqn (2.32) at higher organic phase loadings in inert diluents is the molecular association of the alkylammonium salts, EHL, formed in the organic phase. With few exceptions, the salts of most of the acids considered here are cationic aggregation colloids that form micelles of a variety of sizes, shapes, and properties.[66] The process of micellar aggregation in inert organic solvents used as diluents for the amine extractants is a stepwise formation of oligomers, where the extent (number of aggregated units) and the degree (size of aggregated units) of aggregation, factors well understood, depend on the characteristics of the amine, the acid radical, and the organic solvent[66,67] Generally speaking, there is a considerable similarity between what is required of a good extractant and the surface-active properties, and thus aggregation, of the salt it forms. Thus, any quantitative treatment of extraction data for which Eqn (2.31) does not hold must take into account the aggregation process:

$$q\mathrm{EHL} \leftrightarrow (\mathrm{EHL})_q \qquad \beta_q = \frac{[(\mathrm{EHL})_q]}{[\mathrm{EHL}]^q} \tag{2.33}$$

where q depends on the chemical nature of the salt, its concentration, the nature of the diluent, and temperature.[66]

4.2. Extraction Characteristics

Extractive capacity of tertiary amines exceeds that of the primary and secondary ones, sometimes significantly.[4,68,69] Aliphatic tertiary amines with less than six carbon atoms per chain and aromatic tertiary amines are poor extractants.[63,70,71] Detailed study[4] on the extraction of lactic acid by a variety of amines of different classes in a number of diluents has been found that primary alkylammonium lactates either are excessively water soluble at room temperature or exhibit surface-active properties or both. Lactates of secondary aliphatic amines are more stable and organic solvent soluble, although gel formation interferes with phase separation. Secondary ammonium salts, at least the acetates,[72] appear also to be prone to some thermal degradation.

This trend of the extraction power is of course dictated by the basicity of the amine. The proton association constant is the highest for tertiary amines and increases with the number of carbon atoms[10,67] although the nature of the medium also has a marked effect on the magnitude of the proton association constant. Monocarboxylic acids appear to be more extractable[63,68,69] than the dibasic ones under otherwise comparable conditions.

As mentioned above, diluent affects the basicity of the amine and thus the stability of the ion pair formed and its solvation. The stability governs the equilibrium conditions of the acid extraction, especially at low *Z* values where the equilibrium aqueous acid concentration is very low. Under such conditions, polar diluents are more favorable than the zero-polarity, low-dielectric-constant aliphatic and aromatic hydrocarbons. This has been shown to be the case in the extraction of citric,[71,73] lactic,[11] and succinic acids.[63,74,75] Figure 2.7 demonstrates the effect of changing diluent on the extraction of succinic acid.[63] The trend correlates with the solubility parameter and/or electron acceptor properties of the diluent.

The extractability of citric acid[47,71] into Alamine 336 and trioctylamine (TOA) in hexane, toluene, methylisobutyl ketone, cyclohexanone, butanol, and isopentyl alcohol diluents decreases with increasing temperature in the 25–60 °C range. This temperature effect on the distribution ratio can be utilized for the extractive recovery of citric acid.[5,8]

What appears to be a rather extensive study[64] covers citric acid extraction from aqueous solutions of constant, 1.0 mol/L ionic strength into tri-*n*-dodecylamine dissolved in toluene in the 0.015–0.30 mol/L concentration range at 25 °C. From graphical plots of distribution data in the form of *Z* versus the equilibrium amine and acid concentrations and from a least-squares computer program, the existence of several acid–amine species in the organic

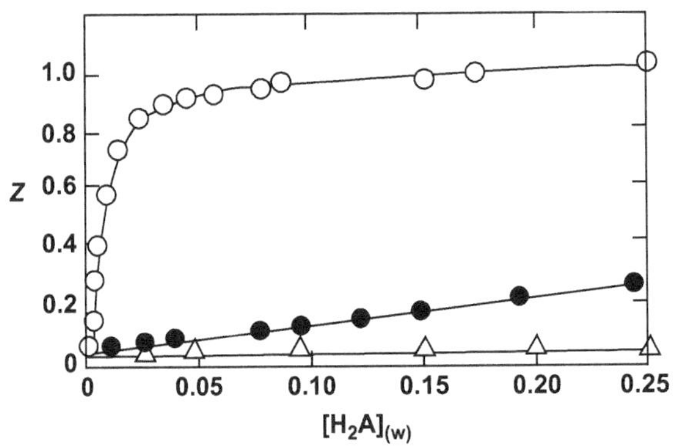

FIGURE 2.7 Plot of *Z* versus. $[H_2A]_w$ for the extraction of succinic acid by 0.05 mol/L solution of tri-*n*-dodecylamine in chloroform (○), benzene (●), and *n*-heptane (△) at 25 °C.

phase was inferred. There is no indication for the formation of species with a 1:1 composition. This is rather strange. The lowest species has a 1:2 ratio and an overall formation constant of 20.1 L^2/mol^2 followed by one with 2:3 ratio and an overall formation constant of 1380 L^4/mol^4 and an aggregated species of a 5:6 ratio with an overall stability constant of $5.9 * 10^8\ L^{10}/mol^{10}$. There is some indication for a micelle with aggregation number of 13. All organic phase species are hydrated, with one, two, or four water molecules, respectively.

In a recent series of papers, Kawano et al.[76–78] have found that the extraction of propanoic acid from its aqueous solutions in the 0.05–2.50 mol/L concentration range by tri-*n*-octylamine in hexane and Amberlite LA-2 in hexane and benzene at 30 °C can be interpreted in terms of the consecutive formation of four acid–amine species with stoichiometries of 1:1, 2:1, 3:1, and 4:1. The reported stepwise formation constants, roughly in the 800–2000 L/mol range, derived via a curve-fitting least-squares computer program, suggest that the model chosen is probably unrealistic. Namely, the absolute values in any one of the three sets of K_{11}, K_{21}, K_{31}, and K_{41} constants are too similar (maximum difference is by a factor of 2) for a meaningful distinction between the four species to be justified.

Most other reports are limited to the identification of the stoichiometry of the acid-amine species in the organic phase with no effort to derive stepwise or overall formation constants. In the context of investigating the extraction of various anionic metal complexes, Pyatnitskii et al.[70,73,79–82] have studied the extraction of citric acid by dioctylamine, tributylamine, tripentylamine, and TOA in several diluents of varying polarity and dielectric constant. In most amine–diluent systems, probably at room temperature the stoichiometry of the amine–acid species is 2:1, and only exceptionally a 1:1 ion pair was identified when a polar ketone or alcohol was used as the diluent. The same 2:1 amine–acid ratio of the extracted species predominates in the tartaric acid[80,83] complex when extracted by tripentylamine or TOA in chloroform.

On the other hand, a 1:1 stoichiometry of the organic species is the prevailing one for malic,[68] lactic,[4,68]and succinic[74,75] acids when extracted by tripentylamine[4,80] or TOA[4,74,75] in chloroform and for succinic acid in several other solvents. In the case of lactic acid extraction[4] more than an equivalent amount of acid can pass into the organic phase.

Kawano et al.[76–79] have recently determined the interfacial adsorption equilibrium constants and the interfacial adsorption area in the process of propanoic acid transfer from an aqueous solution into hexane solutions of TOA and Amberlite 336 at 30 °C. The effects, of both the acid and amine concentrations, on the two adsorption functions have been related to the rate of the acid transfer. The rate-controlling steps are the adsorption of the amine base at the interface, the reaction between the 1:1 ion pair formed and the extra acid molecules, and the desorption or diffusion of the acid-bearing species into the bulk organic phase. The desorption rate constants, which have similar values for the 2:1, 3:1, and 4:1 acid–amine complexes identified by the authors,[65,76–79] differ by as much as 4 or 5 orders of magnitude from that of the simple 1:1 ion pair. The similarity of the other rate constants supports the contention expressed above that not all these complexes may be distinct from one another.

5. EXTRACTION OF STRONG (INORGANIC) ACIDS

To avoid confusion with weak organic acids, strong inorganic acids are denoted by HX. Most strong acids are completely dissociated and both cations and anions are hydrated in aqueous solutions even at pHs as low as 0. The hydration makes them lipophobic and almost insoluble in inert organic solvents. The hydrogen ion is a Lewis acid and is solvated by strong organic (donor or base) molecules, such as those in Table 2.6 (e.g., alcohols, ethers, ketones, esters, amines, phosphoryls, etc.). This results in greater lipophilicity, and the acid becomes more soluble in inert organic solvents. The structure of these solvated hydrogen salts is not well known, but may be represented symbolically by $HE_b^+L^-$, where E refers to the extractant or the solvating solvent and b may have a value of 1–4.

The order of extractability changes with aqueous acidity, but in general follows the order $HClO_4 \approx HNO_3 > HI > HBr > HCl > H_2SO_4$ (see Table 2.6). Since the hydration energies of the acids follow the opposite order, dehydration is an essential step in the solvent extraction process. This order of acids has a practical significance: acids higher can be replaced by the acids lower in the sequence; e.g., HF and HNO_3 are extracted from acidic stainless steel waste solutions into kerosene by addition of H_2SO_4[85] (see Chapter 4).

The extraction of most acids is accompanied by extraction of water. In the extraction of HNO_3 by TBP into kerosene,[86] many different species have been identified, several of which involve hydration. The ratio of acid to extractant is not very predictable. For example, $HClO_4$ apparently is extracted into kerosene with one to two molecules of TBP,[87] HCl into ethylether with one molecule of ethylether, etc.[87] Also, the extracted acid may dimerize in the organic solvent. Assuming that E is almost insoluble in the aqueous phase, the equilibrium reaction can be written in two ways:

1. Using *interface extraction model* we assume that HX reacts with E at the interface:

$$HX_{(w)} + bE_{(o)} \leftrightarrow HXE_{b,(o)} \tag{2.34}$$

Since $[HX]_{(w)}$ and $[E]_{(o)}$ are easily measurable quantities, it is common to define the extraction constant K_{ex} for this model:

$$K_{ex} = \frac{[HXE_b]_{(o)}}{[HX]_{(w)}[E]^b} \tag{2.35}$$

TABLE 2.6 Basicity (Electron-Pair-Donating Tendency) of Some Ions and Molecules

Basicity of some common anions relative to the (hard type) actinide cations $ClO_4^- < I^- < Br^- < Cl^- < NO_3^- < SCN^- <$ acetate$^- < F^-$			
Basicity of some organic molecules Amine compounds			$R_3N^a < R_2NH^b < RNH_2^c < NH_3$
Arsine compounds			R_3As
Phosphine compounds			R_3P
Oxo-compounds		Phosphoryls	$(RO)_3PO^d < R'(RO)_2PO^e < R'_2(RO)PO^f < R'_3PO^g$
		Arsenyls	R_3AsO^h
		Carbonyls	$RCHO < R_2CO(\leq R_2O < ROH < H_2O)^i$
		Sulfuryls	$(RO)_2SO_2^j < R_2SO_2^k < (RO)_2SO^l < R_2SO^m$
		Nitrosyls	$RNO_2^n < RNO^o$
Substitutions causing *basicity decrease of oxo compounds* $(CH_3)_2CH— < CH_3(CH_2)_n— < CH_3— < CH_3O— < ClCH_2—$			

R is an alkyl or aryl group.
[a–c] Tertiary, secondary, and primary amines.
[d] Tri-R phosphate. [e] di-R-R′ phosphonate;
[f] R-Di-R′ phosphinate.
[g] Tri-R′ phosphine oxide.
[h] Arsine oxide.
[i] Ether and hydroxo compounds.
[j] Sulfates.
[k] Sulfones.
[l] Sulfites.
[M] Sulfoxides.
[N] Nitro compounds.
[O] Nitroso compounds.
Source: Ref. 12

2. Using *organic phase reaction model* we assume all reactions take place in the organic phase:

$$HX_{(o)} + bE_{(o)} \leftrightarrow HXE_{b,(o)} \tag{2.36}$$

The equilibrium constant for this reaction is

$$K_{eq,HXE} = \frac{[HXE_b]_{(o)}}{[HX]_{(o)}[E^b]_{(o)}} \tag{2.37}$$

where $K_{eq,HXE}$ is the (organic phase) formation constant of acid–extractant complex.

The distribution ratio of the acid in this system becomes

$$D_{HX} = \frac{[HX]_{(o)} + [HXE_b]_{(o)}}{[HX]_{(w)} + [X^-]_{(w)}} = K_{D,HX} \frac{1 + K_{eq,HXE}[E]^b_{(o)}}{1 + K_a[H^+]} \tag{2.38}$$

Equilibrium measurements do not allow a decision between the two reaction paths.

As reader can see, there are many assumptions, simplifications, and speculations in theoretical considerations of acids extraction. Evaluations of the HNO_3–TBP (in kerosene) interactions are presented below as an example, illustrating the complexity of the extraction considerations. Equations derived by the authors[85] are believed to be valid for extraction of HNO_3 (abbreviated as HX) into 30% TBP in kerosene (abbreviated as E).

1. The formation of an acid monoadduct:

$$H^+ + X^- + E_{(o)} \leftrightarrow EH^+X^-_{(o)} \tag{2.39}$$

The extraction constant is (for simplicity, concentrations are written without charges):

$$K_{ex,1} = \frac{[EHX]_{(o)}}{[H][X][E]_{(o)}} \tag{2.40}$$

2. The formation of a diacid monoadduct:

$$2H^+ + 2X^- + E_{(o)} \leftrightarrow E(HX)_{2(o)} \tag{2.41}$$

and

$$K_{ex,2} = \frac{[E(HX)_2]_{(o)}}{[E]_{(o)}[H]^2[X]^2} \tag{2.42}$$

3. Ion-pair association occurs under strong acid conditions, and the equilibrium constant may be <1:

$$H^+ + X^- \leftrightarrow HX \tag{2.43}$$

$$K_{as} = \frac{[H^+][X^-]}{[HX]} \tag{2.44}$$

4. The distribution of nitric acid is given by

$$D_{HNO_3} = \frac{[EHX]_{(o)} + 2[EHX]_{(o)} + \cdots}{[X^-][HX]} \tag{2.45}$$

5. Furthermore, the dimerization of TBP in the organic phase must be taken into account:

$$2E \leftrightarrow E_2 \quad K_{\mathrm{dim,E}} = \frac{[E_2]_{(o)}}{[E]_{(o)}^2} \tag{2.46}$$

yielding the total concentration of TBP in the organic phase

$$[E_{\mathrm{tot}}]_{(o)} \leftrightarrow [E]_{(o)} + 2K_{\mathrm{dim,E}}[E]_{(o)}^2 \tag{2.47}$$

6. The distribution ratio in terms of only $[H^+]$ and monomeric $[E]_{(o)}$, assuming $[H^+] = [L^-]$ (electroneutrality in the aqueous phase), can then be expressed by

$$D_{HNO_3} = \frac{[E]_{(o)}}{K_{as}}\left(K_{ex1} + 2K_{ex2}[H^+]^2\right) \tag{2.48}$$

Equation (2.48) has been tested,[85] and the results agreed in more than 2300 experiments under varying conditions (see Fig. 2.8). The example illustrates the rather complicated situation that may occur even in such "simple" systems as the extraction of HNO_3 by TBP.

Figure 2.9 shows pH dependence of HCl extraction by TOA in kerosene. Stoichiometric extraction was found at low pH (<4). TOA is a much stronger base than Cl anion, so here we have amine–proton interaction. Equation (2.24) predicts ion-pair formation reaction and a strong drop of pH at $pK_{aE} \approx 3.2$. Apparent basicity of amines, pK_{aE}, is measured by pH of half neutralization,[61] pH_{hn}. Amine of relatively low polarity is converted to a polar ion pair. Polarity is strongly dependent on diluent properties: polar and especially protic diluents stabilize the

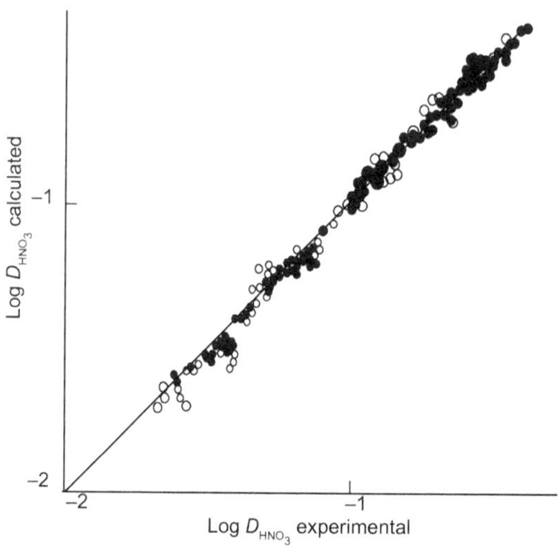

FIGURE 2.8 Test of the equations in example, presented in Ref. 85 for extraction of 0.01–0.5 M nitric acid with 30% TBP in kerosene at temperatures 20–60 °C.

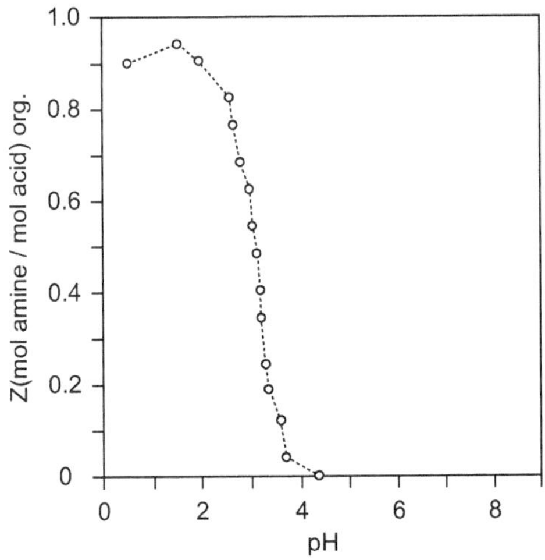

FIGURE 2.9 pH dependence of hydrochloric acid extraction by 0.5 mol/kg TOA in kerosene, initial acid-to-amine molar ratio of 1:1.

ion pair. Results of pH_{hn} measurements for kerosene, xylene, nitrobenzene, TBP, and *n*-octanol show the values equal to 3.50, 3.52, 3.62, 3.88, and 4.16, respectively. More polar diluents improve extraction of strong acids by amines.

6. SUMMARIZING REMARKS

Mechanism of acid extraction by basic extractants or by their solutions is strongly dependent on acid–base properties of both the extracted acid and the extractant. The mechanisms determine the effects of parameters such as pH, acid strength, amine basicity, and diluent properties.

Ion-pair formation is the main extraction mechanism for systems in which the basicity of the amine is higher than that of the anion of the extracted acid, whereas H-bonding of an undissociated acid becomes important only for the extraction of weak acids by weak bases. For extractants, more basic than the anion of the extracted acid, at $pH < pK_{a,E}$, nearly complete protonation takes place to reach stoichiometric extraction through ion-pair formation with dissociated acid molecules.

To explain the anion effect in these systems, modified Gibbs–Donnan model[60] was used. Analysis showed that increasing the anion concentration in the aqueous phase increases the extraction. Proton activity in the organic phase cannot be equal to that in the aqueous phase, unless the activity values of the anions in two phases are equal. If aqueous phase acidity is high enough to maintain high concentration of undissociated acid, above-stoichiometric extraction takes place through H-bonding to the ion pair. In equilibrium with aqueous phases of high pH ($pH > pK_{aH,L}$), if $pK_{a,E} > pK_{aHL}$ ion-pair formation is the dominating mechanism and the degree of extraction is determined by $pK_{aE} - pH$. If $pK_{aHL} > pK_{aE}$, H-bond formation and solvation of undissociated molecules are the dominating mechanisms, and the degree of extraction is determined by $pK_{aHL} - pH$.

Salts of water-immiscible amines (presented as quaternary amines $R_4N^+L^-$) extract acids of similar anion (HA) through H-bonding to the ion pair $R_4NL{:}HL$. In equilibrium with other acids (HX), R_4NX and $R_4NX{:}HL$ may form. The degree of extraction through H-bonding is affected by the pH determining the concentration of the undissociated acid (RL or HX). pH may also affect the anion exchange. If, however, $pK_{a,HL} > pH > pK_{a,HX}$, protonation·of the acid anion becomes the determining factor, and $[R_4N^+X^-] >> R_4N^+L^-]$.

Extraction mechanism determines the effects of various parameters on extraction efficiency and selectivity.

pH effect

Extraction controlled by H-bonding or solvation is mainly determined by the concentration of the undissociated acid[3,51–58] and strongly depends on $pK_{a,HL} - pH$. However; when ion-pair formation is the dominating mechanism, extraction is determined by extractant protonation at the given pH, or by $pK_{a,E} - pH$). Therefore, strongly basic extractants (amines) are efficient in the extraction of carboxylic acids even at pH values much higher than $pK_{a,HL}$.

Extractant basicity

For $pK_{a,HL} >> pK_{a,E}$ the degree of extraction is determined by extractant efficiency for H-bonding and for solvation. Therefore, weak monocarboxylic acids such as propionic and butyric are highly extracted by polar and nonpolar solvents and by amine salts. The contribution of free extractant, if present, is small. However; extraction of strong mineral acids by any extractant or extraction of carboxylic acids by stronger amines with $pK_{a,E} > pK_{a,HL}$ is dominated by the ion-pair-forming mechanism. As a result, it depends on the degree of extractant protonation and thereby on the extractant basicity.

Acid form in the organic phase can be predicted by the difference between the apparent basicity of the extractant determined at extraction of hydrochloric acid, $pH_{hn,HCl}$, (pH half neutralization for HCl) and the pK_a of the extracted acid. As expected, the IR results showed that, in systems where this difference is close to zero, the dissociated acid and the undissociated acid are both present in the organic phase. In systems where the difference is negative, the main acid form in the organic phase is the undissociated one, and where the difference is positive, the acid is mainly dissociated. The apparent basicity of extractants used for calculating the contribution of ion-pair formation is determined by the properties of the extractant, the diluent, and the anion of the extracted acid; $pK_{a,E} = pK_{aE'}\varepsilon_d\varepsilon_a$ where $pK_{a,E'}$ is the specific basicity determined by pH_{hn} and ε_d and ε_a are coefficients for the efficiency of ion-pair solvation by the diluent and for the anion compatibility with the medium, respectively.

The apparent basicity of the extractant, pH_{hn}, was also determined when half-loaded with various monocarboxylic acids instead of hydrochloric acid. Although the basicity sequences of the extractants are similar for the various acids, the apparent basicity was found to be very sensitive to the properties of the anion. In most cases, the pH_{hn} value decreases in the order extractant loaded with trichloroacetic acid >> propionic acid > hydrochloric acid > acetic acid ≈ lactic acid.

At extraction of trichloroacetic acid with aliphatic primary amine (Pimene JMT), the last has higher steric hindrance and lower polarity. This high basicity is explained by the high acidity of trichloroacetic acid and by its low charge density, which reduces the energy lost when the ion leaves the aqueous phase. On the other hand, the high pH_{hn} values for the amine loaded with propionic acid are explained by the high hydrophobicity of the propionate ion, i.e., by the low repulsion of the ion pairs from the organic phase.

Diluent effect

Extraction through ion-pair formation from aqueous solutions of relatively low pH ($pH < pK_{a,E}$) and extraction of acids by relatively strong amines ($pK_{a,E} > pK_{a,HL}$) are strongly dependent on solvation of the ion pair formed. Therefore, extraction is strongly enhanced by polar and particularly by protic diluents. The effects of diluent on extraction through H-bonding (at $pK_{a,HL} << pK_{a,E}$) is governed by the diluent interaction with the undissociated acid and with the free extractant.

Selectivity

The same considerations also provide for prediction of selectivity in the extraction of acids that differ mainly in their acidity.[28,88] On the basis of the theory analysis, the extraction systems can be divided into three classes in terms of extraction selectivity and its dependence on pH:

1) *Systems in which the two acids are extracted mainly by ion-pair interactions.* At relatively low pH ($pH < pK_{a,E}$) and at higher pH for $pK_{a,E} > pK_{a,HL}$, the dominating mechanism is the ion-pair formation and stronger acids are extracted preferentially. In cases where both acids are extracted via ion-pair formation, the acidic species in the organic phases are the conjugate acid of the protonated extractant (e.g., R_3NH^+) bound to the anions of the extracted acids. The acids compete for interactions with the same amine. The extraction of one acid is decreased by the presence of the other acid in the organic phase.

Increasing the pH is equivalent to adding to the system a strong base that neutralizes these acidic species. The selectivity, thus, depends on the acidity of the protonated extractant. This acidity was found to be very sensitive to the properties of the anion such as basicity, hydrophobicity, electron density, and ability to form H-bonds. Other important parameters are the extractant concentration and diluent properties. In many cases, the stronger acid forms a stronger interaction with the amine and is selectively extracted. In such cases, the protonated extractant bound to the anion of the stronger acid acts as a weaker acid than the protonated amine bound to the weaker acid. Thus, as the pH increases, the stronger acid in the system is preferably neutralized, and the selectivity to it increases. Example is the extraction of dichloroacetic and trichloroacetic acids with TOA at $Z < 1$. Extraction studies of some carboxylic acid mixtures by tri-*n*-octylamine in xylene or in 2-ethyl-hexanol[89] or by tri-*n*-octylamine in methylisobutylketone, toluene, and chloroform[90–92] showed that the extractant is a stronger base than both anions of the acids and that the selectivity was for the stronger acid and increased with pH.

However, the acidity of the extracted acid is not the only parameter affecting the selectivity in this kind of interaction. The experiments with lactic and propionic acids and JMT as extractant[28] demonstrate that the hydrophobicity of the anion is a very important parameter influencing the selectivity: propionic acid was preferably extracted although it is the weaker acid.

If total acid concentration in the aqueous phase is high enough to maintain high activity of undissociated acid (or if a salt is present and affects through salting out), the above-stoichiometric extraction takes place. The mechanisms of the above-stoichiometric extraction, H-bonding to the ion pair (a relatively weak base), or solvation prefer weaker acids. Thus, at extraction from dilute solutions, high selectivity is obtained for stronger acids, but this selectivity decreases in more concentrated solutions. At extracted acid to extractant molar ratios considerably higher than 1, preference for the weak acid is observed. On back-extraction of the loaded extractant, however, the weaker acid is "washed out" first as H-bonding is a weaker bond than ion-pair formation. The stronger acid remains in the extractant and can then be recovered in a pure form.

2) *Systems in which both acids are extracted mainly by H-bonding or solvation interactions.* In systems where the extraction is controlled mainly by H-bonding or by solvation, the aqueous solutions are of relatively high pH, where $pK_{a,E}$ is considerably smaller than the $pK_{a,HL}$ of the acids. The acids in these cases are present in the organic phase mainly in their undissociated forms. In such cases, weaker and more hydrophobic acids are preferably extracted. As pH increases (equivalent to the addition of a strong mineral base), the stronger acid in the system is preferably neutralized; thus, the selectivity to the weaker extracted acid increases. The strongest effect is observed around the pK_a of the stronger acid.[93–95] One more example of such interactions is extraction of propionic and lactic acids by tris-2-ethylhexyl amine (TEHA).[28] The results show that propionic acid (the weaker acid) is preferably extracted in the lower pH range, and the selectivity is increased at higher pH, particularly where the pH is higher than the pK_a of lactic acid. Extraction of tartaric and malic acids using a weak extractant TBP in dodecane[96] showed that the weaker (malic) acid was selectively extracted and that the selectivity increased with pH.

Extraction of one acid can be enhanced by the presence of another acid that increases the polarity and the protic properties of the organic phase. This synergistic effect can decrease as the pH is increased, because this extraction mechanism is dependent on the concentration of the undissociated acid.

H-bonding and/or solvation interactions are also the mechanisms for the added acids in the above-stoichiometric range. In such cases, weak and hydrophobic acids are selectively extracted over stronger and less hydrophobic ones. This selectivity can be seen in the experiment where dichloroacetic and trichloroacetic acids are extracted by TOA,[28,88] in which the main acid is dichloroacetic acid.

The values for the mass action constant of H-bond formation, K_H, were calculated for cases in which the IR spectrum showed mainly undissociated acid in the organic phase.[59] The weaker the acid and the higher the hydrophobicity, the higher the K_H. But K_H values are low compared with the pK_a of the acids. Thus, according to the theory, the pK_a value is the main factor that determines the sensitivity of the extraction to pH.

3) *Systems in which one acid is extracted mainly through H-bonding or solvation interactions, while the other is extracted through ion pair formation.* Typically, such systems are formed when the extractant is a stronger base than the anion of one extracted acid (HX), but a weaker base than the anion of the other acid (HL). At elevated pH, the strongest acid is preferably neutralized. This acid might be the undissociated one or the protonated amine bound to the stronger acid. Therefore, the selectivity might reverse direction with increasing pH, as in the case of the hydrochloric and propionic acid extraction by TOA.[28] In such cases, a synergistic extraction can be found. On one hand, the ion-pair interaction (e.g., $R_3NH^+X^-$) is enhanced by the presence of the undissociated acid in the organic phase, HL, which acts as a polar diluent. On the other hand, the ion-pair species forms bonding sites for the undissociated acid, HL. This was evident in the results of Kirsch[90–92] where acetic acid was extracted together with citric or oxalic acid by tri-*n*-octylamine. The distribution ratios of the acids were higher than that predicted from the single acid systems.

Much more complicated are systems in which one of the extracted acids or both are diprotic, and even more. The number of species that can potentially form and compete for the amine and protons (at elevated pH) increases significantly. A few publications are found in this field.[90–92,96] Extractant here is too weak to ionize the second (or more) acidic groups of the polyprotic extracted acid. Thus, the results in these systems are similar to those for monocarboxylic acids. The authors [28,88] studied the extraction selectivity and its dependence on pH in systems of binary mixtures containing at least one dicarboxylic acid. For that purpose, the extractant is chosen to be the strongest base in the examined systems. The authors suggest divalent ion-pair formation and their possible reactions in the range below molar stoichiometry. At stoichiometry and higher concentrations double salts of type $R_3NH^+L^{2-}Na^+$ are formed through neutralization which are converted to divalent ion pair $R_3NH^+L^{2-}HNR_3$. So, extraction of this acid is visibly decreased. In the mixtures of monocarboxylic with dicarboxylic acids of close pK_a, the increased selectivity of monocarboxylic acid may be expected.

At extraction of dicarboxylic acid mixtures if the extractant is a weaker base than the anions, the dominant acid species in the organic phase is the undissociated acid, for all acid-to-amine molar ratios. These species are formed by H-bonding or by solvation interactions. If added, the mineral base raises the pH, neutralizes the stronger acid, and forms the monovalent salt. This salt spontaneously leaves the organic phase, and, as a result, the extraction is reduced significantly at a pH close to the pK_{a1} of the acid down to zero loading.

If extractant is a stronger base than the first anion but a weaker base than the second anion, the dominant acid species in the organic phase is a function of pH and of the acid-to-amine molar ratio. At acid-to-amine molar ratios greater than 1, the undissociated acid is above-stoichiometrically extracted acid in addition to the ion-paired extracted acid. At stoichiometric loading, the monovalent ion pair $R_3NH^+A^-$ is the dominant species. The extraction curve shows a plateau.

If extractant is a stronger base than both anions the chemistry of the extraction is more complicated. The extraction curve has two drop-offs at acid-to-amine molar ratios of 1 and 0.5. Extraction chemistry of systems in which the extractant is a stronger base than the two anions depends also on pH and acid–extractant molar ratio: (a) At acid-to-amine molar ratios greater than 1, the neutralized species are the first acidic group of the above-stoichiometrically extracted acid, and the extraction curve shows a drop-off close to pK_{a1}, until stoichiometric loading is reached; at $pH < pK_{a1}$, the dominant species is $R_3NH^+ \ldots LH^- \ldots HLH$. (b) At stoichiometric loading, which is in the range $pK_{a2} > pH > pK_{a1}$, the monovalent ion pair ($R_3NH^+LH^-$) is the dominant species, and the curve shows a plateau. (c) The additional mineral base above this point neutralizes the stronger acid in the system, which is the second carboxylic acid, $R_3NH^+AH^-$, forming a kind of double salt $R_3NH^+L^{2-}Na^+$. The extracted carboxylic acid forms the divalent ion pair in the organic phase. As a result, the extraction curve shows a second drop-off at the pK_{a2} of the extracted acid, followed by a plateau at an acid-to-amine molar ratio of 0.5 and in the pH range of $pH_{hn} > pH > pK_{a2}$, where the divalent ion pair $R_3NH^+L^{2-} \ldots {}^+HNR_3$ is the dominant species. (d) The addition of mineral base above this point neutralizes the next strongest acid in the system, which is the protonated amine R_3NH^+. As a result, the free amine remains in the organic phase, and the acid, in its divalent mineral salt form, leaves into the aqueous phase. Thus, the extraction curve shows another drop-off at the pH_{hn} of the extractant.

In systems in which divalent ion pairs are formed, the value of the torsion point at $0.5 < Z < 1$ is related to the pK_{a2} of the acid in the organic phase but not equal. For example,[28] at extraction of malic acid by JMT, the torsion point is equal to 4, whereas pK_{a2} of malic acid is 5.1. For the extraction of glutaric acid, the $pK_{a(o)}$ is about 4.7, whereas pK_{a2} of glutaric acid is 5.4. The authors explained that this difference is a function of the difference in the proton activities in the two phases.

In conclusion, the theory, presented in this chapter, is very complicated, speculative, and contradictive in the presentation of different authors. Knowing the $pK_{a.}$ of the acids in the system provides some preliminary understanding for designing extractants. Extractant design should be based on the $pK_{a,E}$ of the extractants, which, particularly in the case of ion-pair formation, is strongly dependent on the diluent and on the anion of the extracted acid. The use of these theories in the development of new processes requires a big number of experimental data.

REFERENCES

1. Kaminski S, Kutermankiewicz M. *Gaz: Cukrow* 1979;**87**:107. Chem Ab 1979;**91**:173351.
2. Jiang Y, Lei Y, Dai G, Li D, Su Y. *Hemaxue Shijie* 1982;**23**(5):133. Chem Ab1983;**100**: 87647w.
3. Kertes AS, King CJ. Extraction chemistry of fermentation product carboxylic acid. *Biotechnol. Bioeng.* 1986;**28**:269.
4. Thyrion FC. Solvent extraction of lactic acid and lactate from aqueous solutions. In: Proceedings of International Solvent Extr. iConference (ISEC'88). 1988. vol. 3, p. 394.
5. Baniel AM. *European patent No. EP 49429*, 1982, Chem.
6. Grinstead RR. *U.S. Patent No. 3,816,524*, Chem Ab 1974;**81**:37253.
7. Colin P. *French Patent No. 1,548.328* (1967), Chem Ab 1970;**72**: 2163r.
8. King CJ, Tung LA. Sorption of carboxylic acid from carboxylic salt solution at pHs close to or above the pKa of the acid, with regeneration with an aqueous solution of ammonia or low-molecular-weight alkylamine. *U.S. Patent 5,132,456*, 1992.
9. Leo A, Hansch C, Elkins D. *Chem Rev* 1971;**71**:525.
10. Marcus Y, Kertes AS. *Ion exchange and solvent extraction of metal complexes.* London: Wiley-Interscience; 1969.
11. Sekine T, Hasegawa Y. *Solvent extraction chemistry.* New York: M. Dekker; 1977.
12. Hanson C, editor. *Recent advances in liquid-liquid extraction.* Oxford: Pergamon Press; 1971.

13. Lo TC, Baird MHL, Hanson C, editors. *Handbook of solvent extraction*. New York: Wiley-Interscience; 1982.
14. Ridberg J, Cox M, Musikas C, Choppin GR, editors. *Principles and practices of solvent extraction*. New York: CRC; 2004.
15. Kovaleva AG. *Ukr Khim Zhur* 1973;**39**:561.
16. Sigematsu T, Tabushi M, Matsui M, Honiyo T. *Bull Chem Soc Japan* 1987;**42**:976 (1%9).
17. Paul RN. *Brit Chem Eng Proc Technol* 1972;**17**:800.
18. Sukhan VV, Pyatnitskii LV, Sakhno AG. *J Anal Chem USSR* 1973;**28**:476.
19. Charykov AK, Pavlovskaya EM. *Russian J Physical Chem* 1981;**51**:593.
20. Zaheeruddin M, Aslam H. *Pak J Sci* 1980;**32**:99.
21. Harris RD, Geankoplis CJ. *J Chem Eng Data* 1962;**7**:218.
22. Nikitin Yu E, Egutkin NL. *Neftekhimiya* 1974;**14**(5):780. Chern Ab 1975;**82**:139208k.
23. Kolossowsky N, Kulikow F, Bekturow A. *Bull Soc Chem* 1935;**38**:460
24. Rao MR, Ramamurty M, Rao CV. *Chem Eng Sci* 1958;**8**:265.
25. Johnson AI, Furter WF, Barry TW. *Can J Technol* 1954;**32**:179.
26. Kovaleva AG. *Ukr Khim Zhur* 1970;**36**:444.
27. Short JF, Eaglesfield P. *Trans Inst Chem Eng* 1952;**30**:109.
28. Canari R, Eyal A. Selectivity in monocarboxylic acid extraction from their mixture solutions using an amine-based extractant: effect of pH. *Ind Eng Chem Res* 2003;**42**:1568.
29. Biancani M, De Filippo D. *Gazz Chim ltal* 1958;**88**:1202.
30. Kojima I, Yoshida M, Tanaka M. *J Inorg Nucl Chem* 1970;**32**:987.
31. Gordon KF. *Ind Eng Chem* 1953;**45**:1813.
32. Marvel CS, Richards JC. *Anal Chem* 1949;**21**:1480.
33. Zaheeruddin M, Khadim-Ul-Fagir. *J Sci Technol (Peshawar, Pak)* 1979;**3**:51.
34. Sekine T, Isayama M, Yamaguchi S, Moriya H. *Bull Chem Soc Japan* 1967;**40**:27.
35. Frumin GT, Ostrovskii MV, Abramzon AA. *Russ J Appl Chem* 1967;**40**:1275.
36. Vignes A. *J Xhirn Phys* 1960;**57**:966.
37. Sporidi G. *German patent no. 1493196*, 1964, Chern Ab 1970;**72**:100046z.
38. Forbes GS, Coolidge AS. *J Am Chem Soc* 1919;**41**:150.
39. Davis MM, Jones P, Patnaik D, Moelwyn-Hughes EA. *J Am Chem Soc* 1951:1249
40. Pavlovskaya EM, Charykov AK, Tikhornirov VL. *Rus J Gen Chem* 1975;**46**:2325.
41. Smith HA, Berman M. *J Am Chem Soc* 1937;**59**:2390.
42. Mottola HA, Freiser H. *Talanta* 1967;**14**:864.
43. Colin P, Moundlic J. *French patent no. 1,494,958*, 1967, Chem Ab 1968;**69**:43459.
44. Shevchenko VB, Renard EV. *Russ J Inorg Chem* 1963;**8**:268.
45. Pagel HA, McLafferty FW. *Anal Chem* 1948;**20**:272.
46. Tsvetkova ZN, Povitskii NS. *Russ J Inorg Chem* 1960;**5**:1364.
47. Wennersten R. In: Proceedings of the international solvent extraction conference, vol. 2, paper 80/63, University of Liege, Liege, Belgium: 1980.
48. Ricker NL, Michaels JN, King CJ. *J Separ Proc Technol* 1979;**1**:36.
49. Pagel HA, Schwab KD. *Anal Chem* 1950;**22**:1207.
50. Grinstead RR. *U.S. Patents No. 3,980,701-4*; *Chem Ab* 1976,**85**:159443–6.
51. Tamada JA. Extraction of carboxyilic acids by amine extractant. Ph.D. Thesis, Lawrence Berkeley Laboratory, University of California; 1989.
52. Tamada JA, Kertes AS, King CJ. Extraction of carboxylic acids with amine extractant. 1. Equilibria and law of mass action modeling. *Ind Eng Chem Res* 1990;**29**:1319.
53. Tamada J, King CJ. Extraction of carboxylic acids with amine extractants. 2. Chemical interaction and interpretation of data. *Ind Eng Chem Res* 1990;**29**:1327.

54. Tamada JA, King CJ. Extraction of carboxylic acids with amine extractants. 3. Effect of temperature, water coextraction, and process considerations. *Ind Eng Chem Res* 1990;**29**:1333.
55. Tamada JA, Kertes AS, King CJ. Extraction of carboxylic acid from aqueous solution by mean of tertiary amine extractants: Equilibria and processing alternative. *Proceedings of International Solvent Extraction Conference (ISEC'88)* 1988;**3**:316.
56. Tung LA. Recovery of carboxylic acids at pH greater than pKa. Ph.D. Thesis, Lawrence Berkeley Laboratory, University of California; 1993.
57. Tung UA, King CJ. Regenerable extraction and sorption of carboxylic acids from aqueous solutions at pH > pK_a. *Proceedings of International Solvent Extraction Conference (ISEC)* 1993;**2**:1010.
58. Yang ST, White SA, Hsu ST. Extraction of carboxylic acids with tertiary and quaternary amines: effect of pH. *Ind Eng Chem Res* 1991;**30**:1335.
59. Canari R, Eyal AM. Extraction of carboxylic acids by amine-based extractants: apparent extractant basicity according to the pH of half-neutralization. *Ind Eng Chem Res* 2003;**42**:145–52.
60. Canari R, Eyal AM. Acids extraction by amine-based extractants: an analysis of the effect of anion concentration in the aqueous phase using the Donnan model. *Ind Eng Chem Res* 2003;**42**:153–9.
61. Eyal AM, Canari R. pH Dependence of carboxylic and mineral acid extraction by amine-based extractants: effect of *pKa*, amine basicity, and diluent properties. *Ind Eng Chem Res* 1995;**34**(4). 1789.
62. Poulad-Meyer, LA. Spectral study of the transfer of acidic or basic compounds from aqueous phases into multicomponent organic extracting media. Ph.D. Thesis, Casali Institute of Applied Chemistry, Jerusalem:The Hebrew University of Jerusalem; 2001.
63. Manenok GS, Korobanova VL, Yudina TN, Soldatov VS. *Russ J Appl Chem* 1979;**52**:156.
64. Vanura P, Kuca L. *Collec Czech Chem Commun* 1976;**41**:2857–77.
65. Kawano Y, Kusano K, Takahashi T, Kondo K, Nakashio F. *Kagaku Kogaku Ronbunshu* 1982;**8**:404.
66. Kertes AS, Gutmann H. In: Matijevic E, editor. *Surf Colloid Sci* 1975;**8**. Chapter 3.
67. Kertes AS. In: Hanson C, editor. *Recent advances in liquid-liquid extraction.* Oxford: Pergamon Press; 1971. p. 15.
68. Moore FL. Anal Chem 1957; 29:1660. Report No. NAS·NS·31 0 1, Washington DC: National Academy of Sciences; 1960.
69. Smith EL, Page JE. *J Soc Chem Ind (London)* 1948;**67**:48.
70. Pyatnitskii V, Tabenskaya TV. *J Anal Chem USSR* 1970;**25**:2060.
71. Wennersten R. *J Chem Tech Biotechnol* 1983;**33B**:85.
72. Ricker NL, Pittman EF, King CJ. *J Separ Proc Technol* 1980;**1**:23.
73. Pyatnitskii IV, Tabenskaya TV, Makarchuk TL. *J Ana Chem USSR* 1973;**28**:484.
74. Vieux AS, Rutagengwa N, Rulinda JB, Balikungeri A. *Anal Chim Acta* 1974;**68**:415.
75. Vieux AS, Rutagengwa N. *Anal Chim Acta* 1977;**91**:359.
76. Kawano Y, Kusano K, Inoue K, Nakashio F. *Kagaku Kogaku Ronbunshu* 1983;**9**:473. Chern Ab 1983; 99:121728w.
77. Kawano Y, Kusano K, Takahashi T, Kondo K, Nakashio F. Extraction equilibria of lower carboxylic acids with long-chain alkylamine. *Soc Chem Eng Jpn* 1982;**8**:404.
78. Kawano Y, Kusano K, Nakashio F. *Kagaku Kogaku Ronbunshu* 1983;**9**:21. Chern Ab 1983; **9**:21–34.
79. Pyatnitskii IV, Harchenko RS. *Ukr Khim Zhur* 1964;**30**:635.
80. Pyatnitskii IV, Tabenskaya TV, Sukhan TA. *J Anal Chem USSR* 1971;**26**:2150.

81. Pyatnitskii IV, Harchenko RS. *Ukr Khim Zhur* 1963;**29**:967.
82. Pyatnitskii IV, Tabenskaya TV. *J Anal Chem USSR* 1970;**25**:815.
83. Kaplunova AM, Eshova NI, Bolshova TA. *J Anal Chem USSR* 1977;**33**:1492.
84. Ahrland S, Liljenzin JO, Rydberg J. The chemistry of the actinides. In: *Comprehensive inorganic chemistry*. London: Pergamon Press; 1973.
85. Jassim TM, Fridemo L, Liljenzin JO. Dissertation. In: Jassim TM, editor. *Co-extraction of pertechnetate with some metal nitrates in TBP-nitric acid systems*. Gothenburg: Chalmers Technical University; 1986.
86. Chiarizia R, Briand A. Third phase formation in the Extraction of Inorganic Acids by TBP in n-Octane. *Solv Extr Ion Exch* 2007;**25**(3):351–71.
87. Eyal A, Hazan B, Bloch R. Recovery and concentration of strong mineral acids from dilute solution through LLX, III. A "temperature swing based process". *Solvent Extr Ion Exch* 1991;**9**(2):223.
88. Canari R, Eyal AM. Selectivity in extraction of lactic, malic, glutaric, and maleic acids from their binary solutions using an amine-based extractant: effect of pH. *Ind Eng Chem Res* 2003;**42**:160–6.
89. Iagirdar GC, Sharma MM. Recovery and separation of mixtures of organic acids from dilute aqueous solutions. *J Sep Process Technol* 1980;**12**:40.
90. Kirsch T, Maurer G. Distribution of binary mixtures of citric, acetic and oxalic acid between water and organic solutions of Tri-n-octylamine. Part II: organic solvent methyl-isobutylketone. *Fluid Phase Equilib* 1998;**142**:215.
91. Kirsch T, Maurer G. Distribution of binary mixtures of citric, acetic and oxalic acid between water and organic solutions of Tri-n-octylamine. Part I: organic solvent toluene. *Fluid Phase Equilib* 1991;**131**:213.
92. Kirsch T, Maurer G. Distribution of binary mixtures of citric, acetic and oxalic acid between water and organic solutions of Tri-n-octylamine. Part III: organic solvent chloroform. *Fluid Phase Equilib* 1998;**146**:297.
93. Anwar MM, Hanson C, Pratt MWT. Dissociation extraction. Part I: general theory. *Trans Inst Chem Eng* 1971;**49**:95.
94. Anwar MM, Hanson C, Pratt MWT, Patel AN. Dissociation extraction. Part II: multi-stage extraction. *Trans Inst Chem Eng* 1973;**51**:151.
95. Robinson RG, Cha DY. Controlled pH extraction in the separation of weak acids and bases. *Biotechnol Prog* 1985;**1**(1):18.
96. Malmary G, Vezier A, Robert A, Mourgues J, Conte T, Molinier J. Recovery of tartaric and malic acids from dilute aqueous effluents by solvent extraction. *J Chem Technol Biotechnol* 1994;**60**(67):647.

Chapter 3

Chemistry of Metal Solvent Extraction

Chapter Outline

Solvent Extraction. DOI: 10.1016/B978-0-444-53778-2.10003-2

1. INTRODUCTION

Solvent extraction of metals is an important technology in hydrometallurgical industry, analytical separations, and liquid waste treatment. Metal ions, cations, and anions are extracted from an aqueous phase into organic phase through reversible chemical reactions, forming organic-soluble neutral complexes. Extractant has usually very low solubility in aqueous phase, but interact with metal ions and obtain high solubility in organic phase complex. So, extraction is described by a heterogeneous chemical reaction on either side of the aqueous–organic interface to the final equilibrium concentrations of the reaction products.

As it was described in previous chapters, chemical reaction may be considered in (1) one of the phases accompanied by the mass transfer of reactants and products to or from another phase and (2) chemical interaction at the interface of the phases and mass transfer of the reactants and products to the bulk phases. The solubility principles and kinetics of the mass transfer were discussed in detail in Chapter 1. Metal extraction is discussed in detail in the references;[1–3] formation and extraction of metal chelates are presented in references.[4,5]

In this chapter, the factors underlying the strength of metal–extractant interactions, their influence for different metals and extracting reagents, other parameters of complexation are reviewed and discussed.

2. METAL EXTRACTION BY CATION EXCHANGERS (ACIDIC EXTRACTANTS)

The chemical reaction of solutes with organic compounds (termed as extractants, or reagents), dissolved in the organic solvent (diluent), is accompanied by the distribution of product complexes between two immiscible phases. Most common classification of extractants of metals are presented in Table 3.1. The extent of complexation for any solute–ligand system is defined by equilibrium constant, which is termed stability or formation constant for metal–ligand interaction.

2.1. Thermodynamic Considerations

Defining M as a metal cation and L as a ligand anion (charges are omitted for simplicity), the complexation reaction steps may be presented as

$$\mathrm{M} + \mathrm{L} = \mathrm{ML} \quad K_1 = \frac{[\mathrm{ML}]}{[\mathrm{M}][\mathrm{L}]} \tag{3.1}$$

$$\mathrm{ML} + \mathrm{L} = \mathrm{ML}_2 \quad K_2 = \frac{[\mathrm{ML}_2]}{[\mathrm{ML}][\mathrm{L}]} \tag{3.2}$$

$$\mathrm{ML}_{n-1} + \mathrm{L} = \mathrm{ML}_n \quad K_n = \frac{[\mathrm{ML}_n]}{[\mathrm{ML}_{n-1}][\mathrm{L}]} \tag{3.3}$$

TABLE 3.1 Classification of Metal Extractants

Extractant type	Compound group and examples	Coordinating atoms	Acidic groups	Chelate ring size
Cation exchangers: carboxylic, sulfonic, phosphoric, phosphonic, phosphinic acids; acidic chelating agents	Carboxylic acid, RCOOH; e.g., perfluorobutyric (C_3F_7COOH), salicylic $C_6H_4(OH)COOH$, cinnamic ($C_6H_5(CH)_2COOH$) acids	O, O	1	4
	Di(alkyl or aryl) phosphoric & phosphinic acids, RR′PO(OH); thioacids	O, O	1	4
	Dithiocarbamate, RR′NC(S)SHxanthate, ROC(S)SH;	S, S	1	4
	Nitrosohydroxylamine, RN(NO)OH; e.g., cupferron ($R = C_6H_5$); hydroxamic acid, RC(O) NHOH	O, O	1	5
	8-Hydroxyquinoline (oxine), C_9NH_6OH	O, N	1	5
	Diphenylthiocarbazone (dithizone), $C_6H_5NHNC(SH)NNC_6H$	S, N or N, N	1	5 or 6
	Diketone, RC(O) CHC(OH)R′; e.g., acetylacetone ($R = R' = CH_3$), HTTAe ($R = C_4SH_3$, $R' = CF_3$)	O, O	1	6
	Mono(alkyl or aryl) phosphoric and phosphinic acid, $RPO(OH)_2$; Di(alkyl or aryl)pyrophosphate,	O, O	2	4
	RP(O)(OH)OP(O)(OH)R′; e.g., dioctylpyrophosphate	O, O	2	6
	Dicarboxylic acids, $R(COOH)_2$	O, O	2	>5 or 2 × 4

(*Continued*)

TABLE 3.1 Classification of Metal Extractants—cont'd

Extractant type	Compound group and examples	Coordinating atoms	Acidic groups	Chelate ring size
Anion exchangers: Polyphenylmetalloid type, polyalkylsulfonium type, polyalkyl-ammonium type, salts of high molecular weight aliphatic amines				
Solvating extractants: Carbon, sulfur–phosphorus-bonded oxygen-bearing extractants, alkylsulfides, ketones, and others				
Mixed extractants: Organic acids & neutral donor extractants, acid-amine mixtures, two acidic extractants, two neutral donor extractants				

where K_n is the stepwise formation or stability constant. Complex formation may proceed up to coordination number (n) of the central metal atom. Complexes have to be saturated by negatively charged ligands, by neutral donor molecules (organic solvents, inorganic salts with donor properties or water). Overall stability (formation) constant β_n:

$$\beta_n = \frac{[ML_n]}{[M][L]^n} = K_1K_2 \ldots K_n = \prod_{i=1}^{n} K_i \quad (3.4)$$

Of course this relation is valid for standard or ideal conditions, when concentrations are equal to their activities. For real conditions concentrations have to be replaced by activities (see Chapters 1 and 2).

The degree of extraction of metal ions depends on the pH of the aqueous phase. The degree of extraction will also vary on the nature of the metal: in the absence of steric effects, an increasing charge to radius ratio of the metal cation will generally be followed by increased extraction. The rules of coordination chemistry and ligand field theory are obeyed.

Metal extraction by acidic reagents is conventionally shown as a stepwise process where the extractant partitions between the organic and aqueous phases, going through the steps of ionization in the aqueous phase, reaction of the metal ion with acid anion until the neutral complex is formed, and then this final complex partitions between two phases.

This description supposes extraction mechanism dependant on chemical reactions in the aqueous phase, requiring a reasonable solubility of extractant in water. If extractant is highly insoluble in water then the reactions are wholly interfacial. In this case description should be replaced by adsorption isotherms relating interfacial concentration of the extractant and metal complex to the bulk phase concentrations (details see Chapter 1). Depending on the acidity of the extractant and the pH range being considered, it may sometimes be appropriate to consider direct reaction between the molecular form of the extractant and the metal cation rather than the extractant anion. The scheme presented first, however, has considerable advantages in describing the chemistry of metal extraction by acid reagents.

There are ligands that can bind to a metal ion in more than one site of basic atoms, such as O, N, S. They are called chelating agents and form metal chelate complexes. Chelates are commonly stronger than analogous nonchelate complexes. Chelation provides extra stability to the metal complex.

The metal ion distribution coefficient increases with increasing aqueous solubility and increasing acidity of the reagent, increasing pH, and also increasing reagent concentration. The value of the formation constant β clearly will affect the value of D, reflecting the specific interactions related to chelating extractants and also other factors such as steric effects from extractant molecular geometry.

Taking all metal species in the aqueous phase into account, the distribution of the metal can be written as

$$D_{\mathrm{ME}} = \frac{[\mathrm{ML}_z]_{(\mathrm{O})}}{\sum\left[\mathrm{ML}_n^{z-n}\right]} = \frac{K_n\beta_n\left[\mathrm{L}^-\right]^z}{\sum\beta_n\left[\mathrm{L}^-\right]^z} \tag{3.5}$$

The distribution ratio depends only on the free ligand concentration, which may be calculated by Eqns (2.24–2.26 from Chapter 2[1]). Most coordinatively saturated neutral metal complexes behave just like stable organic solutes, because their outer molecular structure are almost entirely of the hydrocarbon type, and

1. Eqns (2.24–2.26) from Chapter 2:

$$\log\frac{\left[\mathrm{EH}^+\right]_{(\mathrm{O})}}{[\mathrm{E}]_{(\mathrm{O})}} = \log\left[\mathrm{H}^+\right] - \log K_{\mathrm{a,E}} = \mathrm{p}K_{\mathrm{a,E}} - \mathrm{pH}$$

$$\log\frac{[\mathrm{R_4NX}]_{(\mathrm{O})}}{[\mathrm{R_4NL}]_{(\mathrm{w})}} = \log K_{\mathrm{a,exch}} + \log\left[X^{-1}\right]_{(\mathrm{w})} - \log[\mathrm{HL}]_{(\mathrm{w})}\mathrm{p}K_{\mathrm{a,HL}} - \mathrm{pH}$$

can therefore be extracted by all solvent classes. The rules for the size of the distribution constants of these coordinatively saturated neutral metal complexes are then in principle the same as for the inert organic solutes, described in Chapters 1 and 2. However, such complexes may still be amphoteric due to the presence of electronegative donor oxygen atoms (of the chelating ligand) in the chelate molecule. In aqueous solution such complexes then behave like polyethers rather than hydrocarbons. Narbutt[6] has studied such outer-sphere hydrated complexes and shown that the dehydration in the transfer of the complex from water to the organic solvent determines the distribution constant of the complex.

The degree of extraction of metal ions by organic acids depends on the metal ion basicity and thus in general follows the lyotropic series in the absence of specific extractant metal–ion interactions. The $pK_{a,\,extr}$, when $\log D = 0$, is often used to rank metal ions for a fixed extractant concentration. Several such selectivity series have been published for carboxylic acids and di-2-ethylhexylphosphoric acid.[7,8] As an example extraction with naphthenic acid is shown in Fig. 3.1.

Let us consider the complexation and extraction of Cu(II) by acetylacetone.[9] Acetylacetone can coordinate a metal atom in two ways: in the uncharged keto form (through two keto oxygens) or in dissociated anionic enol form (through the same oxygen). It acts as an acid only in the enolic form.

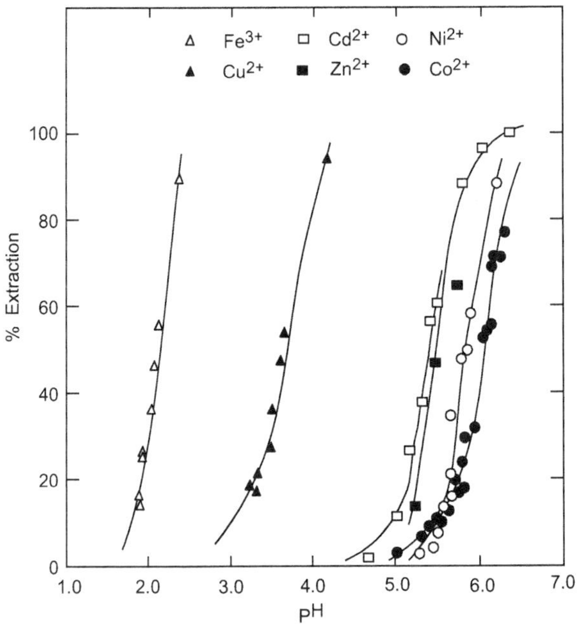

FIGURE 3.1 Extraction of Fe^{3+}, Cu^{2+}, Cd^{2+}, Zn^{2+}, Ni^{2+}, and Co^{2+} with naphthenic acid two additional very weak bonds can be formed perpendicular to the plane; the authors neglect them in this case.

Figure 3.2 shows the extraction of Cu(II) from 1 M $NaClO_4$ into organic phase of acetylacetone (HL) as an extractant in benzene as a diluent at various concentrations of the extractant. Acetylacetone reacts with Cu(II) in aqueous solutions to form the complexes CuL^+ and CuL_2. Because acetylacetone binds through two oxygens, the neutral complex CuL_2 contains two six-membered chelate rings; thus four coordination positions are taken up, forming a planar complex. This complex is usually considered to be coordinatively saturated, but

$$D_{Cu} = \frac{[CuL_2]_{(O)}}{[Cu^{2+}] + [CuL^+] + [CuL_2]} = \frac{K_{DC}\beta_2[L^-]^2}{1 + \beta_1[L^-] + \beta_2[L^-]^2}$$

$$= \frac{K_{DC}\beta_2[L^-]^2}{\sum \beta_n[L^-]^n} \qquad (3.6)$$

In Fig. 3.2 log D_{Cu} is plotted versus $-\log\,[H^+]$ or pH. In the insert of Fig. 3.2 log D is plotted as a function of log $[L^-]$, where $[L^-]$ has been calculated using Eqn (2.26) from Chapter 2 (see Note 1) from pH, $[HL]_O$ organic phase volumes, and K_a. It can be seen that log D_{Cu} is a function of pH at constant $[HL]_{(O)}$ or $[HL]_{(w)}$, while at constant pH the log D_{Cu} depends only on $[HL]_{(O)}$ or $[HL]_{(w)}$.

The distribution curve approaches two asymptotes, one with a slope of 2 and second horizontal (zero slope). From Eqn (3.6) it follows that, at the lowest

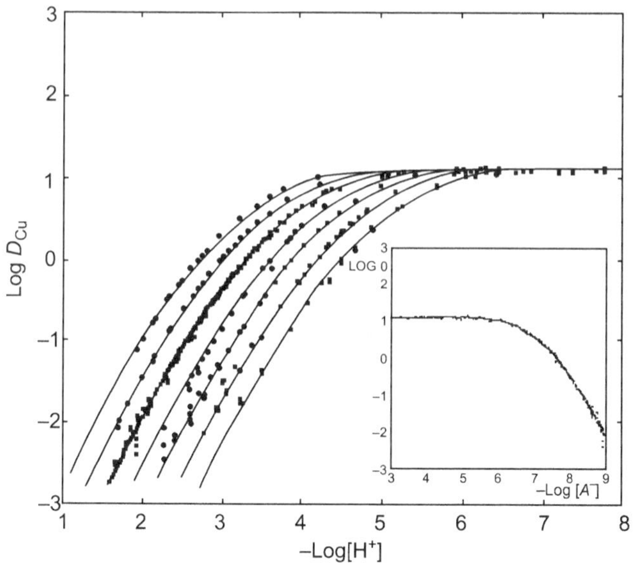

FIGURE 3.2 Extraction of Cu(II) from 1 M NaClO4 into benzene as a function of pH (large figure) and of free acetylacetonate ion concentration (insert) at seven different total concentrations of acetylacetone ([HA]aq 0.05–0.0009 M). *(Source: From Ref. 9 with permission)*

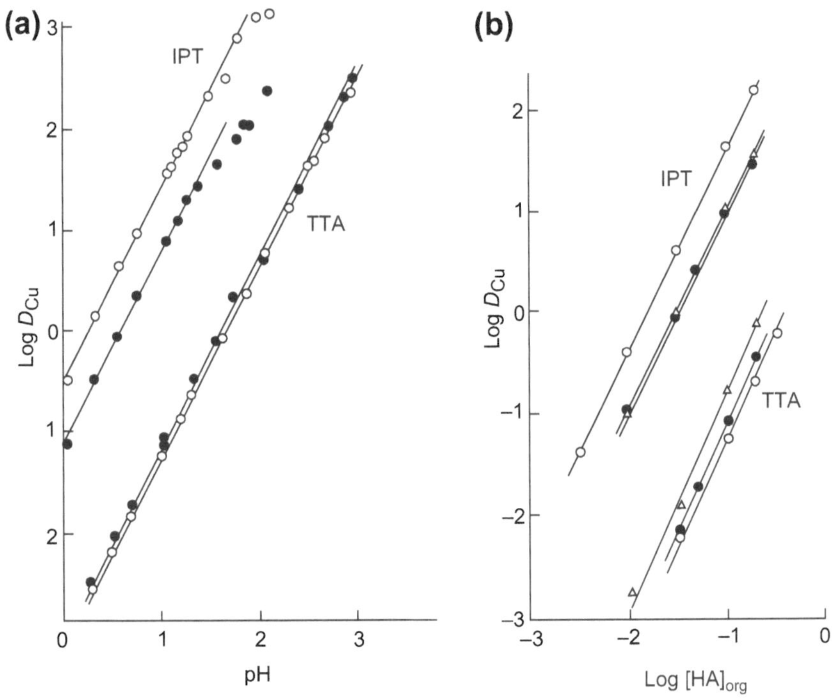

FIGURE 3.3 Distribution of Cu(II) between hexone (△), carbon tetrachloride (●), or chloroform (○) and 0.1 M $NaClO_4$ in the presence of isopropyltropolone (IPT) or thenoyltrifluoroacetone(TTA); (a) as a function of $-\log[H^+]$ at constant $[TTA]_{org} = 0.1$ M; (b) as a function of $[TTA]_{org}$ at constant $[H^+] = 0.1$ M. (*Source: From Ref. 10*)

$[L^-]$ concentration (lowest pH), the concentration of CuL^+ and CuL_2 in the aqueous phase becomes very small:

$$\lim_{L^- \to 0} D_{Cu} = \frac{[CuL_2]_{(O)}}{[Cu^{2+}]} = K_{DC}\beta_2[L^-]^2 \tag{3.7}$$

At the highest L^- concentrations a horizontal asymptote is approached:

$$\lim_{[L^-] \to \infty} D_{Cu} = \frac{[CuL_2]_{(O)}}{[CuL_2]_{(w)}} = K_{DC} \tag{3.8}$$

The horizontal asymptote equals the distribution constant of CuL_2, i.e., K_{DC}. From the curvature between the two asymptotes, the stability constants β_1 and β_2 can be calculated.

This example indicates that in solvent extraction of metal complexes with acidic ligands, it can be more advantageous to plot log D versus log $[L^-]$, rather than against pH, which is the more common (and easy) technique.

For practical purposes it is simpler to use the extraction constant K_{ex} for the reaction

$$M^{z+}_{(w)} + zHL_{(O)} \leftrightarrow ML_{z,(O)} + zH^{+}_{(w)} \tag{3.9}$$

in this case the ML_n^{z-n} complexes in the aqueous phase are neglected. The relevant extraction equations are

$$K_{ex} = \frac{[ML_z]_{(O)}[H^+]_{(w)}}{[M^{z+}]_{(w)}[HL]^z_{(O)}} \tag{3.10}$$

$$D_M = K_{ex}\frac{[HL]^z_{(O)}}{[H^+]^z_{(w)}} \tag{3.11}$$

Thus only one constant, K_{ex}, is needed to predict the metal extraction for given proton and extractant [HL] concentrations. Comparing Eqns (3.6) and (3.11), it is seen that no horizontal asymptote is obtained even at high concentrations of L^-, or HL and H^+. For very large distribution constant of the uncharged complex a straight line with slope $\pm z$ is experimentally observed. Figure 3.4 shows the distribution of Cu(II) between three organic solvents and water in the presence of isopropyltropolone (HITP) or thenoyltrifluoroacetone (HTTA) as a function of pH and [HL].[10] A plot of log D against $-\log[H^+]$ should yield a straight line with slope +2, and a plot against log [HA] should also yield a straight line with slope +2. It indicates that the aqueous phase does not contain any significant concentrations of the complexes CuL^+ yet CuL_2 must be formed in considerable concentrations, otherwise there would be no extraction of Cu(II). The line also corresponds to the asymptote Eqns (3.7, 3.8). The conclusion is that the aqueous phase is completely dominated by Cu^{2+},

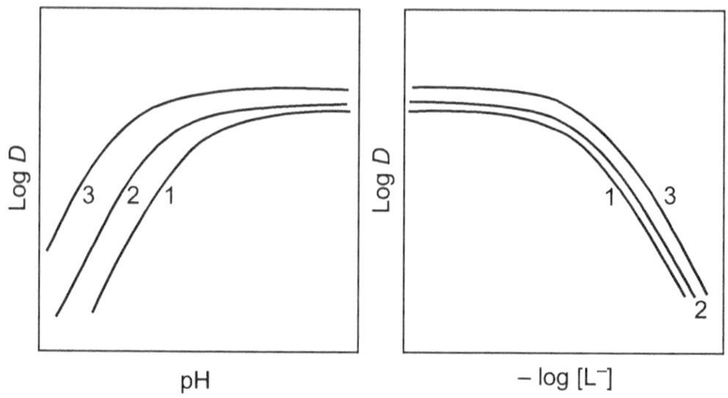

FIGURE 3.4 Log D as a function of pH and $-\log[L^-]$ for the case when both complexes of the form ML_n and $ML_n(LH)_m$ are formed and extracted: 1, 2, and 3 denote different increasing LH concentrations.

while the organic phase contains only CuL_2. This example illustrates a case of considerable analytical importance, especially for the determination of complex formation constants for hydrophilic complexes, when the equilibrium constants for the stepwise metal–organic complexes are of secondary interest. K_{ex} values are tabulated in several reference works.[1–5] K_{ex} is a conditional constant and only valid provided no other species are formed besides the extracted one.

When both the organic and the aqueous phases are dominated by the same uncharged complex over a large concentration region, the distribution constant K_{DC} of the neutral complex ML_z, can be directly measured. Otherwise K_{DC} can be estimated or calculated from K_{ex}.

Differences between series for individual acids may be due to specific interactions, steric factors in relation to extractant geometry, or experimental error as the pH range covering some of the metals given in these series is very small. The variation of degree of extraction with the aqueous solubility of the extractant [Eqn (3.6)] is partly offset by the solubility of the resultant metal complex also change. Variation of the aqueous phase anion can affect the selectivity order due to the metal–anion interactions in the aqueous phase. Provided no metal-inorganic anion complexes are extracted, the distribution coefficient will be decreased by such interactions as follows:

$$\log D = \log K_{DC} + z\text{pH} + z\log[\text{HL}]_{(O)} - \log \sum_{0}^{i} \beta_i[\text{ML}]_i \qquad (3.12)$$

In general, the aqueous interaction would be expected to increase in the order $ClO_4 < NO_3^- < Cl^- < CO_3^- < SO_4^-$. Equation (3.12) applies to a neutral ligand when nonextractable complexes such as ammonia are formed. The degree of extraction increases with increasing acidity of the extractant in the general order: carboxylic acids < alkylphosphoric acids < sulfonic acids.

Obviously, chelating extractants cannot be ranked in the regular series arising from specific interactions with compounds, dominating in the system. However, the effect of increasing acid strength within such classes of reagents can be demonstrated for hydroxyoximes where chlorination of 4-nonyl-2-hydroxybenzophenone oxime in the 3 position changes of the pK_a for copper from 0.96 to 0.00,[11] whereas the acid dissociation constants for the antiisomers are 11.5 and 10.5, respectively. A similar effect, of course, exists within each class of simple acids, and pK_a for copper with lauric acid is reduced from 4.1 to 3.4 when *o*-bromolauric acid is used.[12]

2.2. Stoichiometry and Structure of Extracted Species

The most popular way to study the stoichiometry of the extracted species has been the use of *slope analysis*. This relies on plots of $\log D$ versus pH at constant extractant concentration and also $\log D$ versus log[HL] or $\log[L^-]$ at constant pH. However, the use of Eqn (3.11) for this purpose is based on the following simplifying assumptions:

1. No polymeric species are formed in either phase.
2. Activity factors are essentially constant, so equilibrium constants based on concentration can be used.
3. Only uncharged species are extracted.
4. The formation of intermediate nonextractable complexes can be neglected.
5. No adduct formation between complexes and undissociated extractant molecules or the organic diluent or modifier takes place.
6. All hydrolysis reactions are insignificant.

It is thus very unusual for the simple extraction Eqn (3.9) and the derived Eqn (3.11) to account completely for metal extraction by acidic extractants. Assumption 2, however, can be complied with as nearly as possible by using constant ionic strength media and low metal concentrations. Thus, the activity coefficients are constant and the free organic extractant concentration will also remain essentially constant, provided $[HL]/z$ at $[M^{z+}] >> 1$. Assumption 3 is normally valid. Assumption 1 is seldom valid, as it has already been noted that many of the extractants dimerize to an extent dependent on the nature of the organic diluent. Monoalkylphosphoric acids polymerize and sulfonic acids form micelles, usually with a micelle number of 7. The extraction equation has to reflect this behavior if correct interpretations are to be drawn from slope analysis. The formation of intermediate nonextractable complexes with the extractant as in assumption 4 ($ML_2^{(z-1)+}$, $ML_2^{(z-2)+}$, etc.) will reduce the distribution coefficient and give rise to a curved plot of log D versus log [HL] or log D versus $[L^-]$. The plot of log D versus pH will not be affected unless charged species are extracted. The effect of adduct formation and hydrolysis on the form of the extraction plots can be seen in Fig. 3.3. In the latter the log D – pH plots show an increase in D at increasing [HL] but have a common plateau that is numerically equal to P_{ML}. The log D – log$[L^-]$ plots for all HL concentrations coincide. When adducts are formed, log D_{max} increases with increasing [HL] and the log D – log$[L^-]$ plots are no longer coincident (see Fig. 3.5). Adduct formation usually occurs when the coordination requirements of the extracted metal remain unsatisfied after formation of the neutral ML_z species; thus the water molecules in the remaining coordination positions are replaced by extractant molecules through direct coordination to the central metal ion. Addition of diluent molecules is also possible when sufficiently polar diluents such as alcohols and ether are used. When extractants such as carboxylic acid and dialkylphosphoric acids are involved, the adduct formation is also believed to involve ring formation through the H-bonded adduct.

Behavior when hydrolytic species are involved in the reaction is illustrated in Fig. 3.6. This is also representative of cases where nonextractable complexes are formed in the aqueous phase with ligands other than the extractant. The

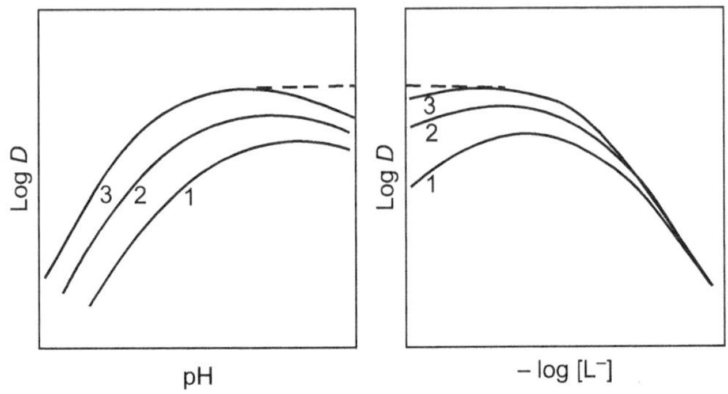

FIGURE 3.5 Log D as a function.of pH and $-\log[L^-]$ for the case when extractable complexes of the form ML_n and nonextractable complexes of the form $M(OH)_p$ are formed: 1, 2, and 3 denote different increasing LH concentrations.

maximum distribution coefficient will depend on the extractant concentration and increases with increasing concentration until formation of the non-extractable complexes can be neglected. Polymerization of the extracted metal complexes is also common with carboxylic[7] and organophosphoric acids[1] but less so with chelating extractants. This phenomenon results in an increase in log D with increasing metal ion concentration.

Various attempts have been made to model the effects of variables. Computer programs have been written on this basis to analyze extraction data and determine the likeliest species present in the organic phase. However, it is quite clear from the foregoing discussion that slope analysis in such complex situations is fraught with difficulty and should be used only in conjunction with physical measurements (e.g., infrared, ultraviolet, and Raman spectroscopy) to obtain an unambiguous description of the extraction system. An excellent critique of the slope analysis technique has been presented by Marcus.[13]

Stoichiometric studies have been carried out for all types of acidic extractants. Generally the chelating extractants form readily defined species, whereas for the acidic nonchelating extractants, a variety of species tend to coexist in the organic phase, with their relative importance depending on the system variables. Extensive data on specific systems are available in Refs. 1,5,14,15. Thermodynamic data, formation constants, and so on, for metal complexes with carboxylic acids, alkylphosphoric, phosphonic, phosphinic acids, and sulfonic acids have been published by Marcus et al.,[16,17] because the acidic chelating extractants behave in a relatively simple manner, considerable interest has been shown in the application of regular solution theory to predict their distribution behavior in solvent extraction systems. Considerable success in this regard has been achieved for the β-diketones, and an excellent review of the use of theory in various systems, including β-diketones, 8-quinolinol, carboxylic acids, and similar reagents has been published in Ref. 18.

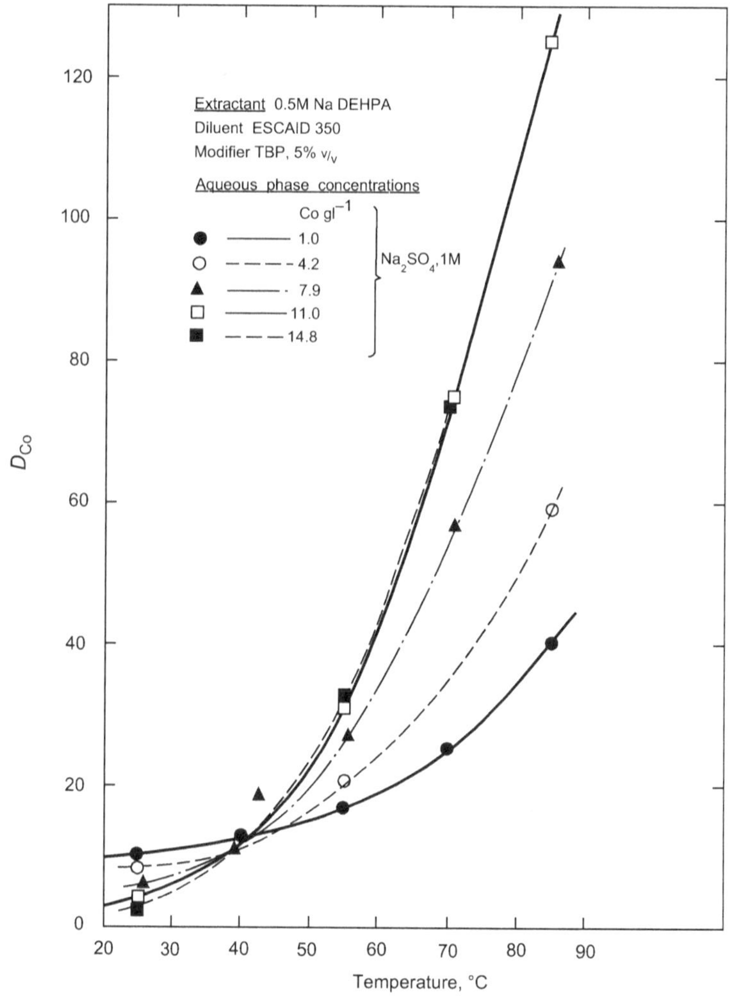

FIGURE 3.6 Effect of temperature on extraction of Co^{2+} from sulfuric acid solution by sodium salt of DEHPA.

2.3. Properties of Extractants

For a simple system, the distribution of the extractant will depend on the value of the thermodynamic partition constant, the pK_a value of the acidic group, and the aqueous pH. The partition constant will depend on the relative solubilities of the extractant in the organic and aqueous phases, and in general P_{HL} will increase with increasing molecular weight in any homologous series and will also depend on the nature of the diluent. This simple picture can be complicated by (1) polymerization or aggregation of the extractant in the organic phase, (2) direct interaction or adduct formation between the extractant and the diluent, (3) formation of acid hydrates soluble in the aqueous phase, (4) amphoteric behavior of the extractant, mostly found for chelating extractants such as 8-quinolinol, and (5) polymerization or aggregation of the extractant and the ionized extractant anion in the aqueous phase.

For carboxylic acids, the major complications are dimerization of the acids in the organic phase and the presence of singly charged anion molecule dimers, as well as doubly charged dimeric species in the aqueous phase. The mathematical treatment of such effects on the distribution of carboxylic acids has been fully discussed in Ref. 7, as has the aqueous and organic phase chemistry of importance in solvent extraction systems. Quantitative information on distribution and dimerization constants of carboxylic acids is given in Ref. 14.

The corresponding behavior of acidic organophosphorus compounds has been reviewed in Ref. 1. A number of equilibria relating to polymerization, hydrate formation, and so on, have to be assumed in the distribution of these acids between the organic and aqueous phases. The degree of polymerization again depends on the nature of the diluent, and it is found that the monoalkyl dihydrogen phosphoric acids $ROPO(OH)_2$ are more highly polymerized than the dialkyl monohydrogen phosphoric acids. The acids are also amphoteric, extracting mineral acids under certain conditions, and thus the equilibria associated with this behavior have to be included. H-bonded interactions between the diluent and other donor compounds such as tri-*n*-butylphosphate (TBP) and alkylamines[19] are also possible. H-bonding interactions are, in fact important in all these systems and most of the interactions are likely to be competitive. The combination of extractant with a neutral donor or another extractant can be of considerable significance in metal ion extraction and can give rise to the phenomena of *synergism* and antag*onism*, which will be discussed later. Data on distribution, dissociation, dimerization constants, and other parameters are available from the literature.[1,14,16]

Sulfonic acids are even more highly aggregated than alkylphosphoric acids and tend to form micelles even at very low organic phase concentrations. The presence of micelles thus dominates the whole extraction chemistry of sulfonic acids. The solvent extraction behavior of these acids is reviewed in Refs. 1,20,21. The physical chemistry of aggregation and micellization of carboxylic, sulfonic acids, and their salts in organic solvents has also reviewed.[22] Great care must be taken when examining the literature for values of dimerization constants, critical micelle concentrations (CMC), and similar parameters. Anhydrous organic solution data must be treated cautiously, for it is certain that in a solvent extraction situation the presence of water will produce hydrates and, through competitive H bonding with the extractants, materially affect the degree of polymerization and observed CMC values.

The chelating extractants tend to behave in a more regular fashion than the organic acids. Nevertheless, hydroxyoximes dimerize in the organic phase to an extent that depends not only on the aromaticity of the diluent, but also on the hydroxyoxime type. The presence of higher aggregates has recently been established,[23] and the degree of aggregation has considerable significance. A full description of the physical properties and extraction data for proprietary hydroxyoxime reagents has been given by Ashbrook.[24,25]

The β-diketones, probably the most widely studied group of chelating extractants, undergo enolization, which is an additional factor affecting extractant behavior and distribution. Dissociation and distribution constants for a variety of diketones between carbon tetrachloride and 0.1 M sodium perchlorate solution have been provided by,[14] who have also reviewed the extensive literature on this class of extractants.

Amphoteric behavior is shown by some of these chelating acid extractants; thus 8-quinolinol-based reagents and α-hydroxyoximes are known to extract acids and anionic or neutral metal complexes from aqueous solution. This behavior is related to the presence of weakly basic atoms in the extractant molecule, the chemistry of which is similar to that of anion exchangers and solvating reagents. Such behavior must be borne in mind when metal extraction over a wide range of acidities is considered.

2.4. Influence of Temperature

Variation of temperature affects the extraction chemistry, as instanced by the cobalt–nickel extraction system with di-2-ethylhexylphosphoric acid (DEHPA).[26] As temperature increases the extracted cobalt species changes from a hexa coordinated dihydrate to a tetrahedral anhydrous species with a consequent increase in the distribution coefficient (Fig. 3.6); increasing cobalt concentration at constant temperature causes polymerization of the organic phase species with consequent increase of distribution coefficient (Fig. 3.7). The nickel complex does not show this behavior at all, and this difference is the basis of a separation of cobalt from nickel in sulfate solution by DEHPA.

3. METAL EXTRACTION BY ANION EXCHANGERS (ION PAIR FORMATION)

Metals that react with inorganic ligands to form negatively charged complexes can be extracted into organic solvents with large organic cations in a process referred to as *liquid anion exchange*.

3.1. Equations for Liquid Anion Exchange Extractions

The discussion on the properties of the amine salts serves to explain why, in metal amine extractions, it is difficult to obtain simple mathematical relations that agree well with the experimental data. While there is no difficulty, in principle, in obtaining reasonably good values for the formation of the negatively charged metal complexes in the aqueous phase, there is a major problem in defining the organic phase species, which may consist of free amine [RN],

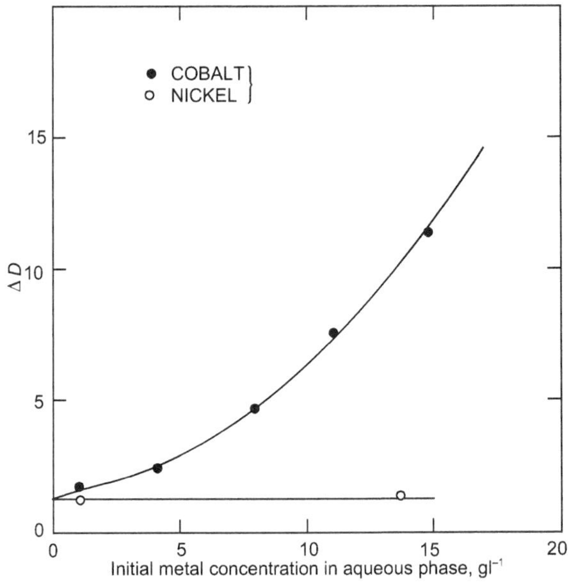

FIGURE 3.7 Influence of initial metal ion concentration on distribution coefficient at 55 °C.

monomeric [RNHL], and polymeric $[RNHL]_n$ amine salt, and several extracted metal complexes $[(RNH)_p(ML_p^{n-})]$. A contributing difficulty in practice is the need to use high ligand concentrations, $[L^-]$, in the aqueous phase in order to obtain the negatively charged complexes (see Fig. 3.8). In general the metal reactions with a monobasic anion are presented by Eqns (3.3, 3.4).

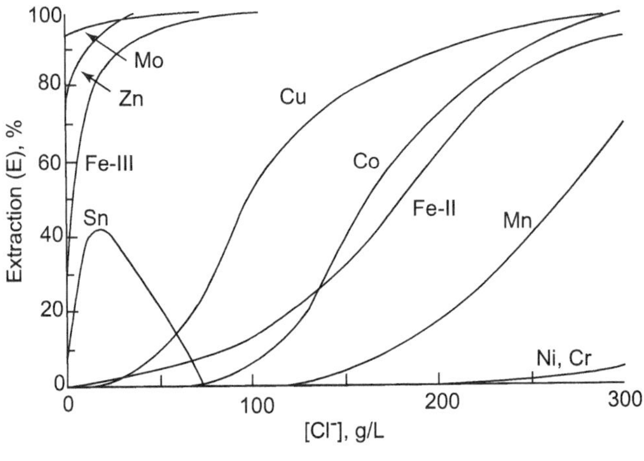

FIGURE 3.8 %Extraction of various metals versus chloride anion concentration from a dissolved stainless steel scrap chloride solution at pH = 2 with organic phase, contained 25% trioctyl amine, 15% dodecanol in 60% kerosene.

When $z - n = q$ is negative, a negatively charged metal complex has been formed, which can be extracted

$$ML_n^{-q} + qRNH^+L^-_{(O)} \leftrightarrow \left([RNH^+]_q^q[ML]_n^{-q}\right)_{(O)} + qL^- \quad (3.13)$$

And/or

$$ML_n + R_4NL \leftrightarrow RN^+ML^- \quad (3.14)$$

We define the extraction constant by

$$K_{ex} = \frac{\left[[RNH]_q^{+q}\right]_{(O)}\left[L_n^-\right]^q}{\left[ML_n^{-q}\right][RNHL]^q_{(O)}} \quad (3.15)$$

It must be assumed that the aqueous phase contains all the stepwise complexes ML_n^{z-n}. Thus, the distribution ratio is

$$D_M = \frac{\left[(RNH)_q^{+q}ML_n^{-q}\right]_{(O)}}{\sum\left[ML_n^{z-n}\right]} = K_{ex}\frac{\beta_q\left[L^-\right]^q[RNHL]^q_{(O)}}{1+\sum\beta_q\left[L^-\right]^q} \quad (3.16)$$

The distribution of M depends on both the free amine salt in the organic phase and the concentration of free ligand, L^-, in the aqueous phase until all metals in the aqueous phase are bound in the ML_n^{-q} complex. At constant amine concentration, Eqn (3.16) indicates that a plot of D_M versus $[L^-]$ would have a linear slope q if the denominator of Eqn (3.16) is $<<1$; i.e., the metal species in the aqueous phase are dominated by the uncomplexed metal ion M^{z+}. At higher $[L^-]$ concentrations, where the ML_n^{-q} complex begins to dominate in the aqueous phase, the D_M value becomes equal to $K_{ex}[RNHL]_{q,org}$. Eqns (3.16) and (3.5) show that the S-shaped curves result for metals with large K_{ex} values. In a plot of D_M versus $[RNHL]_{org}$ a straight line of slope q is obtained only at constant $[L^-]$. From such measurements q, K_{ex}, and β_p can be evaluated.

The interaction of amine salts and quaternary ammonium ions with metal anionic complexes is mainly electrostatic; thus, as the degree of extraction is due to the extent of ion pair formation, selectivity will depend on charge, ion size, and extent of complex formation in the aqueous phase. Selectivity is generally achieved by varying the ligand concentration in the aqueous phase.

Stoichiometry for acidic extractants has traditionally been determined by slope analysis techniques, plotting the distribution coefficient of the metal of interest as a function of the aqueous concentration and extractant concentration. Interpretation of these results can be difficult because of hydrolytic polymerization of aqueous phase complexes when oxyanions are concerned and the degree of aggregation of the amine salt and amine salt metal complex in the organic phase. Extensive computer studies of distribution data have been carried out.[27] Again, the use of slope analysis or numerical analysis alone for the determination of stoichiometry must be done cautiously unless there are other appropriate physical chemical data.

Third-phase formation, that is, the splitting of the organic phase into two parts, is common in these systems and occurs through solubility problems relating to aggregation. This almost always results in a light organic phase rich in diluent and lean in amine and a dense organic phase rich in amine salt–metal complex. Third-phase formation is eliminated by addition of diluent *modifiers* such as long-chain aliphatic alcohols that carry out specific solvation of the amine salts and metal complexes through either dipole–dipole interactions or H bonding.

Extractable anionic species are found with several different types of ligand. For example, some metals occur as oxyanions that are extractable by amines, such as ReO_4^-, TcO_4^-, and MnO_4^-. The group VI metals chromium, molybdenum, and tungsten also occur as oxyanions, but here the system is more complex in that aggregation and proton association with the anionic complexes can also occur. The anions derived from mineral acids are also known to form complexes with metals. The extent of complexation of metals by these anions, as before, is as follows:

$ClO_4^- < NO_3^- < Cl^- < SO_4^-$; thus the extraction of metals by quaternary ammonium ions follows this ligand order.

The nitrate ion tends to form only weak complexes with metal ions in aqueous solution, but a few elements can be extracted in the form of nitrate complexes with amine salts. These elements tend to be tetra- and hexavalent actinides and tervalent lanthanides, the heavier transition elements of Groups IV and V, and some of the platinum elements.[28,29]

For example, at extraction of trivalent actinides by the tertiary amine salt, trilaurylmethylammonium nitrate, TLMA * HNO_3, in *o*-xylene from 0.01 M nitric acid solutions of various $LiNO_3$ concentrations the authors[27] have found that the amine was monomeric only at very low concentrations ($\leq$0.1 M in the organic phase) but at higher concentration formed both dimers and trimers. Using trace concentrations of Ce(III) and Am(III) a log–log plot of D_M against the nitrate ion activity, $m\gamma_\pm = [LiNO_3]^{1/2}$, had a slope of approximately 3 (see Fig. 3.9b). From Eqn (3.15) this slope corresponds to the *q*-value of 3 when the aqueous phase is dominated by the free metal ion, which is not an unreasonable assumption at low nitrate concentrations.

In plots of log D_{Am} against $[TLMA * NO_3]_{org}$ at different nitrate concentrations, the curves in Fig. 3.9a had straight slopes of 1.5–1.8 at low concentrations of TLMA * NO_3, but bending at higher concentrations, was explained by the formation of polymeric amine species. If Eqn (3.15) is valid, these slopes correspond to the number of TLMA * HNO_3 groups attached to the extracted Am species. Thus, the authors[27] described this complex as $\{(TLMA * NO_3)_n\}_2 * Am(NO_3)_3$ for which $n = 1$ only at the very low amine concentrations. Thus, for $n = 1$ the complex could be written as $(TLMA)_2$ $Am(NO_3)_5$. The Ce and other Am complexes would have similar configurations.

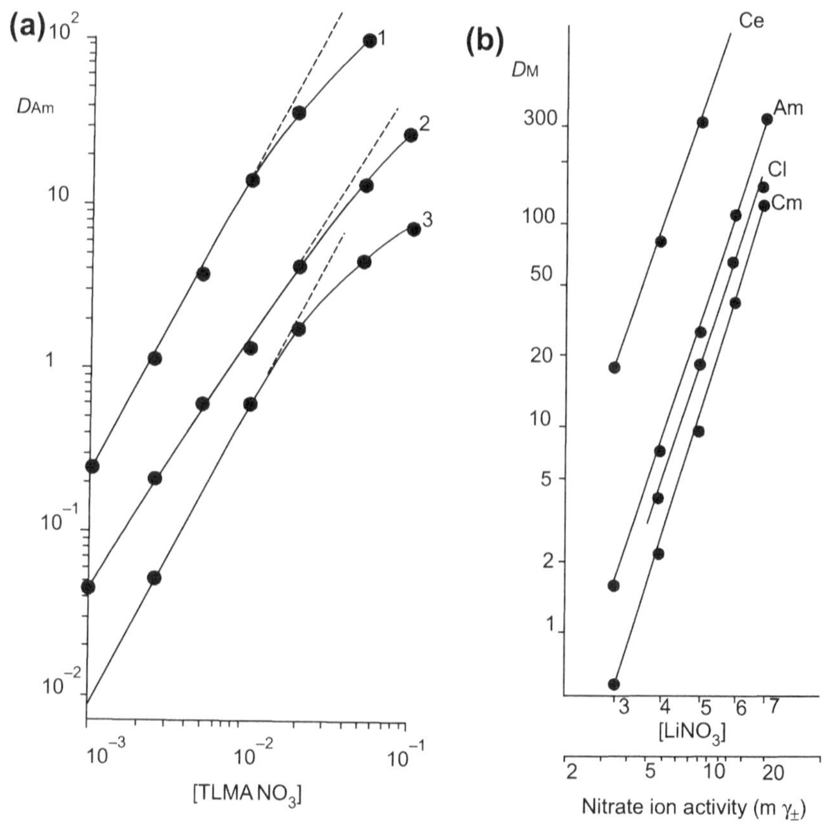

FIGURE 3.9 Distribution ratio of M^{3+} ions between the trilaurylmethyl ammonium nitrate (TLMA) in *o*-xylene and aqueous phases of varying $LiNO_3$ concentrations. (a) As a function of $TLMA{*}NO_3$ concentration at 1–7 M, 2–5 M, 3–3 M $LiNO_3$. (b) Extraction of Eu(III) and tree actinide(III) ions at 0.1 M $TLMA{*}NO_3$ in *o*-xylene and varying aqueous salt concentrations. *(Source: From Ref. 27)*

The order of extraction of anionic metal halides seems independent of the class of amine, although in general the order is quaternary > tertiary > secondary > primary. This order is the reverse of that for hydrochloric acid extraction, and an explanation of this in terms of back donation between chlorine atoms and the amine protons has been proposed by Diamond.[30] The polyvalent metals exhibit the highest distribution ratio, followed by the divalent transition metals.[1] Frequently, extraction from alkali chloride media is preferred to mineral acid, presumably because of competition in the latter case by extraction of the mineral acid itself. The selectivity order of metals follows the tendency to form stable anionic chlorospecies, thus Au(III) > Ga(III) > Fe(III) > In(III). However, the stoichiometry of the extracted complex often indicates the formation of ionic aggregates.

The sulfuric acid systems are complicated by the presence of the sulfate-bisulfate equilibrium. Thus, the extraction of metal sulfate complexes by amine salts is strongly pH dependent since the bisulfate ion (which predominates at low pH) will compete for extractant with the metal sulfate complex. The distribution data for metal sulfates exhibit a maximum at acidities of about 0.1 M, above which the extraction rapidly decreases as the bisulfate concentration increases. Extraction also depends strongly on the class of amine with the order primary to secondary to tertiary being followed. This order is a reversal of that found for both nitrato and chloro species but the same as that for sulfuric acid extraction. The elements that can be extracted are very similar to those given for the nitrate systems.[28]

Some other systems that form anionic metal complexes have been studied, e.g., thiocyanate; carboxylates, in particular acetate, formate, and oxalate; and phosphate.[14]

3.2. Properties of Liquid Anion Exchangers

Hydrogen ions can be solvated by strong donor molecules like TBP, sometimes leading to the extraction of a solvated hydrogen salt, EH^+L^- and form the complex, which could be considered as the complex ion pair. The amines form stronger adducts with hydrogen ions; in fact, they are so strong that they remain protonated while exchanging the anion. A classical reaction is the formation of the ammonium ion NH_4^+ when NH_3 is dissolved in water.

Anion exchangers in solvent extraction are generally protonated forms of primary, secondary, and tertiary high molecular weight amines and quaternary compounds. The form of the extractant in both phases and its distribution behavior must be known in order to understand an extraction process. For anion exchangers, it is essential to elucidate formation and nature of the quaternary salt and its relationship with the free base. The below examples are taken mainly from the well-known quaternary ammonium compounds and amine salts that are the most widely used in commercial processing.

The organic amines (RN) have a nitrogen atom N attached to a large organic molecule R usually containing >7 aliphatic or aromatic carbon atoms and give a high distribution coefficient when distributed between an aqueous electrolyte and an immiscible organic phase. They are highly soluble in organic solvents (diluents), almost insoluble in water. In contact with an aqueous phase containing HL, the amine base RN reacts with the acid HL to form RNH^+L^-, but extracts with an excess amount of acid HL (over the 1:1 HL:RN ratio) into the organic solvent, and also with additional water. A feature of amine salts and ammonium compounds is their ionic character, which is retained under both acidic and basic conditions.

As it was described in Chapter 2, four types of organic amines exist: primary amines RNH_3^+, secondary $R_2NH_2^+$, tertiary R_3NH^+, and quaternary R_4N^+. The hydrocarbon chain R is usually of length C_8–C_{12}, commonly a straight aliphatic chain, but branched chains and aromatic parts also occur. Primary, secondary, and tertiary amines require protonation before they can act as anion exchangers and thus also have acid extracting properties in the free-base form.

Quaternary ammonium compounds that contain five or less carbon atoms are readily water soluble. High molecular weight salts are only sparingly soluble in water and are assumed to behave as strong electrolytes.

The practical concentration of amine in the organic solvent is usually less than 20%; at higher concentrations the amine salt solutions become rather viscous. At low concentrations, quaternary ammonium salts are present in aqueous solution as monomers. While in more inert diluents they easily polymerize to form aggregates and then micelles, and at higher concentrations a third phase.

The degree of aggregation depends on the properties of the diluent and the nature of the ammonium cation and anion. In toluene, alkylammonium chloride salts increase in degree of aggregation in the following order:[31] tertiary alkylammonium < secondary < primary.

The CMC is dependent on the nature of the solute molecule, the length of the hydrocarbon chains, the temperature, and the presence and concentration of other electrolytes.[1] For example, in xylene $(TLA*HBr)_n$ aggregates with $n = 2$, 3, and 30 have been identified, and in other systems aggregation numbers above 100 have been reported. $TLA*HNO_3$ is mainly trimeric in *m*-xylene at concentrations 0.002–0.2 M, but larger aggregates are formed at higher amine concentrations. These aggregates seem to behave like monofunctional species, each extracting only one anionic metal complex. The aggregation can be reduced, and the third phase formation avoided, by using aromatic diluents and/or by adding a *modifier*, usually another strong Lewis base (e.g., octanol or TBP). Such additions often lead to considerable reduction in the K_{ex} value.

The base strengths of aliphatic amines as measured by the values of K_{ex} vary widely according to the diluent used for the measurements. Increasing substitution of the nitrogen by alkyl groups results in an increase in base strength from primary to tertiary amines at least in the gas phase,[2] in solution the increasing steric hindrance by the bulky alkyl groups impedes solvation of the cation–anion ion pair and thus reduces the stability of the salt relative to the free amine. Many workers have also found that the value of K increases with increasing concentration of the amine salt, and this is attributed to aggregation of the salt in the organic phase.

In general the amines extract metal complexes in the order tertiary > secondary > primary. Only long-chain tertiary and, to a smaller extent, quarternary amines are used in industrial extraction, because of their suitable physical properties: trioctylamine (TOA, 8 carbons per chain) and trilauryl amine (TLA, 12 carbons per chain) are the most frequently used. For simplicity we abbreviate all amines by RN, and their salts by RNH^+L^-. The tertiary and quaternary amine bases are viscous liquids at room temperature and infinitely

TABLE 3.2 Equilibrium Constants for Ion Pair Formation with Trioctyl Amine

Anions	Solvent	$K_{ex,ip}$ Ion pair formation
F^-	Toluene	3.0
Cl^-	Toluene	5.9
Cl^-	Carbon tetrachloride	4.0
Cl^-	Benzene	4.1
Br^-	Toluene	8.0
NO_3^-	Toluene	6.6
NO_3^-	Carbon tetrachloride	5.0
SO_4^{2-}	Carbon tetrachloride	6.7
SO_4^{2-}	Benzene	8.3

Source: From Ref. 1

soluble in nonpolar solvents, but only slightly soluble in water. The solubility of the ion pair RNH^+L^- in organic solvents depends on the chain length and on the counterion, L^-: the solubility of TLA*HCl in wet benzene is 0.7 M, in cyclohexane 0.08 M, in $CHCl_3$ 1.2 M, and in CCl_4 0.7 M. Nitrate and perchlorate salts are less soluble, as are lower molecular weight amines. Table 3.2 gives the equilibrium constants $K_{ex,am}$ of ion pair formation for trioctylamine in various solvents. Although the ion pairs are only slightly soluble in water, they can exchange the anion L^- with other anions, X^-, in the aqueous phase. (Note that we use L^- to indicate any anion, while X^- is used only for inorganic anion.)

$$RNH^+L^-_{(O)} + X^- \leftrightarrow NH^+X^-_{(O)} + L^- \tag{3.17}$$

The equilibrium constant ($K_{ex,exch}$ for *ex*traction, ex*ch*ange) for this reaction increases in the order $ClO_4^- < NO_3^- < Cl^- < HSO_4^- < F^-$. From Chapter 2 it follows that the formation constants for the metal complex ML_n^{z-n} usually increases in the same order. Therefore, in order to extract a metal perchlorate complex, very high ClO_4^- concentrations are required; the perchlorate complex is easily replaced by anions higher in the sequence.

All negatively charged metal complexes can be extracted by liquid anion exchangers, independent of the nature of the metal and the complexing ligand. Liquid anion exchange has extensive industrial application, and many examples are given in later chapters. Lists of extraction values are found in other works.[1,16]

Water uptake by the organic phase is proportional to the amine salt concentration and decreases with increasing anion size. The large amines do not seem to be hydrated very much in common diluents. The degree of hydration of the free base increases in the following order: tertiary < secondary < primary.[2] It would appear that the amount of water coextracted with the anion on amine salt formation is dependent on the ionic character of the amine salt and on the nature of the diluent. Thus, the tertiary amine hydrochlorides are monohydrates in benzene and toluene, but less water is coextracted into chloroform or carbon tetrachloride, whereas cyclohexane shows the greatest amount of coextracted water. This obviously relates to the relative ability of the diluent to solvate the amine salt ion pairs. The extraction of water by tri-*n*-octylamine and several of its salts in two diluents was studied.[32] In benzene the amount of water extracted by the amine salt always exceeded that extracted by the free base and increased in the following order: acetate < perchlorate < nitrate < chloride < bisulfate < sulfate.

The effect of coextracted water on the degree of extraction is not fully clear. The authors[1] conclude that water plays only a minor role, but others[33] have shown that water can have a significant effect on the aggregation of free tri-*n*-octylamine in benzene. Although the degree of aggregation of tertiary amine sulfates is considerably less than that of tertiary amine bisulfates, it is not clear whether this is due to the greater amount of water associated with the sulfate form or the increased opportunities for H bonding with the bisulfate system. Certainly, the greater degree of aggregation found for secondary amine sulfate salts compared with tertiary amine sulfate salts has been directly attributed to the opportunities for extensive H bonding in the former, which is not possible in the case of the latter.

These effects have been studied by use of a variety of techniques, such as light scattering, vapor-phase osmometry, vapor-pressure lowering, cryoscopy, infrared spectroscopy, and the two-phase titration technique. Computer analysis of the distribution data has also been widely used.

Organic-phase properties can change during the extraction process. For example, dielectric constant can change markedly with loading increase when amine comprises a significant proportion of the organic phase.

In addition to alkylammonium compounds several other polyalkyl or polyphenyl onium compounds such as quaternary arsenium and sulfonium salts, have been studied as anion exchangers. The properties of these reagents are very similar to those discussed above. More information may be found in Refs. 1–3,14.

4. EXTRACTION THROUGH FORMATION OF H-BONDING AND SOLVATING COMPLEXES

4.1. Extraction of Metal Complexes

Extraction by solvation requires the transfer of a neutral species from the aqueous to the organic phase by solvation of the hydrogen ion of a complex metal acid species (e.g., $HFeCl_4$) or by solvation of the metal ion of a neutral salt species. If the neutral metal complex is coordinatively unsaturated, it forms $ML_z(H_2O)_x$ in the aqueous phase, where $2z - x$ (L being bidentate) equals the maximum coordination number. In the absence of solvating organic solvents, this complex has a very low distribution constant. Obviously, if water of hydration can be replaced by organic molecules E, then the result is a more lipophilic adduct complex ML_zE_b.

Extraction by solvating extractants depends on the extent of complex formation in the aqueous phase, as in the case of the basic extractants. Depending on the ligand, several types of such adducts exist: (1) type ML_zE_b, where L and E are different organic structures; (2) type MX_zE_b, where MX_z is a neutral in organic compound (salt); and (3) type $ML_z(HL)_b$, where L and HL are the basic and neutral variant of the same molecule (so-called *self-adducts*).

4.1.1. Formation of Metal – Organic Complexes of Type ML_zE_b

Extraction is by solvation of either the central metal atom of the complex or the proton in the case of formation of a complex acid species. The metal complex can be written as

$$[ML_z]_{(O)} + bE \leftrightarrow [ML_zE_b]_{(O)} \tag{3.18}$$

or

$$HML_{z+1} + pE_{(O)} \leftrightarrow \left[(HE_b)^+(ML_{z+1})^-\right]_{(O)} \tag{3.19}$$

In the extraction of these metal complexes, as for acid extraction, the solvating agent will replace primary and/or secondary waters of hydration, thus rendering the complex soluble in the organic phase. The degree of extraction of a metal by a solvating extractant should depend on a number of factors, including the nature and concentration of the anionic coordinating ligand L, which, in turn, influences the type of metal complex formed; the degree of hydration of these aqueous metal complexes; and the relative strength of the water–metal and extractant–metal bonds, leading to a competition between water and the extractant for the solvation sites. The authors[1] have attempted to relate these factors to the electropositivity of the metal. The electrovalency requirements of strongly electropositive metals can be satisfied by the strongly polar organophosphorus reagents and hence are readily extractable in an unhydrated form. A less strongly electropositive metal is usually extracted by the same reagent in a partially hydrated form, a similar situation to the extraction of a strongly electropositive metal with the less polar carbon–oxygen extractants. This is explained by the persisting attraction of the metal for water. Similarly, it would be expected that mixed water–extractant complexes would be formed by extraction of the less electropositive metals by carbon–oxygen extractants. Here it is postulated that the extractant cannot compete successfully with water for the primary coordination sites and that extraction occurs by replacement of the secondary hydration sphere by extractant molecules H bonding to water in the primary sphere.

4.1.2. Extraction of Mineral Salts

The extraction of mineral salts is generally less complicated than the extraction of mineral acids. Metal salts with monodentate univalent anions like Cl^-, ClO_4^-, SCN^-, and NO_3^- are strongly hydrated in the aqueous phase and have quite small, if any, solubility in inert solvents. In order to extract these acids, they must either form an adduct with a strongly basic extractant like TBP or TOPO, or be in solvating solvents such as ethers, ketones, alcohols, or esters. Examples of extracted metal salt adducts are: $Br_3(EtO)_b$, $PaCl_3$(MIBK), $UO_2(NO_3)_2(TBP)_2$, $Co(ClO_4)_2$(Octanol)$_b$, etc. (*b* is uncertain). Tables of the extraction of a large number of metal salts by solvating solvents or commercial adduct formers dissolved in kerosene are given in Ref. 1.

It has been shown[34] that monomeric metal hydroxides can be extracted by strong donor molecules; e.g., $Ln(OH)_3(TOPO)_b$ into $CHCl_3$, where *b* is 2–3. Under favorable conditions, the D_{Ln} value may exceed 1, though the fraction of hydroxide is quite low. When the solvent is a good solvater, the determination of the solvation number *b* is difficult, unless the dependence of the extractant concentration on the solvent can be obtained. Solvation numbers can be obtained in mixtures of a solvating extractant and an inert diluent like hexane. Further, in these systems the extraction of the metal commonly requires high concentrations of salt or acid in the aqueous phase, so the activity coefficients of the solutes must betaken into account.

Figure 3.10 shows the extraction of Zn(II) from aqueous thiocyanate (X^-) solutions into 0.001 M TOPO (E) in hexane; the aqueous phase is 1.0 M $Na(SCN)^-$, or $Na(ClO_4)^-$ at pH around 5.[35] Zn(II) is known to form a number of weak $Zn(SCN)_z^{z-2}$ complexes in the aqueous phase. The uncharged one is assumed to accept TOPO to form the complex $Zn(SCN)_2(TOPO)_b$, where $b = 1$ or 2.

Assume that the reaction between the neutral complex and the solvating molecule takes place at the interface (to assume the reaction to take place in the organic phase would be unrealistic, as the zinc thiocyanate is insoluble in hexane); thus the extraction reaction is

$$ZnX_2 + bE_{(O)} \leftrightarrow ZnX_2E_{b,(O)} \qquad (3.20)$$

for which we may define an equilibrium constant $K_{ex,bE}$.

Because more than one solvated species may be extracted, the distribution ratio becomes

$$D_{Zn} = \frac{[ZnX_2E]_{(O)} + [ZnX_2E_2]_{(O)} + \cdots}{[Zn] + [ZnX] + [ZnX_2] + \cdots} = \frac{\beta_2[X_2]\sum K_{ex,bE}[E]_{(O)}^b}{1 + \sum \beta_n[X]^n} \qquad (3.21)$$

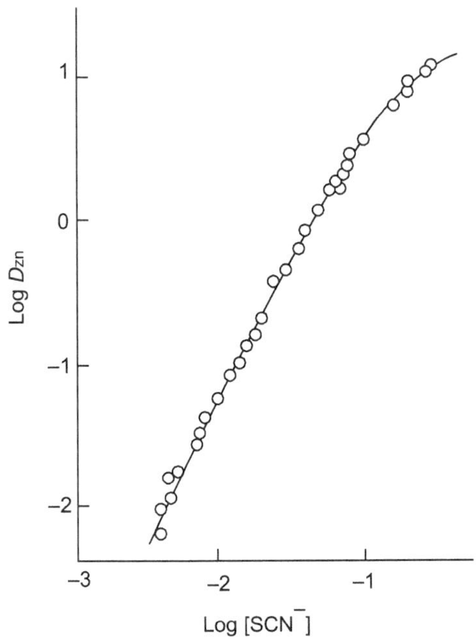

FIGURE 3.10 Distribution ratio of Zn(II) when extracted from 1 M $Na(SCN*ClO_4)^-$ into 0.001 M TOPO in hexane, as a function of aqueous SCN^- concentration. The following equilibrium constants were obtained: $\beta_1 = 3.7, \beta_2 = 21, \beta_3 = 15, K_{ex}$ 2.5 * 10^7 fo $b = 2$. *(Source: From Ref. 35)*

The solvation number b can be determined from the dependence of D on $[E]_{(O)}$ while [X] is kept constant. From the slope of the line in Fig. 3.10 at low SCN^- concentrations, it follows that $b = 2$; thus only one complex is identified: $Zn(SCN)_2(TOPO)_2$. The authors were able to calculate the formation constants β_n from the deviation of the curve from the straight the line at constant $[E]_{(O)}$, assuming b constant. With these equilibrium constants, the line through the points was calculated with Eqn (3.21).

The use of carbon–oxygen-bonded extractants for metals is well established, especially in halide and nitrate systems. Details of the various systems can be found in this chapter and in other sources.[1,14,15,17,28] Generally, it is found that ketones are better extractants for metal species than ethers because of their greater electron donor properties and thus are more widely used. Among these ketones, methyl isobutyl ketone has been used in the laboratory, but its low flash point has limited its commercial use to separations such as niobium and tantalum, where only a small-scale plant is needed to satisfy demand.

Most of the work with phosphorus-oxygen donor extractants has been concerned with halide and nitrate systems. In contrast to the extraction of acids by these reagents, the neutral metal salts are usually extracted into the organic phase in an anhydrous form. Once again, very extensive studies have been made of these reagents in the extraction of metal salts,[1,14–16] including physicochemical studies of the nature of extracted species in both solution and the solid state. A few general conclusions can be made concerning these systems. First, the nature of the diluent used with the reagent will affect the extraction behavior. Dilution itself affects the physical properties of the organic phase and also the water content and the activity of the reagent. Moreover, extraction decreases with increasing diluent polarity, probably as a result of completion for the phosphoryl oxygen by the various potential H-bonding molecules in the system. However, in most cases the distribution ratios cannot be correlated with physical properties of the diluents. Obviously, the ability of the diluent to solvate the extracted metal complex is important, and in these systems third-phase formation is fairly common in particular. Chemical stability of the reagents is another important factor; ester linkages can undergo hydrolysis, leading to production of acidic species. This, of course, changes the nature of the extractant and may produce a system completely different (i.e., extraction by an acidic reagent) from that originally proposed. Purification of organic phases and avoidance of degradation are thus important in these systems.

4.1.3. Extraction by $ML_z(HL)_b$-type Complex (Self-Adducts) Formation

Hydration only occurs in neutral complexes that are coordinatively unsaturated by the organic ligand. The hydrate water reduces the extractability of the complex. In the absence of strong donor molecules, which can replace this hydrate water, there is still a chance for the undissociated acid to replace the water, leading to a *self-adduct* according to the reaction

$$ML_z(H_2O)_w + bHL \leftrightarrow ML_z(HL)_b + wH_2O \tag{3.22}$$

This competition between the formation of a complex with HL or with H_2O is observed as an increased extraction with increasing HL concentration. However, stoichiometrically the complexes $ML_z(HL)_b$ and H_bML_{z+b} are equivalent. Formally the former is a self-adduct and the latter is an ion pair. Thus, Eqn (3.22) could be rewritten as

$$ML_z(H_2O)_w + bHL \leftrightarrow M(H_2O)_{w-2}L_{z+b}^{-b} + bH^+ + (w-2b)H_2O \tag{3.23}$$

assuming the HL can replace $2H_2O$ through its bidentate structure. All of the hydrate water can be displaced by the organic ligand, with formation of negatively charged chelate complexes.

Chemical equilibrium experiments, e.g., distribution ratio measurements, cannot distinguish between these two types of complexes; however, they may be identified by fingerprinting techniques like NMR, IR, or X-ray structure determinations.

Existence of similar complexes like ML_zE_b support the existence of self-adducts. The case of promethium(III) acetylacetone is an interesting illustration of this problem.

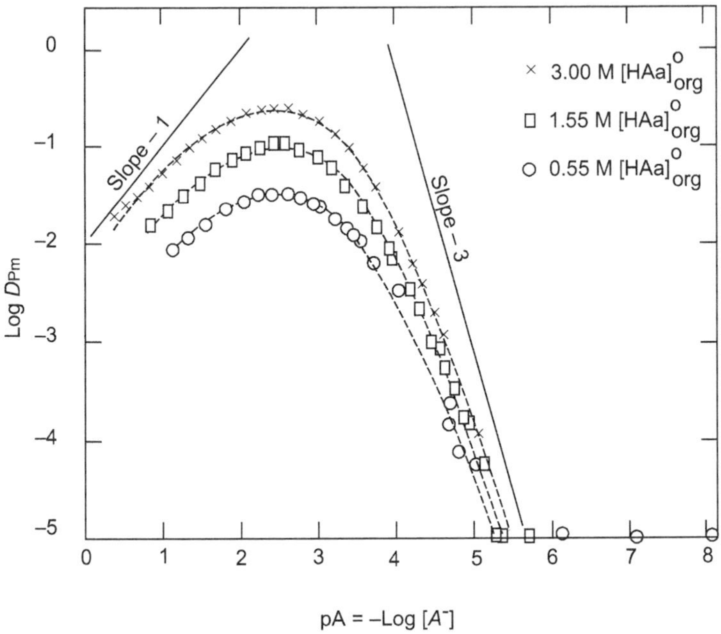

FIGURE 3.11 Extraction of Pm(III) by acetylacetone (HLa) from 1 M $NaClO_4$ into benzene at three different original concentrations of HLa in the organic phase. $pL = -\log[La^-]$ is calculated. The analysis of the system yielded the constants $\log \beta_1 = 5.35$, $\log \beta_2 = 9.20$, $\log \beta_3 = 13.22$, $\log \beta_4 = 14.06$, $K_{ad1} = 7$, $K_{ad2} = 3$, and $K_{DC} = 0.008$. *(Source: From Refs. 36,37)*

Figure 3.11 shows the distribution ratio log D_{Pm} for promethium in the system acetylacetone (HLa), benzene/1 M $NaClO_4$ as a function of $-\log [L^-] = pL$ at various starting concentrations, $[HL]_{O,org}$, in the organic phase. The coordination number of Pm(III) with respect to oxygen is reported to be 8 or 9. In the aqueous phase all stepwise complexes up to PmL_4^- can therefore be expected to form. The vacant coordination sites may be filled with water (forming a hydrate) or undissociated acetylacetone (forming a self-adduct). The last assumption is supported by the fact that a large number of complexes of type LnL_3E_b are known. Omitting water of hydration, the distribution ratio becomes

$$D_{Pm} = \frac{K_{DC}\beta_3[L]^3\left(1 + K_{ad1}[HL]_{(O)} + K_{ad2}[HL]^2_{(O)} + \cdots\right)}{\sum \beta_3[L]^3} \tag{3.24}$$

where K_{DC} refers to the distribution of uncharged PmL_3 between the benzene and the aqueous phase

$$K_{DC} = \frac{[PmL_3]_{(O)}}{[PmL_3]_{(w)}} \tag{3.25}$$

$K_{ad,b}$ is the equilibrium constant for the self-adduct complex formation in the organic phase

$$K_{ad,b} = \frac{[PmL_3][HL_b]_{(O)}}{[PmL_3]_{(O)}[HL]^b_{(O)}} \tag{3.26}$$

Thus in Eqn (3.24) at constant $[HL]_{(O)}$ the curvature of the extraction curve and its position along the $[L^-]$ axes is only caused by varying $[L^-]$, i.e., the aqueous phase complexation. The numerator refers to the organic phase species only, and is responsible for the position of the extraction curve along the D-axis; thus at three different constants $[HL]_{(O)}$, three curves are expected to be obtained, with exactly the same curvatures, but higher up along the D-scale with higher $[HL]_{(O)}$ concentration; as can be seen in Fig. 3.12. The asymptote with the slope of -3 at high pL (i.e., low $[L^-]$), fits Eqn (3.24), when Pm^{3+} dominates the denominator (i.e., the aqueous phase), while the asymptote with slope of $+1$ fits the same equation when the aqueous phase is dominated by PmL_4^-. Between these two limiting slopes, the other three PmL_n complexes are formed in varying concentrations. A detailed analysis of the curves yielded all equilibrium constants see in Refs. 36,37.

The maximum distribution ratio for the Pm–HL system (D_{Pm} about 0.1) is reached in the pH range 6–7. It is well known that the lanthanides hydrolyze in this pH region, but it can be shown that the concentration of hydrolyzed species

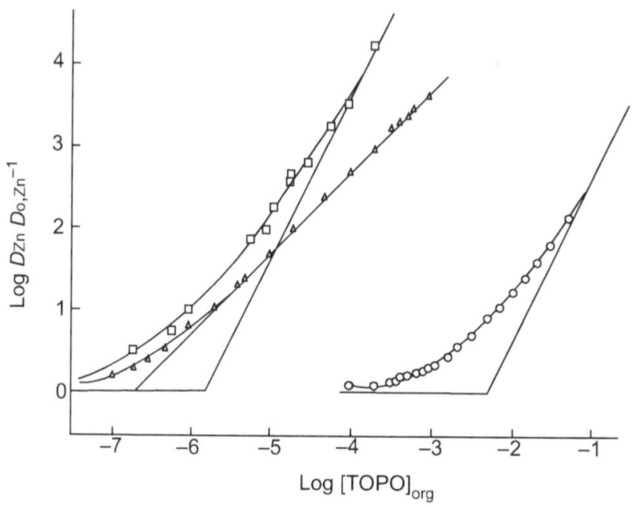

FIGURE 3.12 Extraction of Zn(II) by mixed complex formation, from 1 M $NaClO_4$ into carbon tetrachloride containing the complexing extractants acetylacetone (□), trifluoroacetone (○), or hexafluoroacetone (△) as a function of the concentration of neutral extractant, trioctyl phosphine oxide (E). The curves are fitted using the constants: log K_{ad1} = 3.07 (AA), 6.70 (TFA), 7.0 (HFA), and K_{ad2} = 4.66 (AA), 7.4 (TFA), 11.6 (HFA). *(Source: From Ref. 37)*

is <1% of the concentration of PmL_n species for the conditions of Fig. 3.13, and can be neglected.

Self-adducts are rather common, and have been identified for complexes of Ca, Sr, Ba, Ni, Co, Zn, Cd, Sc, Ln, and U(VI) with acetylacetone, thenoyltrifluoroacetone, tropolone, and oxine. Table 3.3 lists some self-adduct constants. The fact that the constants vary with the organic solvent indicates that the

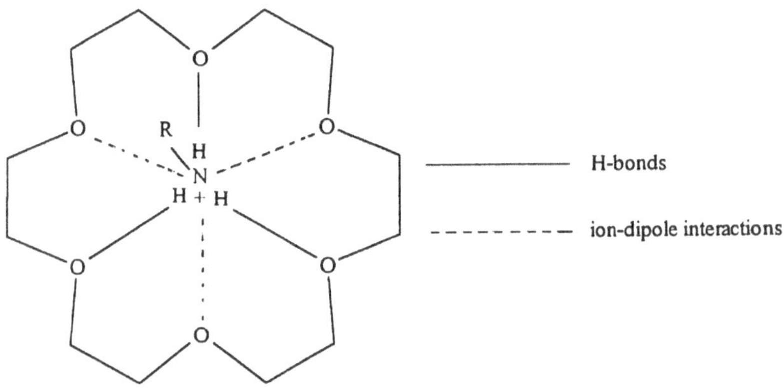

FIGURE 3.13 Scheme of bonding fixation places in $18C_6$ system.

TABLE 3.3 Constants for the Formation of Some Metal complexes (Self-Adducts), according to Eqn (3.21)

Metal ion M_z	Acidic extractant HL	Organic solvent	Log K_{ad}
Zn(II)	Isopropyltropolone, IPT	$CHCl_3$	1.72
Zn(II)	Isopropyltropolone, IPT	CCl_4	1.9
La(III)	Acetylacetone, AA	C_6H_6	2.6
Eu(III)	Thenoyltrifluoroacetone, TTA	$CHCl_3$	0.56
Eu(III)	Thenoyltrifluoroacetone, TTA	CCl_4	0.5
Eu(III)	Isopropyltropolone, IPT	$CHCl_3$	2.1
Eu(III)	Isopropyltropolone, IPT	CCl_4	2.0
U(VI)	Acetylacetone, AA	$CHCl_3$	1.16
U(VI)	Thenoyltrifluoroacetone, TTA	C_6H_6	0.50

Source: Ref. 1

self-adduct reaction probably occurs in the organic phase. Table 3.3 contains both very inert (e.g., CCl_4), polar ($CHCl_3$), and π-bonding solvents (C_6H_6).

4.2. Properties of Extractants

From Table 3.1 it is seen that all oxygen-bearing organic solvents will act in the way, as organophosphorus compounds such as the triesters of phosphoric, phosphonic, and phosphinic acids; alkyl phosphites; amine oxides; organic sulfoxides; and organic sulfides. However, the solvating power depends strongly on the basicity of the reagent. In Ref. 1 the authors point out, oxygen-containing solvents are very weakly basic and in general tend not to form oxonium salts. In the carbon-bonded oxygen donor systems (e.g., ethers and ketones), water usually forms an essential part of the complex, forming a H-bonded bridge between the solvating agent and the solute. However, in the organophosphorus systems, water is often eliminated from the organic phase. Whereas, ethers, ketones, and so on are all electron-donating compounds, alcohols are amphoteric and exhibit both donor and acceptor properties. The relevant physical chemistry of carbon-bonded oxygen donors and the extraction of acids with these reagents have been reviewed in Ref. 1.

A number of the extractants are appreciably miscible with water; hence ternary phase diagrams are required to understand the extraction process. As the molecular weight of extractant increases, this solubility decreases.

Slope analysis for determination of the stoichiometry of extracted species was studied by Widmer.[38] The author also shows how solvent extraction can be used to determine association numbers and association orders of the extracting species.

Extraction by phosphorus-bonded oxygen solvents has been extensively studied and is of particular interest in the atomic energy field. The most well known and most used organophosphorus ester is TBP. With regard to the physical chemistry of this type of solvent in organic solution, nonideal behavior is the rule. Hydrogen bonding type and dipole–dipole interactions are common, and both self-association and extractant–diluent associations complexes are well known for this class of extractant. Equilibrium data for such interactions have been tabulated.[16]

Data on aqueous solubilities and hydrates of phosphorus-bonded oxygen donor solvents is abundant, particularly for TBP. Because of the increasing polar nature of the phosphoryl group, the solubility of neutral organophosphorus compounds in water decreases in the following order: phosphine oxides > phosphinates > phosphonates > phosphates. Solubility also decreases with increasing of alkyl chain length. Like the carbon oxygen donors, these phosphorus oxygen compounds extract mineral acids from aqueous solutions.

In general, extraction power increases with increase in the number of carbon–phosphorus bonds. Excellent bibliographies are available,[1,14,15] and equilibrium constant data are tabulated,[16] including data on the extraction of water and the hydration of the extractants. Other oxygen-containing neutral extractants that have received some attention recently are amine oxides and sulfoxides; much of this work has been carried out and presented in Ref. 39.

5. EXTRACTION THROUGH MIXED COMPLEX FORMATION

Mixed complexes can often be formed in the solvent extraction of metal ions. This is the basis of *synergism*, wherein the degree of extraction of metal by the mixed system exceeds the combined effects of the individual components, that is, a stability constant effect associated with the mixed complex. It was formerly thought that an opposite effect called *antisynergism* existed, but it is now clear that the destruction of synergism is due merely to mass-action effects or changes in physical properties of the organic phase, associated with the predominance of one of the reactants in the organic phase. Thus, it is expected that mixed-complex systems will obey the same rules as single-extractant systems. The compounds added to the organic phase to produce mixed complexes will also partition between both phases just like the primary extractant and may also interact with all species present in both phases. Organic phase interactions between the two organic reactants are of extreme importance as they will materially affect the metal extraction behavior of the system. The thermodynamics of mixed-complex formation is of special interest, and mixed-complex phenomena are discussed in the following paragraphs.

5.1. Acid Extractants

Mixed-complex formation in acid extractant systems can occur in several ways. By far the most common is the addition to the systems of neutral donor molecules such as alcohols, organophosphorus compounds, nitrogen contain ligands such as amines, and acids where the compound remains undissociated. Mixed complexes are also formed between two acidic extractants with both exchanging protons in the extraction reaction. Thus, a wide variety of complexes and behavior arise, and it is common to find the same extractant acting as an acid toward one metal ion in a complex and as an unionized donor molecule toward another. The determination of formation constants and a general algebraic description of mixed-complex formation in acidic systems has been discussed by Dyrssen.[40]

5.1.1. Acid Extractants and Neutral Donor Molecules

The formation of mixed complexes in these system can be described in two ways:

$$M^{z+}_{(w)} + zHL_{(O)} + bE_{(O)} \leftrightarrow ML_zE_{b,(O)} + zH^+_{(w)_{zb}} \tag{3.27}$$

$$K_{ex} = \frac{[ML_zE_b]_{(O)}[H^+]^z_{(w)}}{[M^{z+}][HL]^z_{(O)}[E]^b_{(O)}} = D_M\frac{[H^+]^z_{(w)}}{[HL]^z_{(O)}[E]^b_{(O)}} \tag{3.28}$$

and

$$[ML_z(HL)_b]_{(O)} + qE \leftrightarrow [ML_z(HL)_{b-q}E_q]_{(O)} + [(HL)_q]_{(O)} \tag{3.29}$$

$$K_{ex} = \frac{[ML_z(HL)_{b-q}E_q]_{(O)}[(HL)_q]_{(O)}}{[ML_z(HL)_b]_{(O)}[E]^q_{(O)}} \tag{3.30}$$

Thus the distribution of the metal, D_M, is shown to depend on the concentrations of proton and HL_{org} to the power z of the charge of the metal ion, and on the concentration of E to the power of b (i.e., number of extractant molecules in the extracted complex).

Equation (3.20) describes an addition mechanism, whereas Eqn (3.22) describes a substitution mechanism. Clearly, the addition mechanism will mostly occur when the central metal atom in the complex ML_n is coordinately unsaturated. However, this is not always the case, and an increase in coordination number of the central metal atom to accommodate the neutral ligand is also found. It would seem, therefore, that in all cases neutral ligand bonds directly to the central atom. Kertes[16] has listed some general rules for the stability of mixed complexes: (1) stability increases with increasing base strength of the neutral ligand, in the absence of steric factors; (2) change of the diluent can materially affect complex stability, although not stoichiometry; and (3) the stronger the original complex, the less will be the tendency to form a mixed-complex species.

Without considering the independent parameters, it is often difficult to understand why K_{ex} varies strongly, and it may be impossible to predict improvements of the system. A good example is the extraction of Zn(II) by β-diketones and TOPO. Figure 3.12 illustrates the extraction of Zn(II) from 1 M $NaClO_4$ into carbon tetrachloride by β-diketones (HL) and TOPO as an additional donor.[41,42] At log $D_{Zn}/D_{0,Zn}$ plotted as a function of log $[E]_{(O)}$., the distribution ratio proceeds from almost zero, when almost no adduct is formed, towards a limiting slope of 2, indicating that the extracted complex has added two molecules of E to form ZnL_2E_2. From the curvature and slope the $K_{ad,b}$ values were determined.

Formation of solvated species in extraction systems can be considered as a special case of mixed-complex formation, where the neutral donor molecule is the undissociated acid. The formation of other mixed complexes from the solvated species by substitution can be readily understood. Substitution is also obvious in the replacement of water in hydrated chelated molecules by neutral donor compounds. It is less easy to understand the formation of species such as

$ML_z(HL)_b$ in the extraction of metals with dimeric extractants such as carboxylic acids or dialkyl monohydrogen phosphoric acids. The acid dimer may interact in the organic phase with neutral ligand according to the equation

$$(HL)_2 + 2E \leftrightarrow 2(HL * E) \tag{3.31}$$

But no proof exists that (HL*E) species actually take part in the extraction process. Indeed, this would be highly unlikely as the extraction reactions generally take place at the interface, and for acidic systems, there is considerable evidence that the interfacial species are monomeric. It would thus be unlikely for the species LH*E to be present at the interface and thus be active in the extraction reaction. Nevertheless, interaction between neutral ligands and the extractant can have a profound effect on system behavior; e.g., as the concentration of neutral ligand is increased, the associated mass-action effects can very from synergistic to a considerable reduction in the degree of metal ion extraction. A corresponding effect on reaction kinetics will occur, as discussed later.

In metal chelate extraction by far the most studied reagents are the β-diketones, which can form mixed complexes with a wide variety of neutral ligands.

Of particular interest is the recent discovery that metal complexes of alkylated 8-quinolinol and also hydroxyoximes can form adducts with neutral donor molecules, for although this type of complex with 8-quinolinol is recorded,[43,44] little was known about the behavior of hydroxyoximes in this situation. Addition of carboxylic and dialkylphosphoric acids to metal–hydroxyoxime systems produced mixed complexes and, for some metals, a considerable amount of synergism; thus synergistic extraction by mixtures of an aliphatic α-hydroxyoxime (LIX63, Henkel Corporation) and various carboxylic and phosphoric acids has been widely reported.[12,45,46] Investigations of nickel extraction showed that addition of the carboxylic acid to the four-coordinate orange $(LIX63)_2Ni$ complex produced a green six-coordinate mixed complex.[47] Spectral studies showed that the hydroxyoxime chelate structure remained intact and the acids were acting as neutral donor molecules. It is also shown that no interaction between the two types of donor molecule occurred. An interesting feature of this mixed system was that the rate of extraction of nickel was very slow.[48] Other systems showing similar mixed complexes are aromatic β-hydroxyoximes and carboxylic[49]; dialkylphosphoric and aromatic sulfonic acids[50] for copper, nickel, and cobalt; and alkylated 8-quinolinol and carboxylic acids for nickel,[49] cobalt,[51] zinc, and cadmium.[52] In this latter system, interaction between the carboxylic acid and the 8-quinolinol has been found[53] that materially affects the interfacial properties of the system and thus permits the extraction of cobalt without simultaneous oxidation to cobalt(III). Although addition of nonyl phenol to hydroxyoxime systems does not seem to cause mixed complex formation, the two reagents do interact to cause a decrease in copper extraction with increasing nonyl phenol concentration and a decrease in the rate of extraction.[54,55] There is some evidence for a similar effect with nonyl phenol and alkylated 8-quinolinol mixtures, but these reagents do show a small synergistic effect accompanying mixed complex formation in the extraction of copper.[56] Obviously, this is a case where the observed effect will depend critically on the relative amounts of chelating agent and neutral ligand in the organic phase.

In the carboxylic acid extraction system, mixed complexes form on addition of TBP salicylic acid mixtures.[7] Little synergism, however, is found with either system. Synergism was found in the extraction of cesium with mixtures of 4-phenylvaleric acid and *4-sec-butyl-2-(a-*methylbenzyl) phenol:[57] here the phenol behaves as a neutral solvating agent at pH values below 11. Some data on mixed complex stability constants and adduct formation with neutral ligands have been compiled.[17]

Mixed-complex formation and synergism in dialkylphosphoric acid systems have been reported extensively, with particular interest in the addition of neutral organophosphorus compounds. Once again the stability of the mixed complex increases with increasing donor properties of the molecule; thus for neutral organophosphorus compounds the series, in order of effectiveness, is $(RO)_3PO < R(RO)_2PO < R_2(RO)PO < R_3PO$. Interaction between the extractant and the neutral ligand is again important in the determination of the chemistry of these systems. Mixed extractant systems containing alkylphosphoric acid have also been studied with other neutral donors such as aliphatic and aromatic alcohols.

Mixed complexes in sulfonic acid, neutral ligand systems have been reported[58] for the extraction of zinc by dinonyl-naphthalene sulfonic acid (DNNSA) and its sodium salt with trioctylphosphine oxide (TOPO), TBP, or 2-ethylhexanol. No mixed complexes were formed with the systems containing the free acid, and the observed decrease in zinc extraction on addition of the neutral ligands can be attributed to interaction between DNNSA and the ligands. As expected, TOPO interacts more strongly with DNNSA than does TBP. A mixed complex was formed between zinc, the sodium salt of DNNSA, and TOPO, and synergism occurred, but neither TBP or 2-ethylhexanol had any effect.

5.1.2. Two Acidic Extractants

In some mixed β-diketone systems both diketones combine in satisfying the charge and coordination requirements of the central metal ion. This type of mixed-complex formation does not necessarily lead to synergism. Differentiation must be made between systems wherein mixed-complex formation is produced through adduct formation[59] and by mixed-chelate formation. It has been suggested[60] that the stability of the mixed chelate species is achieved only by a combination of a weak and strong β-diketone. Similar phenomena for iridium and europium complexes with a β-diketone and β-isopropyltropolone have been reported.[61] Recently, the formation of mixed complexes of lanthanide and actinide cations with dinonylnaphthalene sulfonic acid (DNNSA) and di-2-ethylhexylphosphoric acid (D2EHPA) has been reported.[62] The degree of complexation with D2EHPA and the dimensions of the complexed ion are shown to be the important factors for their inclusion in the DNNSA micelles. The core volume of the DNNSA micelles is about the size of the trivalent metal ion D2EHPA complex, but inclusion of the neutral D2EHPA metal complex does not occur. Instead, only positively charged species are included in the DNNSA micelle, and the charge neutralization is achieved by loss of protons. Interfacial tension changes noted in these systems correlate bulk changes in metal ion extraction behavior. Yet again, the importance of interfacial phenomena in solvent extraction systems is emphasized.

5.1.3. Acid–Base (Amine) Mixed Extractants

Mixed extractant systems containing acids with bases such as alkylamines have also been studied. Various names were used for these extractants, including amine salts, mixed extractants, mixed ionic solvents, binary extractants, acid–base-coupled (ABC) extractants. A full description of these systems may be found elsewhere,[1,14,15,63–66] together with a useful compilation of stability constant data.[16] Several potential applications of acid-base mixed extractants in salt extraction were studied and were found to possess attractive properties. The efficiency and selectivity of metal salt extraction by acid-base extractants and their dependence on extractant composition, aqueous phase composition, pH, and temperature have been reviewed and analyzed.[67,68] Extraction is efficient and selective. Compared with liquid cation exchangers, acid-base extractants are less sensitive to acidity in the aqueous phase and extract alkali and alkaline earth metals better. Both the cation and the anion are extracted; therefore, no acid or base addition is required for pH adjustment or for stripping. Extraction is reversible and provides for back-extraction of the extracted salt by water.

Dialkylphosphoric acid-alkylamine systems are unlikely to form mixed complexes as mixtures of these reagents evolve heat during the formation of the amine alkylphosphate salts. A large decrease in the extraction of metals can be expected in such systems. However, if the pH of the aqueous phase is so high that the amine is in the free-base form, adducts similar to those reported for the carboxylic acids could be formed. Similarly, when the alkylphosphoric acid is in a salt form, e.g., the sodium salt, adduct formation with amines should be possible.

The authors[67] have shown that addition alkylphosphoric acids to tri-*n*-octylamine depressed the extraction of sulfuric acid through strong interaction between the organic acid and base. Their conclusion that extraction of inorganic species such as UO_2SO_4 and $PtCl_6^{2-}$ by alkylamines should be depressed by addition of alkylphosphoric acids was supported by a study of the extraction of $PtCl_6^{2-}$ with tri-isooctylamine in the presence of di-*n*-butylphosphoric acid.[68] Similarly, at low sulfuric acid concentration the extraction of uranium from

UO_2SO_4 by a mixture of tri-*n*-octylamine and di-2-ethylhexylphosphoric acid was depressed by formation of the alkylphosphate amine salt. However, as the sulfuric acid concentration was increased synergistic extraction of uranium was found[69–71] that was attributed to the formation of a mixed complex $(R_3NH)^+(UO_2SO_4HR_2)^-$. A similar behavior was discovered[72] in the extraction of hafnium by 2-ethylhexylphosphoric acid and tri-*x*-octyl-amine. In this case the mixed complex had the structure $(R_3NH^+)_2[Hf(SO_4)_2R]^{2-}$. The authors showed that similar mixed complexes were formed in the extraction of uranyl sulfate by amine–monoalkylphosphoric acid mixtures. However, the decrease in extraction found in earlier studies was absent in this system because of the weaker acid strength of the monoalkylphosphoric acids reducing the interaction with the amine in the organic phase. It is clear, therefore, that amines can act in mixed-complex systems both as adducts and extractants similar to the acid extractants such as carboxylic and alkylphosphoric acids.

The lower the stability of the metal diketones and the greater the basicity of the donor atom, the greater is the stability of the adduct formed. Large exothermic enthalpy changes usually characterize these systems.[42] Metal β-diketonates also form mixed complexes with alkylamines and alkylamine hydrochlorides. The inclusion of the amine hydrochloride in the complex is unusual; researchers[73] have postulated the formation of AmT_3R_3NHCl, AMT_3R_3NHT, and $AmT_3NHClHT$ in the extraction of americium by a mixture of tri-*n*-octylamine (R_3N) and thenoyltrifluoroacetone (HT). Researchers[74] have also found adduct formation with copper β-diketonates and alkylamine and alkylamine hydrochlorides. Visible and infrared spectra suggest that the alkylamines bond directly to the copper through the nitrogen atom while the amine hydrochloride binds directly through the chlorine atom. Full discussions of these systems are available.[1,14,15]

5.2. Mixed Two Amines Extractants

Mixtures of two amines have been shown to produce mixed amine complexes in the extraction of uranium from sulfuric acid,[75] with a synergistic effect. *This* behavior is analogous to the formation of mixed chelates with acidic extractants.

No information on mixed-complex formation of metals with amine salts and neutral donors is available.

5.3. Mixtures of Solvating Extractants

The role of solvating extractants and neutral ligands as donor compounds in acidic extraction systems has already been discussed. Systems with two neutral ligands do exist, although not much information is available; Kertes[15] has given a short account. Clearly, H bonding is of prime importance as also is nonideality in these mixtures. Synergism seems to occur in these systems when one of the extractants has a much higher dielectric constant than the other, or when one is a donor capable of H bonding and the other is an acceptor molecule. Diluent extraction interaction may also be important with organophosphorus reagents, and Nishimura et al.[75] have shown that H bonding of chloroform to TBP and tri-*n*-octylphosphine oxide appreciably depressed the extraction power of these extractants. The study showed a greater hydrogen acceptor strength for the phosphine oxide than the alkyl phosphate. A similar effect was found for alkylamines, but they were much poorer hydrogen acceptors in spite of their greater proton affinity.

6. EXTRACTABLE COMPLEXATION OF MONOVALENT METALS

There are a few types of extractable metal complexes that do not fit well into the previous classifications: monovalent metals that form extractable complexes with large organic monobasic anions, and, conversely, monovalent inorganic anions that form extractable complexes with large organic cations. They are treated as a separate homogenous class of extractants.[2] The large monovalent ion pair, tetraphenylarsonium tetraphenylborate (Ph_4As^+ Ph_4E^-), which has been suggested as a reference in solvent extraction, also belongs to this class. Of some importance is the extraction of alkali ions by tetraphenylborate, by tetraphenylantimonium ions, and of perrhenate by tetraphenylarsonium ions. Because most of the volume of these complexes is taken up by organic groups, the complexes are highly lipophilic and, therefore, extractable into organic solvents without additional solvation.

From these systems having limited applications, only the crown ethers offer some interesting extraction systems. The discovery of crown ethers[76] gave impetus to the birth of host-guest complex chemistry. The ethers form cage compounds ("clathrates"); i.e., the metal cation fits into a cage formed by the crown structure. Because the cage can be designed to fit almost any ion of a certain size, rather selective extractions are possible with these type of extractants. The 18-crown-6 core is most perfectly suited for binding, all oxygen atoms in a crown-6 ring seemingly interact with the ammonium moiety. The thorough analysis of data on influence of the structural variations have led to the following scheme of guest fixation by 18C6 systems (see Fig. 3.13).

This pattern, which emerge primarily from a purely chemical approach, was further proved by numerous X-ray crystal analyses[77] and molecular mechanics–molecular dynamics calculations.[78] In summary, 18C6 and its analogs bind protonated amines by three H bonds connecting alternate oxygen with the three guest protons; the other three oxygen contributing to stabilization of the complex through ion–dipole interactions $O–N^+$. Various computational techniques suggest that the last term "weighs" two to three times less than H bonds.

Very interesting is the selectivity of extraction toward different amines. The association COnstants for primary amines slightly decrease or remain nearly constant upon increasing the length of the radical. In combination with increasing ion pair distribution constants, this leads to overall enhancing extraction. Extraction is sensitive to the steric effect at the ammonium center.

Crowns do discriminate secondary and tertiary amines. The effect is due to their reduced number of protons capable to H-bonding: protonated primary amines form three normal H bonds with oxygens of 18C6, the secondary ones bind through four less favorable bifurcated H bonds with the C_2 conformation macrocycle, and in tertiary amines single proton do not form H bonds at all, only electrostatic interactions take place. This selectivity characteristics found numerous applications in biochemical and pharmaceutical separation technologies. In general the wide variety of complicated compounds and among them alkali, alkali earth and transition metal ions may be extracted with crowns.

7. EXTRACTION WITH AQUEOUS BIPHASIC SYSTEMS

Aqueous biphasic systems (ABS) consist of two immiscible phases formed when certain water-soluble polymers are combined with one another or with certain inorganic salts [e.g., $(NH4)_2SO_4$, K_3PO_4] in specific concentrations.[79,80] They are suitable for carrying out separations of various solutes, such as biomolecules, metal ions, and particulates. In an ABS, the major component in each of the two phases is water, and because of this nondenaturing environment, these systems have been widely employed in biological separations.

The combination of two mutually incompatible but water-soluble polymers or the combination of certain water-soluble polymers and specific inorganic salt solutions, produces two immiscible aqueous phases.[81,82] A number of different water-soluble polymers can be utilized in ABS. As an example see Fig. 3.14. An even wider variety of polymer/salt combinations exist, usually with Na^+, K^+ or NH_4^+ salts of mono-, two- or trivalent anions such as OH^-, CO_3^{2-}, and PO_4^{3+}.[80]

ABS can be used extensively for concentration, separation, and extraction of cations and anions spanning the periodic table from various salt solutions. Chaotropic anions (TcO_4^- I^-, Br^-, etc.) can be extracted by the polyethyleneglycol (PEG)-rich phase without an extractant,[81] and most cations can be extracted into the PEG-rich phase only by utilizing extractants (complexants). Inorganic complexants such as SCN^- and heavy halides are effective for the soft transition metal and main group ions[83] (see Fig. 3.15), and complexing dyes like Arsenazo III, Xylenol Orange, and Alizarin Complex one work very well for the extraction of certain lanthanides and actinides.[84] Crown ethers show promising distribution ratios and selectively increasing for the group 1 and 2 cations.[85] However, better ligands, designed specifically for use in ABS, are needed.

For adaptation of the ABS techniques to chromatographic methods the aqueous biphasic extraction chromatography technique (ABEC) and ABEC resins were developed.[86,87]

The applications of ABSs and ABEC resins indicate excellent potential. These systems hold much promise for solving a variety of practical problems

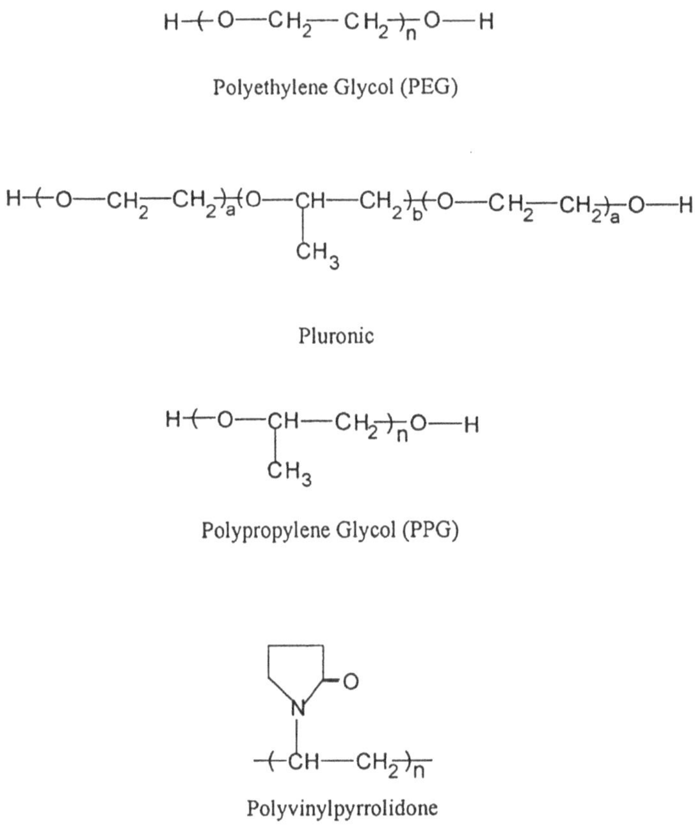

FIGURE 3.14 Water-soluble polymers used for metal ion separations in aqueous biphasic systems.

related to using water-soluble extractants. One of the disadvantages of the ABS, the loss of phase-forming components due to solubility in the opposite phase, can be overcome by using ABEC resins. Selectivity and stripping of the metal ions still need improvement for several possible applications. Selectivity may be improved by the design of new metal-specific ligands for ABS applications, and stripping will depend on the characteristics of these new ligands.

In order to move PEG-ABS and ABEC for metal ion separations out of the laboratory and into the commercial field, predictive tools for the partitioning behavior observed are necessary. In addition, it will be necessary to introduce traditional solvent extraction scientists to the many variables associated with this technology. Both of these steps will require a better understanding of the major factors governing metal ion partitioning in PEG–ABSs and onto ABEC resins. This work has begun, but this is only the beginning of the research effort directed toward a true understanding of the process involved.

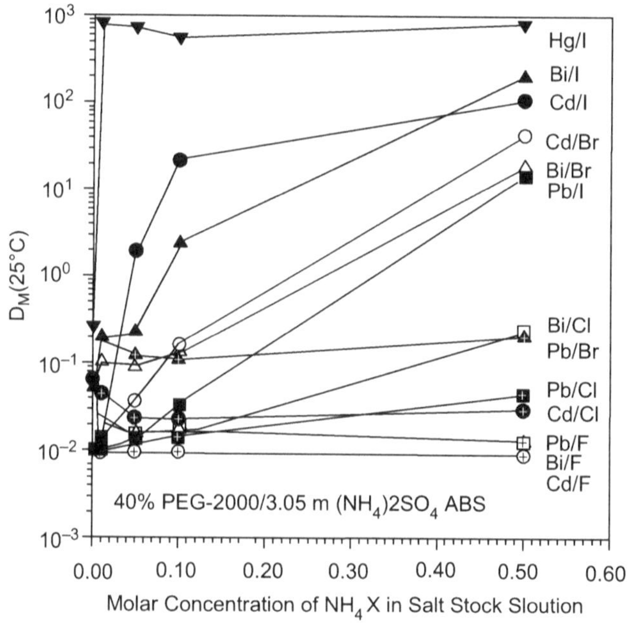

FIGURE 3.15 Distribution ratios of bismuth, cadmium, lead, and mercury versus molar concentration of NH_4X in the salt stock solution (X are Cl, F, Br, I anions). *(Source: From Ref. 81 with permission)*

REFERENCES

1. Marcus Y, Kertes AS. *Ion exchange and solvent extraction of metal complexes*. New York: Wiley-Interscience; 1969.
2. Rydberg J, Cox M, Musikas C, Choppin G, editors. *Solvent extraction principles and practice*. New York, London: CRC Press; 2004.
3. Ritcey GM. *Solvent extraction: principles and applications to process metallurgy*. New York, London: CRC Press; 2006.
4. Stary J. *The solvent extraction of metal chelates*. New York: Pergamon Press; 1964.
5. Zolotov YA. *Extraction of chelate compounds*. London: Humprey Science Publication Ann Arbor; 1970.
6. Narbutt J. Liquid-liquid partition and hydration of cobalt (III) acetylacetonate and cobalt (III) monothioacetylacetonate. *J Phys Chem* 1991;**95**:3432.
7. Flett DS, Jaycock MJ. Extraction of metals by carboxylic acids. In: Marinsky JA, Marcus Y, editors. *Ion exchange and solvent extraction*, vol. 3. New York: Marcel Dekker; 1973. p. 1.
8. Krasovic F, Klofutar C. Solvent extraction of copper(II) with ortho-hydroxyoximes. *J Inorg Nucl Chem* 1965;**97**:349.
9. Liljenzin JO, Stary J, Rydberg J. In: Kertes AS, Marcus Y, editors. *ISEC'68, Solvent Extraction Research*. New York: John Wiley & Sons; 1969.
10. Sekine T, Dyrssen DJ. Solvent extraction of metal ions with mixed ligands. *Inorg Nucl Chem* 1964;**26**:1727.

11. Carpenter JC, Christie PG, et al., editors. *Extractive metallurgy of copper*, vol. 2. New York: AIME; 1976. p. 1025.
12. Cox M, Flett DS. Kinetics of nickel extraction by α-hydroxyoxime/carboxylic acid mictures. In: *Proceedings of the International Solvent Extraction Conference* (ISEC'71), 1971, vol. 1, p. 204 (Soc. Chem. Ind., London).
13. Marcus Y. Gibbs energies of transfer into aqueous alcohols. *Pure Appl Chem* 1969;**20**:85.
14. Sekine T, Hasegawa Y. *Solvent extraction chemistry*. New York: Marcel Dekker; 1977.
15. Kertes AS. Thermodynamics of solvent extraction processes: synergic adducts of copper and zinc diketonates with bidentate nitrogen bases. In: Hanson C, editor. *Recent advances in liquid liquid extraction*. London: Pergamon Press; 1971. p. 15.
16. Marcus Y, Kertes AS, Yanir E. *Equilibrium constants of liquid-liquid distribution reactions, Part 1, Organophosphorus Extractants, IUPAC, Additional Publication*. London: Butterworths; 1974.
17. Marcus Y, Kertes AS, Yanir E. *Equilibrium constants of liquid-liquid distribution reactions, Part III. Compound forming Extractants, Solvating Solvents and Inert Solvents. IUPAC*. London: Pergamon Press; 1977.
18. Irving HM. Ligand-exchange chromatography of racemates. In: Marinsky JA, Marcus Y, editors. *Ion exchange and solvent extraction*, vol. 6. New York: Marcel Dekker; 1974. p. 139.
19. Liem DH. Acidic organophosphorus extractants. *Acta Chem Scand* 1972;**26**:191.
20. Markovits CY, Choppin GR. Solvent extraction of non-ferrous metals: a review. In: Marinsky JA, Marcus Y, editors. *Ion exchange and solvent extraction*, vol. 3. New York: Marcel Dekker; 1973. p. 1.
21. King Industries, Inc., Norwalk, Connecticut, SYNEX Liquid Ion Exchange Reagents, Product Bulletin and Bibliography.
22. Kertes AS, Gutmann H. Decarboxylation of 6-nitrobenzisoxazole-3-carboxylate ion in dichloromethane: the possible role of reverse micelles. In: Matijevic E, editor. *Surface and colloid science*, vol. 8. New York: Wiley-Interscience; 1975. p. 193.
23. Hanson C, Hughes MA, Whewell RJ. The hydrophilic lipophilic balance of hydroxyoximes. *J Appl Chem Biotechnol* 1978;**28**:426.
24. Ashbrook AW. Chelating reagents in solvent extraction processes: the present position. *Coord Chem Rev* 1975;**16**:285.
25. Ashbrook AW. Commercial chelating solvent extraction reagents. III. Oximes: spectra, structure and properties. *Hydrometallurgy* 1975;**1**:5.
26. Flett DS, West DW. *Solvent extraction in hydrometallurgy: the role of organophosphorus extractants Complex Metallurgy'78*. London: IMM; 1978. p. 49.
27. Van Ooyen J, Dyrssen D, Liljenzin JO, Rydberg J, editors. *Solvent extraction Chemistry (ISEC 1966)*. Amsterdam: North-Holland; 1967.
28. Ishimori T, Nakamura E. Japan atomic energy research institute, JAERI, 1047, 1963; JAERI 1062 (1964–1965).
29. Seeley FG, Grouse DJ. *J Chem Eng Data* 1971;**16**:393.
30. Diamond RM. In: Dyrssen D, Liljensin JO, Rydberg J, editors. *Solvent extraction chemistry*. Amsterdam: North-Holland; 1967. p. 349.
31. Grinstead RR, Davis JC. Base strength of amine-amine. Hydrochloride. *J Phys Chem* 1968;**72**:1630.
32. Roddy JW, Coleman CF. The extraction of water by tri-n-octylamine. *J Inorg Nucl Chem* 1969;**31**:3599.
33. Muhammed M, Szabon J, Hogfeldt E. Extraction with long-chain amines. *Chem Scripta* 1974;**6**:61.
34. Cecconie T, Freiser H. Liquid–liquid extraction and separation of lanthanum(III). *Anal Chem* 1990;**62**:622.

35. Moryia H, Sekine T. Studies of the liquid-liquid partition systems. *Bull Chem Soc Jap* 1971;**44**:3347.
36. Rydberg J, Albinsson YJ. Configurational stability of 2-benzoylcyclohexanone. *Solv Extr Ion Exch* 1989;**7**:577.
37. Albinsson Y, Rydberg J. Configurational stability of 2-benzoylcyclohexanone. *Radiochim Acta* 1989;**48**:49.
38. Widmer HM. In: *Proceedings of the International Solvent Extraction Conference* (ISEC'73), 1973, Vol. 1, p. 37 (Soc. Chem. Ind., London).
39. Mikhailov VA. In: *Proceedings of the International Solvent Extraction Conference* (ISEC'77), 1979, vol. 1, p. 52 (Can. Inst. Min. Met., Montreal).
40. Dyrssen D. *Proc Symp Coord Chem*. p. 707. Tihany: Akademini Kiado; 1964. Budapest.
41. Ahrland S, Liljenzin JO, Rydberg J. The chemistry of the actinides. In: *Comprehensive inorganic chemistry*. London: Pergamon Press; 1973.
42. Kassierer EF, Kertes AS. Synergic extraction of some lanthanides. *J Inorg Nucl Chem* 1972;**34**:3209.
43. Schweitzer GK, Neil RB, Clifford FR. Effects of the anion on the extraction of zinc with 8-hydroxyquinoline. *Anal Chim Acta* 1965;**33**:514.
44. Spigorelli JC, Meloan CF. *Sep Sci Techn* 1974;**9**:79.
45. Joe EG, Ritcey GM, Ashbrook AW. Polymer-supported phosphorus-containing ligands for selective metal ion complexation. *J Metals NY* 1966;**18**:18.
46. Flett DS, West DW. In: *Proceedings of the International Solvent Extraction Conference*, (ISEC'71), 1971, vol. 1, p. 214 (Soc. Chem. Ind., London).
47. Flett DS, Cox M, Heels JDG. In: *Proceedings of the International Solvent Extraction Conference* (ISEC"73), 1974, vol. 3, p. 2559 (Soc. Chem. Ind., London).
48. Flett DS, Cox M, Heels JDG. *J Inorg Nucl Chem* 1975;**37**:2533.
49. Nyman BG, Hummelstedt L. In: *Proceedings of the International Solvent Extraction Conference* (ISEC"73), vol. 1 1974, p. 669 (Soc. Chem. Ind., London).
50. Moore RH, Partridge JA. Battelle Pacific Northwest Laboratories Report No. BNWL*SA-4476, 1972.
51. Lakshmanan VL, Lawson GJ. The behaviour of Lix, 63. *J Inorg Nucl Chem* 1973;**35**:4285.
52. Harrison G, Lakshmanan VL, Lawson GJ. Solvent extraction of zinc(II) and copper(II) with mixtures of LIX. *Hydrometallurgy* 1976;**1**:339.
53. Flett DS, Cox M, Heels LDG. Solubilization versus microemulsification of extractant molecules. *J Inorg Nucl Chem* 1975;**37**:2197.
54. Hanson C, Hughes MA, Preston JS, Whewell RJ. Purification and acid-base properties of hydroxyoxime extractants. *J Inorg Nucl Chem* 1976;**38**:2306.
55. Dalton RF. In: *Proceedings of the International Solvent Extraction Conference* (ISEC"77), 1979, vol. 1, p. 40 (Can. Inst. Min. Met., Montreal).
56. Flett S, Spink DR. In: *Proceedings of the International Solvent Extraction Conference* (ISEC"77), 1979, vol. 2, p. 496 (Can. Inst. Min. Met., Monttreal).
57. Zingaro RA, Coleman CF. *J Inorg Nucl Chem* 1967;**29**:1287.
58. Wang SM, Li NC. *J Inorg Nucl Chem* 1966;**28**:1091.
59. Woo C, Wagner WF, Sands DE. *J Inorg Nucl Chem* 1972;**34**:307.
60. Newman L, Klotz. Applications of mixed-ligand complexes in analytical. In: Dyrssen D, Liljensin JO, Rydberg J, editors. *Solvent extraction chemistry*. Amsterdam: North-Holland; 1967. p. 128.
61. Sekine T, Dyrssen D. Solvent extraction of rare-earth metal ions with thenoyltrifluoroacetone. *J Inorg Nucl Chem* 1967;**29**:1489.

62. van Dalen A, Wijkstra J, Gerritsma KW. Liquid-liquid extraction of complexed metal ions by inclusion in dinonyl naphthalene sulphonic acid micelles. *J Inorg Nucl Chem* 1978;**40**:875–81.
63. Eyal AM. Acid extraction by acid-base-coupled extractants. In: Marinsky JA, Marcus Y, editors. *Ion exchange and solvent extraction*, vol. 13. New York: Marcel Dekker Inc; 1997. p. 31–93.
64. Schmuckler G, Harel G. Extraction of salts by mixed liquid ion exchangers. In: Marinsky JA, Marcus Y, editors. *Ion exchange and solvent extraction*, vol. 13. New York: Marcel Dekker Inc; 1997. p. 1–29.
65. Eyal AM, Bressler E, Bloch R, Hazan B. Extraction of $ZnSO_4$ by acid base couple extractants. *Ind Eng Chem Res* 1994;**33**:1067.
66. Eyal AM, Kogan L, Bressler E. Extraction of metal salts by mixtures of water-immiscible amines and organic acids (acid-base couple extractants). 2. theoretical treatment and analysis. *Ind Eng Chem Res* 1994;**33**:1076.
67. Neuman L, Klotz P. Synergistic effect of tri-n-octylamine on the solvent extraction of cerium(III) by thenoyltrifluoroacetone. *Inorg Chem* 1972;**11**(9):2154–9.
68. Ke CH, Li NC. Synergic extraction of cobalt(II) by thenoyltrifluoroacetone. *J Inorg Nucl Chem* 1969;**31**:1383.
69. Deptula C, Minc S. *Nukleonika* 1965;**10**:421.
70. Deptula C. *J Inorg Nucl Chem* 1967;**29**:1097.
71. Deptula C, Mine S. *J Inorg Nucl Chem* 1967;**29**:159.
72. Liem DH, Sinegribova OA. *Acta Chem Scand* 1971;**25**:301.
73. Deptula C. *J Inorg Nucl Chem* 1970;**32**:277.
74. Rinelli G, Abbruzzese C. In: Jones MJ, editor. *Geology, mining and extractive processing of uranium*. London: IMM; 1977. p. 20.
75. Nishimura S, Ke CH, Li NC. Adducts of copper(II) β-diketone chelates with tertiary amines. *J Phys Chem* 1968;**72**:1297.
76. Pedersen CJ. Brief overview on anion binding organic ligands. *J Am Chem Soc* 1967;**89**:7017.
77. Trueblood KN, Knobler CB, Lawrence DS, Stevens RV. Polysulfonylamines. *J Am Chem Soc* 1982;**104**:1355.
78. Gehin D, Kollman PA, Wipff G. Anchoring of ammonium cations to an 18-crown-6 binding site: molecular mechanics and dynamics study. *J Am Chem Soc* 1989;**111**:3011.
79. Walter H, Brooks DE, Fisher D, editors. *Partitioning in aqueous two-phase systems. Theory, methods, uses, and applications to biotechnology.* Orlando, FL: Academic Press; 1985.
80. Rogers RD, Fiteman MA, editors. *Aqueous biphasic separations: biomolecules to metal ions.* New York: Plenum; 1995.
81. Albertsson PA. *Partition of cell particles and macromolecules*. 3rd ed. New York: John Wiley & Sons; 1986.
82. Rogers RD, Zhang J. New technologies for metal ion separations. In: Marinsky JA, Marcus Y, editors. *Ion exchange and solvent extraction*, vol. 13. New York: Marcel Dekker Inc; 1997. p. 141–93.
83. Rogers RD, Bond AH, Bauer CB. In: Logsdeil DH, Slater MJ, editors. *Proceedings of ISEC'93*, vol. 3. London: Elsevier; 1993. p. 1641.
84. Rogers RD, Bond AH, Bauer CB. Toward unusual aluminum-oxygen compounds. *Sep Sci Technol* 1993;**28**:139.
85. Rogers RD, Bauer CB, Bond AH. Crown ethers as actinide extractants in acidic aqueous biphasic systems: partitioning behavior in solution and crystallographic analyses of the solid state. *J Alloys Compd* 1994;**305**:213–4.
86. Rogers RD, Bond AH, Griffin ST, Gorwitz EP. Extraction of strontium from acidic nitrate media using a modified PUREX solvent. *Solvent Extr Ion Exch* 1996;**14**:919.
87. Rippel G, Alattiani E, Szepesy L. Effect of the characteristics of proteins in hydrophobic interaction chromatography. *J Chromatogr A* 1994;**668**:301.

Chapter 4

Engineering Development of Solvent Extraction Processes

Chapter Outline

1. INTRODUCTION

This chapter is devoted to the engineering development of the solvent extraction processes of interest. An aqueous solution contains a valuable component that is desired to recover, and a number of other components from which the desired component should be separated. Considering the technology the developers have to answer several questions.

1. Concerning liquid system: (a) A suitable solvent system has to be identified, which will extract the desired component selectively from the less-desired

Solvent Extraction. DOI: 10.1016/B978-0-444-53778-2.10004-4

components. What fraction of the desired component can be recovered? (b) Suitable physical and chemical conditions for carrying out the extraction have to be found. (c) Thermodynamic design of the extraction process; the rates of extraction of the desired, and possibly also some of the undesired, components have been determined. Optimal parameters for the extraction process are proposed.

These questions are considered as a result of literature search and laboratory studies and answered mainly in the previous chapters.

2. Flow and mass transfer, thermodynamic design of the extraction—stripping technological scheme: (a) Determination of the phase ratio and the number of separation stages necessary. How many of the undesired components are extracted with the desired component, and what can be done to reduce their concentration? (b) Efficiency of the mass transfer and solvent losses and its recovery. (c) Recovery of the desired component from the extract in a usable form: design of a strip flow and mass transfer conditions.
3. Design of the equipment for technology: Mainly extractor–stripper apparatus is discussed from chemical engineering point of view.
4. Energy-efficient, environmental and economic considerations have to be added to these questions.

The initial bench-scale experimental work in solvent extraction processes for answering some of these questions and determining main process parameters (the temperature, flow rate, acidity, and so forth) is conducted with small apparatus, such as separator funnels. Following the successful completion of these tests, when the best reagent and other conditions for the system have been established, small-scale continuous operations are run, such as in a small mixer-settler unit. The data so obtained are used to determine scale-up factors for pilot plant or plant design and operation.

Pilot plant operations are considered from technological and economic point of view. If few (or no) modifications are required, considerable savings can result on scale-up. On the other hand, problems encountered in a plant operation that has been designed without piloting the process could result in the loss of considerable time and expense. It should be pointed out that the size of a pilot plant can vary from a few gallons per minute of total throughput to several hundred gallons per minute. Normally, one would expect that the size of a pilot plant will be in direct proportion to the size of the final plant.

In this chapter, the engineering aspects of scale-up laboratory studies of an extraction process to industrial applications are presented in short. The choice of material, stability, and other questions of mechanical engineering are not considered in this book.

2. EXTRACTION STAGE

Figure 4.1 gives the basic stages of the general solvent extraction process. The phase containing the desired and the undesired components is the *feed* phase. After passing through the *extraction* stage, it is called the *raffinate.* The raffinate may pass to a *solvent recovery* stage in which any solvent is recovered and recycled. The fresh solvent enters the extraction stage and leaves as an *extract.* The extract may pass through a *scrub stage* to wash out some of the undesired components with a *scrub solution,* and the *scrub raffinate* may be recycled to the feed. The extract or the *scrubbed extract* may pass to the *strip stage* in which a *strip solution* back-extracts the desired component from the extract. The loaded strip solution then passes to *product recovery* in which the desired component is finally recovered, while the *stripped solvent* may either be recycled directly to the extraction stage or may pass through a *regeneration stage* in which the solvent is contacted with a *regenerant* to restore its chemical composition to that desired for efficient extraction.

Not all of these steps take place in every solvent extraction process. Equally, in some systems, solvent recovery may be applied to the scrub raffinate or to the loaded strip solution. Consequently, although this is not a completely general flow diagram, it covers most of the systems likely to be found in practice. Variations on this general system should become apparent during the further discussion.

2.1. Single-Stage Process

As described in the previous chapters, if a solution containing a desired component S is contacted with a solvent, then S distributes itself between the feed solution and the solvent according to Eqn (1.12) in Chapter 1. If the phase volume ratio $\theta = V_{(o)}/V_{(w)}$, then the fraction extracted, E_S may be presented by:

$$E_S = \frac{D_S\theta}{D_S\theta + 1} \tag{4.1}$$

So, the fraction extracted in a single stage is a function of the distribution ratio and the phase volume ratio.

Consider, e.g., a desired component S for which $D_S = 10$ and a contaminant X for which $D_X = 0.1$. Table 4.1 shows the effect of varying the phase ratio on the extracted fractions of S and X. The ratio of the fractions extracted is a measure of the product purity.

This shows how the extent of extraction varies with phase ratio at a constant distribution ratio, and how the variation in the phase ratio affects the product purity. It is evident that it is impossible to achieve both high recovery and a high degree of separation simultaneously in a single stage.

A second phenomenon of great industrial importance is the effect of saturation of the solvent on the product purity. Implicit in the derivation of

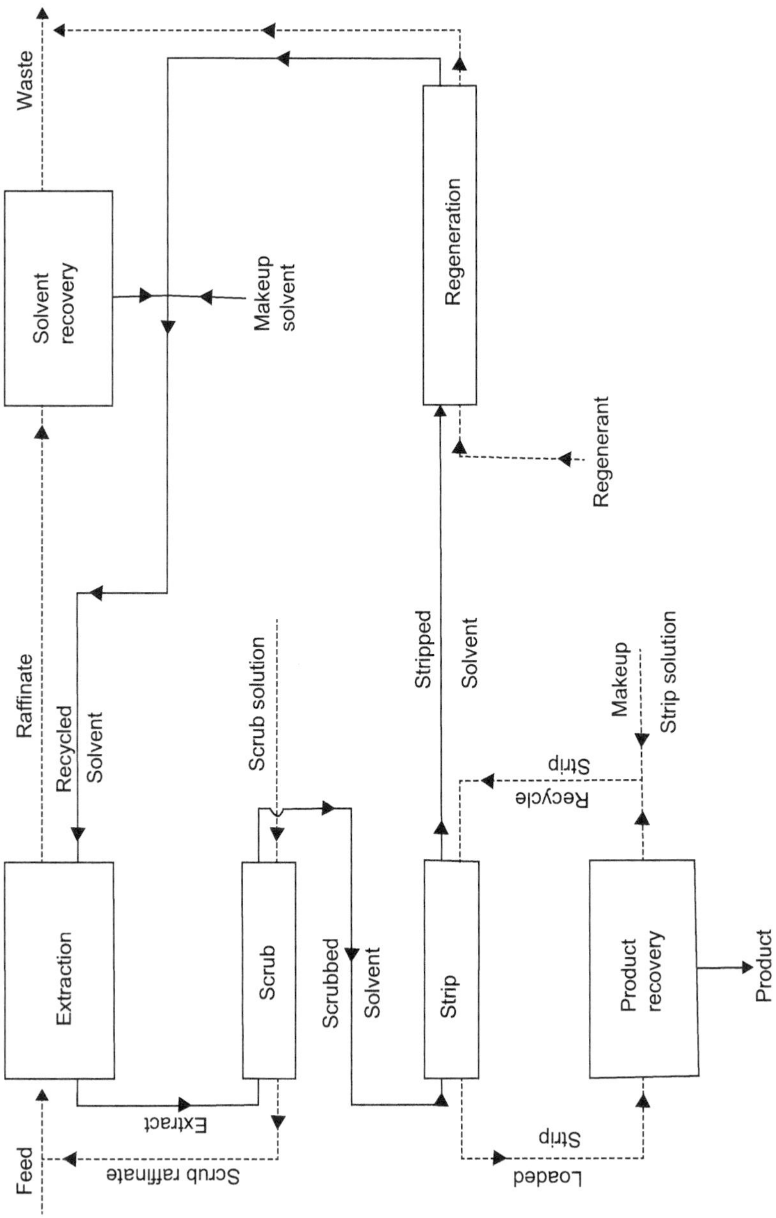

FIGURE 4.1 Basic processes in solvent extraction.

TABLE 4.1 Effect of Phase Ratio in Single-Stage Extraction

	Fraction extracted		
Phase ratio θ	E_S	E_X	Product purity, E_S/E_X
0.1	0.50	0.001	50
1	0.91	0.091	10
10	0.99	0.50	1.98

Eqn (4.1) is the assumption that D is a constant. As discussed in Chapter 1, D is a constant only under ideal and constant conditions (usually at trace concentrations in both phases), but usually changes markedly as the concentrations of the reactants in the two phases vary. At higher concentrations, so much of an extractable component may be extracted that an appreciable fraction of the extractant is bonded to the extracted component: accordingly, in turn, the concentration of the free ligand in the organic phase is significantly reduced.

Industrial practice naturally requires the maximum use of the relatively expensive solvent; therefore, saturation of the solvent, with reduction of the free ligand concentration to a minimum, is a general rule. Thus, a model is needed to quantify the effect of the reduction in D, as the concentration of the extracted species in the organic phase increases. In many extraction systems D is proportional to $[L]^n$, where $[L]$ is the free ligand concentration, and n is determined by the number of ligand molecules per molecule of extracted complex. It may be expressed by the following equation:

$$D_S = [S]_{(o)}/[S]_{(w)} = D^0\{[\mathrm{L}]_{(o),\mathrm{tot}} - m[S]_{(o)}\}^n \qquad (4.2)$$

where D^0 is the distribution ratio at trace concentrations, readily determined in the laboratory research; $[L]_{\mathrm{tot}}$ is the total ligand concentration; and m is the ratio of ligand molecules to extracted molecules S close to saturation. The term in braces on the right-hand side of the equation is the free ligand concentration.[1]

It should be noted that m and n in Eqn (4.2) are sometimes equal, but more often differ in value. This underlines that this equation has no theoretical basis, but is merely a convenient way of representing experimental data. Equation (4.2) can be fitted to a wide range of experimentally determined isotherms, with quite small residual errors over the entire range of concentrations of interest.

When more than one component is extracted, then the free ligand concentration will be reduced by all components in the organic phase. Let us study, e.g., the extraction of two components, A and B, which have similar chemistries of extraction. Assume for both that $m = n = 2$, and that they have D^0 values of 10 and 1, respectively, and the same concentrations in the feed, with the extraction being carried out at a phase ratio $\theta = 1$. The equilibria for the distribution of A and B between the two phases are described using Eqn (4.2):

$$D_A = 10(1 - 2[A + B])^2 \quad (4.2a)$$

$$D_B = 1(1 - 2[A + B])^2 \quad (4.2b)$$

Two mass balance equations:

$$[A]_{feed} = [A]_{(w)} + [A]_{(o)} \quad (4.2c)$$

$$[B]_{feed} = [B]_{(w)} + [B]_{(o)} \quad (4.2d)$$

These four equations are sufficient to solve four unknown variables $[A]_{(w)}$, $[A]_{(o)}$, $[B]_{(w)}$, and $[B]_{(o)}$.

The results of such calculations are shown in Fig. 4.2. It shows that the more extractable specie A competes strongly for the ligand, and the equilibrium curve of A in the presence of B is depressed only slightly below that for the extraction of A alone, whereas the equilibrium curve for the extraction of B in the presence of A is depressed strongly relative to the extraction of free B. Thus, at high concentrations, saturation effects can improve product purity to a far greater extent than the equilibrium isotherms of the individual species would indicate superficially.

We can conclude that single-stage extraction is rarely adopted because of the effect of phase ratio on the fractional recovery and purity, and the effect of saturation on purity.

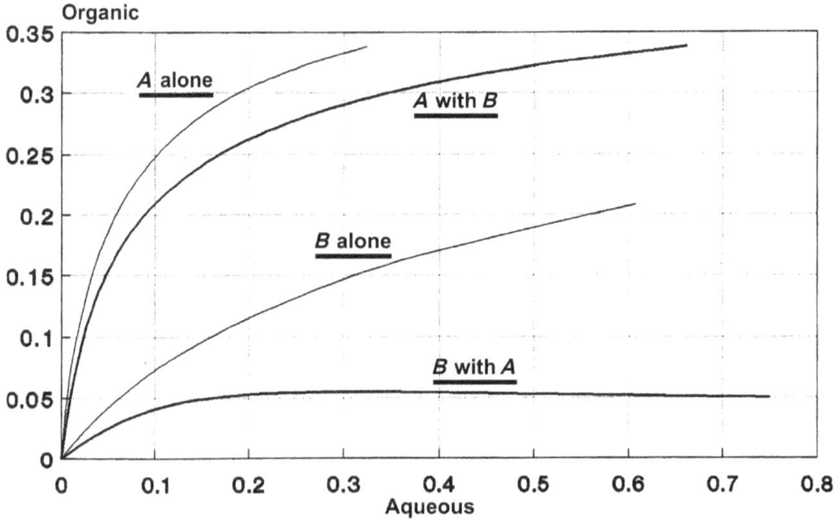

FIGURE 4.2 Extraction of two species simultaneously.

2.2. Stagewise "Cross-flow" Systems

The low degree of extraction can be improved if the raffinate after single-stage extraction could be reextracted with fresh solvent at the same phase ratio. The second extract will show a similar purity. The first and second extracts are combined to give a better overall recovery than in a single stage extraction. For a single, highly extractable component this argument is entirely valid. Performing a mass balance over any one stage it is found that

$$V_{(o)}[A]_{(o)} = V_{(w)}([A]_f - [A]_r) \tag{4.3}$$

which on rearranging becomes:

$$([A]_f - [A]_r)/[A]_{(o)} = \theta \tag{4.4}$$

where $[A]_f$ is the concentration of A in the feed to any one stage; $[A]_r$ is the concentration of A in the raffinate leaving any one stage; $[A]_{(o)}$ is the concentration of A in the organic phase that is in equilibrium with $[A]_r$.

If the equilibrium relationship between $[A]_{(o)}$ and $[A]_r$ is known, the series of Eqn (4.4) governing each individual stage of reextraction can be solved analytically.[1] The properties of the solution can be readily understood graphically. Figure 4.3 is a graphic construction that represents a series of equations such as Eqn (4.4), for the case $\theta = 1$. From $[A]_{f-1}$ a line of slope $1/\theta = -1$ intercepts the equilibrium curve at $[A]_{(o),1}$. A perpendicular from this intercept cuts the x-axis at $[A]_{r-1} = [A]_{f-2}$ from which a further line of slope -1 is

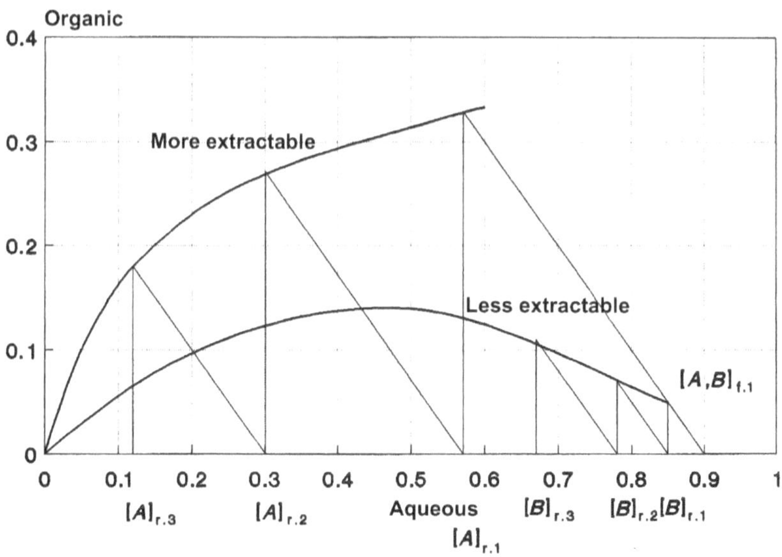

FIGURE 4.3 Cross-flow extraction of one compound.

drawn. Of course, θ may be varied from stage to stage, but there is no particular advantage in doing so. This illustrates that even with a distribution ratio <1, useful recoveries (in our example, over 99%) can be achieved in comparatively few stages by repeated extraction. However, as the number of stages increases, the concentration of the extracted species in the combined extract drops significantly.

The use of four cross-flow extraction stages is clearly preferable to a single extraction using a volume of organic four times that used in each stage of the cross-flow. Equally, of course, the use of more than four extraction stages, each with a proportionately smaller volume, would improve the performance.

The mathematical treatment of a cross-flow cascade is given below. The mass Q_E, which is extracted through the n stages:

$$Q_E = \Sigma V_{n,(o)}[C]_{n,(o)} \tag{4.5}$$

Cumulative recovery:

$$E_{tot} = \frac{Q_E}{\{[C]_f V\}} \tag{4.6}$$

while the average concentration in the combined extracts are:

$$[C]_{av,(o)} = \frac{Q_E}{(nV_{(o)})} \tag{4.7}$$

provided the phase ratio is kept the same in each repeated extraction; if not, $n \times V_{(o)}$ in Eqn (4.7) must be replaced by ΣV_n.

Let us now consider the effect of repeated extraction on the degree of separation that can be achieved from contaminants. Figure 4.4 presents data for two components, three extraction stages, and $\theta = 1$. The equilibria are calculated by solving simultaneously two equations of the form of Eqn (4.2) for the two components. The cross-flow cascade gives a product containing, on average, 0.07/0.26 = 0.27 fraction $[B]_{(O)}/[A]_{(O)}$, whereas an equivalent single stage gives a product containing 0.08/0.23 = 0.35. Thus, the cross-flow cascade gives a significant improvement in product purity relative to a single-stage extraction. This effect clearly arises from the use of a relatively low phase ratio

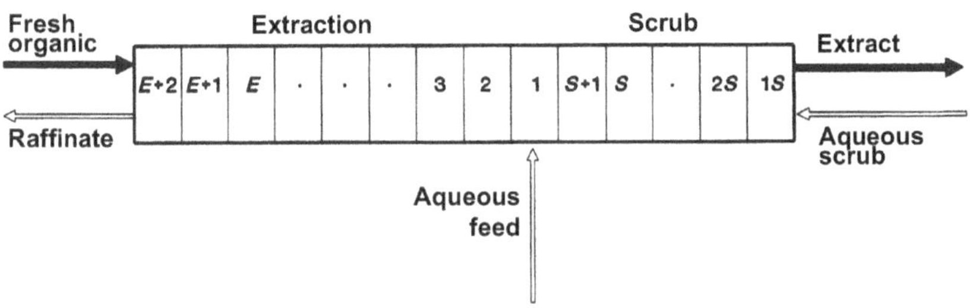

FIGURE 4.4 Countercurrent cascade arrangement.

in each stage of the cross-flow cascade. Unlike the example of Table 4.1, cross-flow recovery of the desired component is actually higher than that of the single stage at the higher phase ratio.

Cross-flow cascades find some use in practice because they can yield high recoveries of a valuable product that has a low distribution coefficient and can provide useful improvements in product purity.

2.3. Countercurrent Extraction

A typical flow arrangement of countercurrent extraction systems is shown in Fig. 4.4 from which it is clear that how the name arises. A mass balance over any one stage

$$V_{(o)}([A]_{(o),i} - [A]_{(o),i+1}) = V([A]_{r,i-1} - [A]_{r,i}) \tag{4.8}$$

And after rearrangement:

$$\theta = \frac{V_{(o)}}{V_{(w)}} = \frac{([A]_{r,i-1} - [A]_{r,i})}{([A]_{(r),i} - [A]_{(r),i+1})} \tag{4.9}$$

If the cascade is designed with the organic concentration increasing from stage to stage and a phase ratio $\theta = 1$, the performance is explained in Fig. 4.5. Here, the calculation has been carried out from the organic feed end (from the aqueous feed end is also possible). From $[A]_{r,3}$ a vertical line is drawn to intersect the equilibrium curve, from which point a line of slope -1 is drawn to reach the point $[A]_{r,2}$ on the axis. A vertical line from this point reaches the equilibrium curve at point *E*, and again a line of slope -1 is drawn, but on this occasion it is only carried to point G, because there is a contact between

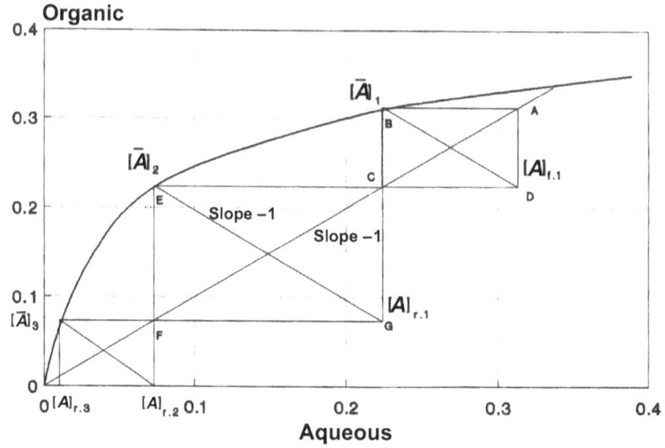

FIGURE 4.5 Countercurrent extraction of one component A.

aqueous concentration $[A]_{r,1}$ and organic, not of zero concentration, but of concentration $[A]_3$. As this calculation continues, it is found impossible to proceed much beyond the point marked *A* on the diagram because the cascade has too little solvent to perform at a phase ratio $\theta = 1$. Considering the rectangles *ABCD* and *CEFG* in Fig. 4.5, the diagonals *BD* and *EG* have the same slope. The diagonals *AC* and *CF* thus have the same slope, and *AF*, therefore, is a straight line. The point *A* has the coordinates $[A]_{f,1}$, $[A]_{(o),1}$ and the point *F* has the coordinates $[A]_{r,2}$, $[A]_{(o),3}$; Thus, the slope of the line

$$AF = \frac{1}{\theta} = \frac{([A]_{(r),1} - [A]_{(r),3})}{([A]_{(r),1} - [A]_{r,2})} \tag{4.10}$$

which is identical with Eqn (4.9), except it represents the mass balance over two stages.

Mass balance lines such as *AF* are known as *operating lines* in countercurrent cascades. Clearly, an operating line cannot cross an equilibrium line. Wherever an operating line approaches an equilibrium line, a *pinch point* is the result—the number of stages needed to achieve the desired degree of extraction approaches infinity.

In Table 4.2 the calculations for a single component in a countercurrent cascade are presented with a value of $\theta = 2.47$, in the direction, from the feed end. Comparison with the results of Table 4.1 clearly indicates that the countercurrent cascade offers a greater overall extraction in fewer stages and can yield an extract of higher concentration.

Figure 4.6 is drawn as an equilibrium isotherm on the basis of the data of Table 4.2. Here, there is an operating line and a series of "steps" between the operating line and the isotherm. These steps are entirely equivalent to the lines establishing the mass balances for each stage. For instance, the horizontal line *AB* represents $[A]_{f,1} - [A]_{r,1}$, whereas the vertical line *BC* represents $[A]_{(o),1} - [A]_{(o),2}$. This graphic of an extraction isotherm, an operating line, and the stepwise evaluation of the number of stages is known as a *McCabe–Thiele diagram.*[2] Historically, it found great application in a variety of mass transfer operations, *from* gas adsorption, through distillation, to solvent extraction. However, the

TABLE 4.2 Extraction of a Single Component A in a Countercurrent Cascade

	Aqueous phase		Organic phase			
Stage no	Feed	Raffinate	Feed	Extract	E	E_{tot}
1	0.9	0.495	0.201	0.364	0.45	0.45
2	0.495	0.056	0.023	0.201	0.89	0.94
3	0.056	0.0025	0.00	0.023	0.96	0.997

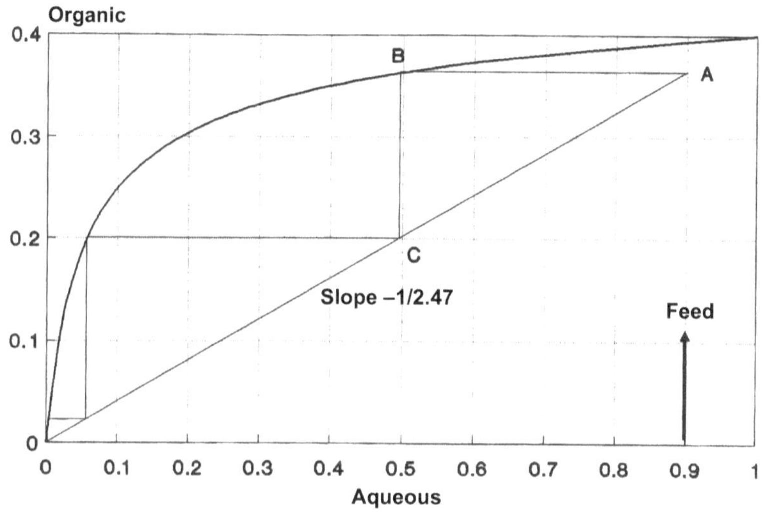

FIGURE 4.6 Extraction isotherm for three-stage countercurrent extraction cascade (McCabe–Thiele diagram).

advent of modern computational techniques has made it largely redundant, as it is often easier, and certainly more accurate, to calculate the cascade directly. It is not sufficient to view the isotherm as fixed. Quite small changes in aqueous composition or in phase ratio can change the isotherms and cause dramatic effects on the performance of a countercurrent cascade.

Now consider the extraction of two species simultaneously in a countercurrent cascade. To obtain the equilibria requires solving two equations of the type of Eqn (4.2) simultaneously, whereas determining the concentrations within the cascade taking into account a mass balance such as Eqn (4.9) at each stage. Table 4.3 presents such calculations for the same two equilibria as used before.

The behavior of the system is very counterintuitive. The coupling of the two equilibria leads to wild swings in the aqueous concentrations. The presence of high concentrations of the less extractable species so disturbs the equilibria that the more extractable component has a distribution ratio of about 4 even in the third stage of the cascade, at which its distribution ratio might have been expected to approach the infinite dilution value of 10. In spite of this, three stages suffice to recover over 99% of the more extractable component, but the product is not particularly pure indeed, in the final extract, the less extractable component is nearly 35% of the total extracted.

It is observed that although the more extractable component is well behaved, the less extractable is not. This is a fairly general finding in industrial practice for which there are coupled equilibria with components that have similar extraction behavior-graphic methods suffice only for the most extractable components, and even then, only if the real equilibrium for that component is used. For similar reasons, highly selective solvents are usually sought.

TABLE 4.3 Extraction of Two Components (A and B) in a Countercurrent Cascade. Phase Volumes Ratio = 3.96

	Aqueous phase		Organic phase			
Stage no	Feed	Raffinate	Feed	Extract	E	E_{tot}
1. More extractable component A						
1	0.9	0.242	0.061	0.227	0.73	0.998
2	0.242	0.031	0.007	0.061	0.87	0.27
3	0.031	0.0018	0	0.007	0.94	0.032
2. Less extractable component B						
1	0.9	1.287	0.218	0.120	−0.43	−0.43
2	1.287	1.11	0.173	0.218	0.14	−0.23
3	1.11	0.424	0	0.173	0.62	0.53

2.4. Scrubbing

When a high-purity product is sought, the same philosophy of saturating the aqueous phase with the unwanted component is used in scrubbing the extract. This entails some loss of the more extractable species from the extract, but the scrub raffinate is usually returned to be mixed with the aqueous feed; hence, the more extractable component is not lost. For example, let us compute the cascade, designed to recover over 95% of the more extractable component at 99% purity. The feed was approximately equivalent to that fed to the countercurrent cascade of Table 4.3, but was more concentrated to allow dilution by the scrub raffinate. In the scrub, $\theta = 8.295$, or a flow of 0.422 volumes of aqueous phase per 3.5 volumes of organic phase. This aqueous volume is then added to 0.578 volumes of fresh feed to give $\theta = 3.5$ in the extraction section.

It has to be stressed that this type of cascade is rather sensitive to changes in many of the operational parameters, including phase ratio. It must be operated under carefully controlled conditions for optimal results. Such control is well within the capability of modern methods, which will extend not only to flow rate and temperature control, but also often to compositional control.

In a similar vein, such cascades are best computed by calculating stagewise from the raffinate end of the extraction section and from the extract end of the scrub section, and adjusting the values of e in each section so that conditions at the center of the cascade match. Attempts to calculate from one end of the whole cascade to the other, or to calculate outward from the feed point, are interestingly unstable.

The scrubbing techniques were developed in the radiochemical industry, where total product impurities measured in parts per million (ppm) were first required on a tonnage basis.

In conclusion of this section we have to stress the following:

1. The fraction of the desired component that can be recovered is determined not only by its inherent extractability, but also by the phase ratio and the presence of competing extractable components.
2. The volume of solvent needed for a particular duty depends on the system adopted for the extraction; in general, a single-stage extraction requires more solvent than a cross-flow cascade, which in turn needs more than a countercurrent cascade.
3. The purity of the product is determined not only by the inherent selectivity of the solvent system for the component that is sought, but also on the phase ratio, the concentration of the contaminants, and the performance of any scrubbing of the extract.

3. STRIPPING ORGANIC SOLUTIONS

The general principles established for extraction apply to stripping, although in order to accomplish the strip as efficiently as possible, distribution ratios have to be significantly less than unity. Stripping can equally be done in single stage, cross-flow, or countercurrent. As in extraction, the same mass balance limitations on phase ratio apply. Let us design a stripping countercurrent cascade with organic feed concentration of 0.300, assuming >95% strip and a final aqueous phase of <2. The maximum phase ratio is $\theta_{max} = 2/(0.95 \times 0.350) = 7.02$. (Maximum in this case because stripping is in the reverse direction to extraction.) Results presented in Table 4.4 calculated for the phase ratio of 6.55. Four stages are needed. The strip raffinate is nearly saturated and has a concentration of 1.71. The stripped organic contains only 4% (0.012/0.30) of the component *A* in the extract, so the stripping efficiency is 98%. The less extractable (more readily stripped) component *B* was stripped in two stages to remove it completely from the organic.

Note that the stripped extract still contains traces of the desired component. This is a slight nuisance as the stripped extract will, in the normal course of events, be recycled to the extraction stages where it will reduce the extraction efficiency in the first stage. This practice is fairly common in industrial practice, as it is more economic to lose a little of the desired component in the extraction stages than to have too dilute a strip raffinate by seeking a total strip. The purpose of seeking a concentrated strip solution is to reduce the energy required to recover the product from the strip solution. In the case of metal salts, precipitation, electrolysis, direct reduction, and a host of other techniques may be used to generate the final product. In the case of the extraction of organic compounds, distillation, crystallization, or similar separation methods are used.

TABLE 4.4 Countercurrent Stripping of Two Component, *A* and *B* Phase Ratio = 6.55

	Aqueous phase		Organic phase	
Stage no	Feed	Raffinate	Feed	Extract
1. Less strippable component A				
1	0.924	1.965	0.300	0.141
2	0.511	0.924	0.141	0.078
3	0.304	0.511	0.078	0.046
4	0.187	0.304	0.046	0.029
5	0.118	0.187	0.029	0.018
6	0.076	0.0118	0.018	0.012
2. More strippable component B				
1	0.001	0.020	0.003	0.000
2	0.000	0.001	0.000	0.003

In each case, the more concentrated the strip solution, the less energy is required to recover the desired components.

3.1. Regeneration of Organic Phase

Sometimes, an undesired component is not stripped. It then builds up in the organic phase until it interferes seriously with the extraction. In this case, the solvent is given a more vigorous strip to regenerate its performance by removing the contaminant. A small sidestream may be bled continuously from the recycled organic phase and regenerated, or the contaminant may be allowed to build up in the extract for some time before the solvent is regenerated on a batch basis. Because of its vigorous nature, regeneration of the organic phase can be expensive, but sometimes the contaminant is very valuable—platinum, gold, and cobalt complexes have acted as contaminants on occasions, and their recovery has paid for the regeneration.

4. EXTRACTION EFFICIENCY

Solvent extraction is a kinetic process. The key variables in determining the rate of extraction are as follows:

1. The driving force of the system
2. The interfacial area

3. Specific resistances in the interfacial region, particularly any slow interfacial reactions

The rate is also affected by diffusion through the bulk liquids, but in general industrial practice there is sufficient turbulence to ensure that the bulk phases are well mixed. All these are discussed in detail in Chapter 1.

There is some control over interfacial area, though not too much flexibility is available because a very large interfacial area is associated with very fine droplets or very thin films. This may result in excessive loss of solvent by entrainment.

A little may be done with the displacement of the system from equilibrium (the average displacement is maximal in a countercurrent cascade). The driving force can be increased by reducing the slope of the operating line, i.e., increasing the phase ratio, but this is generally not economical. Very little can be done with interfacial resistances once the extraction system has been chosen, although renewal of the surface during bubble coalescence and dispersion assists in overcoming some forms of this resistance. Thus, the industry either has to make the time of contact long enough to ensure that equilibrium is essentially attained or it has to accept the inherent inefficiency of a single stage, and employ more stages than would otherwise be needed. In practice, it employs the latter strategy.

4.1. Efficiency in a Single Stage

The stage efficiency, S_E, of extraction of component A is

$$S_E = \frac{[A]_f - [A]_r}{[A] - [A]_{r.e}} \tag{4.11}$$

where $[A]$ is the concentration of component A in one phase, $[A]_r$ is the raffinate (or extract) concentration actually achieved, and $[A]_{r,e}$ is the raffinate (or extract) concentration in that phase that would have been obtained at the phases, reached equilibrium.

It is clear from inspection of Fig. 4.7 that the value of S_E is not changed by the phase on which the calculation is based. The difference between $[A]_{r,e}$ and $[A]_r$ is clear from Fig. 4.7. It is important to note that S_E varies with the choice of phase for calculation.

4.2. Efficiency in a Cascade

The concepts of Section 4.1 are readily extended to determine the efficiency of a cascade. For example, the results of calculations identical to those of Table 4.2 but with a 75% stage efficiency are given in Table 4.5. Comparison with the earlier data for equilibrium conditions shows that an extra stage is needed. In spite of this, the final raffinate is significantly higher than that was achieved at 100% extraction efficiency. The influence of the efficiency value is clearly greater if the solutions are more dilute.

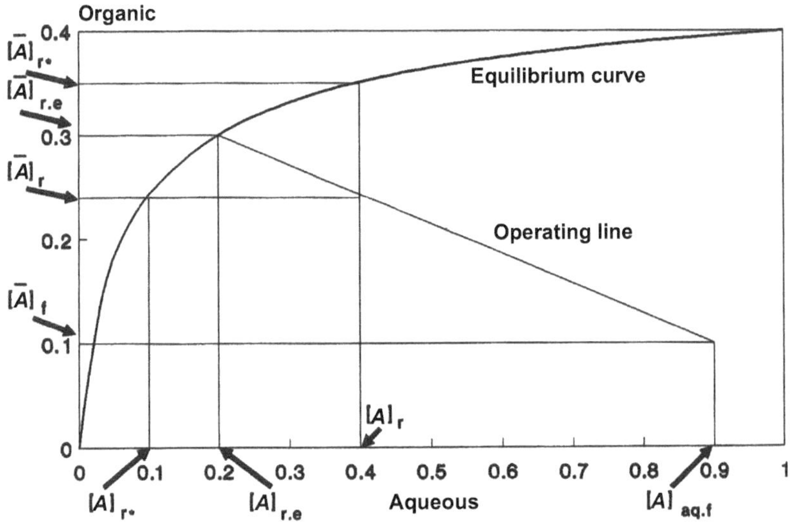

FIGURE 4.7 Stage efficiency calculations.

TABLE 4.5 Extraction of Component *A* with 75% Stage Efficiency

	Aqueous phase		Organic phase	
Stage no	Feed	Raffinate	Feed	Extract
1. Cross flow, phase ratio 1:1				
1	0.900	0.624	0.332	0.368
2	0.624	0.375	0.261	0.332
3	0.375	0.179	0.149	0.261
4	0.179	0.068	0.060	0.149
5	0.068	0.023	0.002	0.060
2. Countercurrent, phase ratio 2.5:1				
1	0.900	0.569	0.226	0.360
2	0.569	0.199	0.076	0.226
3	0.199	0.058	0.019	0.076
4	0.058	0.016	0.002	0.019

In multicomponent systems where there is mutual interference in extraction by several components, the efficiency shrinkage comes on top of the other reductions in the equilibrium curve, and for this reason there is stress in such systems on achieving high efficiency.

5. EQUIPMENT DESIGN FOR CONTINUOUS EXTRACTION-STRIPPING PROCESSES

The industry usually prefers continuous operations: it is generally simpler to control automatically, and it makes better use of labor. To obtain a large transfer area between two phases dispersion of one of the phases into drops have to be made during the mixing. It may be achieved in a construction of type of a gravitational column or a mechanically mixing apparatus. As its name suggests, the last comprises some means for mixing the two phases.

5.1. Dispersion and Mixing of the Phases: Types of Mixers

5.1.1. Agitated Extractors

In modern industry gravitational columns are practically not used because they are inefficient. Extraction techniques use mixing, stirred tanks, and settlers together in so-called mixer-settlers. The mixer usually consists of some form of impeller or propeller in a tank that usually contains some things to prevent the mixed liquids from swirling and thus reducing the efficiency of mixing. The simple stage can be calculated from the data, discussed above. Several stages are connected in series to mixer-settler batteries. Figure 4.8 illustrates two stages of a mixer-settler cascade.[3] The aqueous feed enters the first stage, where it is mixed with partially loaded extract from the second stage. The mixed phases pass to the settler, where they are separated under gravity. The fully loaded extract overflows an upper weir and passes to scrubbing and/or stripping. The first-stage raffinate overflows a lower weir and is pumped to the second stage, where it is mixed with the solvent feed. After settling again, the final raffinate passes to treatment before discharged as waste.

There is physicochemical criterion that is important in industrial mixing, and that is ensuring the correct phase is dispersed in the other. Sometimes mass transfer is more rapid if one phase is the dispersed phase rather than the other. Alternatively, the dispersed phase is chosen because, by definition, it will not contain droplets of the continuous phase. In this way the dispersed phase, after settling, will not entrain the continuous phase and entrainment losses from the settler will be reduced. Whatever the reason for choosing the dispersed phase, it is important to ensure that the mixer will keep that phase dispersed during operation, as changes in the dispersed phase, i.e., phase inversion, can cause considerable operating problems.

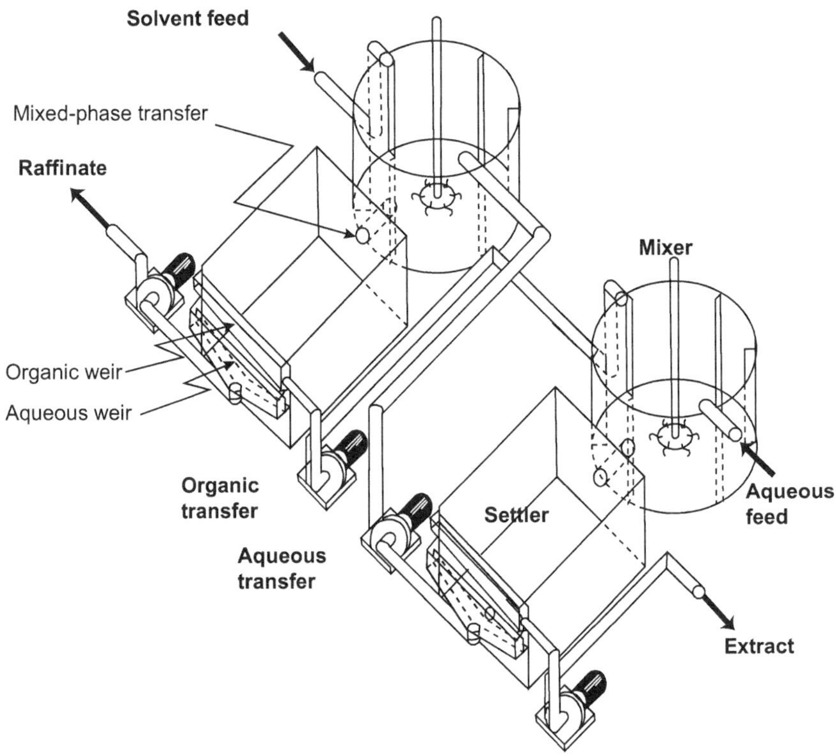

FIGURE 4.8 Layout of two stages of a mixer-settler cascade.

Figure 4.9 shows sections of some frequently used stirred countercurrent columns. All of them contain horizontal stator baffles that divide the column to successive sections. Each section has a mixer, all of which is fixed to a central shaft. Stirred extractors differ according to the kinds of mixers and of stator baffles. The mixers are dispersing one phase into droplets and the stator baffles are support to the separation of the liquids in each stage and provide the countercurrent of the liquids from stage to stage.

Usually, the continuous phase is the phase present in greater volume. It is possible to run for long periods with the greater volume phase dispersed, but phase inversion is always a risk in such circumstances. To overcome this risk, where it is desired that the lesser volume phase is continuous, then a portion of that phase may be recycled from the settler back to the same mixer to ensure that within each stage it is the greater volume, even if it is the lesser volume phase.

In large-scale operation, the volumetric flow of the phase to be dispersed is so large that it becomes necessary to disperse that phase into the mixed phases. Otherwise "blobs" of the dispersed phase will act locally as the continuous

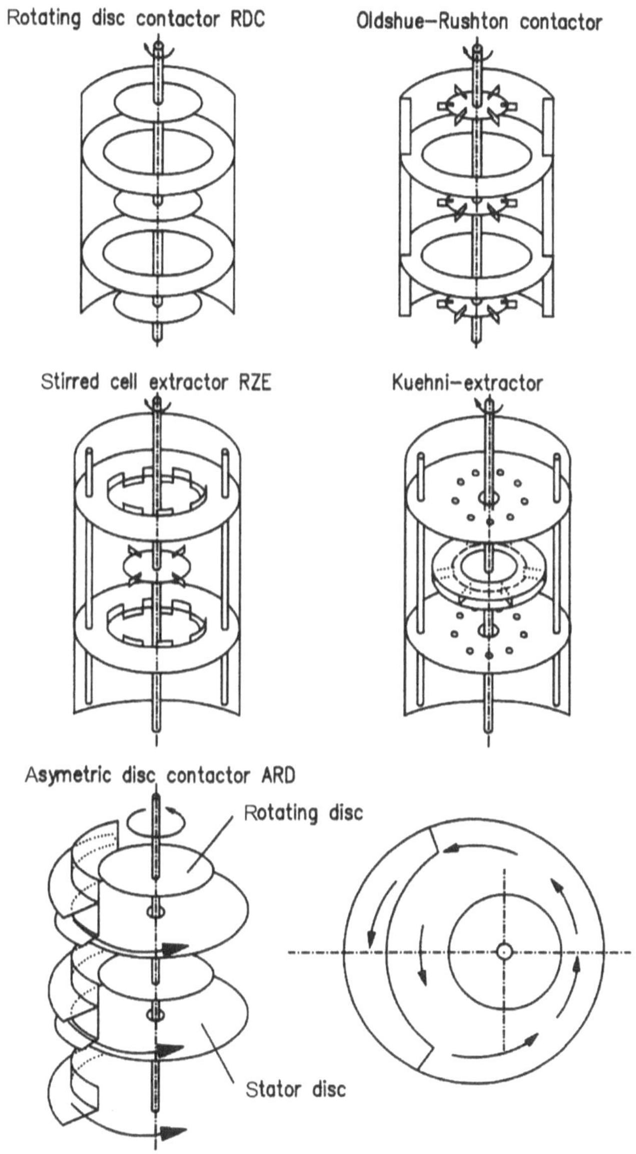

FIGURE 4.9 Sections of several types of stirred countercurrent columns.

phase, and the intended continuous phase will be dispersed in the blobs before the shear forces in the mixer break them up. This can lead to excessive entrainment losses.

Many other methods for mixing have found use. In a design offered by Lurgi,[4] the phases are mixed in an axial flow pump, and then passed down a relatively long pipe where the turbulence of flow keeps the phases mixed while the extraction takes place. In another design, the individual phases are pumped and then joined and passed through a static mixer. There are no particular physicochemical reasons for preferring one type of mixer to another. One mixer will work as well as another when they provide the following:

1. Adequate interfacial area for the extraction, without creating such small droplets that they will not settle efficiently.
2. Sufficient residence time for the desired degree of extraction to take place.

In some mixer-settler designs, the impeller is arranged both to mix the two phases and to provide the necessary energy to transfer the phases from one stage to the next, in which case it is known as a pump-mix mixer. The head required to move a phase from one stage to the next is small, so the impeller need not be efficient as a pump. Nevertheless, the design of impellers for the dual purpose of both mixing and pumping is more of an art than a science. Moreover, in full-scale operation it has been found difficult to start up cascades of pump-mix mixers and achieve equilibrium rapidly. Accordingly, this design has primarily found use in small-scale applications in the nuclear and pharmaceutical industries.

5.1.2. Pulsing Mixers

Pulsing means that the whole liquid content of a sieve tray column is continuously pushed up and down by a piston that moves to and fro (left side of Fig. 4.10) or the whole plate package is moved up and down (right side of Fig. 4.10). These types of columns show about the same efficiency if the plate geometry is identical. It also shows the principal interface in the upper part of the column so that the lighter phase is dispersed.

5.1.3. Centrifugal Mixer-Settler

Centrifugal extractors have cylindrical sieve trays arranged concentrically round the shaft. Sometimes they are called rotating columns. Popular industrial version is the Podbielniak-extractor, which was developed and used first for penicillin extraction.[5] It had a horizontal shaft.

In Fig. 4.11, a simple centrifugal extractor with a vertical shaft is presented.[6] The rotor is covered with a glass disk. It brings to processes with high-speed camera. Two phases are feeding into and away with the aid of rotating sealing devices at shaft. The heavy phase is fed at the rotor center and conveyed away by centrifugal force. The light phase is fed at the periphery of the rotor and displaced inside by the heavy phase. This causes a countercurrent flow of the phases. Controlling the value of the light-phase discharge, the degree of filling rotor with heavy and light phases can be defined (i.e., location of the interface).

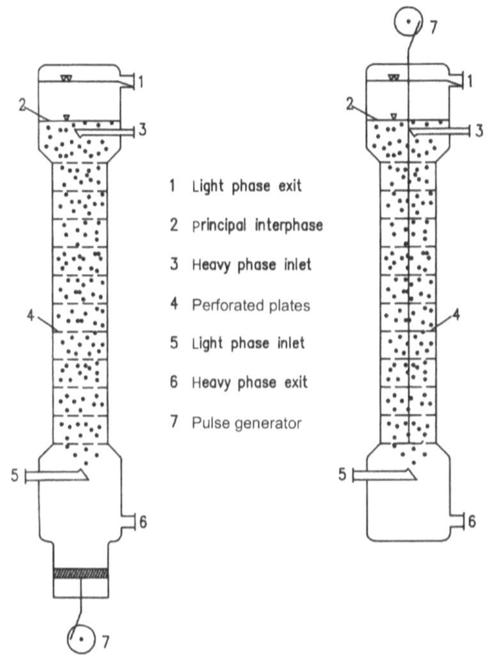

FIGURE 4.10 Sieve tray columns with pulsing of the whole liquid content (left) or of the whole plate package (right, Karr column).

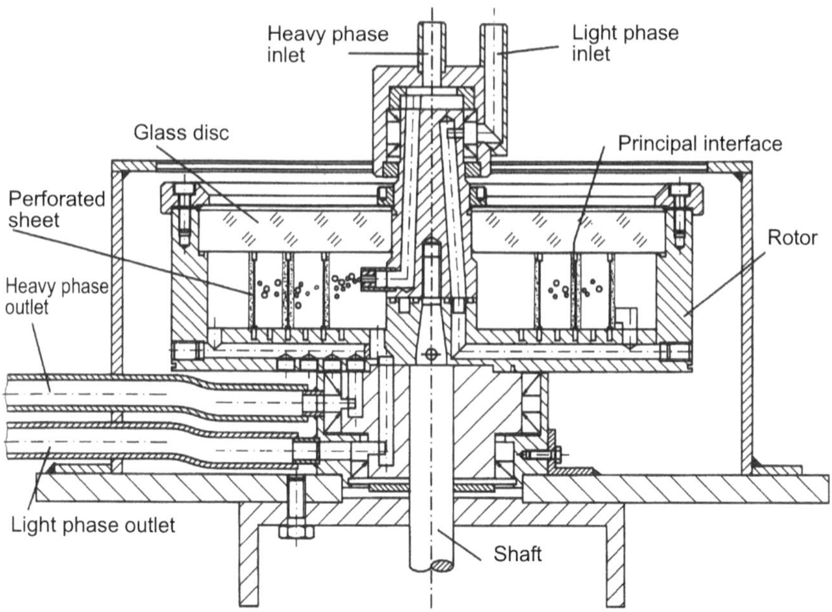

FIGURE 4.11 Experimental centrifuge with cylindrical dual-flow perforated sheets arranged concentrically round shaft (*Source: From reference* 6).

One more interesting example is the Robatel four-stage centrifugal extractor.[7] The heavy and light phases are introduced near the bottom and top, respectively. They flow countercurrently and leave the extractor at opposite ends. The extractor consists of a rotating bowl divided by baffles into horizontal compartments. The stationary central shaft carries mixing disks that run through each compartment and serve to mix and pump two phases to the settling chamber of the stage.

The flow or mixing rates and mass transfer of the phases for every type of extractors partly considered in the above sections but mainly are special and related more to mechanical engineering and not considered in this book.

5.2. Separation of the Phases

To separate the dispersed phases the settlers are used. Settlers tend to be less varied in their design. They typically comprise a relatively large, shallow tank, rectangular in shape, with an inlet for the mixed phases at one end and two outlets for the separated phases at the other. Various devices are used to introduce the mixed phases gently into the settler, and to control the flow of the mixed phases while they separate, but these do not change the basic principle of separation under gravity. The minimum value of the residence time τ of the drop phase in the settler is calculated from the volumetric flow rate of the dispersed phase, V_d:

$$\tau = \frac{V}{V_d} \tag{4.12}$$

where V is the settler volume.

Literature contains many proposals for modeling the coalescence and settling time.[8] But it is a very complicated task with many different conditions hinder the drainage. Especially surfactants, which occur in every technical extractant, cannot yet be evaluated by mathematical models. Surfactants may lead to a drastic increase of the coalescence time.

For a practical settler design the next considerations are used. The level of the heavy phase outlet within the settler controls the level of the interface (Fig. 4.12).

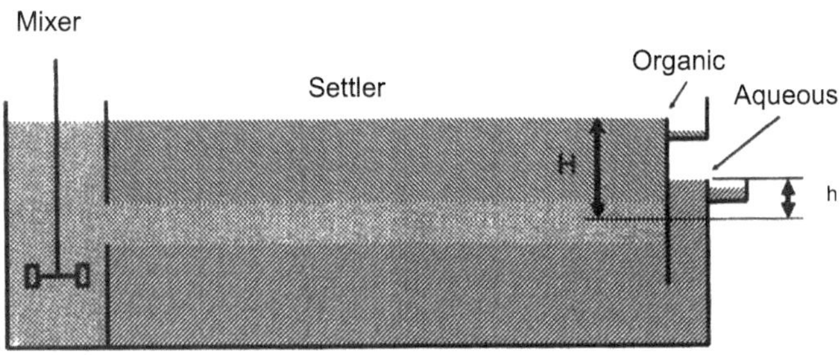

FIGURE 4.12 Control of interface level.

At the interface, the static pressure due to light phase above the interface is determined by the density of the light phase, ρ_0 and the depth of the light phase, H. Similarly, the static pressure above the interface in the overflow leg is determined by the density of the heavy phase, ρ, and the height h of the weir above the interface. The pressure must be the same at the same elevation, so $h \times \rho = H \times \rho_0$, or

$$H = h \times \left(\frac{\rho}{\rho_0}\right) \tag{4.13}$$

Because the difference between the densities of the two phases is often small, and may vary from one stage to the next, H can vary strongly with h. Thus, it is often necessary to make the height of the weir adjustable. Similarly, if there are significant differences in flow rate, then the depth of the liquid overflowing the weir has an effect on h and thus on the height of the interface. The importance of these considerations is that the shallower the settler, the more difficult the interface control. Shallow settlers are desired because they reduce the inventory of the solvent. However, it is possible to make the settler so shallow that the interface control can be lost. In one case, large, shallow settlers suffered from the effects of wind pressure, which caused such massive oscillations in the level of the interface that light phase often passed over the heavy phase weir.

Another way to determine the height H of the dense dispersion is by the simple formula:[9]

$$H = C\left(\frac{V_\mathrm{d}}{S}\right)^{\mathrm{y}} \tag{4.14}$$

where factors C and exponent y combine all unknown parameters and are determined by experiments in pilot settlers and S is the settler's cross-sectional area.

The introduction of the mixed phases into the settler has been found to be important if clean separation is to be ensured. "Picket fences," vertical plates set at an angle to the flow from the mixer into the settler, have been used to calm the flow and ensure the spreading of the mixed stream across the width of the settler. Various packings (Fig. 4.13) have been employed to aid settling, but under industrial conditions they are liable to clog with adventitious material or "third phases."

Baffles (Fig. 4.13) placed across the settler, of progressively lower height from entrance to overflow, have been employed to hold back the mixed phases to permit them to separate. The mixed phases will spread rapidly right across a settler unless there is a baffle to hold them back. There have been many reports of "wedges" of mixed phase in small-scale settlers. These are never seen in industrial practice because considerations of pressure at a given point, as were used above to determine the height of the interface, show that a wedge is inherently unstable.

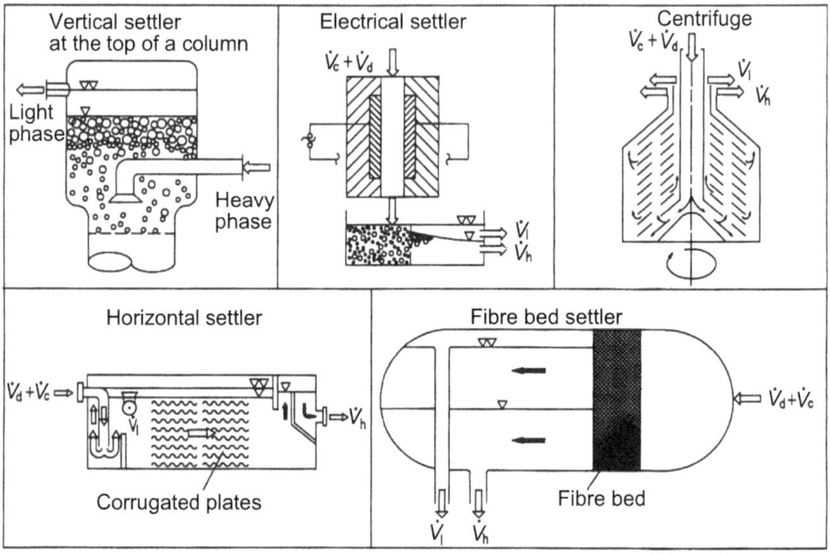

FIGURE 4.13 Settlers for separation of liquid–liquid dispersions; V_h, volume flow of heavy phase; V_l, volume flow of light phase (*Source: From reference* 10).

In one mixer-settler design (Fig. 4.13), the mixed phases flow down a shallow trough placed over the settler, which gives them an opportunity to coalesce and separate before entering the settler. In this way, the capacity of the settler is markedly increased, with a concomitant reduction in the inventory of solvent required for a given duty.

Electrical fields and centrifuges are used to support coalescence in special cases.[10,11]

The manner in which individual mixer-settler stages can be linked together to form countercurrent cascades is illustrated in Fig. 4.8. If each stage is on the same level, then some form of pump must be provided to move each phase from one stage to the next. As indicated earlier, it is sometimes convenient to use the mixer for this duty. Another arrangement has the individual stages set at different elevations, so that one phase (usually the phase with the greatest flow rate) can gravitate from one stage to the next. In circular cross-section mixers, this usually takes the form of vertical baffles mounted on the wall of the mixer.

5.3. Differential Extraction

There is no need to operate in countercurrent cascades. Consider, e.g., pumping the heavy phase to the top of a tower and letting it flow down against drops of the light phase rising upward. Assuming transfer from the heavy to the light phase if the tower were high enough, at the top there would be saturated light phase and at the bottom, depleted heavy phase. Clearly, the concept of a stage has no meaning in such a tower. Instead, we deal with differential transfer units, which are a measure of the change in concentration per unit of difference in concentration (recall that the rate of extraction is determined largely by the difference between the actual and the equilibrium concentration of a solute, or "driving force"). At each point in the tower, a component A has an actual concentration $[A]$ and an equivalent equilibrium concentration $[A]_e$ (Fig. 4.14). Then the *number of transfer units (NTU)* required for the extraction is given as first approximation.[1]

$$\mathrm{NTU} = \int_{[A]_r}^{[A]_f} \frac{\mathrm{d}A}{([A] - [A]_e)} \tag{4.15}$$

where $[A]_f$ and $[A]_r$ are the feed and raffinate concentrations of A, respectively.

Driving force $[A] - [A]_e$ of extraction process in the differential tower is presented schematically in Fig. 4.14.

This integral clearly depends on the slope of the operating line. If the operating line approaches closely to the equilibrium curve, then the driving force approaches zero and the inverse becomes very large. (The same as in the case with stagewise systems.) That is, when the operating line is close to the equilibrium curve and there is a pinch point, then the NTU becomes large.

Equation (4.15) refers to a single phase, and $[A]_e$ is related to $[A]$ at each point via the phase ratio (or operating line, which is the same). The NTU could be calculated for the extract phase instead of the aqueous phase, i.e., either from the difference between $[A_R]$ and $[A_R]_e$ or from that between $[A_E]_e$ and $[A_E]$. In simple cases the equation may be integrated directly. Otherwise, it is necessary to perform the integration numerically or graphically.

Once the NTU has been found, the height of the tower is determined. The height of each transfer unit (HTU) is determined by physical parameters such as

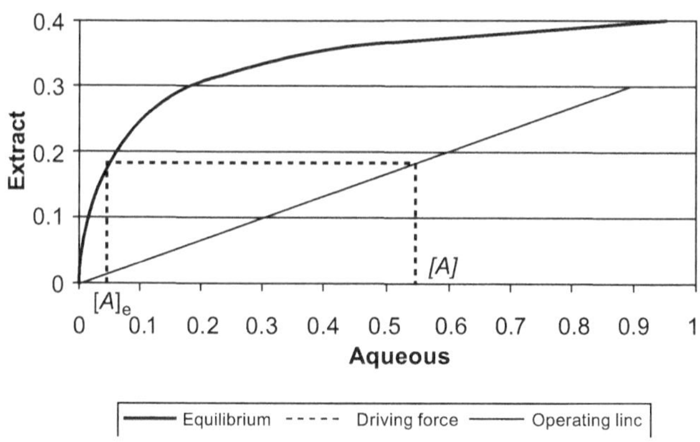

FIGURE 4.14 Illustration of the determination of the driving force from equilibrium and operating lines.

the droplet size, the flow patterns in the tower, and the effect of any packing. These parameters affect the rate of mass transfer.

Very often the rate of mass transfer cannot be estimated from first principles. In this case, it is necessary to estimate the height by determining the NTU achieved and then dividing the actual height of the column employed by the NTU, i.e.:

$$\mathrm{HTU} = \frac{H}{\mathrm{NTU}} \tag{4.16}$$

where H is the height of the column employed.

HTU is subjected to the effects of both radial and axial mixing, and these are not readily quantified, so scale-up of columns of this kind is often not based on fundamentals, but rather on correlations determined from detailed studies of several systems in the particular design of the column chosen.

Physically, towers designed for countercurrent contact can be open, but more usually contain some form of packing or plates. The material of the packing is chosen so that one phase wets it preferentially, thus increasing the surface area for mass transfer. Similarly, the plates are designed to break up droplets and increase the surface area. In addition, the contents of the tower may be agitated either by an internal agitator or by pulsing the fluids. The energy imparted by agitation or pulsation breaks up the droplets of the dispersed phase.

When the equilibrium curve is relatively linear, the driving force does not vary greatly down the length of the column, and the NTU approaches the number of McCabe–Thiele theoretical stages. In this case, it is reasonable to speak of the number of stages in the column, and to calculate a *height equivalent to a theoretical stage (HETS)*. However, if the equilibrium curve and the operating line are far from parallel, the number of theoretical stages becomes a poor measure of the column's performance, and the NTU should be used.

6. SOLVENT LOSSES

All considerations through above theoretical chapters were based on assumption that the two phases, the aqueous and the organic, are totally immiscible. For many systems this is a reasonable approximation, but for some systems mutual miscibility must be taken into account, particularly when the primary solute is organic.

The basic principles remain unchanged: the primary difference is the choice of a consistent basis for calculation, such as a solvent-free basis. Graphic techniques based on triangular coordinates provide approximate answers, but modern computational techniques are preferred.

Some consideration should be given at this point to prevent loss of the organic phase in the aqueous raffinate. This loss can arise by either solubility in the aqueous phase or by entrainment of droplets not fully settled. The solvent lost in this way can offer a finite environmental hazard and be an economic cost on the process.

Clearly, the primary duty is good engineering practice. Often, however, additional security is provided in the following forms:

1. Additional settler capacity for final raffinate.
2. Extraction of residual organic phase using a third diluent, from which it is later separated, typically by distillation.
3. Coalescence on a solid wetted preferentially by the organic phase.
4. Flotation with an air in the presence of surfactants.

7. ECONOMICAL CONSIDERATIONS

The need for pilot plant operations is considered by one school of thought to be an unnecessary expense in the development of a solvent process. The rationale here is that money and time spent in piloting the process can be used to modify the full-scale plant, if necessary. Thus, if few (or no) modifications are required, considerable savings can result. This approach is probably satisfactory if the process is very similar to an existing process and sufficient data are available on which to base the plant design and operation. The practical experience of the plant design engineers and operators is also of considerable importance in such cases. On the other hand, problems encountered in a plant operation that has been designed without piloting the process could result in the loss of considerable time and expense. Of course, almost any process or plant can be made to work, even though the basic data and design are poor. The costs, however, may be astronomical. It should be pointed out, however, that the size of a pilot plant can vary from a few gallons per minute of total throughput to several hundred gallons per minute. Normally, one would expect that the size of a pilot plant will be in direct proportion to the size of the final plant.

8. PROBLEMS WITH SCALE-UP TO INDUSTRIAL SYSTEMS

From the previous sections it is evident that the experimental results have to be proven with pilot experiments. Pilot extractors are usually much smaller than large-scale industrial equipment. For example, pulsed columns for pilot experiments have diameter up to 30–250 mm, whereas industrial columns have up to 2500–3000 mm and more.[12] Can pilot data be used for the development of industrial apparatuses? This is the problem for engineers.

It depends much on the experience of the developer, its careful analysis obtained from laboratory and pilot experimental data. Nevertheless, some preliminary calculations have to be supplemented by the large-scale experiments, including the following:

1. Determination of solubility in triangular diagram and their dependences on pH, temperature, and salt content of the raffinate.

2. Determination of the ratio of solvent to raffinate flow rate.
3. Measuring the separation time to evaluate the coalescence behavior.
4. Measurement of the physical properties of the feed and product phases.

Obtained data have to be carefully analyzed comparing with pilot and laboratory experimental results.

In reference [13] original decision was proposed: a steady-state simulator for industrial-scale processes. It performed well during process development, design, and optimization. The main problems were partial mass transfer (nonideal stages), backmixing caused by entrainment, and decrease in solvent activity due to contamination or change in the concentration of the active ingredients. The effect of the entrainment is introduced directly into the mathematical model. The differences between the theoretical and real data were taken into account when calculations and deviations were expressed through stage efficiency coefficient. Either the Murphree efficiency or the mass transfer factor (MTF) was used. The entrainment is measured directly in every stage. The stage efficiency coefficient is calculated periodically from plant data. All the data are fed as input to the simulator. With the modifications, the simulator was used successfully for the $CaBr_2$ extraction from the sea water brine of the Dead Sea plants.

REFERENCES

1. Lloyd FJ. Principles of industrial solvent extraction. In: Rydberg J, Cox M, Musikas C, Choppin GR, editors. *Solvent extraction principles and practice*. New York, Boca Raton: Society of Chemical Industry; 2004.
2. Lo TC, Baird MHI, Hanson C, editors. *Handbook of solvent extraction*. New York: Wiley-Interscience; 1983.
3. Warwick GCI, Scuffham JB. The Davy–McKee mixer–settler. In: Lo TC, Baird MHI, Hanson C, editors. *Handbook of solvent extraction*. New York: Wiley-Interscience; 1983. p. 287–97.
4. Muller E, Stonner HM. In: Lo TC, Baird MHI, Hanson C, editors. *Handbook of solvent extraction*. New York: Wiley-Interscience; 1983. p. 311–8.
5. Treybal RE. *Liquid extraction*. 2nd ed. New York: McGraw-Hill; 1963.
6. Otillinger F, Blass E. *Chem Eng Technol.* 1988;**11**:312.
7. Brunner KH, Hemfort H. *Adv Biotechnol Processes* 1988;**8**:1–50.
8. Hartland S, Jeelani SAK. *Chem Eng Sci* 1987;**42**:1927.
9. Ritcey GM. *Solvent extraction: Principles and applications to process metallurgy*. Elsevier; 2006.
10. Blass E, Lobman A, Meon W, Rommel W. *Chem Eng Technol* 1989;**61**:597.
11. Rebelein F, Ein Beitrag zur Tropfenabscheidung aus Flussigkeiten mit Hilfe von Faserbetten. Doctoral Thesis, TU Minchen, 1989.
12. Blass EF. Engineering design and calculation of extractors for liquid-liquid systems. In: Rydberg J, Cox M, Musikas C, Choppin GR, editors. *Solvent extraction principles and practice*. New York, Boca Raton: Society of Chemical Industry; 2004.
13. Grinbaum B, Angel SA. How to fit a simulator to a real liquid-liquid extraction plant. *Process Metallurgy* 1992;**7B**:1235–40.

Chapter 5

Examples of Application of Solvent Extraction Techniques in Chemical, Radiochemical, Biochemical, Pharmaceutical, Analytical Separations, and Wastewater Treatment

Chapter Outline

Solvent Extraction. DOI: 10.1016/B978-0-444-53778-2.10005-6

1. INTRODUCTION

Solvent extraction theory and the engineering design aspects of solvent extraction processes have been considered in the previous chapters. This chapter was devoted to the examples of application of the process in laboratory, scaleup studies and commercial applications for specific separations in different fields of practice.

The initial bench-scale experimental work in solvent extraction processes is conducted for determination of main process parameters (the temperature, flow rate, acidity, etc.) using small apparatus such as separator funnels. Following the successful completion of these tests, when the best reagent and other conditions for the system have been established, small-scale continuous operations are run, such as in a small mixer-settler unit. The data so obtained are used to determine scaleup factors for pilot plant or plant design and operation.

Pilot plant operations are considered from a technological and economic point of view. If few (or no) modifications are required, considerable savings can result on scaleup. On the other hand, problems encountered in a plant operation that has been designed without piloting the process could result in the loss of considerable time and expense. It should be pointed out that the size of a pilot plant can vary from a few gallons per minute of total throughput to several hundred gallons per minute. Normally, one would expect that the size of a pilot plant will be directly proportional to the size of the final plant.

In this chapter examples, published during the last decade, since the publication of the last books in the field[1,2] of specific solvent extraction processes in the laboratory and/or scaleup practice are introduced.

2. EXTRACTION IN HYDROMETALLURGY (METALS EXTRACTION)

The use of solvent extraction in hydrometallurgy extends to a wide range of metals from different feeds including low-grade ore, scrap, waste leachates, and dilute aqueous solutions. The technology was used first in Nuclear Technology (see Section 3). Then, some high-value metals (e.g., precious group metals, PGM) solvent extraction technologies became commercially viable and were developed and used. After development of selective chelating acidic reagents (extractants), the solvent extraction was able to compete with classical separation–concentration operations in hydrometallurgy.

Liquid–liquid extraction of metals is widely used in analytical chemistry for preparation and concentration of samples, which is discussed in Section 5. The theory of metal extraction was analyzed in detail in Chapter 3, and novel vision in Chapter 9. In this section, some recent publications in the field are considered.

2.1. Metal Ion–Extractant Complexes Formation

1. Metal cations: Cu^{2+}, Ni^{2+}, Co^{2+}, etc.
2. Complex metal anions, e.g., $[UO_2(SO_4)_3]^{4-}$, $[Mo_8O_{26}]^{4-}$ and others
3. Complex metal cations, e.g., $[MoO_2]^{2+}$
4. Neutral metal species such as $UO_2(NO_3)_2$

2.2. Types of Metal Extractants

2.2.1. Chelation Extractants

$$2RH + [M^{2+}\ SO_4^{2-}] = R_2M + [2H^+\ SO_4^{2-}]$$

Properties, advantages, and disadvantages are as follows:

1. They operate on a hydrogen ion cycle.
2. Some operate well with both acid and ammonia leach solutions.
3. They tend to be more selective than either ion-pair or organic acid extractants.
4. They tend to be kinetically slower than organic acid and ion-pair extractants.
5. They have good operating characteristics with respect to phase separation (see Table 5.1).
6. They are usually more difficult to produce.

Most common application is in the extraction of copper (over 3 million tons annually), Ni, and Pd.

2.2.2. Organic Acid Extractants

$$4(RO)_2\,PO_2\,H_{org} + Zn^{2+} = [(RO)_2PO_2]_2\,Zn[(RO)_2PO_2H]_{2org} + 2H^+$$

Properties, advantages and disadvantages are as follows:

1. They show less selectivity than chelating extractants.
2. They can promote stable emulsions with basic solutions.
3. They operate on a hydrogen ion cycle.
4. Their kinetics of extraction are usually quite rapid.
5. Some of these reagents are easy to produce and some are difficult. Highly variable selling prices depending on the reagent.

Most significant application is in the extraction of Zn and Co (see Table 5.2).

2.2.3. Ligand Substitution Extractants

$$[PdCl_4^{2-}] + 2R_2S_{org} = [PdCl_2(R_2S)_2]_{org} + 2Cl^-$$

Properties, advantages, and disadvantages (see Table 5.3) are as follows:

1. Metal extraction is not normally sensitive to pH.
2. At times remarkable selectivity can be achieved.
3. Kinetics can be rapid or slow depending on the metal and the ligand.
4. These reagents can be relatively inexpensive or very expensive depending on the complexity.

2.2.4. Neutral or Solvating Extractants

$$[2(RO)_3PO]_{org} + [UO_2(NO_3)_2]_{aq} = [UO_2(NO_3)_2\,2(RO)_3PO]_{org}$$

Strength: $R_3PO > (RO)_3PO > R_2CO > ROH > R_2O$:

Trialkylphosphine oxides > trialkylphosphates > ketones > alcohols > ethers

TABLE 5.1 Commercial Chelating Metal Extractants

Reagent class	Structure	Extractants
α-Hydroxyoximes	$R^1C(NOH)CH(OH)R^2$	5,8-Diethyl-7-hydroxydodecan-6-oxime (LIX63)
β-Hydroxyaryloximes	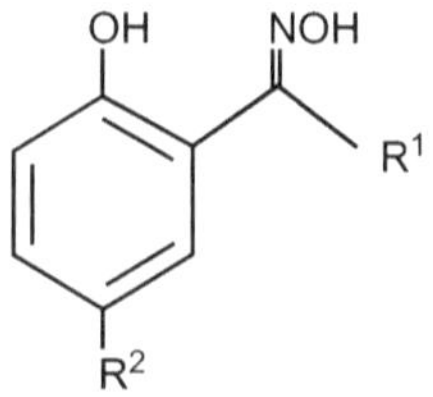	
R^1 = phenyl	$R^2 = C_9H_{19}$ LIX65 N + LIX63(1%)	LIX65 N LIX64 N
$R^1 = CH_3$	$R^2 = C_9H_{19}$	LIX84
R^1 = H	$R^2 = C_9H_{19}$	P1
	P1 + nonyl phenol	P5000 series
	P1 + tridecanol	PT5050
	P1 + ester modifier	M5640
	$R^2 = C_{12}H_{25}$	LIX860
	LIX84 + LIX860	LIX984
	LIX980 + tridecanol	LIX622
8-Quinolinol	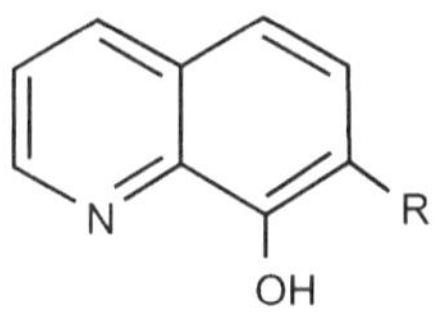	
R = 5,5,7,7-tetramethyl-1-octenyl- (pre-1976) = 4-ethyl-1-methyloctyl- (post 1976)		Kelex 100
R = unknown side chain		Kelex 108
R = unknown saturated alkyl		LIX26
β-Diketones	$R^1COCH_2COR^2$	
	$R^1 = R\text{-}C_6H_5$ $R^2 = CH_3(CH_2)_5$	LIX54

TABLE 5.2 Commercial Acidic Metal Extractants

Reagent class	Structure	Extractants
Carboxylic acids	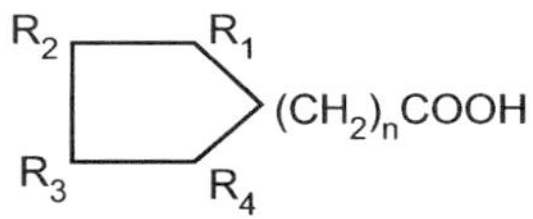	Naphthenic acids
	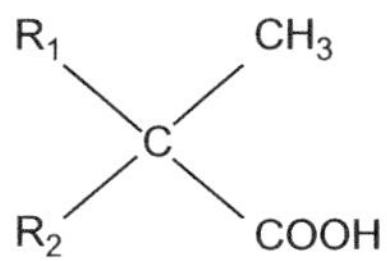	Versatic acids
		$R_1 = R_2 = C_5$ = Versatic 10
		$R_1 = R_2 = C_4{-}C_5$ Versatic 911
Phosphorus acids	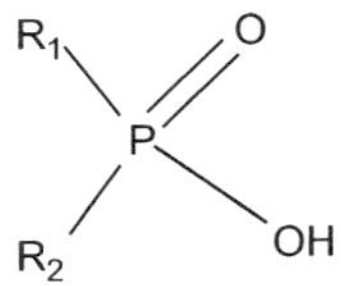	
Phosphoric acids	$R_1 = OH$; $R_2 = C_4H_9CH(C_2H_5)CH_2O$	Mono-2-ethylhexylphosphoric acid
	$R_1 = R_2 = C_4H_9CH(C_2H_5)CH_2O$	Di-2-ethylhexylphosphoric acid (DEHPA)
	$R_1 = R_2 = p\text{-}CH_3(CH_2)_6CH_2C_6H_5O$	Di-*p*-octylphenylphosphoric acid (OPPA)

Phosphonic acids	$R_1 = C_4H_9CH(C_2H_5)CH_2$; $R_2 = C_4H_9CH(C_2H_5)CH_2O$	2-Ethylhexylphosphonic acid mono-2-ethylhexyl ester (PC88A), (P-507)
Phosphinic acids	$R_1 = R_2 = C_4H_9CH(C_2H_5)CH_2$	Di-2-ethylhexylphosphinic acid (P-229)
	$R_1 = R_2 = CH_3(CH_2)_3CH_2CH(CH_3)CH_2$	Di-2,4,4-trimethylpentyl-phosphinic acid (CYANEX 272)
Thiophosphorus acids	$R_1R_2P(S)OH$	
	$R_1 = R_2 = CH_3(CH_2)_3CH_2CH(CH_3)CH_2$	Di-2,4,4-trimethylpentyl-monothiophosphinic acid (CYANEX 302)
	$R_1R_2P(S)SH$	
	$R_1 = R_2 = CH_3(CH_2)_3CH_2CH(CH_3)CH_2$	Di-2,4,4-trimethylpentyl-dithiophosphinic acid (CYANEX 301)
Sulfonic acid	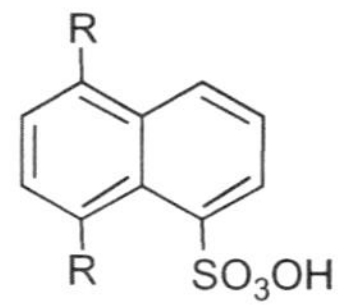 $R = C_9H_{19}$	5,8-Dinonylnaphthanenesulfonic acid DNNSA

TABLE 5.3 Commercial Solvating Metal Extractant

Reagent class	Structure	Extractants
Carbon–oxygen extractants		
Amides	$R^1CONR^2_2$	
	$R^1 = CH_3$	N503
	$R^2 = CH_3(CH_2)_5CH(CH_3)$	
	$R^1 = R^2$ unknown	A101
Ethers	$CH_3(CH_2)_3OCH_2CH_2OCH_2$-$CH_2O(CH_2)_3CH_3$	Dibutylcarbitol (Butex)
Ketones	$(CH_3)_2CHCH_2COCH_3$	Methyl isobutylketone (MIBK)
Phosphorus–oxygen extractants		
Phosphorus esters	$R^1R^2R^3PO$	
	$R^1,R^2,R^3 = CH_3(CH_2)_2CH_2O$	Tri-*n*-butylphosphate (TBP)
	$R^1,R^2 = CH_3(CH_2)_2CH_2O$, $R^3 = CH_3(CH_2)_2CH_2$	Dibutylbutylphosphonate (DBBP)
	$R^1,R^2 = CH_3O$, $R^3 = CH_3$	Dimethylmethylphosphonate
Phosphine oxides	$R^1R^2R^3PO$ $R^1,R^2,R^3 = CH_3(CH_2)_6CH_2$	Tri-*n*-octylphosphine oxide (TOPO) CYANEX 921
Phosphorus–sulfur extractants		
Phosphine sulfides	$R^1R^2R^3PS$	
	$R^1,R^2,R^3 = CH_3(CH_2)_6CH_2$	CYANEX 471X
Carbon–sulfur extractants	R^1R^2S	
	$R^1,R^2 = C_6H_{13}$	Dihexylsulfide
	$R^1,R^2 = C_7H_{15}$	Diheptylsulfide

Properties, advantages, and disadvantages are as follows:

1. Normally exhibit fast extraction and stripping kinetics.
2. Extract neutral metal complexes.
3. In general are not selective.
4. In combination with other extractants, can act as synergists, antagonists, or phase modifiers (see Table 5.4).

TABLE 5.4 Commercial Ion-Pair Metal Extractants

Reagent class	Structure	Extractants
Primary amines	RNH_2	Primene JMT
	$R = (CH_3)_3C(CH_2C(CH_3)_2)_4$	
Secondary amines	R^1R^2NH	
	$R^1 = C_9H_{19}CH = CHCH_2$	
	$R^2 = CH_3C(CH_3)_2(CH_2C(CH_3)_2)_2$	Amberlite LA-1
	$R^1 = CH_3(CH_2)_{11}$	
	$R^2 = CH_3C(CH_3)_2(CH_2C(CH_3)_2)_2$	Amberlite LA-2
	$R^1,R^2 = CH_3(CH_2)_{12}$	Adogen 283
	$R^1,R^2 = CH_3CH(CH_3)CH_2(CH_2)_6$	HOE F2562
Tertiary amines	$R^1R^2R^3N$	
	$R^1,R^2,R^3 = CH_3(CH_2)_7$	Trioctylamine
		Alamine 300
	$R^1,R^2,R^3 = C_8{-}C_{10}$ mixture	Alamine 336
		Adogen 364
		Hostarex A 327
	$R^1,R^2,R^3 = (CH_3)_2CH(CH_2)_5$	Tri-isooctylamine
		Adogen 381
		Alamine 308
		Hostarex A 324
	$R^1,R^2,R^3 = (CH_3)_2CH(CH_2)_7$	Tri-isodecylamine
		Alamine 310

(Continued)

TABLE 5.4 Commercial Ion-Pair Metal Extractants—cont'd

Reagent class	Structure	Extractants
	$R^1,R^2,R^3 = CH_3(CH_2)_{11}$	Adogen 363
		Alamine 304
	$R^1,R^2,R^3 = CH_3(CH_2)_{12}$	Adogen 383
	$R^1,R^2,R^3 = C_{28}H_{57}$	Amberlite XE204
	$R^1 = CH_3(CH_2)_7$, $R^2 = CH_3(CH_2)_9$	
	$R^3 = CH_3(CH_2)_{11}$	Adogen 368
Quaternary amines	$R^1R^2N(CH_3)^+Cl^-$	
	$R^1,R^2,R^3 = C_8–C_{10}$ mixture	Aliquat 336
		Adogen 464
		HOE S 2706

2.2.5. *Ion-Pair Extractants*

$$[R_3N]_{org} + [HL]_{aq} = [R_3NH^+L^-]_{org}$$

$$a[R_3NH^+L^-]_{org} + [M^{z+}X^{a-}_{z+a}]_{aq} + [(R_3NH)_aMX_z]_{org} + a[L^-]_{aq}$$

Properties, advantages, and disadvantages are as follows:

1. Amines extract anionic metal complexes.
2. Extraction with amines usually takes place from acidic to near-neutral solutions with the best pH range dependent upon the specific system.
3. Generally, amines require the use of aromatic diluents and/or require that a phase modifier, normally a long chain alcohol, be added to the hydrocarbon diluent in order to prevent the formation and precipitation of organic insoluble amine acid or amine metal complexes.
4. The kinetics of extraction and stripping are usually rapid.
5. Quaternary and tertiary amines are more selective than secondary or primary amines.
6. There are a wide variety of amines that are available and/or can be synthesized (see Table 5.4).

2.3. Modifiers

1. Organic compounds with the ability to donate and/or accept a hydrogen bond which are soluble in the diluent. Alcohols, esters, ketones, alkylphenols, phosphine oxides, phosphate esters, ethers, and nitriles.
2. When used with amines, modifiers help solubilize the amine acid and amine metal ion pair complexes
3. When used with oximes for copper extraction, modifiers allow the oxime to be stripped with less acid.

Modifiers that accept a hydrogen bond tend to increase stability of the oxime and also increase selectivity over iron. Modifiers that donate a hydrogen bond to an acceptor have the opposite effect.

2.4. Synergistic Mixtures

A mixture of D2EHPA with the oxime Reagent LIX 63 gives a blend that extracts certain metals at lower pH than either extractant by itself. This is the only example of many reagent combinations that show a synergistic behavior. For many other examples, see Preston (1983)[6,7] The LIX 63/D2EHPA combination was investigated as a Ni extractant but the poor stability of the LIX 63 to hydrolysis made the combination too costly.

Kinetics can be accelerated by the addition of phase transfer catalysts, e.g., protonated amines.

Synergistic mixtures are likely to be the future for "new reagents" instead of actually developing expensive reagents that must then go through the regulatory process. Synergism optimization is possibly difficult to control because of the differential losses, and the incentive will be to develop and optimize a mix for a particular deposit.

2.5. Diluents

1. Solubilize the extractant and metal–extractant complex.
2. Have a low viscosity and density in order to assist phase separation and reduce entrainment.
3. Should be free of objectionable components to minimize crud formation.
4. Should be chemically stable to circuit conditions.
5. Should have a flash point 25 °C or higher above the highest operating temperature in the SX circuit.
6. Should be insoluble in the aqueous phase.
7. Must have low evaporation losses.
8. Should not be detrimental to metal extraction/stripping chemistry.
9. Should be readily available at a low cost

2.6. Copper Extraction

2.6.1. Laboratory Scale Research

Hydroxyoxime Copper Extractants[8]

Solvent extraction of copper with acidic extractants (see Table 5.1) met some problems with separation of iron(III) and some other heavy metals. In common, selectivity of some heavy metals, extracted by acidic extractants may be considered from Fig. 5.1. Development of chelating extractants (see Table 5.2) led to the design of the commercial reagents specifically for copper extraction. These were an aliphatic α-hydroxyoxime (LIX 63, Henkel) and β- hydroxyoxime based on benzophenone (LIX 64). LIX 63, with pH_{50} values of. 3–4, was unable to extract copper from commercial leach liquors without alkali addition, also this reagent extracted iron(III) at lower pH values than copper. These disadvantages were largely overcome by LIX 64 that extracted copper selectively from iron in the pH range 1.5–2.5, but with slow kinetics of extraction.

It became evident that increasing the extractant strength increases the acid strength required for stripping, which reduces the overall copper transfer when interfaced with a normal tankhouse spent electrolyte. Two approaches have been proposed to overcome these difficulties. ICI with their Acorga reagents have showed that addition to the hydroxyoxime of nonyl phenol (P5000 series)

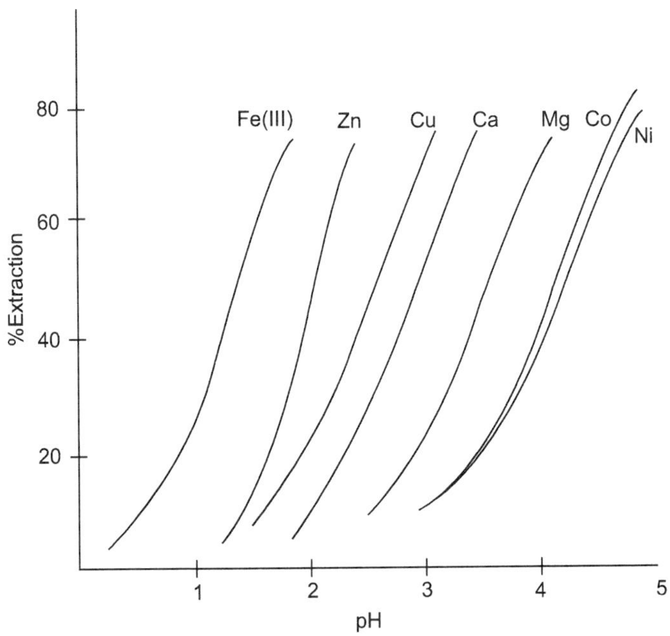

FIGURE 5.1 pH dependence of copper ion by DEHPA. *(Source: From Lo et al.[4])*

or tridecanol (PT5050) improves the stripping performance, with little effect on extraction.[9,10] The M series of Acorga), produced by adding modifiers such as long-chain alkyl esters, improved performance by reduction of entrainment and crud formation. Second approach was developed by Henkel, which involves a 1:1 mixture of an aldoxime and ketoxime, combining the extractive strength, selectivity, and fast kinetics of the akloxime with the proved hydrolytic stability and stripping performance of the ketoximes. These were second generation of commercial copper extractants.

Quinoline-Based Copper Extractants[11]

These are Kelex 100 (Sherex) and LIX 34 (Henkel), extractants derived of 8-quinolinol and 8-alkylarylsulfonamidoquinoline, respectively. They were developed for selective separation of copper from iron(III). The separation was based on different extraction rate of copper and iron. Separation factor larger than 1,000 were obtained. They have an advantage to extract copper from its high concentration and high acid concentration feeds in comparison with hydroxyoximes. Disadvantages of these extractants are easy protonation (taking acid at stripping) and high cost that are overweighed for commercial use.

β-Diketone Copper Extractants

These extractants, e.g., LIX 54, cannot compete with hydroxyoximes for extraction from acidic media but have some advantages at extraction from ammonia leach solutions (they do not transfer ammonia). The stripping is possible with low acidity and high copper concentration solutions. Using LIX 54 copper extraction from printed circuit board etch liquors was commercialized successfully.

Solvating, Nitrogen-Based Extractants

Chloride hydrometallurgy led to the interest in extractants, operating with high selectivity and a wide range of metals and metalloide but without pH control. An extractant, E, based on a pyridine-carboxylic ester derivative, readily extract

$$CuCl_2 + 2E_{org} + CuCl_2E_{2\ org} \tag{5.1}$$

An example of this kind extractant is Acorga CLX 50. Unlike the amine extractants, this compound does not protonate easily and provides excellent selectivity for metals, existing as chloroanions in hydrochloric acid solutions. But extraction becomes significant at more than 8 M chloride concentration.[12]

2.6.2. Industrial Scale Copper Extraction

Addition of a small amount of LIX 63 to the LIX 64 solved the problems arising with the LIX 63. The mixture, LIX 64 N, for many years was the extractant of

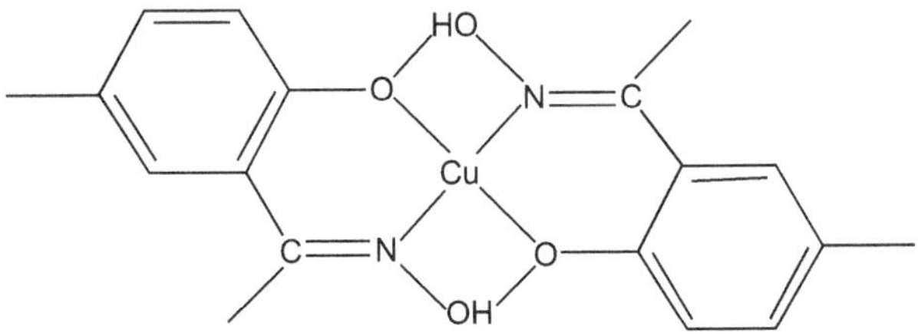

FIGURE 5.2 Structure of bis(5-methylsalicylaldoximato)copper(II).

choice for commercial copper extraction from acid solution. A thermodynamic model, using a computer program, has been developed to determine the extraction and stripping isotherms of copper. Developed by reagent suppliers, the program is distributed to their customers. Similar programs are used by Cognis and Cytec.[5] The extracted complex had a 2:1 oxime/metal stoichiometry, with the two oxime molecules linked by hydrogen bonds to give an overall planar configuration (see Fig. 5.2). The copper could be stripped with sulfuric acid (150–200 g/L). Commercial success of extractants lead to development of modifications: LIX-65 N LIX-70, SME 529 Shell (methyl chain), P50 ICI Acorga, which improved both efficiency and kinetic performance.[1–4] These references contain detailed descriptions of flowsheets and operational arrangements of commercial copper solvent extraction plants.

The second-generation reagents used in the important copper extraction plants includes PhelpsDodge, in New Mexico and Arizona, Codelco-El in Chile, Roxby Downs in Australia and others.

2.7. Cobalt–Nickel Solvent Extraction[13,14]

Cobalt and nickel are very similar in chemical behavior, which causes problems in their separation. Any separation based on solvent extraction will be difficult unless some differences in chemical behavior can be exploited. Such differences include a higher rate of water exchange of the hexahydrated divalent cation for cobalt than for nickel, which might result in the development of a kinetic, rather than thermodynamic, separation. In addition, although the redox potentials of cobalt(II)–cobalt(III) and nickel(II)–nickel(III) couples are similar for the hexa-aquoion, on complexation, cobalt(II) is much more readily oxidized than nickel(II). Also, once formed, the cobalt(III) complexes are relatively inert to substitution. Finally, in the divalent state, cobalt has a strong tendency to form tetrahedral complexes in concentrated electrolytes (e.g., $CoCl_4^{2-}$), whereas in such aqueous solutions nickel(II) retains hexacoordinating (e.g., $Ni(H_2O)_6Cl_2$). These general differences in chemical behavior have been exploited to provide the solvent extraction processes currently used or proposed for cobalt–nickel separation. All these processes remove cobalt from nickel solution, as no commercial flow sheet has yet been devised to remove nickel selectively from cobalt.

Cobalt and nickel can be separated by several solvent extraction processes, the choice depending on the total flow sheet requirements, on the ratio of the metals in the feed solution, and on the required purity of the products.

The various systems are as follows:

1. Extraction of cobalt from chloride media by amines.
2. Extraction of cobalt by organophosphorus acids, especially CYANEX 272, from acidic solutions.
3. Extraction of nickel after prior oxidation to cobalt(III) using hydroxyoximes.
4. Extraction of nickel by mixtures of simple oximes and carboxylic acids.

2.7.1. Laboratory Scale Cobalt–Nickel Extraction

Cobalt–Nickel Anion Exchange Extractants

As noted above, cobalt(II) in strong electrolyte solutions readily forms tetrahedral anionic complexes that can be extracted into an immiscible organic phase by amines or quaternary ammonium compounds (see Table 5.4). The most commercially important aqueous phase ligand is chloride, although thiocyanate has also been studied by INCO Canada.[13] Nickel/cobalt ratios of 2,000:1 in the nickel raffinate and a cobalt product with a 1,000:1 ratio over nickel are readily achievable. In chloride medium, the degree of formation of $CoCl_4^{2-}$ and, hence, the level of extraction depend on the chloride concentration is shown in Fig. 5.3. Separation for more than 104 is achievable at chloride ion concentration in aqueous phase 8–10 mol/dm^3. Stripping may be achieved by water washing of the loaded organic phase. Maximum cobalt extraction was observed at hydrochloric acid concentration 9 mol/dm^3. Extraction depends

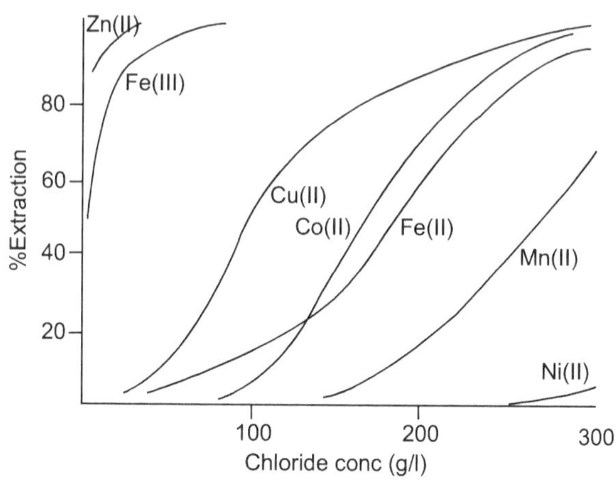

FIGURE 5.3 Extraction of metal chlorocomplexes by amines. *(Source: From Ritcey[2])*

also on the concentration and type of extractant in the following order: tertiary > secondary > primary amines.

Co–Ni Extraction by Acidic Extractants

Simple acidic extractants, such as carboxylic or sulfonic acids, have similar values of formation constants and are not used in cobalt–nickel separation. Also, the rate of nickel extraction is very slow compared with cobalt and, in addition, atmospheric oxidation of cobalt(II) to three valent station is readily occurred, which prevents normal stripping processes.[14] A similar oxidation of the loaded organic phase has been observed with substituted quinolines and *β*-diketones. The oxidation of cobalt(II) in the organic phase can be minimized by the addition of donor molecules, such as the solvating extractants, such as tri-*n*-butylphosphate (TBP), tri-*n*-octylphosphine oxide (TOPO), and carboxylic acids. In addition to stabilizing cobalt(II), various degrees of synergism can be obtained, enhancing the separation of nickel from cobalt.

Although separation factors of up to 150 have been observed with the system LIX 63–carboxylic acid, but again, there are problems with the slow rate of nickel extraction. By using a nonchelating oxime instead of LIX 63, separation factors of about 200 have been obtained,[8] without any reduction in the rate of nickel extraction or oxidation of cobalt. The structure of the extracted complex for both metals is $MR_2(oxime)_2$.[6,7] The problem in continuous operation with this system is that the acid strip solution tends to hydrolyze the oxime.

Earlier it was noted that nickel and cobalt could be extracted by carboxylic and sulfonic acids, with nickel being extracted at the lower pH. However, with alkylphosphorus acids, a selectivity reversal is observed, with cobalt being favored under acid conditions. The cobalt–nickel separation factor has been shown to depend upon metal concentration, reagent structure, diluent, temperature, and the presence of a diluent modifier. Thus, with increasing cobalt concentration the color of the organic phase changes from pink to blue, indicating a change from octahedral coordinated species (e.g., $[CoR_2(HR)_2]$) to an unsolvated polymeric tetrahedral species $[CoR_2]_n$. In contrast, nickel remains octahedrally coordinated, with a structure $[Ni(R_2)HR_x(H_2O)_{4-x}]$, maintaining a green color in the organic phase. This behavior was first noticed for di-2-ethylhexylphosphoric acid (DEHPA).[15] With phosphonic acid ester (PC88A, Daihachi) and a phosphinic acid (CYANEX* 272, Cytec), further studies[16] have shown a reagent dependence on the cobalt–nickel separation factor, increasing from 14 (DEHPA) to 280 (PC88A) and 7,000 (CYANEX 272), at pH 4 and 258 °C. Other factors that increased the separation include a change from aliphatic to aromatic diluents and an increase in temperature. Introduction of diluent modifiers, such as the solvating extractants TBP[17] or isodecanol, reduces the separation factor by depolymerizing the system and partially solvating the extracted complex to form octahedral species. A problem arises from oxidative degradation of the diluent by cobalt, which depends on the aromatic content of the diluent, on the temperature, and on cobalt loading.[18] This degradation can be eliminated by an introduction of antioxidants such as nonyl phenol. In addition to the diluents, oxidation of hydroxyoximes and 8-quinolinols has also been reported.

Solvent extraction has become important at development of hydrometallurgical processes for producing high-purity nickel products from nickel lateritic ores. The ammonia leach leaves the iron as an insoluble residue that is removed by filtration. The filtrate is contacted with a ketoxime reagent (e.g., LIX84I, Cognis) to extract the nickel:[19]

$$Ni(NH_3)_x^{2+}\ 2HR_{org} \leftrightarrow NiR_2 + 2NH_4^+ + (x-2)NH_3 \qquad (5.2)$$

where R is oxime.

However, this will also extract cobalt(II) under the same conditions, and once extracted the cobalt is oxidized to cobalt(III) and cannot be stripped. Therefore to avoid this problem, the cobalt is oxidized prior to extraction. So, nickel can be selectively extracted from a feed solution (9 g/dm^3 Ni and 0.3 g/dm^3 Co) leaving about 0.01 g/dm^3 Ni in the raffinate. Nickel is subsequently stripped from the loaded organic phase with strong ammonia.[20]

One of the main disadvantages of this flow sheet is that the cobalt(III) causes some oxidation of the reagent and a process of reoximation is required.[20] Also, ammonia transfers to the organic phase, which again represents an operating loss. However, the introduction of solvent extraction raises the nickel content of the products to >99.5% nickel as compared to 85–90% nickel with the flow sheet that did not use solvent extraction.[19]

An alternative route to cobalt/nickel separation following ammonia leaching in the Caron process is to produce a mixed Co/Ni carbonate product. This can then be dissolved in sulfate media and processed using one of the acidic extractants.[18]

2.7.2. Industrial Scale Solvent Extraction of Co–Ni

Some examples of industrial use of solvent extraction separations in the treatment of Co–Ni hydrometallurgy are presented below.

Bulong Process

The Bulong operation is situated near Kalgoorlie in Western Australia. In 2000, the plant produced 9000 t/a nickel and 729 t/a cobalt. Following the pressure acid leach, the leachate is neutralized with lime to precipitate residual iron, chromium, and aluminum. The clarified filtrate is contacted with CYANEX 272, containing TBP to inhibit third phase formation, at pH 5.5–6.0 and 508 °C to extract the cobalt, zinc, and manganese and maximize the rejection of nickel. Then the nickel is extracted from the raffinate with a carboxylic acid (Versatic 10) at pH 6.2–7.0.[21] The coextracted calcium and magnesium are scrubbed. The nickel is finally stripped from the organic phase with recycled electrolyte from the nickel electrowinning circuit. Coextracted impurities are removed initially from the cobalt circuit by sulfide precipitation of cobalt, copper, and zinc from the loaded strip liquor. The sulfide cake is then redissolved in acid and the zinc extracted with DEHPA, the copper removed by resin ion exchange and the cobalt electrowon. DEHPA has been proposed commercially to separate cobalt–nickel with either column contractors or mixer-settlers,[19] and PC88A has been incorporated into a flow sheet.[20]

Problems associated with this flow sheet include cross contamination of the organic phases and aqueous solubility losses of the Versatic acid.[21]

Cawse Process[8]

The Cawse mine produced 8500 t/a nickel and 1900 t/a cobalt in 2000 at the lowest cost production in the world.

Purification of the pressure acid leachate from treatment of the Cawse limonitic ore involves the precipitation of cobalt and nickel hydroxides with magnesium oxide following prior neutralization with limestone to remove iron and other impurities. The hydroxide cake is selectively leached with ammonia, rejecting any iron, manganese, or magnesium but redissolving any copper and zinc in the precipitate. The cobalt is then oxidized to cobalt(III) prior to extraction, to prevent coextraction with nickel with the hydroxyoxime LIX 84I (Cognis). The nickel is stripped with dilute sulfuric acid and electrowon. Copper and some cobalt build up in the organic phase and these levels are controlled by including a bleed stream, from which the copper is removed with a highly concentrated acid strip and cobalt by stripping in the presence of a reducing agent. Another major problem is the oxidative degradation of the oxime extractant by the cobalt(III) requiring the addition of a reoximation step to maintain extractant capacity. The cobalt in the raffinate is recovered by sulfide precipitation.

Murrin-Murrin Process

The Murrin-Murrin mine is also near Kalgoorlie in Western Australia. This is the largest of the three projects, with a first-phase production of 45,000 t/a nickel and 3000 t/a cobalt with the expansion to 115,000 t/a nickel and 9000 t/a cobalt.

The pressure acid leachate, following removal of the insoluble residue, is treated with hydrogen sulfide under pressure to precipitate nickel and cobalt and reject impurity elements. The sulfides are then dissolved in a pressure oxygen leach to produce a sulfate liquor that is treated with CYANEX 272. Initially, zinc is removed at low pH followed by a pH adjustment and cobalt extraction. Any coextracted nickel is scrubbed from the loaded organic phase with a dilute cobalt sulfate solution and the cobalt stripped with dilute sulfuric acid. Both the cobalt and nickel are recovered from solution as powders by hydrogen reduction that also produces ammonium sulfate as a by-product. The main problem in this flow sheet has been the crystallization of nickel ammonium sulfate because of the high nickel concentration (100 g/dm^3) in the feed to extraction. The problem was solved by preloading the extractant with nickel prior to extraction by contact with the nickel sulfate raffinate.

Several industrial scale flow sheets were developed recently. The anion-exchange process has been used commercially in Norway (Falconbridge Nikkelwerk) and Japan (Sumitomo) for cobalt–nickel separation.[22] Extraction of metals from chloride solutions into kerosene containing trioctyl amine (TOA) has found application in the production of pure Co and Ni metals from their ores. The aqueous phase contains the Co and Ni in addition to other metals in chloride solution, where the chloride concentration goes to 300 g/L (i.e., about 8.5 M). Negatively charged metal chloride complexes in the aqueous phase are formed at these concentrations are extracted by an amine like TOA, or by its amine salt, $TOAH^+Cl^-$. Fe(III) forms much stronger complexes with Cl^- ions, so, it is much more easily extracted. The complex formation from M^{z+} to MCl_{z-n}, where $z - n \leq -1$, proceeds in the stepwise manner described in Chapter 3. For example, Fe(III) forms the complex $FeCl_4^-$, while cobalt forms $CoCl_4^{2-}$.

Bulk extraction followed by selective separation in smaller circuits will be more popular. An example of bulk extraction is that of Goro,[23] using Cyanex 301 for Co + Ni coextraction, followed by stripping with chloride and selective separation of Co from Ni using an amine.

2.8. Precious Group Metals (PGM) Extraction

Chemical complexity of precious elements and the difference in the raw material feed leads to the different flow sheets. The traditional route required multiple precipitation and redissolution steps to provide the desired product purity, which, with the consequent wash streams, involved increased recycling and retreatment. This locks up considerable amounts of treated elements, which places an economic constraint on the whole production. Sometimes the operation may last a year between dissolving a sample of scrap at the start of the flow sheet and the emergence of the pure elements for marketing. During this time, the value of the elements is locked up in the plant.

To understand these complexities of the production technologies, it is necessary to outline some of the fundamental chemistry of the precious metals. These elements are noble metals and, as such, can be dissolved only with great difficulty. The usual leaching agent is hydrochloric acid, with the addition of chlorine to increase the solution oxidation potential. This strong chloride medium results in the almost exclusive formation of aqueous chloroanions, with, under certain circumstances, the presence of some neutral species. Very seldom are cationic species formed in a chloride medium. However, these elements do possess a range of easily accessible oxidation states and, with the possibility of a number of different anionic complexes that are dependent on the total chloride concentration; this provides a very complicated chemistry.

A summary of the most important chloro complexes found in these leach solutions is given in Table 5.5, from which the mixed aquochloro and polynuclear species have been omitted. The latter are found especially with the heavier elements.

The chloro-species also differ in their rates of substitution, which, in general terms, are slower than those of the base metals and follow the order: Au(I), Ag(I) >> Pd(II) > Au(III) > Pt(II) > Ru(III) >> Rh(III) > Ir(III) > Os(III) >> Ir(IV), Pt(IV), with the states from Rh(III) being termed inert. Thus, kinetic factors tend to be more important, and reactions that should be possible from thermodynamic considerations are less successful as a result.

On the other hand, the presence of small amounts of a kinetically labile complex in the solution can completely alter the situation. This is made even more confusing in which the basic chemistry of some of the elements has not been fully investigated under the conditions in the leach solutions.

TABLE 5.5 Common Chloro-Species of the Precious Group Metals

Metal	Oxidation state	Major chloro-species	Comments
Silver (Ag)	I	$AgCl$	Insoluble
		$AgCl_2^-$	High HCl concentration
Gold (Au)	I	$AuCl_2^-$	Very stable
	III	$AuCl_4^-$	
Platinum (Pt)	II	$PtCl_4^{2-}$	Conversion IV to II slow
	IV	$PtCl_6^{2-}$	Most common species, kinetically inert
Palladium (Pd)	II	$PdCl_4^{2-}$	Most common
	IV	$PdCl_6^{2-}$	Conversion II to IV difficult
Iridium (Ir)	III	$IrCl_6^{3-}$	Both species stable
	IV	$IrCl_6^{2-}$	Conversion IV to III easy
Rhodium (Rh)	III	$RhCl_6^{3-}$	
Ruthenium (Ru)	III	$RuCl_6^{3-}$	Complex equilibria between III and IV depends on redox potential and chloride concentration
	IV	$RuCl_6^{2-}$	
Osmium (Os)	III	$OsCl_6^{3-}$	
	IV	$OsCl_6^{2-}$	Os(IV) more stable than Ru(IV)

Consequently, a solvent extraction process to separate the precious metals must cope with a wide range of complexes in different oxidation states, which vary, often in a poorly known fashion, both in kinetic and thermodynamic stability. Therefore, different approaches have been tried and different flow sheets are produced.

Because gold and silver can be easily separated from the remaining precious metals, they tend to be removed from the flow sheet at an early stage. In addition, significant quantities of gold and silver exist, in either primary ores or waste materials that are not associated with the platinum group metals. Silver often occurs as a residue from the chloride leaching operation and so is removed first, followed by gold, which can be extracted as the tetrachloroauric(III) acid, $HAuCl_4$, by a solvating reagent.

Notice that none of the flow sheets uses solvent extraction exclusively. Because the aqueous chemistry of osmium and ruthenium is very complex, most operators remove these elements by distillation of the tetraoxides, MO_4. Also, it has been advantageous to use ion exchange to separate and concentrate rhodium. The various extraction routes for individual elements are discussed in the following subsections.

Thermodynamic model has been developed[24] for the precious metal complexes extracted from an ammonium copper thiosulfate solution.

2.8.1. Extraction of Gold

Leaching solution of gold is $AuCl_4^-$, and all the processes rely on the extraction of this ion or the parent acid by ion-pair or solvating extractants, respectively. For example, dibutylcarbitol (diethyleneglycol dibutylether) by INCO,[25] 4-methyl-2-pentanone (methyl isobutylketone; MIBK),[25] long-chain alcohols (the most effective: 2-ethylhexanol)[26] Recovery of gold is best achieved by chemical reduction with, e.g., iron to give an impure product for subsequent refining.

Gold, like silver, being classified as a soft acid, should prefer interaction with soft donor atoms like sulfur. Therefore, it is no surprise to find that extraction with the trialkylphosphine sulfide, CYANEX 471X, is very efficient and provides excellent separation from elements such as copper, lead, bismuth, and other metalloids found in anode slimes.[26] Stripping can be achieved using sodium thiosulfate that allows the separation of recovery and the reduction to the metal.

Amine salts can also be used to extract anionic species, but, in general, they are less selective for gold. However, trioctylamine (TOA) was used[27] to extract gold from chloroanions such as copper, iron, tin, and zinc. The selectivity is attributed to the presence of TOA-HNO_3 in the organic phase. Coextracted metals are scrubbed with dilute nitric acid, and gold is recovered by a hydrochloric acid–thiourea solution. The same authors also used *N*,*N*′-di-*n*-butyloctanamide as a solvating extractant. The protonated forms of Kelex 100 and

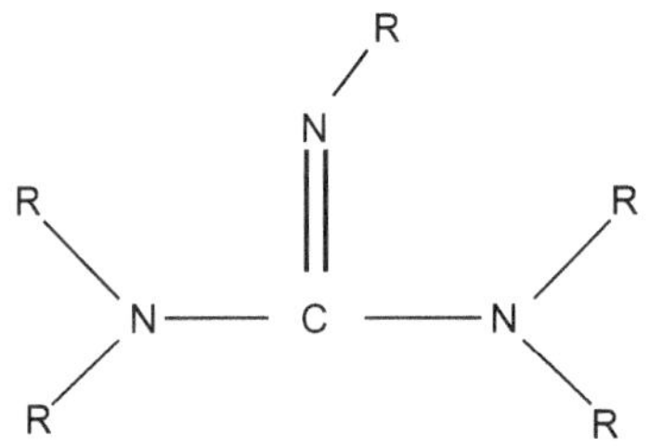

FIGURE 5.4 Structure of LIX 79.

LIX26, both acting as ion pair extractants, have been proposed[28] for gold extraction.

Alkaline cyanide solutions are another common lixiviant for gold, giving complex cyano species in solution. These anionic complexes may be extracted from aqueous solution by amines as ion pairs.[29,30] Recently, another extractant, LIX79 (trialkylated guanidine see Fig. 5.4) has been developed specifically to extract gold and silver from these media.[31,32] The reagent extracts both metals rapidly from solutions in the pH range 9–11.5 (gold) and 9–10 (silver). At higher pH values the metals are stripped from the organic phase as the guanidine becomes deprotonated.

Environmental concern over the use of cyanide leaching has stimulated the development of several alternative leaching processes. The gold as an anionic complex, $[Au(S_2O_3)_2]^{3-}$ may be extracted by amines, at pH < 8.0, or solvating reagents, e.g., TBP (pH >10)[33] or di(2-ethylhexyl)dithiophosphoric acid.[34,35] It was also found that a mixture of these reagents gave some synergism over the pH range 9–11. Similarly, the use of thiourea as an alternative to cyanide for the recovery of gold and silver has several advantages, including improved kinetics and lower toxicity of the reagents.

2.8.2. Extraction of Silver

Up to date, silver is not treated by solvent extraction from leachates, but there are some processes from the mixtures with silver. Cytec reagents with donor sulfur atoms are used to extract this "soft" element. For example, tri-isobutylphosphine sulfide (CYANEX 471X) extracts silver from chloride, nitrate, or sulfate media selectively from copper, lead, and zinc.[36] Silver is recovered from the loaded organic phase by stripping with sodium thiosulfate, and the metal recovered by cementation or electrolysis. Silver can also be extracted from chloride solution by a dithiophosphinic acid (CYANEX 301).[37]

2.8.3. Extraction of Palladium and Platinum

These elements are present in ores and recycled feeds, such as catalyst wastes, and as an intermediate bulk palladium–platinum product from some refineries. Two common oxidation states and stereochemistries are used for separation of elements: square planar palladium(II) and octahedral platinum(IV). Palladium(II) has the faster substitution kinetics.

Two general directions are considered: selective extraction of metals with different reagents or coextraction of the elements followed by selective stripping. Flow sheets with di-*n*-octyl sulfide or a hydroxyoxime as selective extractants are used to remove the palladium, followed by tri-*n*-butylphosphate (TBP; INCO) and an amine (MRR) to recover platinum. The extraction of palladium by the sulfide[38] is slow with 2–3 h to achieve equilibrium. This inhibits commercial flow sheet design. Tertiary amines can act as a kinetic accelerator, reducing equilibrium times to minutes, but no information has been published on the effect of the amines on selectivity of palladium–platinum separation. High loading of the sulfides is possible, and stripping is easy with aqueous ammonia. Neutralization of the strip solution with hydrochloric acid gives $Pd(NH_3)_2Cl_2$ as product.

The aqueous raffinate from extraction is acidified to 5–6 mol/dm^3 with hydrochloric acid to optimize platinum extraction by the solvating extractant TBP. The coextraction of iridium is prevented by reduction with sulfur dioxide, which converts the iridium(IV) to the not extractable (III) species. The species H_2PtCl_6 are extracted by TBP and stripped by water.[17] Metal is recovered by precipitation with ammonia as $(NH_3)_2PtCl_2$.

The second selective extractant (MRR) uses a hydroxyoxime to remove palladium.[8] The actual compound used has not been specified, but publications refer to both an aliphatic α-hydroxyoxime and an aromatic β-hydroxyoxime. The α-hydroxyoxime LIX63 has the faster extraction kinetics, but suffers from problems with stripping. For the β-hydroxyoxime, a kinetic accelerator in the form of an amine (Primene JMT) has been proposed. The precise mode of operation of this accelerator is unknown, but it may be a similar process to that proposed for the sulfide system (i.e., a form of proton-transfer catalysis where the protonated amine extracts $PdCl_4^{2-}$ which then reacts with the hydroxyoxime in the organic phase). Good loading of palladium and palladium–platinum separation factors of 103–104 are found. Stripping of the palladium is easy with hydrochloric acid, and $(NH_3)_2PdCl_4$ is obtained by precipitation. The raffinate from palladium extraction is treated to remove osmium and ruthenium. Stripping of the platinum is difficult because of the high-distribution coefficient, but techniques that have been successful include the use of strong acids, such as nitric or perchloric, to break the ion pair, or alkali to deprotonate the amine salt.

Influence of residual gold on the solvent extraction-based recovery of cyanide from more traditional industrial cyanidation liquors was presented in Woollam and Grant (2008)[39] Some authors observed a significant increase in the rate of acid catalyzed degradation of the hydroxyoxime LIX 84I in the palladium solvent extraction circuit when the reagent concentration was increased. However, a reduction in strip acid concentration was found to alleviate this problem sufficiently, while leaving phase disengagement and overall strip circuit efficiency at acceptable levels.

The phosphine sulfide CYANEX 471X has been proposed as an alternative to the foregoing reagents for palladium–platinum separation. Here, the faster extraction of palladium affords separation factors of 10:1, with a 10-min contact time. The chemistry of the process is similar to that of the alkyl sulfides. Various other separations have been published; e.g., TBP will extract platinum(IV) chloroanions preferentially to palladium(II).

An alternative general process involves coextraction of palladium and platinum, followed by selective stripping. A novel amino acid extractant, made by the reaction between chloracetic acid and Amberlite* LA-2, a secondary amine, is used to extract the two elements from the leach liquor. However, in the given flow sheet, no selective stripping is employed, both elements being stripped by hydrochloric acid. The resulting chloroanions are then separated by using di-*n*-hexylsulfide to extract $PdCl_4^{2-}$. Other coextraction-selective strip routes that have been proposed include the following:

1. Extraction with a secondary amine, followed by stripping of the palladium by hydrazine, and bicarbonate to strip platinum by deprotonation of the amine ion pair.
2. Extraction with a tertiary amine, Alamine* 336, followed by palladium stripping with thiourea and platinum with thiocyanate. Both of these systems suffer from problems with stripping and metal recovery from the strip solutions. Results of testing series of polyamines based on the tris(2-aminoethyl)amine (tren) platform as extractants of palladium(II), platinum(II), rhodium(III), and ruthenium(III)[40] for the chloro complexes show some hopes. This study helped to confirm the importance of water in some complex formation structures. New tripodal amido and urea group-based anion-binding ligands as potential extractants for $[PtCl_6]^{2-}$ from acidic chloride solutions are presented in Fig. 5.5.[41] Much greater selectivity for the complex anion over chloride was reported compared to model reagents such as trioctylamine. The efficient extraction of soft metal ions, including silver(I) and palladium(II), onto various copolymers of *N*-isopropylacrylamide and thioethers was reported.[42] Some potential for enhanced selectivity may encourage further development of these systems. On the base of secondary amine, Amberlite LA-2 system a computer simulation of the solvent extraction of platinum from a chloride matrix was developed[43] and proved on several flow sheet options to be simulated. The results of the selected configuration were validated in a subsequent mini-pilot plant trial.
3. Extraction with 8-quinolinol derivatives, which can act as either a chelating acid or, after protonation, as an anion extractant. It was observed[44] that palladium extraction was favored at low acidity, but platinum was extracted best at high acidity (see Fig. 5.6). Here, no chlorine atoms are retained in the complex, because a bidentate chelating ligand is used, and the system is easy to strip. Fast kinetics are observed with a 3–5-min contact time to reach equilibrium at ambient temperature. The extraction conditions can be easily

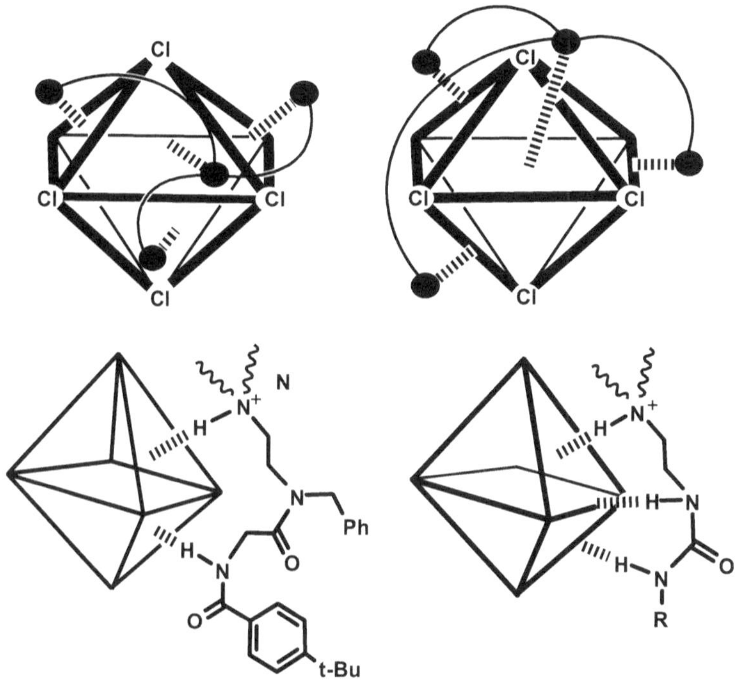

FIGURE 5.5 Possible binding modes for the amide and urea arms of the tripodal reagents used as potential extractants for $[PtCl_6]^{2-}$ from acidic chloride solutions. *(Source: From Tasker et al.[41] with permission)*

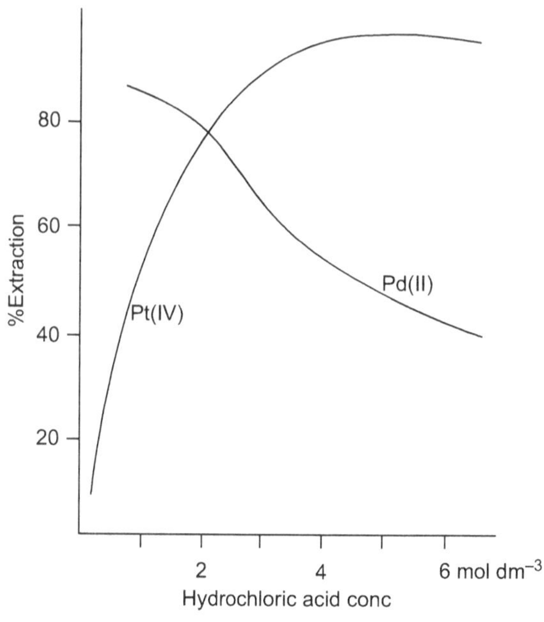

FIGURE 5.6 Extraction of platinum(IV) and palladium(II) by LIX26 (5%), isodecanol (5%) in Solvesso (90%). *(Source: From Du Preez and Preston[44])*

adjusted and good selectivity over base metals, such as iron and copper, is found. Differential stripping is possible from the loaded organic phase by variation of the acidity of the strip solution: dilute solutions were used to release platinum and more concentrated for palladium. Separation factors >200 can be obtained.

4. The use of N-alkylamides that selectively extract platinum from strong acid media, presumably by ion pair formation between the protonated amide and the chloroanions.[45] Stripping is easy with dilute acid or water, which deprotonates the amide and releases the metal anion. The process favors metals in high oxidation states [e.g., platinum(IV) or iridium(IV)], so that reduction of any iridium must precede the extraction of platinum and palladium, as with other systems. The platinum–palladium separation factors depend on the amide structure, but are rather small, about 25–55, so that scrubbing of the loaded organic phase is required to give a good separation.

The rapid and highly selective recovery of palladium(II) from hydrochloric acid solutions in the presence of platinum(IV) using sulfur-containing monoamide and diamide compounds was presented recently.[46] The results compared well to the slower separation achieved using more conventional dialkylsulfide extractants. The effect of hydrochloric acid concentration on the kinetics and equilibrium data for the extraction of palladium(II) with novel pyridine carboxamide and phosphonium ionic liquid systems was reported.[47]

2.8.4. Extraction of Indium and Rhodium

In the commercial flow sheets, these elements are left in the aqueous raffinate after platinum and palladium extraction. Indium can be extracted in the +IV oxidation state by amines or TBP. The separation from rhodium is easy, but the recovery of iridium may not be quantitative because of the presence of non-extractable iridium halocomplexes in the feed solution. The authors proposed coextraction of iridium, platinum, and palladium by a tertiary amine and the selective recovery of the iridium by reduction to Ir(III).[48] Iridium can also be separated from rhodium by substituted amides.

Rhodium is recovered by precipitation from the raffinate after ion exchange, but solvent extraction with an amine followed by thiourea stripping has been suggested.[49] The problem in these systems is that extraction is favored by low acidity and low chloride concentrations so under the conditions found in the commercial flow sheets rhodium extraction would be low.[50,51] In addition, stripping from the organic phase is difficult with the possibility that the rhodium remains in the organic phase as the dimer, $Rh_2Cl_9^{3-}$.

Extraction with 8-hydroxyquinoline extractants has also been studied. Low extraction (maximum 40%) and stripping problems were observed.[52] A critical review discussing the flow sheets and their problems in more detail has been published.[53]

2.9. Lanthanide Extraction

The lanthanide group of elements (see Table 5.6) is very difficult to separate by traditional methods because of their similar chemical properties. The techniques originally used, like the precious metals, included multiple fractional recrystallizations and fractional precipitation, with many recycle streams to achieve reasonably pure products. Such techniques were unable to cope with the demands for significant quantities of certain pure compounds required by the electronic industry.

The lanthanides occur in a number of ores, but as the relative concentration of individual elements differ according to the source, several extraction schemes have been developed to take advantage of this natural variability. Ion exchange and solvent extraction separations are based on the steady decrease in size across the lanthanide elements, which result in an increase in acidity or decrease in basicity with increasing atomic number. This causes a variation in

TABLE 5.6 Properties of Lanthanide Elements

Element	Atomic number	Atomic mass	Ionic radius (Ln^{3+}) (Å)	
Lanthanum (La)	57	138.9	1.061	Light earths
Cerium (Ce)	58	140.1	1.034	Light earths
Praesodymium (Pr)	59	140.9	1.013	Light earths
Neodymium (Nd)	60	144.2	0.995	Light earths
Promethium (Pm)	61	–	0.979	Light earths
Samarium (Sm)	62	150.4	0.964	Light earths
Europium (Eu)	63	152.0	0.950	Light earths; Middle earths
Gadolinium (Gd)	64	157.3	0.938	Middle earths; Heavy earths
Terbium (Tb)	65	158.9	0.923	Middle earths; Heavy earths
Dysprosium (Dy)	66	162.5	0.908	Middle earths; Heavy earths
Holmium (Ho)	67	164.9	0.894	Middle earths; Heavy earths
Erbium (Er)	68	167.3	0.881	Heavy earths
Thulium (Tm)	69	168.9	0.869	Heavy earths
Ytterbium (Yb)	70	173.0	0.858	Heavy earths
Lutetium (Lu)	71	175.0	0.848	Heavy earths
Yttrium (Y)	39	88.9	0.88	Heavy earths

formation coefficient of any metal–extractant complex, allowing preferential binding to an ion exchange resin, or extraction of the complex into an organic phase. The variations in formation coefficients and hence, separation factors, between adjacent elements, are small; so many equilibria and thus a large number of extraction units are required. Even then, separation is possible only by recycling both the exit streams. This configuration, rarely used outside lanthanide production, is essential to the preparation of pure products or the separation of adjacent elements.[54] The bank of mixer-settlers is fed close to the midpoint, with a mixture of components, and by returning the exit streams to the process, 100% reflux is obtained. The components are force-fed to build up their concentration, and the reflux is continued until the desired degree of separation is achieved. Then, the operation is stopped and the individual mixer-settlers emptied to give the required products. Figure 5.7 gives a typical separation of lanthanum, neodymium, and praseodymium from nitrate media, using TBP as extractant. The number of extraction units can be varied according to the products required. For example, lanthanum can be concentrated in the aqueous raffinate and neodymium in the organic extract in holding tanks external to the mixer-settlers, thereby decreasing the number of units required.

It should be noted that yttrium, although not a member of the lanthanide series, often accompanies them in ores. The charge/size ratio of yttrium places it between dysprosium and holmium, although, depending on the extraction system, it can behave as either a light or heavy element.

Ore leaching agent is one of the common mineral acids. The choice of extractant depends on the leaching acid used. For example, nitric acid as a leaching allows the use of solvating extractants (e.g., TBP), whereas the carboxylic or organophosphorus acids as extractants may be used with either hydrochloric, nitric, or sulfuric acids as leaching agents. Occasionally, mixtures of two extractants will improve the separation of individual elements.

Lanthanides are also found as minor components in other ores, particularly in association with uranium or in phosphate rock. These are often coextracted

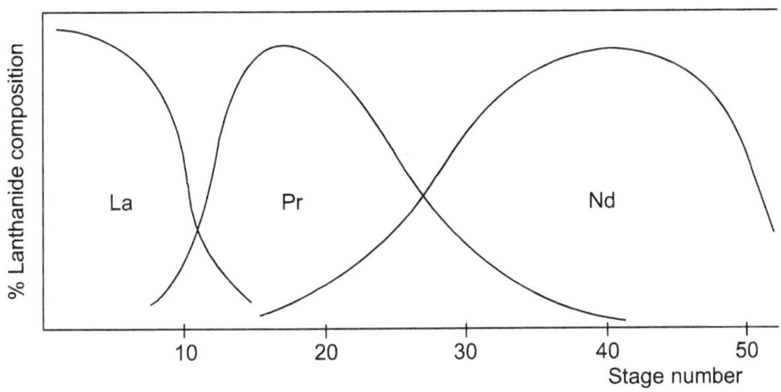

FIGURE 5.7 Concentration profiles for mixer-settler bank under total reflux: system TBP (50%) from nitrate medium. *(Source: From Brown and Sherrington[54])*

with the major product and can be economically recovered from the waste streams resulting from the uranium or phosphoric acid extraction.

2.9.1. Solvating Extractants for Lanthanides

The extraction of the lanthanides from nitrate media with TBP depends on several factors including pH, nitric acid concentration, ionic strength, and concentration of the elements in the aqueous phase:

$$Ln(NO_3)_3 * xH_2O + 3TBP + \{Ln(NO_3)_3 * 3TBP\}_{org} + xH_2O \qquad (5.3)$$

The situation is complicated by the ability of TBP to extract nitric acid as HNO3 * xTBP, which then competes with the lanthanides, so the process is optimized in terms of high pH, but low nitric acid concentration. Increased ionic strength has the advantage of salting out the aqueous phase complexes into the organic phase, whereby neutralization of excess nitric acid increases the distribution coefficient. The total concentration of lanthanides in the feed affects the separation factors between adjacent elements; the separation factor decreasing as the concentration rises (see Table 5.7). In contrast, the separation factors between yttrium and lanthanides increase with dilution, so that a process with a dilute feed can be used to produce purified yttrium. A commercial flow sheet has been designed to produce pure lanthanum using 50% TBP, the lanthanum remaining in the raffinate after extraction. The organic phase is stripped with water, and the aqueous solution concentrated and extracted with the same organic phase to separate praseodymium and neodymium. A typical profile of the mixer-settler bank is described above.

Other solvating organophosphorus extractants have been studied to a small extent and these show some processing advantages with, e.g., an organophosphonate showing higher separation factors than TBP.

TABLE 5.7 Separation Factors of Adjacent Lanthanides at Extraction by TBP from Nitrate Solutions

Aqueous phase concentration (Ln_2O_3 g dm^{-3})	Separation factors					
	Sm/Nd	Gd/Sm	Dy/Gd	Ho/Dy	Er/Ho	Yb/Er
460	2.26	1.01	1.45	0.92	0.96	0.81
430	–	–	1.20	0.96	0.65	–
310	2.04	1.07	1.17	0.94	0.82	–
220	1.55	0.99	1.08	0.89	0.78	–
125	1.58	0.82	0.92	0.83	0.72	–
60	1.40	0.78	0.89	0.77	0.70	0.63

Source: From Ref. 37

2.9.2. Amine Extractants

Amines provide an alternative method of extraction from nitric acid media, extracting the nitrato-complexes of the lanthanides as ion pairs. Because of competition from nitric acid, the process has to operate above pH 2, usually with metal nitrates as salting out agents. The extraction also depends on the amine structure, with asymmetric compounds giving the highest distribution coefficients. Again, separation from yttrium is favored, with separation factors in the range of 5–20. An interesting process to produce pure yttrium from a mixed lanthanide feed depends on this element behaving as a heavy lanthanide in nitrate media and as a light lanthanide in thiocyanate solution.

2.9.3. Acidic Extractants

Both carboxylic (e.g., Versatic acid) and organophosphorus acids have been used commercially to extract the lanthanides. The extraction follows the formation of the metal–extractant complex (5.4) and so depends on the pH of the feed.

$$Ln^{3+} + 3H_2L_2 \quad \leftrightarrow \quad LnL_3 * 3HL + 3H^+ \tag{5.4}$$

Organophosphorus acids, of which DEHPA has received the most attention, are more powerful extractants than Versatic acid and extract the heavy lanthanides even at low pH values. This gives problems with stripping, and the heaviest elements, ytterbium and lutetium, require 20% hydrofluoric or 50% sulfuric acid as strip solutions. Variation of pH thus allows the separation of heavy, including yttrium, from light elements. Separation factors between adjacent elements follow the order: phosphinic >> phosphonic > phosphoric > carboxylic acids, although few data are available forthe first two extractants (see Table 5.8). Organophosphonic and organophosphinic acids potentially offer some more processing advantages.

In addition, with organophosphorus acids, higher separation factors are found from chloride than from sulfate or nitrate media. Thus, parameters that can be used to design a flow sheet include type and concentration of extractant, pH of the feed solution, nature of anion (organophosphorus acids), and nature and concentration of the strip solution. 0.1 mol/dm^3 DEHPA solution completely extracts all the elements above europium at pH = 2 and, as noted earlier, this makes the heavy elements difficult to strip. Increasing the acidity of the chloride solution decreases the separation factor values. A similar, but smaller, dependency of separation factors on acidity has also been found with sulfate systems.

Several different flow sheets have been developed around DEHPA, which, as discussed earlier, vary according to the precise composition of the feed material. As an example, typical process using nitric acid leach of a xenotime ore is present below. The leach liquor is first treated with a DEHPA solution to extract the heavy lanthanides, leaving the light elements in the raffinate. The loaded reagent is then stripped first with 1.5 mol/dm^3 nitric acid to remove the elements from neodymium to terbium, followed by 6 mol/dm^3 acid to separate yttrium and remaining heavy elements. Ytterbium and lutetium are only

TABLE 5.8 Separation Factors for Adjacent Lanthanides Using Acidic Extractants

	Ce/La	Pr/Ce	Nd/Pr	Sm/Nd	Eu/Sm	Gd/Eu	Tb/Gd	Dy/Tb	Ho/Dy	Er/Ho	Tm/Er	Yb/Tm	Lu/Yb
Versatic 911[a]	3.00	1.60	1.32	2.20	–	1.96	–	–	1.17	1.28	–	–	–
DEHPA[a]	–	–	1.40	4.00	2.30	1.50	5.70	2.00	1.80	2.40	3.40	–	–
DEHPA[b]	5.24	1.86	1.32	9.77	2.14	1.99	–	–	–	–	–	–	–
Phosphonic acid[b]	13.8	3.47	1.51	13.80	2.45	1.86	6.92	3.23	2.17	2.82	2.63	3.71	1.74
Phosphinic acid[b]	16.98	1.86	1.32	13.80	2.14	1.32	6.02	2.82	1.74	2.63	3.47	2.63	1.51

[a] *Source: Ref. 54*
[b] *Source: Ref. 55 (All organophosphorus extractants used in this work had 2-ethylhexyl substituents)*

partially removed; hence, a final strip with stronger acid, as mentioned earlier, or with 10% alkali is required before organic phase recycle. The main product from this flow sheet was yttrium, and the yttrium nitrate product was further extracted with a quaternary amine to produce a 99.999% product.

The rate of extraction of the lanthanides by the 2-ethylhexyl ester of 2-ethylhexylphosphonic acid (e.g., PC88A, P507) is slower than for DEHPA but the higher distribution coefficients allow extraction from more dilute solutions. Also, as acidic stripping is easier than with DEHPA, the reagent has been used commercially to separate lutetium from terbium and ytterbium.[56] Recently emphasis has shifted to the use of organophosphinic acids for the extraction of lanthanides[57] with CYANEX 272, providing significant advantages over other organophosphorus acid extractants for the heavy lanthanides with greater average separation factors and easier stripping. Furthermore, a 1:1 mixture of CYANEX 272 and a carboxylic acid gives significant synergism for the heavy lanthanides with higher loadings (0.2 mol/dm^3) than either CYANEX 272 (0.08 mol/dm^3), DEHPA, or the phosphonic acid ester (P507) (0.17 mol/dm^3).[44] In addition, stripping from the synergistic mixture with either hydrochloric or nitric acid is easy. The latter enables the production of high-purity products free from chloride ions.

Extraction with versatic acid takes place by a cation-exchange mechanism with maximum extraction close to the pH of hydrolysis. No preference is shown for a particular anion in the feed solution, and separation factors are less dependent on the concentration of elements in the feed than with TBP. The high cerium–lanthanum separation factor indicates that versatic acid would be a better extractant than TBP to produce pure lanthanum. The small separation factors still require both extractions to be performed under total reflux in a large (56-stage) mixer-settler train.

3. SOLVENT EXTRACTION IN NUCLEAR CHEMISTRY AND TECHNOLOGY

Nuclear chemistry and technology include the following:

1. Technology of production of fuel for nuclear reactors (U, Th).
2. Reprocessing of the irradiated spent fuel for extraction of U, Np, Pu, transplutonium elements.
3. Treatment of the nuclear waste.

All technological flow sheets contain solvent extraction processes for separation, concentration, purification of the radioactive elements as products and/or wastes.

There is a large bibliography of nuclear chemistry and technology. The reader can refer to some handbooks,[58–61] encyclopedias,[62,63] books,[64–67] published after 2000, and a huge number of articles, published in special journals such as *Nuclear Technology, Journal of Nuclear Materials*, etc. The objective of this subsection provides a brief introduction to publications in application of solvent extraction processes and technology in the nuclear field during the last decade.

Because of secret problems, sometimes it is difficult to distinguish between laboratory or pilot studies and industrial scale publications, so this was not separated in the text below.

3.1. Solvent Extraction in Nuclear Fuel Processing

The actinide elements are "hard acid cations" and their bonds are principally ionic; thus, ionic interactions are the main driving force for bond formation (see Fig. 5.8). Consequently, oxygen donors form complexes with the actinide ions much better than the "type B ligands."[61] Trivalent actinides have slightly more covalency than the lanthanides.[68] This has led to the search for new extractants containing more covalent donor atoms, such as sulfur or nitrogen.

3.1.1. Uranium Production

Industrial-scale processes for recovery of uranium from ore leach liquors were developed at the U.S. Oak Ridge National Laboratory.[69] The process used for recovery of uranium from its ores depends on the nature of the ore. All the processes include a leaching step that solubilizes the metal. Solvent extraction is used most frequently for the recovery and purification of uranium from the leaching liquors.[70,71]

Two extractants are used to recover uranium from sulfuric acid leach liquors: di-(2-ethylhexyl)phosphoric acid (DEHPA) in the Dapex process and commercial mixtures of trioctyl to tridecylamines (e.g., Adogen 364 or Alamine 336) in the Amex process. The chemical principles on which these processes are based are given in Table 5.9. The Amex process is more widely

Oxidation state (ion)	AC	Th	Pa	U	Np	Pu	Am	Cm	Bk	Cf	Es	Fm	Md	No
I (M^+)													+	
II (M^{2+})							+			+	+	+	++	*
III (M^{3+})	*	++	+	++	++	++	*	*	*	*	*	*	*	++
IV (M^{4+})		*	++	++	++	*	++	++	++	+				
V (MO_2^+)			1 *	++	*	++	++							
VI (MO_2^{++})				*	++	++	++							

FIGURE 5.8 Oxidation states of the actinide elements: * most stable ions in aqueous solutions; ++ oxidation states observed in aqueous solutions; + unstable ions observed only as transient species..

TABLE 5.9 Chemicals for the Amex and the Dapex Processes for U(VI) Recovery from H_2SO_4 Leach Liquors

	Amex	Dapex
Extraction	$UO_2^{2+} + SO_4^{2-} + 2(R_3NH)_2SO_4 \leftrightharpoons (R_3NH)_4UO_2(SO_4)_3$ R_3N = Alamine 336 or Adogen 364 (trialkylamines with alkylgroups with 8 to 10 carbons)	$UO_2^{2+} + 2(HDEHP)_2 \leftrightharpoons UO_2(HDEHPDEHP)_2 + 2H^+$
Stripping	*Acidic stripping* $(R_3NH)_4UO_2(SO_4)_3 + 4HX \leftrightharpoons 4R_3NHX + UO_2^{2+} + 3HSO_4^- + H^+$ HX = HCl or HNO_3 *Neutral stripping* $(R_3NH_4)_4UO_2(SO_4)_3 + (NH_4)_2SO_4 + 4NH_3 \leftrightharpoons 4R_3N + UO_2(SO_4)_3^{4+} + 6NH_4^+ + SO_4^{2-}$ *Alkaline stripping* $(R_3NH)_4UO_2(SO_4)_3 + 7Na_2CO_3 \leftrightharpoons 4R_3N + UO_2(CO_3)_3^{4-} + 4HCO_3^- + 3SO_4^{2-} + 14Na^+$	*Alkaline stripping* $UO_2(HDEHPDEHP)_2 + 4Na_2CO_3 \leftrightharpoons UO_2(CO_3)_3^{4-} + 4NaDEHP + H_2O + CO_2\uparrow + 4Na^+$ (TBP must be added to avoid the third-phase formation owing to low organic phase solubility of NaDEHP.) *Acidic stripping* $UO_2(HDEHP)_2(DEHP)_2 + 2H^+ \leftrightharpoons 2(HDEHP)_2 + UO_2^{2+}$

used than the Dapex process because of the greater selectivity of trialkylamines than DEHPA for uranium in H_2SO_4 solution.

High-grade pitchblende ores are leached with nitric acid to recover uranium. Extraction of uranium from nitrate solutions is usually performed with TBP. TBP-based solvents are used in several areas of the nuclear industry, especially for reprocessing of spent nuclear fuels and for refining the uranium product of the Amex and Dapex processes. Extraction of uranium by TBP solvents is described in the next subsections.

The concentration of uranium contained in phosphate rocks (50–200 ppm) is higher than that in seawater (see below). Even though economic recovery of uranium from phosphate rock is difficult, several phosphoric acid plants include operation of uranium recovery facilities. Various processes[72,73] have been used for uranium extraction from phosphoric acid solution; their main features are listed in Table 5.10. The DEHPA-TOPO process is increasingly preferred over others because of the stability of the extractant and the well-understood chemistry of the process. Today, phosphoric acid for use in the pharmaceutical or food industry must be free from uranium and other harmful metals. The processing of phosphate rock for production of fertilizers may be an important source of uranium because uranium removal may become compulsory to avoid its dissemination into the environment from the use of phosphate-based fertilizers.

Hydrometallurgy of copper ores sometimes results in waste solutions containing appreciable amounts of uranium. The uranium bearing aqueous raffinate from copper extraction is usually a dilute sulfuric acid solution. Uranium can be recovered using the same technique as described above for sulfuric acid leach liquors. A typical example is uranium production at the Olympic Dam mine in Australia, where the copper ore bodies are estimated to contain a total of over a million metric ton of uranium.

Estimates of the price of recovery uranium from seawater have shown that the cost is several orders of magnitude higher than for recovering U from high grade ores. High separation factors are needed for the recovery of U as the seawater contains many other elements. However, recent research on calixarenes[74] has showed that this type of macrocycle can achieve very high separation factors between U(VI) and other cations if the number of functional groups on the calixarenes corresponds to the number of ligands needed to saturate the coordination sphere of the UO_2^{2+} ion in its equatorial plan. Figure 5.9 shows the structure of two lipophilic *p-tert*-butylcalixarenes(6), bearing carboxylate or hydroxamate groups on the lower rim, which have been considered as extractants. The best functional group for the calixarene(6) is the hydroxamic group, CONHOH, probably because, at the pH of seawater, neutral U(VI) species are formed in the organic phase; with carboxylate groups the lower pK_a of R–COOH favors anionic species. The influence of chlorides in the feed solutions is studied in Soldenhoff et al.[75]

TABLE 5.10 Main Industrial Processes for the Recovery of U from Wet Phosphoric Acid

Extractant	Mechanism of extraction	Stripping	Remark
Octylpyrophosphoric acid (H_2OPPA)	$U^{4+} + 2H_2OPPA \leftrightharpoons U(OPPA)_2 + 4H^+$ (high distribution ratios)	In 25% H_2SO_4 + 10–20% HF: $U(OPPA)_4 + 4HF \leftrightharpoons 2H_2OPPA + UF_4$	Cheap but unstable extractant. Difficulties in stripping.
Octylphenyl orthophosphoric acids (mono plus di) H_2MOPA + HDOPA	$U^{4+} + (HDOPA)_2 + H_2MOPA \leftrightharpoons (DOPA)_2\ MOPA + 4H^+$ (high distribution ratios)	Oxidation to U(VI) : $U(DOPA)_2MOPA + 1/2O_2 + 3H_2O \leftrightharpoons UO_2^{2+} + H_2MOPA + (HDOPPA)_2 + 2OH^-$	Cheap extractant. Difficulty of stripping. Phase disengagement problems.
HDEHP-TOPO	$UO_2^{2+} + 2(HDEHP)_2 + TOPO \leftrightharpoons UO_2(DEHPHDEHP)_2 + 2H^+$ (rather low distribution ratios)	Reduction to U(IV) in H_3PO_4 or use of $(NH_4)_2CO_3$: $UO_2(HDEHP\ DEHP)_2TOPO + 3(NH_4)_2CO_3 + 2NH_3 \leftrightharpoons 4NH_4DEHP + TOPO + UO_2(CO_3)_3^{4-} + 4NH_4^+$	Rather expensive extractant. Must oxidize U(IV) to U(VI) before extraction.

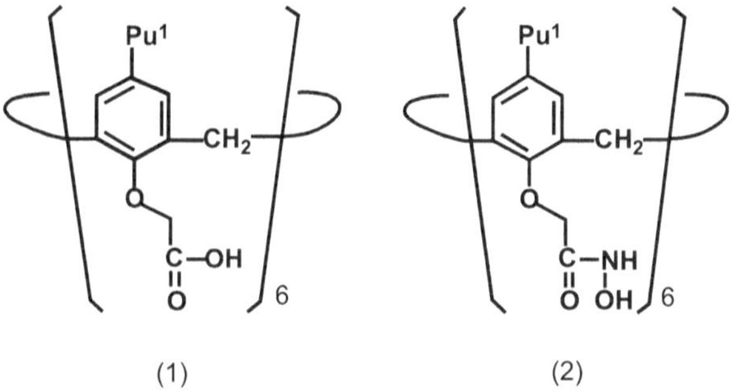

FIGURE 5.9 Formula of developed calixarene(6) selective extractants for U(VI) extraction from sea water: (1) Carboxylate functions; (2) hydroxamate functions. *(Source: From Jha et al.*[73]*)*

3.1.2. Thorium Production from its Ores

Thorium, which consists mainly of the isotope ^{232}Th, has often been proposed as a secondary source of nuclear energy.[58–62] Irradiation of ^{232}Th with neutrons produces ^{233}Th, a beta-emitter that decays to fissionable ^{233}U. The world reserves of ^{232}Th are about three times larger than those of U. However, because the U nuclear fuel cycle was developed first and is easier to use for nuclear energy production, the thorium cycle has not received as much attention. However, thorium must be considered as a future energy source. Fission of thorium produces less hazardous waste, because the amount of long-lived transuranium isotopes is about 100 times less than in the uranium cycle.

The purification of thorium takes place by TBP extraction after the dissolution in nitric acid. When thorium is dissolved in sulfuric acid, purification is achieved by extraction with long-chain alkylamines.[76]

Recently, supercritical fluid extraction[77–79] was successfully tested for recovery of uranium and thorium.

3.2. Spent Nuclear Fuel Recycling (Reprocessing)

The initial objective for reprocessing irradiated nuclear fuels[58–63] was to recover and purify ^{239}Pu for military applications. However, with the primary goal of commercial nuclear power reactors, reprocessing of the irradiated fuels was considered as the best use of the available uranium resources. Reprocessing also allowed recovery of valuable fissionable nuclides that could be recycled in fast breeder reactors.

3.2.1. Purex Process Recycling

To separate the actinide cations, advantage can be taken of their general chemical properties.[59] The different oxidation states of the actinide ions lead to ions of charges from +1 (e.g., NpO_2^+) to +4 (e.g., Pu^{4+}) (see Fig. 5.8), which allows the design of processes based on oxidation reduction reactions. In the Purex process, e.g., uranium is separated from plutonium by reducing

TABLE 5.11 Composition of Solution After Dissolution of Spent Fuel from Several Reactor Types in Nitric Acid

Fuel type characteristics	Graphite gaseous reactor	Pressurized water reactor	Fast breeder reactor
U(VI) (mol dm^{-3})	1.05	1.05	0.85
Pu(IV) (mol dm^{-3})	0.00275	0.01	0.22
HNO_3 (mol dm^{-3})	3	3	3
Zr (mol dm^{-3})	0.0013	0.0095	0.0184
Ru (mol dm^{-3})	0.0005	0.0058	0.02
Fission products (kg per metric ton)	4.16	35	87
Transplutonium elements (mol dm^{-3})		0.0002	0.0028

extractable Pu(IV) to nonextractable Pu(III). Under these conditions, U(VI) (as UO_2^{2+}) and also U(IV) (as U^{4+}), if present, remain in the TBP phase.

Irradiated UO_2 is dissolved in nitric acid, resulting in a solution with the approximate composition listed in Table 5.11. Nitric acid is preferred over other strong inorganic acids because it is the least corrosive acid for the stainless steel equipment used in large-scale reprocessing of spent nuclear fuels. The solution is treated by the Purex process. Most of the existing plants use some variation of the Purex process. Typically, the extractant composition (percentage TBP, diluent) and the extraction equipment (i.e., pulse columns, mixer-settlers, etc.), vary from plant to plant. However, the upper concentration limit is 30%TBP to prevent a phase reversal due to the increased density of the fully loaded solvent phase.

Three purification cycles for both uranium and plutonium are used. High levels of beta and gamma radioactivity are present only in the first cycle, in which 99.9% of the fission products are separated. The other two cycles, based upon the same chemical reactions as the first cycle, obtain additional decontamination and overall purity of the uranium and plutonium products. Distribution ratios of the actinides in largest concentration and valence state, and some important fission products, which form the basis for the design of the Purex process flow sheet, are shown in Table 5.12.

A typical flow sheet for the first cycle of the Purex process is shown in Fig. 5.10. The main steps are as follows:

1. Coextraction of 99.9% of Pu and U and some 90% of ^{237}Np from 3 to 4 mol/ dm^3 HNO_3 solution into 30 vol% TBP, diluted with a mixture of aliphatic

TABLE 5.12 Distribution of Various Metallic Species Between 30% TBP–Dodecane, 80% Saturated with U(VI), and HNO_3 Solutions at 25 °C

Metal[a]	3 mol dm^{-3} HNO_3	1 mol dm^{-3} HNO_3	0.01 mol dm^{-3} HNO_3
U(IV)	3.3	1.5	0.8
Np(VI)	0.8	0.4	–
Np(V)	Negligible	Negligible	Negligible
Np(IV)	0.12	0.04	Hydrolyzes
Pu(IV)	0.7	0.18	Hydrolyzes
Zr(IV)	0.01	0.0015	Hydrolyzes
Ru	0.0013	0.006	–
Ce(III)	0.0004	0.004	
Sr(II)	Negligible	Negligible	Negligible
Cs(I)	Negligible	Negligible	Negligible

[a]*Tracer level concentrations except for uranium.*

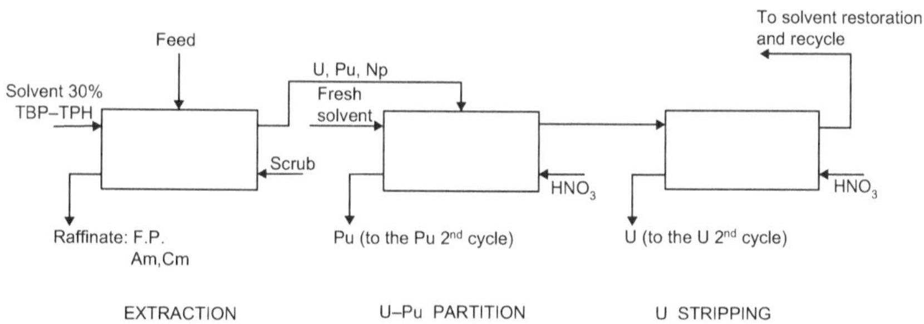

FIGURE 5.10 Flow sheet for the first purification cycle in the Purex process (TPH is hydrogenated tetrapropylene, a commercial branched dodecane).

hydrocarbons or, occasionally, with pure dodecane. Fission products and trivalent actinides (Am, Cm) remain in the aqueous raffinate.

2. U–Pu partitioning: U(VI) remains in the TBP phase when Pu(IV) is reduced to Pu(III) and stripped (back extracted). Several reducing agents [e.g., U(IV), Fe(II), and hydroxylamine nitrate ($NH_2OH–HNO_3$)] can be used; all have their advantages and disadvantages. Reduction of Pu(IV) to Pu(III) in the U–Pu partition step is a process that has to be carefully controlled, because inadequate reduction can result in accumulation of

plutonium in the partitioning equipment and increase hazards from nuclear criticality:

a) U(IV) solutions have to be prepared at the plant site and usually have to be stabilized against oxidants by addition of a nitrite scavenger such as hydrazine (N_2H_4), which can lead to the formation of highly explosive hydrazoic acid (HN_3). Although the use of U(IV) as reducing agent has the advantage of adding no new salts to the high-level waste, it does increase the solvent load and can lead to third phase formation. As it follows the uranium stream in the Purex process, it is not suitable in the second and third cycles, where Pu is purified from traces of U.

b) Instead other reducing agents, such as ferrous sulfamate or hydroxylamine, are preferred.[80] Ferrous sulfamate, $Fe(SO_3NH_2)_2$, rapidly reduces Pu(IV) to Pu(III) and the sulfamate ion reacts with HNO_2 to prevent the HNO_2-catalyzed oxidation of Pu(III). However, ferrous sulfamate adds some undesirable inorganic salts to the aqueous high-level waste.

c) Hydroxylamine is a kinetically slow reducing agent and has to be used in conjunction with other reagents, such as U(IV) and hydrazine.

d) Electrolytic reduction of Pu(IV) to Pu(III) appears to be a very useful method of reduction. However, there is no large-scale plant experience, and a nitrate scavenger, such as hydrazine, is often necessary.

3. Uranium stripping: Dilute HNO_3 solutions at 45–50 °C are used to remove uranium from the TBP phase. Traces of the fission products ruthenium and zirconium are eliminated in the second and third cycles of the Purex process. Also, in the second and third cycles, neptunium and the last traces of plutonium are removed from the uranium product.

3.2.2. Other Reprocessing Flow Sheets

Plutonium purification processes that were based on extraction by trilaurylamine [TLA] were developed in France[81] and were used for a time in the French UP2 Purex plant. The spent fuel pins are dissolved in acid and "reprocessed" to recover energy values (i.e., unused uranium and other fissionable isotopes, notably plutonium).

A range of polycyclic molecules has been developed and tested for the separation of actinides from lanthanides. The method is potentially useful in a spent nuclear fuel treatment flow sheet, where the resistance of reagents to ionizing radiation is clearly an issue.[82] The use of diglycolamide derivative of TODGA/TBP for the recovery of trivalent actinides from a PUREX raffinate was reported in Modolo et al.[83] Another PUREX raffinate treatment strategy involves the DIAMEX process developed by the French Atomic Energy Commission which employs a malondiamide as the extractant. Results of the successful recent tests[84,85] on the DMDOHEMA malonamide based step of this process, including palladium separation, were reported. Influence of diluent on the extraction behavior of U(VI) with benzyloctadecyldimethylammonium chloride (BODMAC) was reported in sano et al.[86] Neptunium co-recovery with uranium and plutonium was also tested.[87]

3.2.3. Reprocessing of Irradiated Thorium Fuels

Natural thorium spiked with ^{233}U, ^{235}U, or ^{239}Pu has been used as fuel in several large test reactors. After irradiation, thorium fuels contain Pa, ^{233}U, Np, and fission products. The Thorex solvent extraction process is generally used to reprocess spent Th-based fuels. As in the Purex process, the solvent is TBP diluted in an appropriate mixture of aliphatic hydrocarbons. Figure 5.11 shows the Thorex process flow sheet used for reprocessing high-burn-up thorium fuel.[88]

Several difficulties arise from the use of TBP in the Thorex process. Thorium (IV) is the major element present, and the loading capacity of a 30% TBP–dodecane solution is considerably lower for the tetravalent actinides than it is for U(VI); thus, a larger organic/aqueous phase ratio must be used to achieve suitable recovery of the thorium. Distribution ratios of Th(IV), Pa(V), and Zr(IV) are such that a good separation of Th(IV) is possible only at low aqueous feed acidity. Two extraction-strip cycles are necessary with fuels containing high quantities of hydrolyzable fission products. In the first cycle, carried out with acidic aqueous feed, thorium and uranium are separated from the bulk of the fission products. In the second cycle, performed with an acid-deficient feed solution, the final purification of uranium and thorium is achieved.

If reprocessing of spent Th fuels is needed in the future, many drawbacks of the Thorex process can be avoided by using *N*,*N*-dialkylamides instead of TBP.

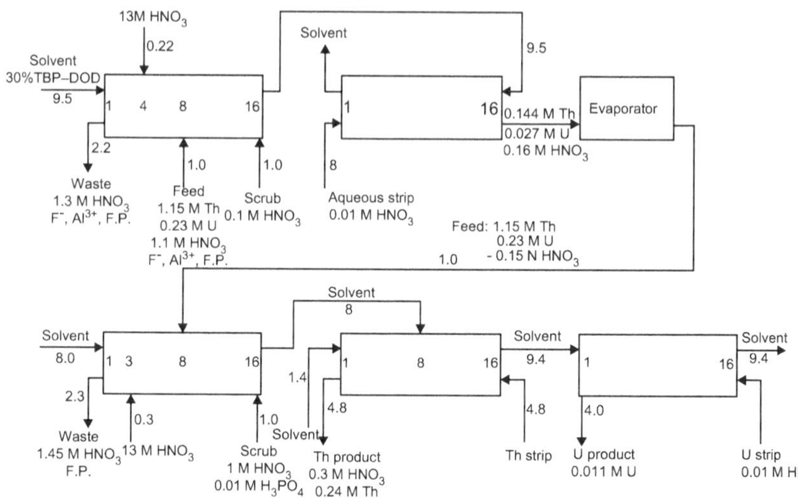

FIGURE 5.11 Two-stage acid Thorex process for highly irradiated fuels. Numbers in the frames indicate stage number, numbers on the lines indicate flow volumes relative to the feed volume (DOD and FP are dodecane and fission products).

From studies of Pu(IV)–U(VI) separation with amides, it is known that U(VI)–Th(IV) separation in acidic medium is easier than in TBP. The advantages of amides over phosphates could probably be considered more freely for Thorex than for Purex plants because no Thorex plant has been built yet. The extraction behavior of Pa(V) has not been yet investigated but in view of the general chemical properties of the two extractants, smaller distribution ratios are expected with the *N,N*-dialkylamides.

3.3. Nuclear Waste Management

Several kinds of radioactive wastes are produced by the nuclear industry, including those classified as high-level (HLW), low-level (LLW), and transuranic (TRU) wastes.[89] Typically, the solid and liquid wastes that contain $3.7 * 10^6$ Bq/kg (100 nCi/g) or more of radionuclides, with a half-life ($t_{1/2}$) greater than 20 years must be eventually disposed in immobilized forms such as vitrified glass in deep geological repositories. Conversely, low-level wastes, which contain $<3.7 * 10^6$ Bq/kg of radioactive elements can qualify for disposal in less expensive near-surface facilities.

Human safety, environmental protection, and cost of treatment must be considered with primary importance during the evaluation of waste treatment options. The primary goal of this effort is to reduce the actinide and fission product activities to levels which will allow the disposal of the majority of the radioactive waste to low-level waste (LLW). This would require removal of approximately 98.1%, 99.5%, and 99.98% for the TRUs, ^{137}Cs and ^{90}Sr, respectively. For economic reasons, therefore, there continues to be great interest to develop efficient and practicable solvent extraction processes for nearly quantitative extraction of all actinides, and other radionuclides.

In waste management programs, it might be desirable to isolate the minor actinides (Np, Am, Cm) and recycle them. These actinides appear together with the trivalent lanthanides that have undesirable nuclear properties (e.g., high neutron reaction cross sections), and thus it becomes necessary to remove the trivalent lanthanides. Unfortunately, the trivalent lanthanides behave chemically like the trivalent actinides, because their similar ionic radii decrease with the atomic number in each series and overlap, resulting in complexes of very similar strength, leading to very small separation factors.

After the separation of the actinides from the high-level waste, it is desirable to remove certain other fission products from the nuclear wastes. Some Cs^+, Sr^{2+} are low-charged cations that react well with macrocyclic ligands (e.g., crown ethers, calixarenes).[89] Research to synthesize and investigate the properties of macrocyclic ligands for application in nuclear waste treatment has been an active international effort.

Several solvent extraction technologies have been developed for the separation of the transuranic elements (TRUs), ^{137}Cs, and ^{90}Sr from actual acidic radioactive waste solution in a centrifugal contactor pilot plant. The solvent extraction technologies include the Transuranic Extraction (TRUEX) process; the Strontium Extraction (SREX) process; a chlorinated cobalt dicarbollide (ChCoDiC) process; a chlorinated cobalt dicarbollide/polyethylene glycol process; a phosphine oxide (POR) process and a universal solvent extraction process. For each of these solvent extraction processes, considerable experimental development and flow sheet modeling were performed. Subsequent immobilization processes for the high and low-level waste products from the separations are also being developed, including trivalent americium and curium, from all HNO_3 waste solutions.[89,90]

3.3.1. TRUEX Process

TRUEX process was developed for the removal of TRUs from acidic radioactive wastes. The process uses octyl(phenyl)-*N*,*N*-di-isobutyl carbamoyl methyl phosphine oxide (CMPO) as the active extractant, tributyl phosphate (TBP) as a phase modifier, and a hydrocarbon diluent. Trivalent lanthanides follow Am^{3+} and Cm^{3+} when the latter are stripped with dilute HNO_3; Anion-exchange resin and solvent extraction procedures have been used for separating uranium and technetium. The Truex process (see Fig. 5.12) has been

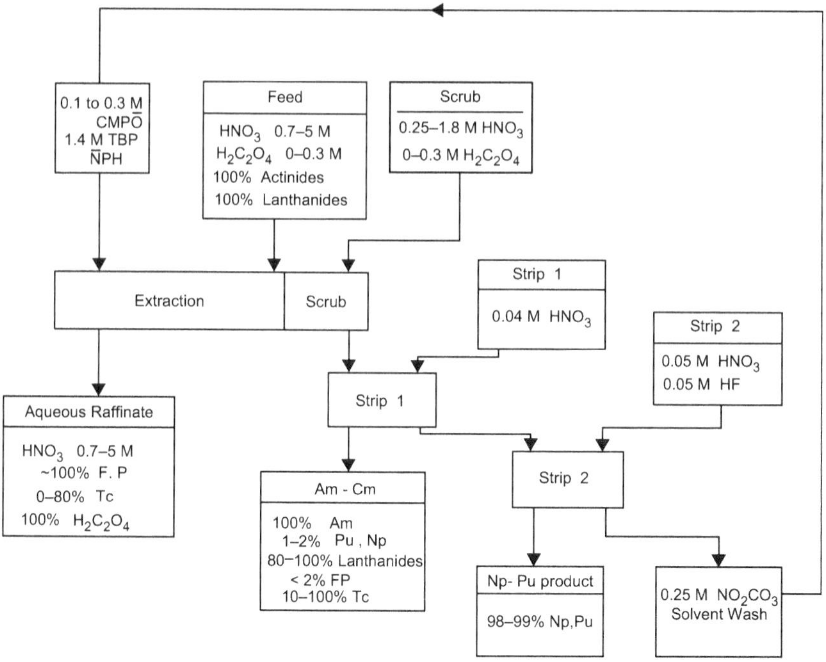

FIGURE 5.12 Truex process flow sheet for use with nitric acid wastes (NPH is a mixture of aliphatic hydrocarbons).

extensively and successfully tested.[90] Laboratory batch tests with simulated and actual waste solutions, as well as countercurrent pilot-scale flow sheet tests with simulated waste have been performed. The development of two TRUEX flow sheets has been demonstrated using actual tank waste in a centrifugal contactor pilot plant.[91]

3.3.2. SREX Process: Extraction of ^{90}Sr and $^{134,135,137}Cs$ from Acidic Wastes

To facilitate near-surface disposal of nuclear wastes, efficient removal of the $t_{1/2} \geq 30$-year fission product ^{90}Sr is needed. This has stimulated much research in the area of solvent extraction. The remediation of radioactive wastes usually requires the selective removal of Sr in the presence of numerous other ions. The complexation of macrocyclic polyethers, also called crown ethers, for Sr in aqueous solutions has been studied extensively. The crown ether 4,4(5),di-(*t*-butyl cyclohexano)-18-crown-6 (DtBuCH18C6) was determined to be a highly selective and efficient extractant for Sr.[92] TBP used as a phase modifier in a hydrocarbon diluent. The SREX process[93] provides such removal from highly acidic solutions. Batch contacts with simulated waste solutions show that the SREX process solvent is highly selective for strontium. From various contaminants, only barium and technetium coextract with strontium to any extent. Irradiation tests confirm that the SREX process solvent is satisfactorily resistant to radiolysis. Hydrolytic attack is also minimal. The SREX process has been successfully tested in both batch contacts and continuous countercurrent tests with simulated waste solutions. An overall ^{90}Sr decontamination factor (SREX process feed to SREX process raffinate) of 4600 was achieved from Truex wastes. SREX flow sheet has been demonstrated using actual waste solution in a centrifugal contactor pilot plant.[93]

3.3.3. Phosphine Oxide Process

Alkylated phosphine oxides are typically excellent extractants for TRUs in the +3, +4, or +6 valence states from acidic waste solutions.[94] Attachment of different alkyl or aryl substituents to the phosphonate group results in different compounds with variations in selectivity, extraction efficiency, solubility, and radiolytic and hydrolytic stability. A phosphine oxide solvent extraction process consists of an alkylated phosphine oxide in a hydrocarbon diluent. Laboratory batch contact, countercurrent pilot-scale testing with simulated and actual waste solutions has been performed. POR flow sheet has been demonstrated using actual waste solution in a centrifugal contactor pilot plant.[89]

3.3.4. Cobalt Dicarbollide Processes

Chlorinated cobalt dicarbollide (ChCoDiC) has been developed as an extractant for Cs from acidic waste solutions. With the addition of polyethylene glycol

(PEG) to the solvent, Sr can also be partitioned from acidic waste streams. Due to the acid form of cobalt dicarbollide being completely dissociated in the organic phase, a polar solvent such as nitrobenzene, has been used. The ChCoDiC process for the removal of ^{137}Cs and ^{90}Sr from acidic radioactive waste was developed.[95] Laboratory batch contact and countercurrent pilot-scale testing with simulated and actual waste solutions has been performed. ChCoDiC flow sheets (with and without PEG for the extraction of Sr) have been demonstrated using actual waste solution in a centrifugal contactor pilot plant.[96]

3.3.5. Universal Solvent Extraction Process

A single solvent extraction process to remove all or most of radionuclides, was desired instead of combination of processes in separate unit operations. The development of a universal solvent containing ChCoDiC with PEG to remove ^{137}Cs and ^{90}Sr, and a carbamoylmethylphosphine oxide derivative (diphenyl CMPO) to remove the TRUs, was initiated in 1996.[89] Laboratory batch contact and countercurrent pilot-scale tests with simulated and actual waste solutions have been performed. This testing has culminated in the development of a universal solvent extraction flow sheet which has been demonstrated using actual waste solution in a centrifugal contactor pilot plant.[89]

A review and comparison of the existing processes has been published.[97–99]

3.3.6. Bifunctional Organophosphorus Compounds for Waste Treatment

Strong efforts of researchers are devoted to development of novel universal extractants for the waste management. Several classes of reagents have been or are currently being investigated for their ability to extract actinides from both HNO_3 and HCl waste solutions. Bifunctional organophosphorus compounds of the formula $(R'O_p)(RO_m)PO–(CH_2)_n–PO(RO_m)(R'O_p)$ are known to form complexes with metal ions.[95] These compounds include diphosphonates ($m = 1$, $p = 1$), diphosphinates ($m + p = 1$), or diphosphine oxides ($m = p = 0$). Higher distribution ratios for +3 actinides from strong HNO_3 solutions are obtained when $n = 1$, which corresponds to the ideal size of a molecule for chelation of metal ions. Two carbamoyl-organophosphorus extractants have been intensively tested for their usefulness in plant-scale removal of any and all actinides from HNO_3 waste solutions: dihexyl-*N,N*-diethylcarbamoylmethylphosphonate [$(C_6H_{13}O)_2POCH_2CON(C_2H_5)_2$] and octyl(phenyl)-*N,N*-di-isobutyl-carbamoyl-methyl phosphine oxide [CMPO; $C_8H_7(C_6H_5)POCH_2CON(CH_2–CH(CH_3)_2)_2$]. The latter extractant is used in the Truex process.[90,91] Of the many radioactive and inert constituents in waste solutions, only ^{99}Tc and lanthanides coextract with actinides into the Truex process solvent.

The perceived advantages of bifunctional diamides as extractants for actinides in waste solutions include their ease of incinerability and the fact that radiolytic and hydrolytic degradation products are not deleterious to the process. These results were confirmed by extensive work in the last decade[100,101] in an EU collaboration program to develop a process (DIAMEX) for the extraction of actinides. Thirty new extractant molecules were synthesized and rated in order of ability to extract trivalent actinides; with selectivities toward the fission and corrosion products; third phase boundaries as a function of the aqueous nitric acid concentration; and the loading by lanthanides. Several tests using countercurrent separation in mixer-settler or centrifugal extractors with simulated and genuine high-level waste showed that the recovery of Am(III, IV, VI) is quantitative. The back extraction of Am(III) was complete.

These early flow sheets were not designed to strip U(VI) and Pu(IV). The distribution ratios of these ions at low acidities are lower than those measured with CMPO and one can guess that the stripping of Pu(IV) and U(VI) from Diamex will not be a difficult task. In addition, the degradation products of the solvent do not retain the actinides in the organic phase, as happens with the organophosphorus extractants.

Inactive metals in the wastes, such as Fe(III), Mo(VI), and Zr(IV), are retained in the aqueous feed by addition of suitable quantities of oxalic acid As with CMPO, some Ru is extracted. Currently, the extraction of Pd, Tc, Np, and their poor stripping remain a problem for which process modifications are necessary. The flow sheet in Fig. 5.13 has been tested in a battery of 16 centrifugal extractors with a genuine raffinate from a small-scale Purex process at the European Institute for Transuranium Elements (ITU), Karlsruhe.

3.3.7. Monofunctional Organophosphorus Reagents for Waste Treatment

Various monofunctional organophosphorus reagents [e.g., TBP; DBBP (dibutylbutylphosphonate); DEHPA (di-(2-ethylhexyl)phosphoric acid)] were considered for removal of alphaemitters from HNO_3 waste solutions. Typical flow sheets for the use of DEHPA, TBP and DBBP were tested. Many drawbacks were found:[102–104] hydrolyzation of Zr(IV) and Pu(IV), difficulties in stripping tetravalent and hexavalent actinides, etc.

3.4. Solvent Degradation and Regeneration

Routine regeneration of the TBP solvent is an important step in the Purex process. The TBP, as well as the diluents, are degraded by hydrolysis and radiolysis. Degradation of TBP results in products such as dibutylphosphoric acid (DBPA), monobutylphosphoric acid (H_2MBP), and phosphoric acid. Depending upon the type of diluent, hydrolytic and radiolytic attack produces nitrato esters, nitro compounds, carboxylic acids, ketones, and others. The degradation products of TBP are stronger complexants for cations than TBP itself, and their complexes are not very soluble in the organic phase. Elimination

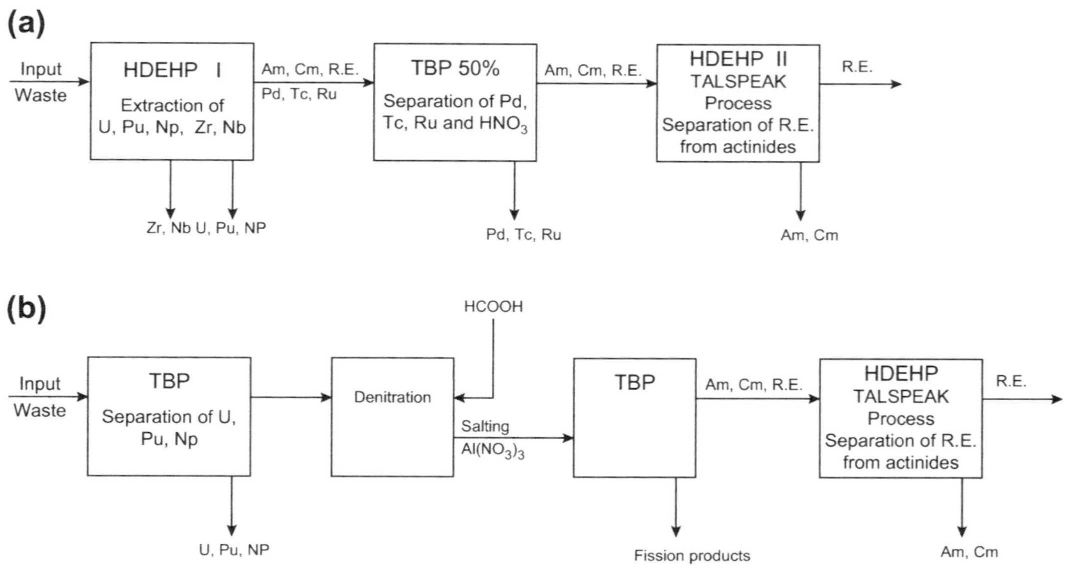

FIGURE 5.13 Schemes for partitioning Purex high-level waste. ((a) data from Ref. 86; (b) data from Ref. 87.)

of the degradation products, particularly the organophosphorus species, prior to solvent reuse is an important factor in the success of the Purex separations.[105]

Degraded TBP solvent is typically cleaned by washing with sodium carbonate or sodium hydroxide solutions, or both. Such washes eliminate retained uranium and plutonium as well as HDBP and H_2MBP. Part of the low-molecular-weight neutral molecules such as butanol and nitrobutane, entrained in the aqueous phase, and 90–95% of the fission products ruthenium and zirconium are also removed by the alkaline washes.

Alkaline washing is not sufficient to completely restore the interfacial properties of the TBP solvent, because some surfactants still remain in the organic phase. Various additional (secondary solvent cleaning) procedures have been proposed, including vacuum distillation,[106] treatment with macroreticular anion-exchange resins,[107] and treatment with activated alumina.[108] These secondary cleanup operations increase solvent quality, and continue to be studied to better evaluate their process and economic advantages. In the French UP3 Purex plant, vacuum distillation is used to regulate TBP concentration and solvent quality with beneficial effects. The latter include retention of metals and use of less concentrated Na_2CO_3 in the solvent wash, which helps to limit the amount of sodium salts that must be vitrified.

3.5. New Developments

3.5.1. Partitioning of Wastes

The fission products contain nearly 50 elements located in the middle part of the periodic system. Some of the fission products are produced in considerable amounts, though their radioactivity is rather low,[109] and they were therefore thought to have a potential value on the mineral market. Some other fission products were considered as suitable for radiation sources (heat sources, radiography, etc.). Thus, processes were designed to isolate these from the rest of the fission products. Today, there is little interest in such applications, but there is renewed interest in separating some key nuclides (e.g., ^{90}Sr, ^{99}Tc, ^{129}I, $^{134,135,137}Cs$), which have important medical applications. However, the prime interest comes from the desire to generate wastes that could be more easily and safely stored in deep geological repositories. In the most favorable situation, they could be safely stored in relatively inexpensive near-surface engineered facilities. The separation of the remaining actinides (traces of Pu, Np, Am, Cm, etc.) and of the most hazardous long-lived fission products is referred to as partitioning. The partitioned radionuclides would either be transmuted (e.g., actinides, ^{99}Tc, ^{129}I) in nuclear reactors or accelerators to form stable or short-lived nuclides, or stored (^{90}Sr and ^{137}Cs) in hardened engineered near-surface facilities for 300–500 years to allow them to decay to negligible levels.

Solvent extraction is thought to be a key technique for partitioning and such partitioning could improve the economics of final waste handling. The current conventional approach (see Fig. 5.14a) to partition radionuclides from Purex

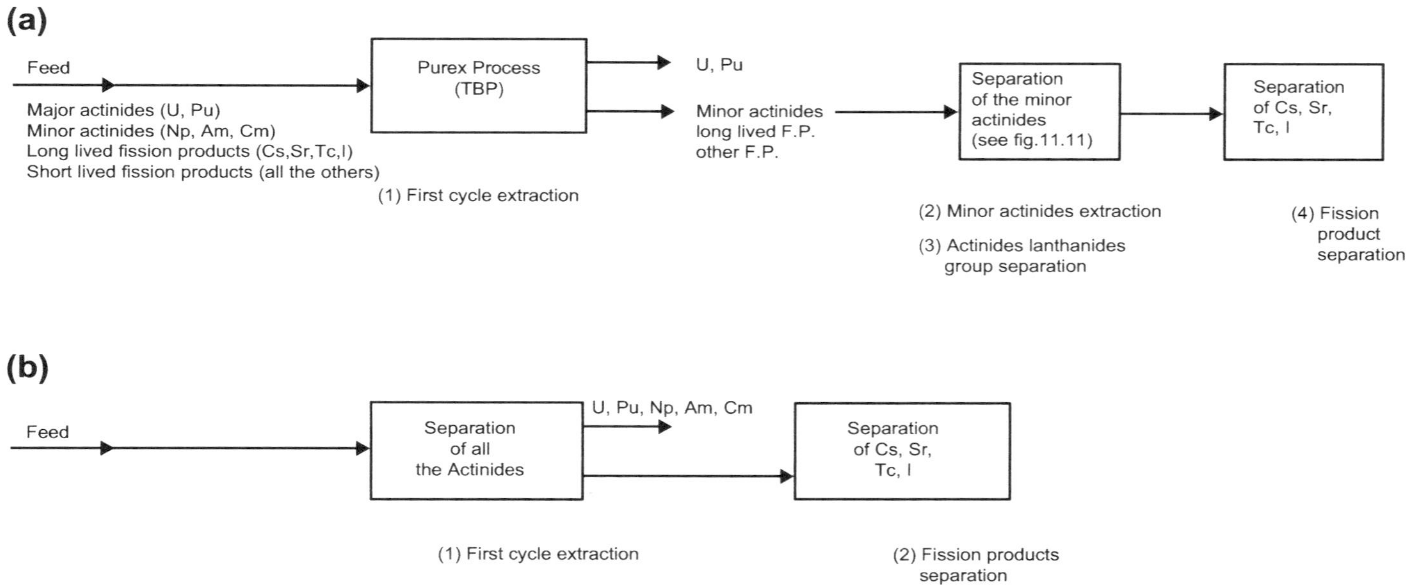

FIGURE 5.14 Concepts for partitioning radionuclides from Purex process high-level waste: (a) today's technology; (b) tomorrow's technology. *Note:* Numbers in parentheses indicate separations that must be performed remotely.

process acidic high-level waste involves a sequential series of separation processes to remove selected nuclides or groups of radionuclides. An example of this approach is the United States "clean use of reactor energy" (CURE) concept[110] that utilizes the Truex process to remove actinides, ^{99}Tc, and lanthanides; the Srex process to remove ^{90}Sr from the raffinate from the Truex process; and precipitation of cesium phosphotungstate from the Srex process raffinate by addition of phosphotungstic acid. Results of batch contacts to test the Truex and Srex process portions of the Cure concept were discussed previously. The cesium phosphotungstate precipitation process has been used extensively on a plant scale at the U.S. Hanford site.

The Cure concept and the Japanese OMEGA concept[111] are quite similar, although they will likely employ different reagents. For example, for removal of actinides from Purex process high-level waste, Japanese scientists are investigating the properties of di-isodecylphosphoric acid. The Japanese Omega concept also involves removal of valuable rhodium and palladium from the high-level waste.

The Cure and Omega concepts developed in the last decade but they have a number of potential disadvantages including the need to handle the highly radioactive waste in three or more extraction cycles. Computer simulations improve the processes applications.[112]

Improved partitioning processes that will accomplish the separations in fewer steps are highly desirable (as an example, see Fig. 5.14b). For example, a solvent extraction process that provides for quantitative recovery of both ^{90}Sr and ^{137}Cs in a single extraction cycle is very much needed. In addition, a desirable goal is to find an extractant that will extract (III), (IV), and (VI) actinides, and reject the lanthanides to the aqueous raffinate. One such program is the SPIN (solvent partition–incineration) program[113] for separation of the trivalent actinides from the trivalent lanthanides. This separation is necessary because nuclear incineration of the actinides requires that the neutron-consuming lanthanides be removed.

Recent investigations[114] indicate that certain soft donor ligands containing nitrogen and sulfur atoms extract trivalent actinides better than trivalent lanthanides from nitric acid solutions. However, the particular extractants studied then were not considered suitable for plant-scale use because of poor chemical stability. However, these results encourage further investigations of soft donor ligands for use in preferential extraction of (III) actinides, as well as (IV) and (VI) actinides, from HNO_3 solutions. The search for good extractants remains a challenge for the future.

3.5.2. Trivalent Actinide–Lanthanide Group Separations

Considerable interest of researchers has been attracted methods for the improvement separation of the minor actinides and some of the fission products from HLW. The goal was to optimize the formula of the soft donors (N and S)

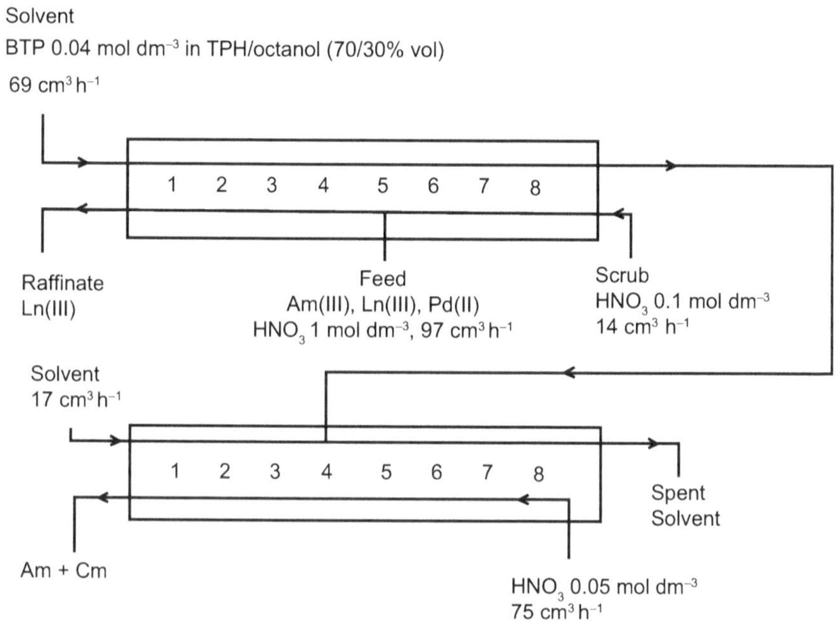

FIGURE 5.15 Flow sheet for the hot test of SANEX 3 process using n-Pr-BTP.

extractants in order to use them in large-scale separations. Many laboratories were involved in this work; a large experimental material describing behavior of these soft donor extractants in countercurrent separations was obtained. Altogether 78 nitrogen donor molecules were synthesized. Among them are the tridentate nitrogen donors from the family of the 2,6-bis-(5,6-dialkyl-1,2,4-triazine-3-yl)-pyridine.[115]

The trivalent actinide–lanthanide separation processes were named SANEX and four chemical systems have been explored. As an example, in Fig. 5.15 is shown schematically a flow sheet based on 2,6-bis-(5,6-dialkyl-1,2,4-triazine-3-yl)pyridine family of solvents. In Table 5.13, some experimental data of the tests of SANEX 3 process, employing this extractant with concentration of 0.04 mol/dm^3 in a mixture of 70% TPH/30% Octanol, which is able to selectively extract trivalent actinides from 1 mol/dm^3 HNO_3.[116]

The decontamination factors for Eu and Gd are rather low,[117] and kinetic or phase disengagement problems are probably responsible for the poor separation.

Sulfur donors belonging to the dialkyl-thiophosphorous acids[118] were also investigated (SANEX 4), and nine molecules have been considered. Separation factors between Am(III) and Eu(III) as high as 17,000 have been found and confirm the ability of sulfur donors to promote intergroup 5f-4f separations of trivalent ions. Individual sulfur donors and synergistic mixtures were investigated. The feed was an active solution simulating the aqueous output of the Diamex process. Pd(II) and Fe(III) were accumulated in the solvent, while

TABLE 5.13 Decontamination Factors (DF) for the Extraction (Mass in the Feed/Mass in the Actinide Back-Extraction) in the SANEX 3 Hot Test.

Element	% In raffinate	% In actinide strip	% In solvent	DF feed/strip
Am(III)	0.11	98.34	1.71	1.017
Cm(III)	0.15	93.89	8.07	1.065
Y(III)	114[a]	1.8	<0.06	54
La(III)	119[a]	<0.05	<0.05	>2000
Ce(III)	118[a]	<0.03	<0.03	>3800
Pr(III)	120[a]	<0.05	<0.05	>1800
Nd(III)	119[a]	<0.02	<0.02	>6300
Sm(III)	114[a]	0.4	<0.08	246
Eu(III)	112[a]	2.3	<0.37	43
Gd(III)	108[a]	3.1	<0.28	32
RuNO(III)	109[a]	<1	0.66	>100
Pd(II)	0.09	0.97	75.5	103
Fe(II)	6.8	<2	73.4	>50

[a]*Uncertainties of analysis are responsible for mass balances >100%.*

Ru is not extracted. The general properties of the sulfur donors, dithiophosphoric acids, were tested and described in Grüner et al.[99] More chemically stable molecules of bis-(2,4,4-trimethylpentyl)dithiophosphinic acid, CYANEX 301,[119] or bis-(chlorophenyl)dithiophosphonic acid[120] were investigated. Synergistic mixtures of 0.5 mol/dm^3 bis-(chlorophenyl)-dithiophosphonic acid and 0.25 mol/dm^3 TOPO, which permit rather acidic feeds (0.5 mol/dm^3 HNO_3) to be used, were also tested. The experiments showed several problems such as formation of gels, decomposition of the extractant at the high acidities needed for stripping, and accumulation of Fe, Pd, Tc, and Np in the solvent. Low extraction and stripping kinetics prevent these flow sheets reaching the scale up separation performance. However, very high Am(III)–Eu(III) separation factors (e.g., 17,500, were measured) indicate that the sulfur donors have a high potential for trivalent actinide lanthanide group separation and must be considered further.

Recently intensive investigations were undertaken in separation and purification of technetium-99 by solvent extraction techniques.[121–123] This interest is explained by growing demand in this isotope in medicine and biology.

4. SOLVENT EXTRACTION IN ANALYTICAL CHEMISTRY

In early analytical applications, optimal solvent extraction conditions were obtained empirically. Then, general chemical and mathematical descriptions of the processes were developed by many researchers in many countries and have been reported in a large number of publications, among them several monographs and reviews.[124–129]

Solvent extraction techniques are applicable in great range of concentrations (from weightless trace levels of radioisotopes to macro levels of metal ions) for which quantitative separations, therefore, is useful in analytical process that is more concerned with measurements and preparative chemistry as a separation, concentration, and purification steps in many fields such as hydrometallurgy, radiochemistry, biochemistry, medicine, ecology, engineering, etc. Among these, automatic methods of analysis have gained a notable momentum and have motivated the development of a large number of commercial instruments.

Solvent extraction has also played a major role in sample pre- or post-treatment to improve selectivity and sensitivity. Initially simple schemes of liquid–liquid extraction were used as separation methods for the cleanup and preconcentration of samples, mainly because of its simplicity, reproducibility, and versatility. Later, the distribution between two liquid phases was used as an efficient tool in chromatographic separation processes. More recently, the same principles have been implemented using supercritical fluids in preconcentration, cleanup, and column separation schemes. In the last decades, the principles and the properties of solvent extraction reagents were used to develop extraction chromatography techniques, to design different types of membrane electrodes, and to prepare and develop impregnated materials, where the extractant is placed in a solid matrix.

In this subsection the application of solvent extraction techniques in analytical chemistry is shortly presented. Analytical techniques in radiochemistry are presented in Section 3.5. Liquid membrane and liquid membrane electrode analytical techniques are excluded from this chapter because it is presented in more detailed and organized order in Ref. 130.

4.1. Solvent Extraction in Samples Preparation and Pretreatment Techniques

In sample preparation or sample pretreatment steps, there are a number of important operations that may include dissolution of the sample, transformation of the elements into specific inorganic forms, conversion of the analytes into alternative chemical species, separation of the analyte from other chemical species present, and preconcentration. These first stages of the analytical process decisively influence the precision, sensitivity, selectivity, rapidity, and cost.

Liquid–liquid extraction is by far the most popular separation method for the cleanup and preconcentration of samples because it is simple, reproducible,

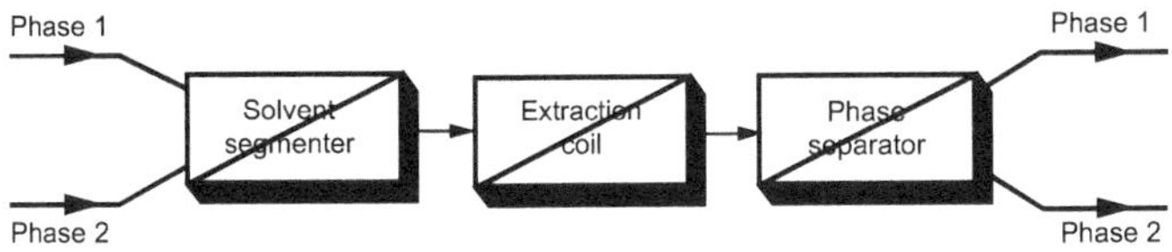

FIGURE 5.16 General scheme of a continuous solvent extraction contactor.

and versatile. There are several ways to achieve these objectives, from the original discontinuous ("batch") and nonautomatic techniques to continuous separation techniques incorporated with automated methods of analysis. The methodologies can be classified into two general types:

Discontinuous methods are performed in conventional separating funnels in one or more steps. Microscale extraction is carried out in autosampler tubes.

Continuous methods can be performed by different procedures such as solvent recirculation, extraction chromatographic techniques, or counter-current chromatography. The introduction of extractors in continuous methods such as segmented flow analysis (SFA), flow injection analysis (FIA), and completely continuous flow analysis (CCFA) are described below. Continuous-flow extraction involves segmenting of an aqueous stream with an organic solvent and separation of the phases using a membrane separator.[131] An interesting possibility is offered by countercurrent extraction, in which gravitational forces retain the organic phase while the aqueous phase is pumped through it. The basic components of a continuous liquid–liquid extractor are shown in Fig. 5.16. As can be seen, the continuous liquid–liquid extraction system consists of three main parts and performs the following functions:

1. It receives the two streams of immiscible phases and combines them into a single flow.
2. It facilitates the transfer of material through the interfaces of segmented flow in the extraction coil, the length of which, together with the flow rate, determines the duration of the actual liquid–liquid extraction.
3. It splits, in a continuous manner, the segmented flow from the extraction coil into two separated phases (phase separator).

4.1.1. Solvent Extraction for Separation Steps

Solvent extraction facilitates the isolation of analyte(s) from the major matrix and/or the separation of the particular analyte from concomitant trace or minor elements. Extraction is usually a fast and simple process, which demands only simple equipment. In practice, the separation/preconcentration step consists either of selectively removing the matrix without affecting the analyte(s) or of isolating the analyte(s) without affecting the matrix. There is a variety of separation/preconcentration methods available.[132,133] The most popular method includes solvent extraction, precipitation, sorption and chromatographic techniques, volatilization, and electrodeposition. The choice is dictated by the sample to be analyzed, the analytes, concentration levels to be determined, and the characteristics of the determination technique. In any case, the incorporation of a separation/preconcentration step increases the analysis time, may result in losses of the analyte(s), and at the same time raises demands both on the purity of reagents used and on the analytical expertize required.

The selectivity of the extraction is expressed by the separation factor S[134] that is derived from the individual distribution coefficients (D_1, D_2) for two species (5.5, 5.6):

$$S_{1/2} = \frac{D_1}{D_2} \tag{5.5}$$

This coefficient gives quantitative information for the separation of both species. Working with mixtures of extractants, additional experimental information is needed and the synergistic coefficient (SC) defined by Tagushi and Goto:[135]

$$\mathrm{SC} = \frac{\log D_{1,2}}{D_1 + D_2} \tag{5.6}$$

where $D_{1,2}$ denotes the distribution coefficient of the species for the mixture of extractants. In analytical chemistry, for an extraction system based on a pH-dependent extraction reaction, as in the case of acidic extractants, the value of SC has been estimated by the following equation:

$$\mathrm{SC} = n\Delta\mathrm{pH}_{50} \tag{5.7}$$

where n is the charge of the metal ion and $\Delta\mathrm{pH}_{50}$ is the difference of pH corresponding to 50% extraction when the total concentration of the extraction system is the same for the single system and for the mixtures.

4.1.2. Preconcentration for Trace Element Determination

Preconcentration is an operation in which the relative ratio of trace components vs. the macro component is increased; it is aimed, typically; for overcoming limited detection characteristics of the instrumental technique. The efficiency of this operation, the preconcentration factor (PF), is defined in terms of recovery as:

$$\mathrm{PF} = \frac{[C_\mathrm{T}]}{[C_\mathrm{T}^0]} \tag{5.8}$$

where $[C_\mathrm{T}]$ and $[C_\mathrm{T}^0]$ are the concentrations of the microcomponent in the concentrate and in the sample respectively. The need for preconcentration of trace compounds results from the fact that instrumental analytical methods often do not have the required selectivity and/or sensitivity. Thus, the combination of instrumental techniques with concentration techniques significantly extends the range of application of the instrument. The chemical techniques used in preconcentration can provide, in many cases, analyte isolation, as well as high enrichment factors. The concept of sample preconcentration prior to determination could apply to many situations other than just concentration enrichment and minimization of matrix effects.

Selection of a preconcentration scheme based on any liquid–liquid extraction step will depend upon on the type of analyte and/or sample matrix. Concentration factors reasonably achieved by solvent extraction have practical limits. However, considering a reasonable practical limit of extracting a 100 cm^3 sample with 1–2 cm^3 of organic phase, a concentration factor of 100–50 can be achieved. One of the limitations of the preconcentration factor in batch solvent extraction is the difficulty of obtaining good separation of the small volume of organic phase. Significant portions of the organic solvent may adhere to the walls of the vessel, requiring repeated washings, which decreases the concentration factor. Thus, to conduct successful analyses for many organic and inorganic compounds at trace concentrations, it is necessary to extract these compounds and use a concentration step prior to analysis. Many of the techniques developed for preconcentration are described in specialized books.[124–128] Proper choice of the extracting solvent can often be the critical step in the procedure.

Preconcentration of Inorganic Compounds

There has been a growing demand for highly sensitive analytical methods of trace components in complicated matrices. Recent advances of new ligands, chelating reagents, and instrumentation have improved detection and specificity. However, it is often necessary to concentrate the trace component with high selectivity. Chelating agents are typically widely used.

Considerable number of methods are not sufficiently specific and so prior separation was required. For example, interference in atomic absorption determinations is common, and so separations of interfering species are frequently required.[136] Group preconcentration of trace elements has the objective of isolating the maximum number of elements in a single step using the minimum number of reagents, with the excess reagents being retained in the aqueous phase. Using appropriate masking agents, suitable choice of pH, the reagents may become highly selective and specific for a particular species. In some cases, exchange techniques in which a less stable metal chelate is the source of the chelating agent may be very useful. Extraction reagents available for this purpose fall into several basic categories: chelating, ion exchangers, solvating, etc. They are described in detail in the previous chapters.

A series of preconcentration techniques based on the use of a solid phase carrying a liquid metal extractant has been developed.[137–139] These solid phases substitute the solvent (liquid phase) carried of the liquid–liquid extractant with a solid phase. Such impregnated materials have been prepared using different types of solid supports: paper, resin beds, gels, polyurethane foams, flat and hollow fiber membranes, clays, silica, etc. The major advantage of most of these solid adsorbents is that the functional group is immobilized on a solid substrate, therefore providing the possibility of either batch extraction of the analytes from solution or using the solid phase in a column.

Preconcentration and Separation of Organic Compounds

A scheme for analysis of organic compounds generally consists of sampling, extraction, prefractionation, and analysis, usually by gas chromatography (GC) or gas chromatography/mass chromatography (GC-MS). Determination of organics at parts per trillion (ppt) levels can be performed by combining sensitive and selective detection with sample preconcentration.[140]

Increasing the degree of preconcentration can make the limits of detection extremely small. Many standard methods of analysis[141–143] often include preconcentration as an integral part of the sampling and extraction procedure. It is especially important for environmental samples where many toxic and carcinogenic compounds are distributed from a wide variety of sources. Because of their serious damaging effects at low levels, to assess their environmental impact it is necessary to achieve the greatest possible analytical sensitivities. In some cases, detection limits as low as a few ppt are required. In addition to the polynuclear aromatic hydrocarbons (PAHs), much attention has been focused on many halogenated pollutants such as pesticides, polychlorinated biphenyls (PCBs), trihalomethanes (THMs), polychlorinated dibenzofurans (PCDFs), and dibenzo-*p*-dioxines.[144–149] Such substances can be found in air, water, and solid and biological samples. Most preconcentration techniques fall into two classes: solvent extraction followed by solvent reduction or sorbent trapping with subsequent solvent elution or thermal desorption. There are many variations of these methods, and they are frequently used in combination.

Solvent extraction reduction is most frequently performed mainly in connection with the extraction of solid and biological samples by liquid partition. Extractions are typically accomplished using a Soxhlet apparatus[150] that provides the benefits of multiple extractions. By repeated distillation and condensation of the solvent, the apparatus allows multiple extractions using the same (small) volume of solvent.

Preconcentration will only give precise results for quantitative work if the initial extraction technique gives high, or at least known and reproducible, recoveries of the desired compounds from the initial sample. Typically, several cycles are needed, and the solvent containing the extracted compounds must be concentrated to a small volume. This is normally carried out by evaporation under reduced pressure. Volatile compounds may be lost in this procedure. Even the very small volume losses on the glass walls of the flask or pipettes may result in significant and nonreproducible component loss.

Accelerated solvent extraction (ASE)[151,152] is a relatively new extraction technique using equipment that holds the sample in a sealed high-pressure environment to allow conventional solvents to be used at higher temperatures. Using a higher temperature without boiling, ASE allows smaller volumes of solvent to be used in a single-stage extraction. Extraction kinetics are also faster, so the entire process is much faster than Soxhlet extraction. After heating, the cell is allowed to cool below the normal boiling point of the solvent and pressure is applied to the cell to force the solvent and extracted materials through a filter.

4.2. Extraction in Analytical Determination Techniques

Extraction may provide an appropriate measured signal for analysis. Examples of this use of solvent extraction are found in spectroscopy, isotope dilution radiometry, and ion-selective electrodes using liquid membranes. In the latter case, electrochemical determinations are possible by controlling the local concentration of specific ions in a solution by extraction.

4.2.1. Methods Based on Spectrophotometric Detection

Only a few element species (e.g., MnO_4^{2-}, CrO_4^{2-}) are capable of absorbing light in the UV-VIS range. Usually, all spectrophotometric methods are based on reactions of analytes with color-forming reagents. The role of the most important organic reagents has been discussed in many publications.[150,153,154] Thousand articles have been published on analytical procedures for chemical elements in which solvent extraction is involved. The scope of this subsection limits us to few examples for each element which are presented in Table 5.14. Chelating reagents are the most popular. Although most of these reactions were initially developed for aqueous phase measurement, in many cases taking into account the low aqueous solubility of the analyte-chromogenic reagents, extraction into an organic phase was used for direct spectroscopic measurements. Among the most important reagents used are the following:[155] Dithizone (diphenylthiocarbazone) is a weak acid insoluble in water at $pH < 7$ but readily soluble in CCl_4 and $CHCl_3$. It reacts with many metal ions to form chelates that can be extracted into the organic phase from which the excess of the green reagent[156] is stripped with dilute NH_3. The most stable dithizonates (Pt, Pd, Au, Ag, Hg, and Cu) are extracted from strongly acid solutions. Other metals (Bi, Ga, In, Zn) are extracted from weakly acid media and some (Co, Ni, Pb, Tl, Cd) from neutral or alkaline media. Some compounds are extracted rapidly (Ag, Hg, Pb, Cd) and others more slowly (Pd, Cu, Zn), while a few (Rh, Ir, and Ru) require prolonged heating to be formed.

Azo dyes contain an azo link between two aromatic rings possessing an orthohydroxy group. The most important reagents include PAN, PAR, and Arsenazo III and generally they offer high sensitivity for the majority of transition metals.

TABLE 5.14 Elements Determined by a Solvent Extraction Procedure

Element	Extractant	Analytical technique	References
Alkali metals	Crown ether compounds	Spectrometry (AAS)	152
Lithium	Dipivaloylmethane	X-ray Fluorescence	153
Potassium	Crown ether compounds	Spectrophotometry (UV-VIS)	154
Rubidium/ Caesium	Nitromethane	Spectrometry (ICP-MS)	127
Copper	Dithizone	Isotope dilution	155
	Pyridine-carboxylic acid derivatives	Spectrophotometry (UV-VIS)	156
	Mixtures of Cyanex 301 and LIX 984 N	Spectrophotometry (UV-VIS)	157
Silver	Diethyldithiocarbamate	Isotope dilution	158
	Dithizone	Spectrophotometry (VIS)	150
Gold	Diphenyldipyridylmethane	Activation analysis	159
	Dithizone	Spectrophotometry (VIS)	150
Beryllium	Micelles-diketones	Spectrometry (ICP-MS)	160
Magnesium	Sulfonamide di-Schiff ligands	Spectrophotometry (VIS)	161
Calcium	Tributylphosphate/CCl4	Spectrophotometry (VIS)	162
Strontium	Polyethylene glycol	Activation analysis	163
Barium	Crown ether compounds	Spectrometry (AAS)	162
Zinc	Dithizone	Spectrophotometry (VIS)	164
	Diethyldithiocarbamate	Spectrophotometry (VIS)	150
	Thioxine	Molecular fluorescence	150
Cadmium	Diethyldithiocarbamate	Polarography	165
	Thioxine	Molecular fluorescence	150
	Pyridyl ketoximes	Spectrometry (Fl AAS)	166
			167
Mercury	Dithizone	Spectrophotometry (VIS)	168
			169
			170

(*Continued*)

TABLE 5.14 Elements Determined by a Solvent Extraction Procedure—cont'd

Element	Extractant	Analytical technique	References
Aluminum	Diisopropyl ether	Spectrophotometry (VIS)	171
Gallium	Tripod phenolic ligands	Activation analysis	172
	Thioxine	Spectrophotometry (UV-VIS)	150
	8-Hydroxyquinoline	Spectrophotometry (UV-VIS)	150
Indium	1-(2-Pyridylazo)-2-naphthol	Polarography	173
	Thioxine	Molecular fluorescence	150
	8-Hydroxyquinoline	Spectrophotometry (UV-VIS)	150
Thallium	2-Propanol	Spectrophotometry (UV-VIS)	174
	8-Hydroxyquinoline	Spectrophotometry (UV-VIS)	150
Boron	Different ligands	Spectrometry (ICP-MS)	175
Lanthanides	Diethylhexylphosphoric acid	Activation analysis	176
Scandium	Mesityl oxide	Spectrophotometry (UV-VIS)	177
Yttrium	Diantipyrylmethane	Complexometry	178
Thorium	Di-dodecylphosphoric acid (HDDPA)	Spectrophotometry (UV-VIS)	179
	8-Hydroxyquinoline	Spectrophotometry (UV-VIS)	150
Uranium	Quaternary amine micelles	Spectrophotometry (UV-VIS)	180
	Diisobutyl ketone	Spectrometry (ICP-MS)	181
	Thenoyltrifluoroacetone	Spectrophotometry (UV-VIS)	150
Germanium	Dibutyl ether	Spectrometry (AAS)	182
Tin	Diisopropyl ether	Activation analysis	183

TABLE 5.14 Elements Determined by a Solvent Extraction Procedure—cont'd

Element	Extractant	Analytical technique	References
Lead	2-Octylaminopyridine	Polarography	184
	Diazadibenzocrown ethers	Spectrophotometry (UV-VIS)	185
	Dithizone	Spectrophotometry (UV-VIS)	186
Titanium	Monolaurylphosphoric acid	Isotope dilution	187
Zirconium	Trioctylphosphine oxide	Spectrophotometry (UV-VIS)	188
Hafnium	Thenoyltrifluoroacetone	Spectrophotometry (UV-VIS)	189
Phosphorous	Butanol	Spectrophotometry (UV-VIS)	190
Arsenic	Acetic acid	Spectrometry (AAS)	191
	Thioxine	Spectrometry (AAS)	192
Antimony	*N*-Benzoyl-*N*-phenylhydroxylamine	Spectrometry (AAS)	193
	Thioxine micelles	Spectrometry (AAS)	194
Bismuth	Dithiocarbamate	Spectrophotometry (UV-VIS)	195
	Thioxine	Spectrophotometry (UV-VIS)	150
	Dithizone	Spectrophotometry (UV-VIS)	150
Vanadium	D2EHPA/TBP	Spectrometry (AAS)	196
	8-Hydroxyquinoline	Spectrophotometry (UV-VIS)	197
	Thenoyltrifluoroacetone	Spectrophotometry (UV-VIS)	150
Niobium	Tetraphenylarsonium salt	Spectrophotometry (UV-VIS)	198
	8-Hydroxyquinoline	Spectrophotometry (UV-VIS)	150

(Continued)

TABLE 5.14 Elements Determined by a Solvent Extraction Procedure—cont'd

Element	Extractant	Analytical technique	References
Tantalum	Methylisobutyl ketone	Spectrometry (AAS)	199
Selenium	Diethyldithiocarbamate	Spectrometry (AAS)	200
Tellurium	*N*-*n*-Octylaniline	Spectrometry (AAS)	201
Chromium	Methylisobutyl ketone admicelles	Spectrometry (AAS)	202
	8-Hydroxyquinoline	Spectrophotometry (UV-VIS)	203
Molybdenum	*N*-Benzoyl-*N*-phenyl hydroxylamine	Polarography	204
	Poly(sulfonylpiperidinyl methylene hydroxide	Spectrophotometry (UV-VIS)	205
	Methylisobutyl ketone	Spectrometry (AAS)	206
Tungsten	Methylisobutyl ketone	Spectrometry (AAS)	207
Manganese	Tetramethylene-dithio-carbamate	Spectrometry (AAS)	208
	8-Hydroxyquinoline	Spectrophotometry (UV-VIS)	150
Technetium	Tetraphenylarsonium chloride	Spectrophotometry (UV-VIS)	209
Rhenium	Aliquat 336	Activation analysis	210
Iron	1,1,1-Trifluoroacetylacetonates	Spectrometry (AAS)	211
	4-Acetyl-5-methyl-1-phenyl-1H-pyrazole-3-carboxylic acid	Spectrophotometry (UV-VIS)	212
Cobalt	Aliquat 336	Spectrophotometry (UV-VIS)	213
	1-Nitroso-2-naphthol	Spectrometry (AAS)	214
	Diethyldithiocarbamate	Spectrophotometry (VIS)	150
Nickel	Dimethylglyoxime	Polarography	215
	Aliquat 336	Spectrophotometry (UV-VIS)	216
	1-Nitroso-2-naphthol	Spectrometry (AAS)	217
	Thenoyltrifluoroacetone	Spectrophotometry (UV-VIS)	150

TABLE 5.14 Elements Determined by a Solvent Extraction Procedure—cont'd

Element	Extractant	Analytical technique	References
Platinum	Diphenyldithourea	Spectrophotometry (UV-VIS)	218
	Diethyldithiocarbamate	Spectrophotometry (UV-VIS)	150
	8-Hydroxyquinoline	Spectrophotometry (UV-VIS)	150
Palladium	Diphenyldithourea	Spectrophotometry (VIS)	219
	2-Methyl-8-quinolinol	Spectrophotometry (UV-VIS)	220
	2,2′-Dithiodianiline	Spectrophotometry (UV-VIS)	221
Rhodium	Diphenyldithourea	Spectrophotometry (UV-VIS)	222
	8-Hydroxyquinoline	Spectrophotometry (UV-VIS)	150
	Thenoyltrifluoroacetone	Spectrophotometry (UV-VIS)	150
Ruthenium	8-Hydroxyquinoline	Isotope dilution Spectrometry (AAS)	223

Chelating dyes include triphenylmethane reagents (e.g., Pyrocatechol Violet, Eriochrome Cyanine R, Chrome Azurol S, Xylenol Orange) and xanthene reagents (fluorones, e.g., Gallein, Pyrogallol Red, phenylfluorone, and salicylfluorone). They form chelates with most metals. Ionic surfactants make it easier to dissociate the protons of chelating triphenylmethane reagents and facilitate reactions with easily hydrolyzable metals (Be, Al, Fe, Se, Ti, Zr), leading to very sensitive but poorly selective methods. Nonchelating dyes include basic triphenylmethane dyes (e.g., brilliant green, malachite green, crystal violet), xanthene dyes (e.g., Rhodamine B, Rhodamine 6 G), azine dyes (e.g., methylene blue), and acid dyes (e.g., eosin, for chemical erythrosin). These are intensely colored and when paired with an oppositely charged analyte ion leads to high sensitivities. Crown ethers are not chromogenic unless they contain a pendant chromogen able to dissociate a proton in a basic medium. The resulting anion interacts strongly with the crown-complexed cation compensating the electric charge. The formation of a zwitterion leads to a hydrophobic extractable species with a considerably shifted absorption maximum compared with the protonated species. This allows the same spectrophotometric determination to be used for a large number of metal ions, provided the appropriate crown compound is used in each case. Another method involves using specific macrocyclic reagents that incorporate a chromophoric group, even if the latter is not involved in the complexation of the metal ion.

4.2.2. Other Detection Techniques (Atomic Absorption, Flame Emission Spectroscopy, X-Ray Fluorescence, Luminescence, etc.)

High selectivity of Atomic Absorption and Flame Emission Spectroscopy makes a separation process superfluous. Preconcentration, especially the removal of the solution medium, is often essential in atomic absorption determinations. Extraction into an organic phase decreases the detection limit by increasing the rate of introduction of the solution into the flame, since most organic solvents have a lower viscosity and surface tension than aqueous solutions. In addition, the sensitivity of the determination may be influenced favorably by changes in flame temperature and composition.[224] Using the combination of solvent extraction-flame atomic absorption, attention must be paid to the problems of vapor toxicity or noncombustibility of the extractant or its diluent. For example, benzene or chlorine containing solvents should always be avoided. Favored extraction systems are solutions of a chelating agent such as oxine (8-hydroxyquinoline), a ketone (methylisobutylketone), or butyl acetate.[150]

The advantages of solvent extraction in combination with atomic absorption apply equally well for flame emission spectroscopy. In addition, the latter analytical method often requires separation of the analyte from a large excess of other components. This may be achieved either by extracting the elements to be determined or by carrying out the spectrometric analysis of an aqueous solution from which the interfering components have been removed by extraction.[225]

Solvent extraction has also been used to enhance the selectivity of polarographic determinations. Such measurements are normally carried out in aqueous solutions, and extraction followed by back-extraction has been widely used. However, it may be unnecessary to perform a back-extraction if the organic extractant phase has a sufficiently high dielectric constant to dissolve sufficient background electrolyte for a voltammetric determination or if the organic phase can be diluted with suitable polar solvents, such as methanol or acetonitrile.[226]

Although X-ray fluorescence is generally used on solid samples, it may be beneficial in some cases to combine the procedure with solvent extraction. In addition to the usual advantages of separation and preconcentration (before the solid sample is prepared), it may be easier to prepare the required solid sample by evaporation or crystallization from an organic extract phase rather than from an aqueous solution.[150] Of special interest is the technique that uses an extraction system that is solid at room temperature, but has a relatively low melting point. Liquid–liquid extraction is performed at a temperature above the melting point and after separation of the phases, the organic melt is allowed to solidify to give a purified concentrate of the analyte.[157]

Solvent extraction is applied in several other analytical procedures. For example, (1) luminescence (fluorimetry), where detection limit of rhodamine complexes of gallium or indium can be increased by extraction;[227] (2) electron spin resonance using a spin-labeled extractant;[228] and (3) mass spectrometry, where an organic extract of the analyte is evaporated onto pure Al_2O_3 before analysis.[229]

4.3. Basic Materials for Solvent Extraction techniques

4.3.1. Solvent-Impregnated Resins

In the last decade, the development of new impregnated materials with chelating and complexing properties has acquired great importance. These materials are prepared principally by the simple immobilization of complexing organic reagents by adsorption onto conventional macroporous polymeric supports (polar and nonpolar). They provide two important advantages over conventional ion exchange resins: the possibility of selecting the functional group to be immobilized and the facility for continuous column operation. Such macromolecular resins containing an extractant within the lattice of the polymer were developed to bridge the gap between the two techniques of solvent extraction and ion exchange and are roughly classified into extractant-impregnated sorbents and Levextrel resins.[135] The former are prepared by soaking a polymeric resin in a diluent containing the extractant, then evaporating the diluent (dry impregnation method) or retaining the diluent (wet method). Levextrel resins are prepared by adding an extractant to the mixture of styrene monomers during bead polymerization with divinylbenzene. This method has increased in importance mainly because of its adaptability to preconcentration, separation, and/or determination of analytes. Additionally, they could be used in continuous-flow systems and in solid phase spectrophotometry by using chemical reactions that occur at interfaces (e.g., solid/liquid or gas/solid). Furthermore, most of the analytes could be analyzed directly on the solid matrix using X-ray spectrometry, neutron activation analysis, molecular absorption, fluorescence spectroscopy, or isotopic dilution methods. Alternatively, analytes may be eluted from the column and the analysis completed on the solution. In this context, atomic absorption (AAS) or inductively coupled plasma (ICP) spectroscopy are mainly used.

The use of impregnated resins in the preconcentration and separation of trace metal ions provides the following advantages:

1. The active part of the resin (complexing ligand) can be selected to be compatible with the nature of the metal ion and the matrix of the sample, and with the analytical procedure to be applied.
2. The impregnation procedures of the complexing molecules are simple.
3. Their structure and composition could be compatible with integrated detection systems, when used in solid phase spectroscopic measurements.

4.3.2. Solid-Phase Spectrophotometry Applications

Solid-phase spectrophotometry (SPS) is a technique based on the preconcentration of the species of interest onto a solid, aided by complexants or other reagents, and subsequent measurement of the spectrophotometric properties of the species in the solid phase.[230] Depending on the spectrophotometric responses of the analytes or analyte complexes, several procedures based on absorbance and fluorescence measurements have been developed. In the first case, the absorbance of a resin containing the analyte fixed as a colored chromogenic species is measured directly. In the second case, solid-phase fluorimetry (SPF), the diffuse reflected fluorescence, is measured. Most procedures using color or fluorescence measurement are based on the addition to the sample solution of a resin impregnated with a highly specific chromogenic agent for the analyte.

Ion exchanger colorimetry has been used as a sensitive and rapid method for vanadium analysis by immobilization of 2[2-(3-5-dibromopyridyl) azo]-5-dimethylaminobenzoic acid onto an ion exchanger resin AG1X2.[230] Solid phase fluorimetry can be useful for the analysis of very dilute solutions in water analysis or trace metal determination; thus a chelating 8-(benzene-sulfonamido)quinoline, immobilized on Amberlite XAD2 support, has been used for the spectrofluorimetric determination of Zn(II) andCd(II).[231]

4.3.3. Applications for Fiber Optic Chemical Sensors

The development of chemical sensors based on optical measurements has grown steadily in importance during the last decade. While a large variety of devices are possible, they share a common feature in multiple applications, i.e., an immobilized reagent phase that changes its optical properties upon interaction with an analyte on either a continuous or reusable basis. Systems in which chelating liquid extractants are immobilized onto solid polymeric supports have been used for chemical sensing. Particular attention has been given to fiber-optic devices for measuring and controlling metal ions and organic compounds in aqueous media. The applications of such devices have covered areas such as environmental applications, industrial process control, and biomedical and clinical applications. Small sensors based on immobilization of acid-base indicators such as bromothymol blue, bromophenol blue, bromocresol purple, phenolphthalein, phenol red, chlorophenol red alizarin, on nonionic macroporous supports, Amberlite XAD-2 and Amberlite XAD-4, have been developed and used for pHmeasurements.[231] These fiber-optic probes provide advantages over conventional electrodes in safety, reliability, applicability, and cost. A sensor based on perylene dibutyrate adsorbed onto Amberlite XAD4 has been characterized in some depth.[232] Fiber optic sensors, particularly fluorescent sensors, have become the object of considerable interest among researchers in recent years. Their performance is based on the change in fluorescent properties of organic reagents immobilized on a solid matrix upon contact with solutions of metals in a continuous system.

4.4. Micelles in Analytical Chemistry

Another of new techniques for extractive preconcentration, separation, and/or purification of metal chelates, biomaterials, and organic compounds is based on the use of surfactant micellar systems. Surface-active agents aggregate in aqueous solutions to form micelles if the concentration in aqueous solutions exceeds the critical micelle concentration (CMC). Dilute aqueous solutions of certain surfactant micelles, when the conditions (i.e., temperature, pressure, and electrolyte concentration) of the solution are changed, have the ability to separate into two isotropic liquid phases: a surfactant-rich phase with a small amount of water (surfactant phase or coacervate phase) and a phase containing an almost micelle-free dilute aqueous solution. This separation is reversible so that on changing the conditions, e.g., cooling, the two separated phases merge to form a clear solution once again. This phenomenon can be exploited for separation and concentration of solutes. If one solute has certain affinity for the micellar entity in solution then, by altering the conditions of the solution to ensure separation of the micellar solution into two phases, it is possible to separate and concentrate the solute in the surfactant-rich phase. This technique is known as cloud point extraction (CPE) or micelle-mediated extraction (ME). The ratio of the concentrations of the solute in the surfactant-rich phase to that in the dilute phase can exceed 500 with phase volume ratios exceeding 20, which indicates the high efficiency of this technique. Moreover, the surfactant-rich phase is compatible with the micellar and aqueous-organic mobile phases in liquid chromatography and thus facilitates the determination of chemical species by different analytical methods.[233] Typical preparation steps for analytical determination are presented in Fig. 5.17.

The most common surfactants for analytical applications are nonionic (polyoxyethylene glycol monoethers, polyoxyethylene methyl-*n*-alkyl ethers, *t*-octylphenoxy polyoxyethylene ethers, and polyoxyethylene sorbitan esters of fatty acids) that demonstrate cloud point behavior with increasing solution temperature, and zwitterionic surfactants (ammonioethyl sulfates, ammoniopropyl sulfates, ammoniopropane sulfonates, phosphobetaine, dimethylalkyl

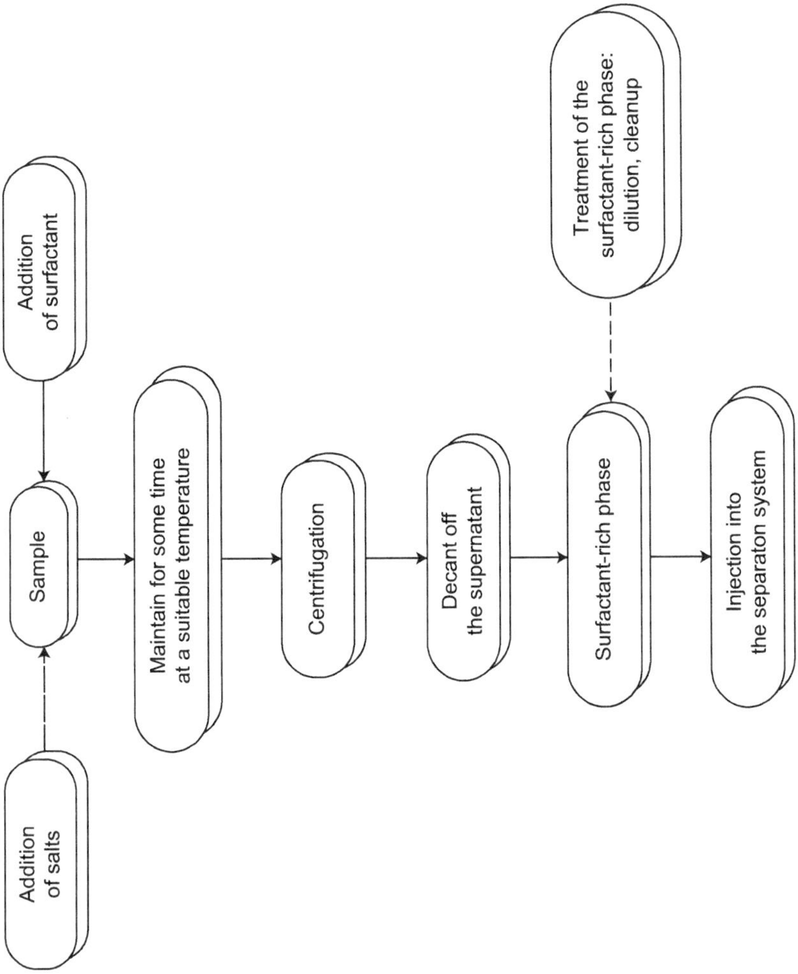

FIGURE 5.17 Steps involved in cloud point extraction (CPE) prior to HPLC, GC, and CE analysis.

phosphine oxides) that show cloud point behavior on decreasing solution temperature. Thus, cloud point temperature depends on the structure of the surfactant and its concentration and range from −0.5 °C to 120 °C.[234,235] In a homologous series of polyoxyethylated surfactants, the cloud point temperature increases with the hydrocarbon chain length and increasing length of the oxyethylene chain. At a constant oxyethylene content in the surfactant molecule, the cloud point temperature is lowered by decreasing the molecular mass of the surfactant and by branching of the hydrophobic group.

The most important advantage of cloud point extraction is that only small amounts of nonionic or zwitterionic surfactants are required and consequently the procedure is less costly and more environmentally benign than other conventional extraction techniques such as liquid–liquid extraction and solid–liquid extraction.[236,237] Moreover, CPE offers the possibility of combining extraction and preconcentration in one step.

Cloud point extraction has been applied to the separation and preconcentration of analytes including metal ions, pesticides, fungicides, and proteins from different matrices prior to the determination of the analyte by techniques such as atomic absorption, gas chromatography, high performance liquid chromatography, capillary zone electrophoresis, etc.

Cloud point extraction has been applied successfully to the extraction of metal chelates for spectrophotometric, atomic absorption, or flow injection analysis of trace metals in a variety of samples.[236–238] Other metal complexes such as $AuCl_4^-$ or thiocyanato-metal complexes can be extracted directly using nonionic surfactants such as polyoxyethylene nonyl phenyl ether (PONPE) or Triton X prior to their determination by ICP or visible spectrometry.[239–241] The main advantages of cloud point extraction of metal ions include the simplicity of the extraction procedure and the possibility of obtaining high preconcentration factors ranging from 10 to 100 allowing the development of analytical methods for determining metal ions at very low concentrations. For example, Pd(II) can be determined in a surfactant phase of Triton X-100 with coproporphyrin III as a complexing agent in a procedure based on phosphorescence at room temperature with detection limits of 20 nmol/dm^3. Some parameters for the cloud point extraction of metals with different surfactant micelles are given in Table 5.15.

The most frequent use of cloud point extraction is for the separation and purification of biological analytes, principally proteins. In this way, the CPE has been used as an effective tool to isolate and purify proteins when combined with chromatographic separations. Most of the applications deal with the separation of hydrophobic from hydrophilic proteins, with the hydrophobic proteins having more affinity for the surfactant-rich phase, and the hydrophilic proteins remaining in the dilute aqueous phase. The separation of biomaterials and clinical analytes by CPE has been described.[234,235,240]

Recently, CPE has been used for sample preparation and preconcentration of organic analytes such as pesticides,[242,243] herbicides, polycyclic aromatic hydrocarbons (PAH),[244] polychlorinated biphenyls (PCBs),[245] and phenols[246]

TABLE 5.15 Cloud Point Extraction of Metals Chelates Using Non-ionic Surfactants Micelles

Metal ion	Ligand[c]	Non-ionic surfactant[d]	Experimental conditions			Reference
			pH	C_F^a	% E^b	
Ni(II)	TAN_c	Triton X-100	7.0	30	Phosphate	234
	PAN	Triton X-100	5–6			234
	PAN	OP[e]	6.0	15–25		235
Ni(II), Zn(II)	PAN	Triton X-114				236
Ni(II), Zn(II), Cd(II)	PAMP	PONPE-7.5				236
Ni(II), Cd(II), Cu(II)	PAMP	PONPE-7.5	5.6			236
Zn(II)	PAN	PONPE-7.5	10	40	Carbonate	236
	QADI	PONPE-7.5	9	40	NH_3	192
Au(III)	HCl	PONPE-7.5			>95	239
Fe(III), Ni(II)	TAC	Triton X-100				236
U(VI)	PAN	Triton X-114	9.2		98	237
U(VI), Zr(IV)	Arsenazo	Tween 40[f]	3		>96	193
Cu(II), Zn(II), Fe(III)	Thiocyanate	PONPE-7.5			72.5–96.8	235

Er(III)	CMAP	PONPE-7.5	20		190
Gd(III)	CMAP	PONPE-7.5	3.3	99.9	190
Cd(II)	PAN	Triton X-114	60		236
Ru(III)	Thiocyanate	Triton X-110	5–10		240
Au(III)	HCl	PONPE-10		>90	239
Ga(III)	HCl	PONPE-7.5		>90	240
Ag(I), Au(III)	DDTP	Triton X-114[f]	9–130		239
Cu(II)	LIX54	Igepal CO-630			236
Cd(II)					241

[a]*Concentration factor.*
[b]*Percent extracted.*
[c]*Nomenclature for ligands: CMAP: 2-(3,5-dichloro-2-pyridylazo)-5-dimethylaminophenol; DDTP: o,o-diethyldithiophosphoric acid; LIX54: dodecylbenzoylacetone; PAN: 1-(2-pyridylazo)-2-naphthol; QADI: 2-(8-quinolazo)-4,5-dipheylbenzimidazole; PAMP: 2-(2-pyridylazo)-5-methylphenol; PAP: 2-(2-pyridylazo)-phenol; TAC: 2-(2-thiazoylazo)4-methylphenol; TAN: 1-(2- thiazoylazo)-2-naphthol.*
[d]*Nomenclature for surfactants: Igepal CO-630: nonylphenoxypoly(ethylenoxy)ethanol; PONPE-7.5: polyoxyethylene(7.5)nonylphenyl ether; PONPE-10: polyoxyethylene(10)nonylphenyl ether; TRITON X: t-octylphenoxypolyoxyethylene ether.*
[e]*OP surfactants refer to (polyethyleneglycol octylphenyl ethers).*
[f]*In this extraction, the concentrated surfactant-rich phase was a solid.*

in environmental samples prior to their determination. The compatibility of the surfactant-rich phase from CPE with micellar or conventional hydroorganic mobile phases allows subsequent determination of the analyte by thin layer chromatography, HPLC,[247,248] micellar electrokinetic capillary chromatography, or CZE.[249] Cloud point extraction is a rapid, simple, sensitive, and efficient sample pretreatment for trace environmental analysis and offers some advantages over conventional liquid–liquid extraction technology in terms of enhanced detection limits from the large preconcentration factors, elimination of analyte losses during evaporation of solvents used in liquid–liquid extraction, and elimination of toxic solvents.[236]

4.5. Liquid Membranes in Analytical Chemistry

The use of liquid membranes (LM) separation techniques in analytical applications has greatly increased in the last 20 years. This field is not in the scope of this book and described in detail in Kislik (2010).[130] Here we describe shortly below the application of LM techniques as ion-selective electrodes.

4.5.1. Ion-Selective Liquid Membrane Electrodes

Ion-selective electrodes (ISE) are described in the Liquid Membrane literature as fixed active functional groups in pores of the membrane support, or Polymer inclusion membranes (PIM), or affinity permselective materials.[130] The principle is incorporation of electron-rich complexing organic compounds as the ionophore together with an appropriate plasticizer and additive into pores of membrane to provide the membrane with the properties of a liquid phase. Ionophores are lipophilic, capable of reversibly binding ions and transporting them across membrane supports. The mode of action of these electrodes is therefore a reversible ion exchange process accompanied by diffusion and migration effects.

Potentiometric sensors, such as ion-selective electrodes, ion-selective microelectrodes, and ion-selective field effect transistors (ISFET) are used in analytical systems. Moreover, the recently introduced ion-selective optodes based on absorbance or fluorescence measurements provide additional sensors for a broad range of monitoring applications.[250,251] In these systems, the active constituent of the liquid membrane is an ion-pairing or complexing agent that is selective for a limited number of specific ions. The reader can find details in our book on liquid membranes.[130]

Many different designs have been suggested to solve the problem of interposing the liquid membrane between the aqueous solutions. The support prevents leakage of the organic solvent into the aqueous solution and at the same time maintains efficient contact with the analyte solution. Schematic design of the electrode is illustrated in Fig. 5.18. The construction of a liquid membrane electrode is rather similar to that of the glass electrode, in that it requires an internal reference electrode. The ion selectivity of a membrane can

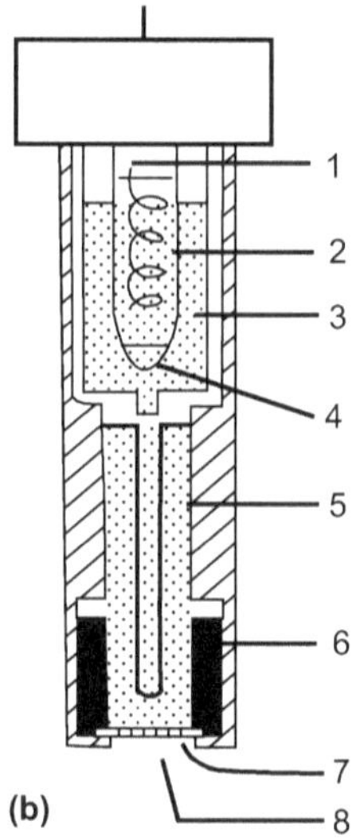

FIGURE 5.18 Schematic diagram of commercial liquid membrane electrode; improved by Szczepaniac and Oleksy[249] version.

be established by measuring the potential difference between two identical reference electrodes. One electrode is immersed in the specimen solution, the other in a reference solution, and the membrane is interposed between them. The composition of solution 2 is constant, and, if both solutions contain monovalent ions, the ion-exchange process can be described as follows:

$$J^+_{\mathrm{w}} + I^+_{\mathrm{mem}} \xleftrightarrow{K_{ij}} J^+_{\mathrm{mem}} + I^+_{\mathrm{w}} \tag{5.9}$$

and the potential difference is given by:

$$\Delta E = E_0 + 0.059 \log(a_i + K_{ij}a_j) \tag{5.10}$$

where E_0 is a constant related to the reference electrodes and the reference solution, a_i and a_j are the activities of ions i and j in the specimen solution, and K_{ij} is the selectivity constant of the membrane.

Equation (5.10) shows that the selectivity behavior of a liquid membrane is specified completely by the membrane selectivity constant, K_{ij}, which in turn is dependent on the equilibrium constant of Eqn (5.11) and on the mobility of ions i and j within the membrane. For the case in which the membrane consists of a neutral carrier,[253] the exchange reaction can be presented as:

$$J^+_{\mathrm{w}} + IS^+_{\mathrm{mem}} \xleftrightarrow{K_{ij}} JS^+_{\mathrm{mem}} + I^+_{\mathrm{w}} \tag{5.11}$$

If the ligand S (neutral carrier) is capable of completely enveloping the cations, the terms K_{ij} and E_0 of the membrane depend only on the ratio of the complex stability constants $\beta_{\mathrm{JS}} = \beta_{\mathrm{IS}}$ of the ions with the ionophore.

Liquid membrane electrodes are not capable of being specific for only one ion in solution. There is always some interference from other ions in solution with the given analyte. The selectivity coefficient provides an indication of the ability of an electrode to measure a particular ion in the presence of another ion. The response of an electrode to an interfering ion can be included in the Nernstian equation:

$$\Delta E = E_0' + \frac{0.059}{z'} \log(a_i + K_{ij} a_j^{z'/z}) \qquad (5.12)$$

where E_0' is a constant emf, including that of the reference electrode and the standard potential of the ion being measured; a_i is activity of species i with charge z'; a_j is activity of interfering ion, j, with charge z; $K_{i/j}$ is selectivity constant of the electrode for i over j.

The electromotive force (emf) of liquid membrane electrodes depends on the activity of the ions in solution. Their performance is similar in principle to that of the glass electrode. To characterize the behavior of liquid membrane electrodes, the linearity of the emf measurements vs. concentration of a certain ion in solution is checked. Additional performance data are the Nernstian slope of the linear range and the pH range over which the potential of the electrode is constant.

As mentioned above, ion-exchange and neutral selective carriers and their properties are described in our previous book.[130] Some liquid membrane electrodes are presented in Table 5.16. They have found wide application in

TABLE 5.16 Ion Exchange Electrodes Widely Commercially Applicable

Ion	Ligand	Application area	References
Li^+	Trioctylphosphine oxide + 255 neutral carrier	Lithium in blood serum and pharmaceuticals	251, 252
K^+	Valinomycin dissolved in diphenyl ether	Potassium in feldspar, urine, blood serum, seawater, vegetables	255
NH_4^+	Nonactin or dinactin	Activity of nitrate reductase, ammonium in mineral water, fruit juice, beer, urine, sewage water	256
Ca^{2+}	Phosphates and phosphonates	Protein-binding studies, calcium in sewage water, mineral water, blood serum, biological fluid	254, 257
NO_3^-	Tributyl octadecyl-phosphonium nitrate	Nitrate in agricultural products, soils, foods	258

biological and technological systems. All electrodes perform well in the concentration range over which the Nernstian slope is maintained, i.e., from 10^{-1} to 10^{-5} mol/dm^3. These electrodes to a certain extent have replaced in both chemical and clinical laboratories the more traditional instrumental methods of analysis, such as flame photometry and atomic absorption spectrometry. There are, of course, many more liquid membrane electrodes, but the availability of satisfactory solid electrodes has greatly restricted their development and practical application.

4.6. Chromatography-Based Analytical Techniques

The term "chromatography" now embraces a variety of processes that are based on the differential distribution of the components in a chemical mixture between two phases. The chromatography is the technique with continuous equilibration between two phases: stationary and mobile. The same fundamental laws of phase equilibrium apply to extraction and chromatography. In principle, the latter is a multiple extraction process in which the mobile phase moves continuously over the fixed phase in a chromatographic column.

4.6.1. Liquid–Liquid Partition Chromatography

The liquid–liquid partition chromatography (LLPC) method involves a stationary liquid phase that is more or less immobilized on a solid support. The analyte is distributed between the two liquid phases. In conventional LLPC systems, the stationary liquid phase is usually a polar solvent and the mobile liquid phase is an essentially water-immiscible organic solvent. On the other hand, in reversed-phase chromatography (RPC), the stationary liquid is usually a hydrophobic solvent, whereas the mobile liquid is the polar solvent, most frequently water or an aqueous mixture with polar organic solvents.[259,260]

In LLPC, solid support is used for the stationary liquid: silica gel, cellulose powder, starch, alumina, silicic acid, etc. The practice of HPLC has greatly simplified the technique in providing column stability for repeated use and for treatment of large volumes.

Reversed-phase chromatography is the predominant technique in HPLC, and chemically bonded silica gel supports are made specifically for the nonpolar stationary phase. In the last decade, as many as 60% of the published LLPC techniques refer to RPC. The reasons for this involve the significantly lower cost of the mobile liquid phase and a favorable elution order that is easily predictable based on the hydrophobicity of the eluate.

The selection of solvents for LLPC is similar to the selection of solvents in liquid–liquid extraction systems. The solid support has little effect upon the selection of the solvent pair, except for the obvious fact that a hydrophilic support for a polar stationary phase requires a hydrophobic mobile phase with the opposite for a reversed-phase system. Both solvents should exhibit good solubility for the solute(s); otherwise the column loading capacity would be too low. Frequently, the separation potential of LLPC columns can be additionally enhanced by using solvent mixtures rather than a single solvent as the mobile phase. In RPC, typical mobile phases are water, aqueous electrolyte solutions, or aqueous mixtures of one or more water-immiscible organic solvents. Water, the most polar of mobile phases employed, is the weakest eluent. For specific purposes, such as the separation of closely related acidic or basic substances, the mobile phase is adjusted by buffers.

A large variety of analytical separation processes has been reported. Table 5.17 demonstrates the range of organic compounds for which LLPC has been applied. The method has been limited essentially to organic compounds.

However, chromatographic separation procedures have been successfully used to separate metals with a combination of a cation exchanger as a stationary phase and a solution of a chelating reagent as a selective mobile phase. Thus chromatographic separations of many of the rare-earth elements: Gd(III), Y(III), La(III), Pr(III), Nd(III), Ho(III), Er(III), Sc(III) from acidic solutions have been achieved using different acidic organophosphorous extractants (e.g., DEHPA, PC88A) as well as neutral organophosphorous extractants (e.g., TOPO, TBP) immobilized onto polymer beads as stationary phases.[261,262] Similar approaches have been used in environmental control and determination of transuranium elements (TUE), especially Pu(IV) and Am(III), in various

TABLE 5.17 Some Applications of Liquid–Liquid Partition Chromatography (LLPC)[260,265–270]

Eluate	Support	Stationary phase	Mobile phase
Aromatics	Alumina	Water	*n*-Heptane
Aliphatic alcohols	Silica acids	Water	Chloroform-CCl4
Fatty acids	Celite	Aqueous sulfuric acid	Butanol-chloroform
Glycols	Celite	Water	Butanol-chloroform
Phenols	Silicic acid	Water	Isooctane
Amino acids	Silica gel	Water	Dichloromethane
Steroids	Silica acids	Methanol/water	Benzene-chloroform
Alkaloids	Silica gel	Water	Acetonitrile-water
Urine	Silica gel	Aqueous sulfuric acid	Dichloromethane
Antibiotics	Silica gel	Water	Dichloromethane
Pesticides	Silica gel	Water	*n*-Heptane

natural samples.[263,264] Recently, new family of resins was developed and intensively studied. These resins are based on the impregnation of Amberlite XAD7 with highly specific reagents for actinide analysis in environmental and biological samples.

4.6.2. Supercritical Fluid Extraction

Supercritical fluids were found to be highly efficient extraction media, chiefly because of their high solvating power, their low viscosities (intermediate between a gas and a liquid), and their low surface tensions that enable their penetration deep into the extraction matrix. Supercritical fluid extraction (SFE) used in isolation is generally not selective enough to separate specific solutes from the matrix without further cleanup or resolution from coextracted species prior to qualitative and quantitative analysis. Consequently, for analytical applications, SFE is usually used in combination with chromatographic techniques to improve the overall selectivity in the isolation of specific solutes. The combined use of SFE with chromatographic techniques is quite recent. The application of SFE for the preparation of samples in the analytical laboratory has received serious attention as a sample preparation step for extracting analytes of interest from a bulk matrix prior to their determination by other analytical methods including chromatographic, spectrometric, radiochemical, and gravimetric techniques.

The need to replace conventional solvent extraction methods of sample preparation with analytical-scale SFE systems comes because the latter are faster, more efficient, have better potential for automation, and also reduce the need for large volumes of potentially hazardous liquid solvents. The potential for applying SFE to a wide variety of environmental and biological samples for both qualitative and quantitative analyses is widely described in reviews.[271–273]

Analytical-scale SFE is most often applied to relatively small samples (e.g., several grams or less). A good solvent for extraction should be selective so that it dissolves the desired analytes to a greater degree than other constituents in the sample matrix. It should be unreactive and stable, preferably nontoxic, noncorrosive to equipment and inexpensive to buy. Many of these requirements are met by supercritical fluids such as carbon dioxide.

Off-Line, On-Line SFE

Analytical-scale SFE can be divided into off-line and on-line techniques. Off-line SFE refers to any method where the analytes are extracted using SFE and collected in a device independent of the chromatograph or other measurement instrument. On-line SFE techniques use direct transfer of the extracted analytes to the analytical instrument, most frequently a chromatograph. While the development of such on-line SFE methods of analysis has great potential for eventual automation and for enhancing method sensitivities,[271–273] the great majority of analytical SFE systems described use some form of off-line SFE followed by conventional chromatographic or spectroscopic analysis.

To perform off-line SFE, only the SFE step must be studied: the extract can be analyzed at leisure by a variety of methods. The final product of an off-line SFE experiment typically consists of the extract dissolved in a few milliliters of liquid solvent, a form that is directly compatible with conventional chromatographic injectors. In contrast, the successful performance of on-line SFE requires coupling step (i.e., the transfer and collection of extracted analytes from the SFE to the chromatographic system), and the final chromatographic conditions have to be controlled. With many on-line approaches the sample extract is committed to a single analysis, and the extract is not available for analysis by other methods. Off-line SFE has the advantages of simplicity and availability of commercial instrumentation, provides extracts that can be analyzed by several instrumental methods. On-line techniques have greater potential for enhanced sensitivity and automation.

The instrumental components necessary include a source of fluid, most often CO_2 or CO_2 with an organic modifier, a means of pressurizing the fluid, an extraction cell, a method of controlling the extraction cell temperature, a device to depressurize the supercritical fluid (flow restrictor), and a device for collecting the extracted analytes.

On-Line Coupling SFE-GC, SFE-SFC

On-line supercritical fluid extraction/GC methods combine the ability of liquid solvent extraction to extract efficiently a broad range of analytes with the ability of gas-phase extraction methods to rapidly and efficiently transfer the extracted analytes to the gas chromatograph. The characteristics of supercritical fluids make them ideal for the development of on-line sample extraction/gas chromatographic (SFE-GC) techniques. Since most supercritical fluids are converted to the gas phase upon depressurization to ambient conditions, SFE has the potential to introduce extracted analytes to the GC in the gas phase. Supercritical fluid chromatography (SFC), in which a supercritical fluid is used as the mobile phase, was introduced.[274,275] SFE directly coupled to SFC provides an extremely powerful analytical tool. The efficient, fast, and selective extraction capabilities of supercritical fluids allows quantitative extraction and direct transfer of the selected solutes of interest to be accomplished to the column, often without the need for further sample treatment or cleanup. Extraction selectivity is usually achieved by adjusting the pressure of the supercritical fluid at constant temperature or, less often, by changing the temperature of the supercritical fluid at constant pressure. SFE coupled with packed column SFC has found many applications in the last decade, particularly in connection with the determination of food compounds, drug residues, herbicide and pesticide residues, and polymer additives.

4.7. Solvent Extraction in Continuous Flow Injection Analysis (FIA)

Flow injection analysis (FIA) is based on the injection of a liquid sample into a continuously flowing liquid carrier stream, where it is usually made to react to give reaction products that may be detected. FIA offers the possibility in an on-line manifold of sample handling including separation, preconcentration, masking and color reaction, and even microwave dissolution, all of which can be readily automated. The most common advantages of FIA include reduced manpower cost of laboratory operations, increased sample throughput, improved precision of results, reduced sample volumes, and elimination of many interferences. Fully automated flow injection analyzers are based on spectrophotometric detection but are readily adapted as sample preparation units for atomic spectrometric techniques. Some books[276,277] and critical reviews[278,279] on FIA methods have been published.

A typical extraction manifold is shown in Fig. 5.19. The sample is introduced by aspiration or injection into an aqueous carrier that is segmented with an organic solvent and is then transported into a mixing coil where extraction takes place. Phase separation occurs in a membrane phase separator where the organic phase permeates through the Teflon membrane. A portion of one of the phases is led through a flow cell and an on-line detector is used to monitor the analyte content. The back-extraction mode in which the analyte is returned to a suitable aqueous phase is also sometimes used. The fundamentals of liquid–liquid extraction for FIA[280] and applications of the technique[281] have been discussed. Preconcentration factors achieved in FIA (usually 2–5) are considerably smaller than in batch extraction, so FI extraction is used more commonly for the removal of matrix interferences.

Chemical reactions between the analyte and a reagent or combination of reagents can be implemented in continuous systems to obtain a reaction product that is capable of facilitating mass transfer and/or continuous detection that results in increased sensitivity or selectivity. These reactions could take place in different locations of the schemes defined previously, e.g.:

1. In a coil after the sample is merged with a reagent stream.
2. In the solvent segmenter (SS) when it is necessary to avoid precipitation of the extractable compound in the aqueous phase before it reaches the continuous extractor.
3. In the extraction coil (EC), when the extraction reaction is based on the formation of metal chelates and the ligand is dissolved in the organic solvent. Here the reaction and the transfer of the chelate take place simultaneously.
4. In a coil located after the extractor using a reagent stream that is merged prior to the detector with the phase containing the analyte.

The derivatizing reactions used in these systems are not substantially different from those described in classical analytical determinations. The most common

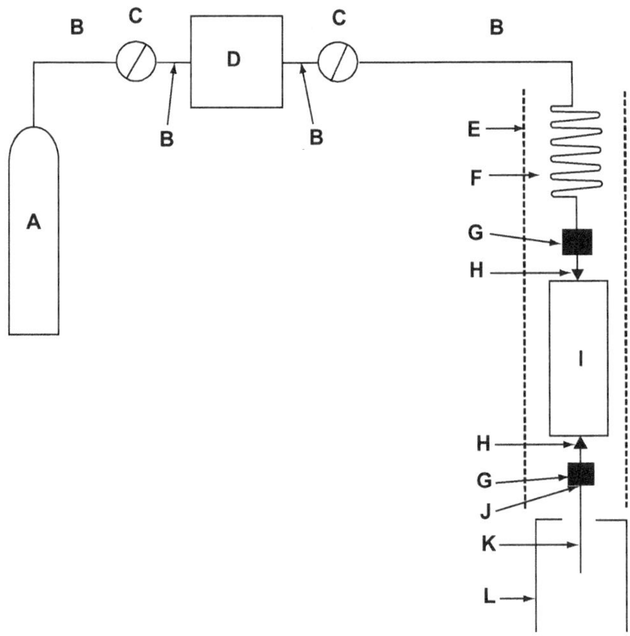

FIGURE 5.19 Schematic diagram of a simple SFE-GC system showing all the required components. (Several manufacturers supply suitable components and specific suppliers are listed only for the reader's convenience.) Components are (A) SFE grade extraction fluid source; (B) 1.5-mm o.d. stainless steel tubing (0.77-mm or smaller i.d.); (C) shut-off valves (SSI model 02–120 or equivalent, Supelco, Bellefonte, PA); (D) SFE pump; (E) SFE cell heater; (F) approx. 0.5-m long coil of 1.5-mm stainless steel tubing for fluid preheater; (G) 1.5-mm tubing union (e.g., Parker or Swagelok brand); (H) finger-tight connectors (e.g., Slip-Free connectors from Keystone Scientific Bellefonte, PA, USA); (I) SFE cell; (J) restrictor connector ferrule (Supelco M2-A, Bellefonte, PA) which is used to replace the stainless steel ferrule in the outlet end of the tubing union "G"; (K) 15–30-mm i.d. fused silica tubing restrictor (Polymicro Technologies, Phoenix, AZ); (L) GC injection port.

are ion-pair formation reactions, where the analyte forms an ion pair directly, e.g., determination of anionic surfactants by using ethyl violet or methylene blue, and the determination of codeine with picric acid and enalapril with bromothymol blue; or after reacting with a reagent to form a bulkier, charged species, e.g., tetracyanatocobaltate or phosphomolybdate. Metal cations are typically transformed into extractable metal chelates by using such common ligands as oxine, dithizone, dimethylglyoxime, etc. Some of the most important applications of the technique are in the determination of surfactants in waters, quinizarin in hydrocarbons, amines in pharmaceuticals, and drugs in urine.[282]

The sample injection may be before or after the continuous extractor device. Figure 20a depicts a manifold for the determination of vitamin B1 in pharmaceuticals,[282] based on the oxidation of thiamine to thiochrome in a carrier of potassium ferricyanide in a basic medium (NaOH). The thiochrome is

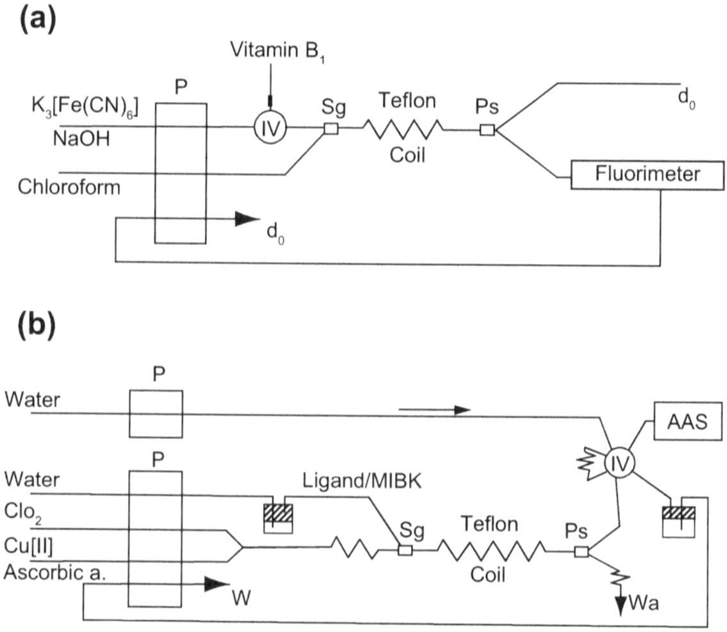

FIGURE 5.20 Typical FIA extraction manifolds: (a) Injection in front of the extractor components; (b) injection after the extraction has been carried out.

continuously extracted into a chloroform stream and the fluorescence of the organic phase is measured continuously.

Figure 5.20b presents the scheme of determination of traces of perchlorate in serum and urine by formation of an ion pair with cuproine-type chelates and subsequent extraction with 6-methylpicolinealdehyde azine dissolved in MIBK.

The sample is continuously added to the system and is mixed with another stream containing Cu(II) and ascorbic acid. This mixture merges with the organic solution in the solvent segmenter. The organic phase stream containing the ion pair emerging from a T-type separator fills the loop of an injector valve. This volume is then injected into a water carrier in which copper is continuously determined and hence perchlorate (0.5–5 ppb) may be indirectly determined.[283]

Activation analysis, isotope dilution radiometry, and liquid scintillation counting are radiometric methods used in analytical chemistry. Solvent extraction techniques are widely used in these methods. Specific methods and equipment are described in Section 4.8.

4.8. Solvent Extraction in Nuclear Analytical Applications

The analytical characterization of radioactive materials for trace and minor metallic constituents is essential in the management and utilization of spent nuclear fuels and associated radioactive waste treatment. These are very complex mixtures and are intensely radioactive. They must be analyzed remotely in a heavy shield box located in a controlled area, which further adds difficulty to the analysis. On the other hand, it facilitates to develop highly sensitive analytical methods or analytical instruments that can accurately and quickly measure the trace amounts of metal elements.

There are two main directions in nuclear analytical chemistry: (1) mass spectrometric analytical methods for samples of long-lived radioactive elements (LLR) such as uranium, thorium, plutonium, etc. at processing of nuclear fuels and reprocessing of irradiated fuels, and (2) radiometric analytical techniques, used for short-lived radioactive (SLR) isotopes. The mass spectrometric detection techniques become more and more popular method for the determination of LLR. At present, inductively coupled plasma atomic emission spectrometry, ICP-AES, analysis is being used as a standard technique for the analysis of uranium, thorium compounds in the fields of nuclear science and technology.

Sometimes, direct analysis of LLR bearing materials was not possible for trace metal elements because of spectral interference and complex atomic emission spectra. Thus, an adequate pre-separation and concentration procedure has been required in advance of the analysis. Mainly conventional solvent extraction methods of sample preparation and analysis are used. Many of these conventional preparative techniques are described earlier.

There are many radionuclides naturally occurring in the environment, including decay products of the isotopes of uranium and thorium, those produced from the cosmic ray reactions, such as ^{3}H, ^{7}Be, ^{10}Be, ^{14}C, ^{26}Al, and ^{129}I. The interest in the determination of these radionuclides mainly comes from the application of them in geochronology and using them as environmental or paleoclimate tracers. Determination of radionuclides, such as ^{222}Rn, ^{226}Ra in air and drinking water is important in the view of radiation protection. Radionuclides applied in industry and hospital may also be released to environment and for the radiation protection purpose, the level of these radionuclides in various environmental and biological samples has to be determined. Radionuclides released from the reprocessing plants can also be used as environmental tracer for the investigation of transport of water mass ($^{134,137}Cs$, ^{99}Tc, ^{129}I) and atmospheric circulation (^{129}I). In decommissioning of nuclear facilities and repository of nuclear waste, invitatory of radioactivity or concentration of various radionuclides in the waste samples need to be determined.

4.8.1. Spectroscopic and Mass Spectrometric Detection

Spectroscopic and mass spectrometric techniques, including Capillary electrophoresis–inductively coupled plasma mass spectrometry, CE-ICP-MS; Nuclear magnetic resonance spectroscopy, NMR; Surface-enhanced Raman spectroscopy, SERS; Ultraviolet–visible spectroscopy, UV-Vis; Extended X-ray absorption fine-structure spectroscopy, EXAFS; Time-resolved laser-induced fluorescence spectroscopy, TRLIF; electrospray ionization mass spectroscopy, ESI-MS all have been used for the quantitative determination of long-lived radionuclides. Sample preparation techniques and determination by spectroscopic techniques, used for the nonradioactive materials are described earlier.

Mass spectrometry is extensively used for the determination of element concentrations in the trace and ultra-trace range by measuring the number of atoms of isotopes of the element. It has also been used for the determination in nonradioactive materials and radionuclides. Mass spectrometric detection techniques such as inductively coupled plasma atomic emission spectrometry (ICP-AES), or mass spectrometry (ICP-MS), accelerator mass spectrometry (AMS), thermal ionization mass spectrometry (TIMS), resonance ionization mass spectrometry (RIMS), secondary ion mass spectrometry (SIMS), and glow discharge mass spectrometry (GDMS) become more and more popular methods for the determination of LLR and allow direct determination of radionuclides in solid samples without any chemical preparation. The analysis is very simple and minimizes the risk from the contamination due to the limited sample handling.

Inductively coupled plasma atomic emission spectrometry (ICP-AES) has been widely selected as the technique of choice to perform the analysis of radioactive materials owing to its simultaneous measurement capability, wide analytical dynamic range, high sensitivity, and relative freedom from matrix interferences.[284,285] At present, because of the high sensitivity, short analysis time and relatively easy operation ICP-AES analysis is being used as a standard technique for the analysis of uranium, thorium compounds in the fields of nuclear science and technology.

Sometimes, direct analysis of LLR bearing materials was not possible for trace metal detection. The limit for many radionuclides is not sufficient and chemical separation of the elements of interest from the matrix is the way for increasing the signal strength and limiting interferences. Severe spectral interference of other elements was found in the intensely complex atomic emission spectra of the actinide matrices. Thus, an adequate pre-separation procedure has been required in advance of the mass spectrometric analysis. Several review articles have addressed the application of mass spectrometric techniques for the determination of radionuclides.[284–289] Many of these techniques are described earlier.

4.8.2. Radiometric Detectors

For great number of radionuclides with half-life of less than 10 years, the radiometric methods are exclusively used for their determination. In these methods, the decay rate (A, number of decays per unit of time) of the radionuclide of interest is measured, the atom number (N) of radionuclide of interest is calculated based on the statistical property of the decay of the radionuclide using its half-life ($T_{1/2}$):$N = A/\gamma = A/(\ln 2/T_{1/2})$. In these methods, the atom numbers of the radionuclide of interest are directly measured.

Different radiometric methods, such as gamma (γ)-spectrometry, alpha (α)-spectrometry, and beta (β) counting are usually used. Three main radiometric detecting methods are shortly presented below.

Alpha Spectrometry

There are several types of detector that can be used for the measurement of alpha-emitting radionuclides; e.g., Frisch grid ionization chambers, proportional counters, plastic- and liquid scintillation detectors and semiconductor detectors. Frisch grid ionization chambers are used for total alpha activity measurements. Semiconductor detectors are normally used for the determination of α-emitting radionuclides in the environmental and waste samples.

Although the resolution of semiconductors for α-spectrometry is good, the relatively small difference in alpha particle energy between some alpha emitters makes it difficult to spectrometrically separate the peaks. This is usually the case with isotopes such as ^{241}Am and ^{238}Pu, ^{237}Np, ^{233}U, ^{234}U, and ^{230}Th. In environmental samples, the concentration of radionuclides is normally very low; it requires a long counting time. Therefore, a good chemical separation of radionuclide of interest from the matrix is required.

The main disadvantage of α-spectrometry is the long analysis time, which is a result of the long chemical separation procedure for the complete separation of the target radionuclide(s) from the matrix as well as from interfering radionuclides, and the very long counting time (1–30 days). However, from the above, it is obvious that normal separation procedures for α-spectrometry is not suitable for fast analysis in emergency situations, which requires getting the analytical results in a shorter time (less than 1 day). Several modified, rapid procedures have therefore been worked out to handle such situations for different radioisotopes in various matrixes.

Gamma Spectrometry

Many α- and β-decay radionuclides also emit γ-rays; they can therefore be determined by γ-spectrometry. A number of detectors can be used for the measurement of γ-rays, such as scintillation counter using NaI, CsF, and ZnS as scintillator, gas-filled counter, and semiconductor detectors based on lithium-doped Si or high purity Ge, CdTe, and GaAs. Mainly Si(Li) and Ge (lithium drift germanium Ge(Li)) semiconductor detectors are widely used for the γ-spectrometry measurement of radionuclides in the biological, environmental, and nuclear waste samples, because of their high energy resolution. The recent development in the γ-spectrometry is the introduction of large-volume Ge crystal that significantly increases the counting efficiency.

Due to the property of electromagnetic radiation, γ-ray can penetrate a long distance in the samples without significant absorption, especially high energy γ-rays. Sample usually does not need any treatment before counting, which makes the analysis quick and easy, and minimizes the risk of contamination during sample preparation. It is therefore that the radionuclides with emission of suitable energy γ-rays are normally measured by γ-spectrometry, such as ^{60}Co, ^{65}Zn, ^{54}Mn, ^{94}Nb, ^{106}Ru, ^{133}Ba, ^{134}Cs, ^{137}Cs, ^{144}Ce, ^{152}Eu, ^{154}Eu, and ^{241}Am. However, the counting efficiency of γ-spectrometry is low (<10% for absolute counting efficiency) and varies with the γ-ray energy, distance of source to the detector and the size of the Ge or Si crystal. The detection limit of γ-spectrometry (>50 mBq) is typically a few orders of magnitude higher than β-counting and α-spectrometry.

Beta Counting

Beta- or electron emitting radionuclides are normally measured by gas ionization detector, such as ionization chamber, gas flow GM detector, or liquid scintillation counters (LSC). GM counter is normally used for radionuclides emitting higher energy electrons due to the absorption of low energy electrons in the thin window separating the sample and gas volume. LSC has an advantage in measuring low-energy electron emitters due to the absence of attenuation when the sample is immersed within the scintillator. LSC can thus be used for the measurement of both high and low energy β-emitters. Determination of beta particle emitters therefore requires chemical separation of the radionuclide of interest from all other radionuclides before counting.

In order to minimize self-absorption in the sample and maintain a high counting efficiency, samples for GM counting needs to be prepared as a thin solid source which can be carried out by the same method as used for the α-spectrometry. The counting efficiency of a GM counter varies from 10 to 70% with the energy of the emitted electrons, thickness of the source, and counter properties. GM counter cannot be used effectively for the determination of radionuclides emitting low energy electrons, such as ^{33}H, ^{14}CI, and ^{241}Pu, because of high absorption in the detector window. LSC has to be used in this case.

LSC has the advantage of none or minimal self-absorption, high counting efficiency, homogeneous distribution of the sample in the scintillation cocktail, relative simple sample preparation procedure and easy standardization using internal or external standard. The quench correction is a drawback of LSC.

4.8.3. Preparative Techniques for Radioactive Elements Analysis

Sometimes, direct analysis of plutonium, americium or uranium bearing materials was not possible for trace metal elements because a severe spectral interference was found in the intensely complex atomic emission spectra of the actinide matrices. Thus, an adequate pre-separation procedure has been required in advance of the ICP-AES analysis. Many separation methods, using solvent extraction and ion exchange chromatography,[284–289] have been proposed for the purification and recovery of actinides or fission products from a variety of radioactive materials in spent fuel reprocessing and associated radioactive waste treatment processes. These techniques and processes were described previously as convenient preparation techniques. They are only realized in hot chambers in radioactive security mode.

Among numerous techniques for radioanalytical purposes, a new family of resins (TVEX) were developed and intensively studied. They are based on introducing an extractant into the polymer matrix during synthesis[264,265] similar to that of polymer inclusion membranes (PIM) in liquid membrane separation technologies.[130] The behavior of uranium, plutonium, and trans-plutonium elements (TPE) using TVEXs with TBP, TOPO, DEHPA, and TOA, etc. has been studied.

Another family of resins was developed[279,280] by impregnation of Amberlite XAD7 with highly specific reagents for actinide analysis in environmental and biological samples.

4.8.4. Radiometric Techniques

Neutron Activation Analysis (NAA)

Neutron activation analysis is the main radiometric method used in analytical chemistry. The sample is irradiated in a neutron flux, and the resulting specific radioactivity of the analyte determined. Although high-resolution gamma spectrometers would seem to eliminate the need for highly selective separation of individual gamma-emitting radioactive elements, group separation still may be necessary.[290] Various separation methods used in NAA such as distillation, precipitation and coprecipitation, electrochemical deposition. Solvent extraction and ion exchange show several advantages. Solvent extraction can be used in combination with activation analysis in two ways: either by separating the analyte from interfering components and preconcentrating before irradiation, or by performing the neutron activation and separating the analyte before the radiometric determination. The latter method has the advantage of eliminating the need for a blank correction.

Good selectivity, quantitativeness, and simplicity of operation have made ion-exchange chromatography (IEC) the leading separation technique employed in the radiochemical version of NAA. Many of the extractants employed in this technique, such as tri-*n*-octylamine and di-(ethylhexyl) orthophosphoric acid, extract ions by anion-exchange mechanism. Chromatographic methods have been widely used in radiochemical NAA for the determination of trace elements in a variety of other matrices such as semiconductor materials, high-purity metals and geological materials. In the analysis of geological materials, a strongly basic anion-exchange resin in the chloride form or newly synthesized chelating resins were used.[288,291]

Sometimes, however, it may appear necessary to remove matrix elements before irradiation. Certain types of samples activate too strongly to permit their safe handling without the use of hot chambers. When the interfering components become highly radioactive and the resulting radioisotopes are long-lived, it is preferable to remove them before irradiation. Another typical case is the determination of very low concentrations of elements in large volumes of solution (e.g., water, wastes), where it is necessary to perform preconcentration and to bring the element(s) in question into the form compatible with the requirements set for irradiation in a nuclear reactor. These convenient preconcentration techniques were described above.

Isotope Dilution Radiometry

In this type of radiometric analysis, a tracer quantity of a radioactive isotope is added to the analyte, which is then partly extracted using a specific extractant. Since the extractant may be considered as reacting totally with the analyte, the ratio of radioactivity in both phases provides the concentration of analyte in the sample.[292]

Liquid Scintillation Counting

Carbon-14 and tritium are radioisotopes with β-emissions of very low energy that are extremely difficult to detect with any form of window counter, due to self-absorption of the β-particles and absorption within the counter window. To reduce the self-absorption losses, it is desirable to mix the active sample homogeneously with the detecting material. This can be done by counting the sample in the gaseous phase. The gaseous activity can then be intimately mixed with the filling gas of any type of gas ionization detector, thus minimizing the effect of β-absorption and resulting in high counting efficiencies.

In this, the radioactive sample and a scintillator material are both dissolved in a suitable solvent, and the resulting scintillations are detected and counted.[292,293] The method is called liquid scintillation counting. If a compound containing an α- or β-emitting isotope is dissolved in a solvent such as toluene, the radioactive emissions result in the formation of electronically excited solvent molecules. If the solution also contains a small amount of a suitable scintillator, the excited solvent molecules rapidly transfer their excitation energy to the scintillator, forming electronically excited scintillator molecules, which then relax by the emission of photons. All scintillator solutions contain: a solvent, a primary solute, the scintillator material; and may contain a secondary solute. The functions of the solvent are to keep the scintillator or solute in solution, and to absorb the decay energy of the radioisotope for subsequent transfer to the solute. Solvents may be effective, e.g., aromatic hydrocarbons such as toluene and xylene; moderate, e.g., many nonaromatic hydrocarbons and poor solvents.

5. APPLICATION OF SOLVENT EXTRACTION IN BIOCHEMICAL AND PHARMACEUTICAL SEPARATIONS

5.1. Application of Solvent Extraction in Biotechnological Separations

The recovery of products from biotechnological processes has traditionally been focused on separation approaches, such as electrophoresis or column liquid chromatography. These methods are difficult to scale up to production levels and often become expensive for medium and low-value products.

Recently liquid extraction has been recognized as a potential method in the primary recovery of fermentation cell culture products, such as carboxylic acids, proteins and amino acids. The separation problem, however, is difficult because the product mixtures are often complex, including cell debris and enzymes. Proteins are not suitable for conventional solvent extraction because of incompatibility with organic solvents, but can be handled in aqueous two-phase systems or by extraction in reverse micellar systems (see below).

Although biotechnological processes often have been stated to be energy efficient in that the reaction temperature is low, it is important to realize that the product concentrations are low and that the product recovery step is often the most energy consuming.

5.1.1. Carboxylic Acids Separation

Citric Acid

Solvent extraction has been proposed as an alternative for the separation of organic and amino acids from fermentation broth.[294] Citric acid can be extracted by both high molecular weight amines and organophosphate solvating reagents. Important criteria have been established for the processing of citric acid by solvent extraction: a distribution coefficient <10 allows for easy stripping with water. Stripping with water is important because, although the citric acid could be recovered with a base, the formation of a citrate salt would require further processing as in the conventional flow sheet, thus removing the advantages of solvent extraction. Extractant system with tri-*n*-butylphosphate (TBP) in Shellsol is shown in Fig. 5.21. A diluent was temperature dependent with extraction decreasing with increasing temperature. Therefore, extraction from the broth at room temperature and water stripping at higher temperature (60–70 °C) provided an efficient process (see Fig. 5.21). The extraction with long-chain amines was hampered by the formation of emulsions and thus poor separation. Adding a modifier, such as octanol, stops emulsion formation. Alamine 336/Shellsol A/octanol gives the best separation.

Acetic Acid

Acetic acid is an important intermediate organic tonnage chemical that may be produced by fermentation. Several technologies have been applied to this separation, including solvent extraction, azeotropic distillation, and extractive distillation. A comparison of economics between azeotropic distillation and solvent extraction combined with azeotropic distillation shows that the solvent

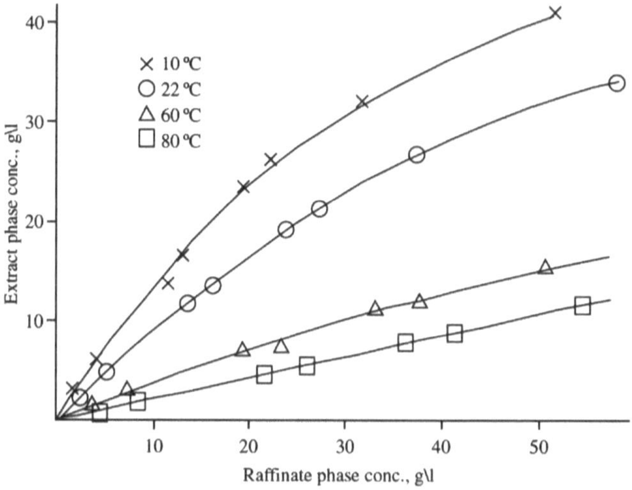

FIGURE 5.21 Isotherms for citric acid extraction with TBP.

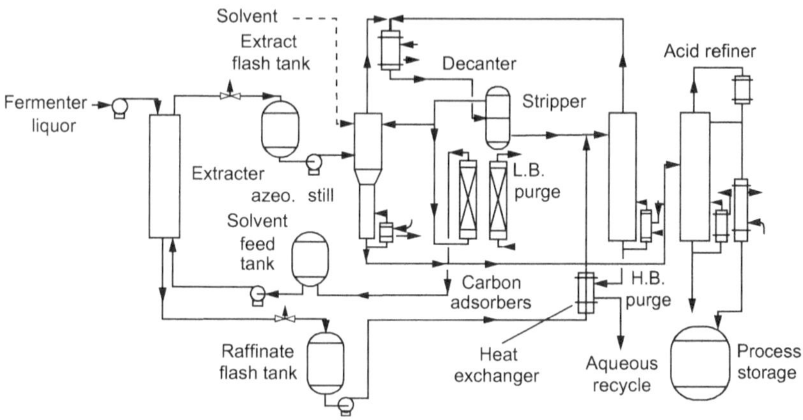

FIGURE 5.22 Acetic acid flow sheet.

extraction successfully introduces savings in plant investment and operating costs.[295] This process flow sheet is outlined in Fig. 5.22.

A number of studies have been carried out on the extraction of other important carboxylic acids from fermentation liquors.[296,297]

5.1.2. Amino Acids

Amino acids, because they contain both carboxyl (–COOH) and amino (–NH_2) groups, can behave as cations at low pH and anions at high pH and have a zwitterion character at intermediate pH values. Although they have no net charge in this intermediate pH range, they are hydrophilic and so their solubility in nonpolar diluents is low. In addition, their extraction with carbon-bonded oxygen donor extractants is poor. Therefore, the most common method for extraction is to convert the amino acid into one of the ionic forms and use an appropriate ion-pair extractant. Thus extraction can be effectively carried out by a cationic extractant (HL), for instance, either a phosphoric acid extractant such as DEHPA,[298] or an organosulfonic acid such as DNNSA, where the extracted species is $(RNH_3COOH)^+$ (A^-).[299] The extent of extraction depends on the relative pK_a values of the two acids, with extraction favored when the extractant has a lower pK_a than the extractant. Thus, glycine ($pK_a = 2.34$) is extracted by DNNSA ($pK_a = 0.68$) over the pH range 1–5, but DEHPA has a higher pK_a value (3.50), and little glycine extraction is found. However, some amino acids with two amino groups, e.g., histidine, form both singly and doubly charged cations and here the singly charged cation can be extracted by DEHPA or even Versatic acid ($pK_a = 4.8$).

In high pH media, the amino acid carries a net negative charge and so can be extracted by an anionic extractant, usually an alkylammonium salt, e.g., tri-*n*-octylamine hydrochloride (R_3NHCl). Here the extracted species is $(R_3NH)^+(RNH_2COO)^-$.[300,301] In both cases, stripping is easy using either dilute alkaline or acidic solutions respectively to regenerate the extracting species.

Of the two extracting systems, the use of an anionic extractant is more common to minimize coextraction of inorganic cations by the organic acids. However, similar problems can also occur with the alkylammonium extractants, especially as the pH of the solution must be sufficiently high to produce an adequate concentration of the amino acid anions, i.e., more than two units above the pK of the amino acid. Thus, coextraction of hydroxyl ions competes with the amino acid anions and in addition lowers the pH of the aqueous phase and reduces the concentration of amino acid anions. To avoid such problems, a large buffer capacity is required in the aqueous phase with the buffer chosen to minimize coextraction of anions.

5.2. Pharmaceutical Separations

Liquid–liquid extraction is extensively used in the pharmaceutical industry for the production of drugs and isolation of natural products.[302] These products are often heat sensitive and cannot be recovered by methods such as atmospheric distillation or evaporation. Owing to competition, little detailed information is available on current commercial operations. A classic example and the best documented, which has encountered problems general for pharmaceutical compounds, is the purification and concentration of penicillin.[303]

5.2.1. Production of Penicillin

The fermentation broth containing penicillin is first filtered to remove mycelium and adjusted to pH 2–2.5 to convert it to the largely undissociated

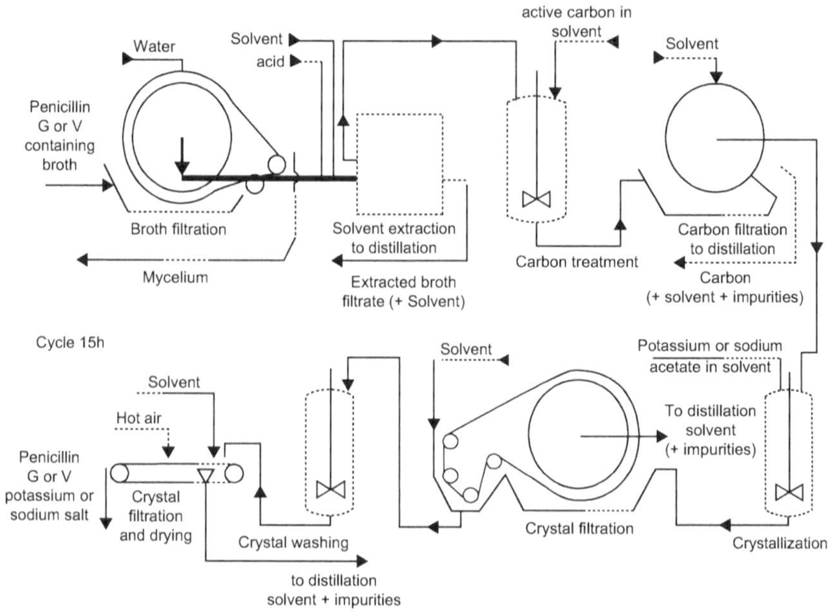

FIGURE 5.23 Purification of penicillin.

penicillinic acid before it is fed to the first extraction stage where it is contacted with, e.g., butyl acetate (see Fig. 5.23). The precise pH used is a compromise between the stability of the penicillin and the partition coefficient.

The partition coefficient increases with pH, but unfortunately, it is accompanied by a rapid decrease in product stability.[304] To minimize the time the product is in the free-acid form, the feed is contacted with the extractant before acidification and the broth/extractant/acid efficiently contacted using an in-line mixer. It is also necessary to carry out the acidification and extraction at low temperature to minimize decomposition.

Extraction takes place in countercurrent centrifugal contactors, e.g., a Podbeilniak extractor to ensure rapid throughput. Penicillinic acid is stable in the organic phase. However, because of entrained acid it is important to convert it to a stable salt as soon as possible. This is achieved by mixing the extract phase with a 2% potassium phosphate solution at pH 6. The potassium penicillin is recovered by filtration and the solvent recirculated. Recovery of the organic phase from the aqueous raffinate is also very important to minimize costs and environmental impact. Butyl acetate, being a low-boiling solvent, can be recovered easily by distillation.

Problems that can arise in the process include the formation of emulsions during extraction from the presence of surface-active impurities in the filtered broth. The effect of these can be minimized by introducing appropriate surfactants that can also reduce the accumulation of solids in the extraction equipment. In addition, other organic impurities are present that can be coextracted with the penicillin. It has been found that a number of these can be removed by adsorption onto active carbon.

Most of the penicillin is used as intermediates in the production of, e.g., cephalosporin, but if it is necessary to produce pure penicillin for pharmaceutical use, it can be purified by reextraction at pH 2–2.5 and further stripping with a phosphate solution at pH 6. This extraction of penicillin is an example of the direct use of a polar organic compound to partition a solute. An alternative process that has been considered is to use organic reagents that form ion pairs with penicillin.[305] Here the authors found that the penicillin anion could be extracted efficiently with a secondary amine (Amberlite LA-2) in the pH range 5–7 where the product is most stable. This type of process is used extensively in hydrometallurgy and can be used to extract both anionic species using cations as shown earlier, or cationic species using organic acid anions. In hydrometallurgy, the system normally uses a hydrocarbon diluent, but for pharmaceutical applications more polar diluents are generally required.

Although the use of such chemically assisted extraction procedures is unlikely to displace the established extraction processes for commercial extraction of penicillin, there are a number of other systems in biotechnology where ion pair formation is used.

6. APPLICATION OF SOLVENT EXTRACTION IN ORGANIC AND BIOFUELS SEPARATION

6.1. Petroleum and Petrochemical Processing

Oil refineries and the petrochemical industry constitute one of the largest application areas for liquid extraction. Other applications in this field are the treatment of lubricating oils and the separation of aromatic and aliphatic hydrocarbons. One more example is the production of high-purity fiber-grade caprolactam. The success of solvent extraction in the petroleum and petrochemical industry is due to careful process integration and also the development of large-scale column extractors.

The development of different processes and solvents for the separation of aromatics from aliphatics has reached a rather stable state. A number of different processes, some of them with capacities of several hundred thousand tons of aromatics per year, are in operation. The more important ones are listed in Table 5.18.

6.1.1. The Union Carbide Process

The process equipment consists mainly of two extraction columns with pulsating trays and four distillation columns according to Fig. 5.24.[306] The feed, with a high content of aromatics, is pumped to the middle of the first extraction column where the aromatics are extracted with the solvent *S*1

TABLE 5.18 Different Processes for Separation of Aromatics from Aliphatics

Process	Extraction solvent	Extraction temperature (K)	Equipment
UDEX	Polyethylene glycols + water	423	Sieve plate column
Union Carbide	Tetraethylene glycol	373	Pulsed plate column
Shell	Sulfolane	393	RDC
DMSO	Dimethylsulfoxide	298	RDC
Arosolvan	*N*-Methylpyrrolidone	308	Vertical mixer-settler column
Formex	*N*-Formylmorpholine	313	Sieve plate column
Carmex	Methyl carbonate	298	RDC

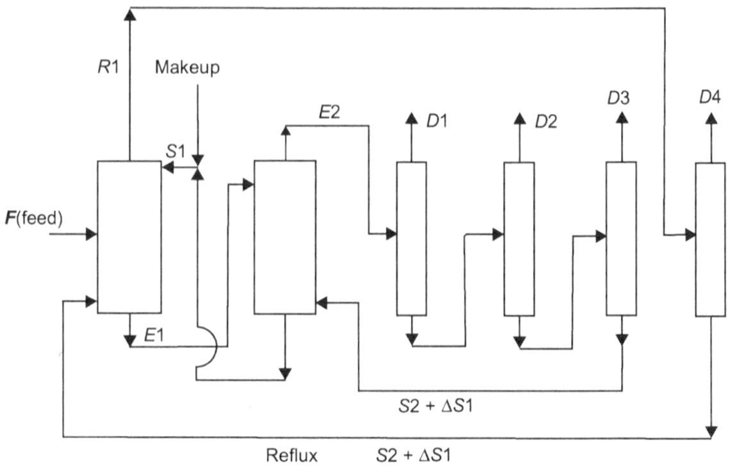

FIGURE 5.24 The Union Carbide process. *(Source: From Somekh*[306]*)*

(tetraethylene glycol). In the lower part, the extracted aromatics are washed with $S2$ (dodecane). The outgoing raffinate phase $R1$ (containing aliphatics, $S2$, and a small amount of $S1$) is distilled in the fourth distillation column. The distillate, $D4$ consists of pure aliphatics, whereas the bottom stream $S2 + DS1$ is recirculated as reflux to the first extraction column. The leaving extract phase $E1$, containing aromatics and $S1$, is fed to the second extraction column where it is washed with more $S2$. The washed organic phase $S1$ is recirculated to the first column, while the extract phase $E2$, containing

TABLE 5.19 Some Operating Conditions for the Union Carbide Process

	First extraction column	Second extraction column
Equipment type	Pulsed plate column	Pulsed plate column
Number of ideal stages	10	7
Extraction temperature	373 K	373 K
Solvent	Tetraethylene glycol	Dodecane
Reflux	Dodecane	–
$S1/S2$	9	1.43
$S1/F$	7	–
Aromatics in feed	–	87 vol%
Recovery of benzene and	–	>99.5%
toluene ($D1$, $D2$)	–	–
Recovery of C8-aromatics ($D3$)	–	>98.5%
Purity of benzene	–	99.9%

aromatics and $S2+DS1$, goes to distillation. The distillates from the distillation columns constitute the pure aromatics, while the bottom stream from the third column, containing $S2+DS1$, is recirculated to the second extraction column. The operating conditions for this process are given in Table 5.19.

6.1.2. Re-refining Process

Solvent extraction is used in a technology for upgrading of low-value, contaminated hydrocarbons into high-value products such as base lubricating oil stock and clean-burning industrial fuels (see Fig. 5.25). Thus, the rerefining process may be used to extract gasoline and diesel from refinery bottoms. To produce fuel, propane is mixed with the refinery bottoms to separate heavy hydrocarbons from middle distillates remaining in the bottoms. These middle distillates can be processed through vacuum distillation to produce gasoline, diesel, and marine diesel.[307]

6.2. Separation of Isomers

Pure isomers are often used as starting products for fine chemicals (e.g., different drugs). The separation of isomers entails great difficulties because they often have boiling points differing by only a fraction of a degree and

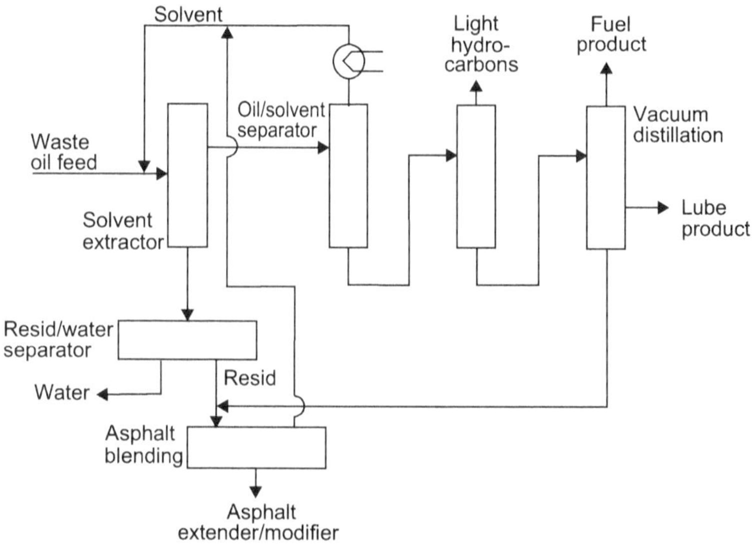

FIGURE 5.25 The re-refining process.

they have closely similar solubilities in many solvents. Solvent extraction processes for the separation of isomers, therefore, have to rely more on chemical reactions than on nonspecific physical interactions between the solute and the solvent.

6.2.1. Separation of m-Xylene from o- and p-Xylene

Xylenes react reversibly to form complexes with boron trifluoride (BF_3) in the presence of liquid hydrogen fluoride (HF). This fact can be utilized in the separation of *m*-xylene from its isomers, because the *m*-xylene complex is much more stable than the others and also more soluble in hydrogen fluoride. After extraction with the HF–BF_3 mixture, *m*-xylene is recovered by vaporizing and recycling the HF and BF_3.[308]

6.2.2. Separation of Alkylnaphthalene Isomers Using Cyclodextrins

Dimethylnaphthalenes are important starting materials for surfactants and polymers. Thus 2,6-dimethylnaphthalene is used to produce 2,6 dimethylcarboxylic acid used in the manufacture of polyesters. The separation of the 2,6- and 2,7-dimethylisomers is difficult as their physical properties are very similar, so other approaches have to be found. Cyclodextrins have a structure where six a-1,4-D(+)-glucopyranose molecules are linked in a doughnut shape (see Fig. 5.26). They can accommodate various organic molecules in the central apolar cavity forming inclusion complexes. Selectivity depends mainly on the

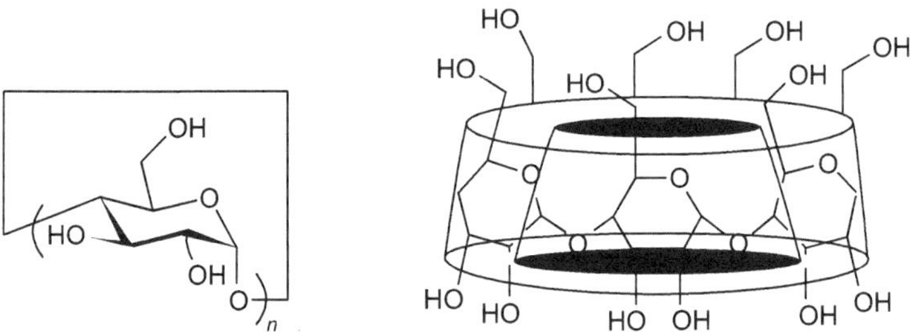

FIGURE 5.26 Structure of cyclodextrin.

size and shape of the guest molecule.[309] The mixture of dimethylnaphthalenes in 1-methylnaphthalene is mixed with an aqueous solution of substituted cyclodextrins and the system allowed to settle into two phases.

The upper phase consists of the oil sample and the lower phase contains the cyclodextrin/2,6-dimethyl isomer inclusion complex. After separating the phases, the aqueous phase is mixed with an organic solvent that dissociates the inclusion complex and extracts the dimethylnaphthalene product. Analysis of the product showed 96% isomeric purity of the 2,6-dimethyl isomer from a 50:50 mixture of 2,6- and 2,7-isomers.[310]

The use of cyclodextrins is still in the development stage and such systems have not yet been used commercially.

6.2.3. *Separation of* m- *and* p-*Cresol Isomers*

These isomers have very close boiling points, but they have appreciable differences in acid strength, as shown by their relative dissociation constants of $9.8 * 10^{-11}$ for *m*-cresol and $6.7 * 10^{-11}$ for *p*-cresol. This is a separation technique for the separation of mixtures of organic acids or bases that exploits the differences in the dissociation constants of the components of the mixture.[311] If a mixture of these bases is partially neutralized by a strong base, there is a competition between the two acids for reaction with the base. *m*-Cresol, the stronger acid, reacts preferentially with the base to form an ionized salt, insoluble in organic solvents. The weaker acid, *p*-cresol, remains predominantly in its undissociated form and, accordingly, is more soluble in an organic solvent. When this process is applied with countercurrent flow of organic solvent and an aqueous base, the mixture of cresols may be separated. The *m*-cresol may be generated from its salt in the aqueous phase by treatment with a strong mineral acid.

6.3. Biofuel Separations by Solvent Extraction

Biomass-based routes to fuels and chemicals have been suggested as alternatives to the fossil-fuel-based routes that are commonly used in industry today. Methanol, ethanol, 1-propanol, 1-butanol, and various other organic compounds can be produced biologically, and thus can be considered as biofuel candidates, as biofeedstocks for production of other chemicals, and as biobased solvents for various applications. For example, ethanol can be used as a fuel, as a solvent in the chemical industry, and as a starting material for many other compounds, including ethylene and its derivatives. Likewise, 1-butanol can be used as a feedstock and has uses as a solvent in many applications such as recrystallization processes used in the pharmaceutical industry.

Furthermore, there is growing interest in 1-butanol as a biofuel. Compared to ethanol, it has a significantly higher energy density (only about 10% less than gasoline). It is also more hydrophobic than ethanol, leading to reduced concerns about fuel system and pipeline corrosion.

Ethanol and 1-butanol are synthesized using fermentation of biomass (sugars, starch, cellulose, etc.) using yeasts or bacteria as the biological agents. Current bioethanol production is based on fermentation with the yeast *Saccharomyces cerevisiae*, though there is also current interest in use of strains of the bacterium *Zymomonas mobilis*.[312] Because the current biological agents have a limited tolerance for ethanol, a typical fermentation broth will have no more than about 16 wt% ethanol.[313]

Improved fermentation agents are being developed[314] with higher ethanol tolerance. Biobutanol has historically been based on fermentation with the bacteria *Clostridium acetobutylicum* or *Clostridium biejerinckii*. These have limited tolerance to 1-butanol and so a typical fermentation broth currently contains about 1.3 wt% 1-butanol.[313] However, significant progress is being made in increasing this tolerance level. For example,[315] have recently described improvements that butanol concentration increased to about 3 wt%.

Furthermore, yeast-based fermentation processes for 1-butanol are now being developed,[316,317] and this may lead to tolerance of much higher levels of 1-butanol, via the same mechanisms a high tolerance for ethanol is achieved. These fermentation processes for bio-based alcohol production ultimately yield a broth consisting mostly of water, the target alcohol, and perhaps other organic compounds. Separation processes are needed for these relatively dilute aqueous solutions.

6.3.1. *Separation of Ethanol*

One of the problems associated with the production of ethanol by fermentation is the inhibition of the yeast activity with ethanol concentrations greater than about 12 wt%. Removing ethanol during fermentation by solvent extraction would allow a greater total product yield from a given batch of feed. The application of in-line extraction of ethanol from the fermentation broth has attracted interest. Problems do arise with direct extraction from broths with the effect of entrained organic solvents on microbial growth and the reduction in mass transfer of the ethanol by the presence of cell debris and other biological material. These problems can be minimized by suitable choice of the extracting

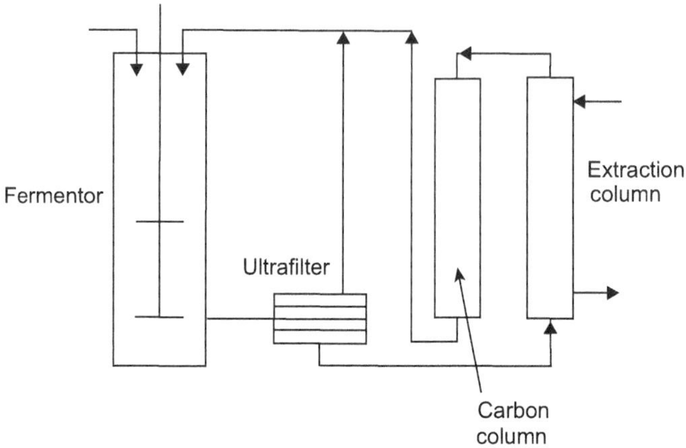

FIGURE 5.27 Extraction of alcohol.

solvent, e.g., long-chain aliphatic alcohols such as butanol, hexanol, etc., and removal of cell debris by filtration.[318] The process operates as a closed loop with the filtered broth circulating through a suitable column extractor before returning to the fermenter via a carbon column to remove residual organic solvent (see Fig. 5.27).

7. SOLVENT EXTRACTION IN RECOVERY OF WASTE AND WASTEWATER TREATMENT

Recently the protection of the environment has become increasingly important for industry with the requirement that the potential impact on the environment is considered for all aspects of industrial processes. Such considerations are supported by environmental legislation that controls all types of emissions as well as the treatment of wastes. Limits for discharge into the aqueous environment have been established. The problems can be minimized by ensuring that the treated solutions are recycled within the plant and that any solutions discharged into the environment are treated appropriately to remove entrained organic compounds and residual toxic metals.

Many technologies could be applied to treatment of these wastes: precipitation, adsorption, electrolysis, more recently developed ion exchange and membrane processes. The flow sheet finally chosen will depend on a number of factors, including the precise composition of the waste, local expertize, disposal options, and of course the overall economics. As a rule, the flow sheet may involve more than one technique. This chapter concentrates on the applications of solvent extraction which is the most appropriate technique for treatment of liquid wastes. Usually, solvent extraction processes have low costs for chemicals and energy. It seems to be good for application in the treatment of both, solid (with leaching) and liquid wastes.

7.1. Solvent Extraction in Metal Waste Treatment

Metal pollution can arise from a number of different sources. The most important metalliferous liquid effluents where solvent extraction could be applied are from the various metal finishing operations: plating, pickling, etching, and the wash waters arising from the cleaning of work pieces. Some sources examples:

1. Mine waters, leachates from solid residues from the processing of primary minerals such as flotation and leach residues, and from landfill;
2. Metal treatment processing including coating (zinc), plating (chromium, nickel, copper), neutralization sludge from the plating industries, pickling (hydrofluoric acid, iron), etching solutions, including rinse waters;
3. Chemical wastes from other industries: tanning (chromium), photographic (silver), electronic (copper, silver, gold, nickel, lead, cadmium)
4. There are more metal-containing solid wastes that may undergo leaching if disposed to land: spent catalysts (cobalt, nickel, vanadium); spent batteries (nickel, cadmium, lithium, lead); combustion ashes, flue dust from steel and brass mills, ash and scrap from industrial production, etc.

In the treatment of waste, the objectives must be:

1. to remove the metals effectively so as to produce a liquid waste stream capable of reuse or finally meeting environmental discharge limits
2. to recover the metals for recycling within the plant, or at appropriate purity for sale
3. to separate other impurities in a form for resale as by-products or safe environmental discharge.

These principles are embodied in the concepts of zero discharge and sustainable technology.

As metal containing wastes differ widely in nature and complexity, selective separation techniques such as solvent extraction will be of increasing importance. To use solvent extraction for metal separation, it is first necessary to have the metals dissolved in an aqueous solution. Starting with a solid waste material, the initial treatment is therefore a leaching procedure to provide an aqueous solution from which the metals may be recovered.

7.1.1. Treatment of Acidic Metal Waste Solutions

It is clear from the previous chapters that extraction by acidic extractants decreases with decreasing pH. In addition, the extraction of each metal ion (M^{z+}) liberates z hydrogen ions (H^+), from the extractant. This means that the pH decreases with extraction, resulting in decreased transfer of metal to the organic solvent. Thus, the more metal in solution and the higher acidity, the less is extracted, no matter how many stages are used.

For example, the extraction of nickel with di-(2-ethylhexyl)phosphoric acid (DEHPA) at a pH below 3 is <10%. If the extraction of nickel is carried out between pH 3–4, the extraction of about 0.3 g/dm^3 Ni^{2+} will decrease the pH to below 3 and the mass transfer of nickel will stop. For copper, the extraction is performed between pH 2–3 and thus, an amount of 3 g/dm^3 Cu^{2+} will decrease the pH from 3 to 2 and again extraction will stop.

To allow a higher extraction capacity (net transfer) and avoid the influence of liberated H^+ ions, simultaneous neutralization with NaOH directed into the mixer is used. Such an example is described later, where nickel is extracted from a cadmium electrolyte. Other means of controlling the pH includes the use of acid extractants in either alkaline (NH_4^+L) or neutral (Na^+L) form.

An increase of the net transfer of zinc was reported[319] for the extraction with DEHPA (50%) in kerosene of zinc from a feed solution containing 40 g dm_3 Zn, emanating from the leaching of steel furnace dust. The resulting raffinate, with a residual Zn concentration of 15–20 g/dm^3, was recycled to leaching. Comparable net transfer conditions were obtained[320] in the modified Zincex process (MZP) for the processing of spent domestic batteries.

Similar high transfer of copper in a solvent extraction procedure has been recently reported.[321] After pressure leaching to liberate copper (60 g/dm^3) under conditions that minimize the free acid (pH = 1.4) and iron in solution, copper was extracted to 10 g/dm^3 in three stages with loading in the organic phase of 12.5 g/dm^3. The situation described is valid in sulfuric acid solutions, when no strong metal complexes are formed.

However with hydrochloric acid the formation of metal chloride complexes makes the situation quite different. For example, the separation of zinc and iron(III) can be achieved at a low chloride concentration, and later the separation of cobalt and nickel is described. The disadvantage of using the chloride system is concerned with stripping. For example, when a zinc chloride complex $ZnCl_4^{2-}$ is extracted with a tertiary amine, the stripped species corresponds to $ZnCl_2$, resulting in an increased chloride concentration and an interruption of the mass transfer. The consequence of this is that it is difficult to achieve concentrated zinc chloride strip liquors.

Several specific examples of metal extraction from the acidic wastes solutions are presented below.

Extraction of Zinc from Weak Acid Effluents

This example describes the extraction of zinc from weak acid solutions. In the manufacture of rayon, rinse waters and other zinc-containing liquid effluents are produced. The total liquid effluent in a rayon plant may amount to several cubic meters per minute with a zinc concentration of 0.1–1 g/dm^3 and pH = 1.5–2. In addition to zinc, the effluent contains surface-active agents and dirt (organic fibers and inorganic sulfide solids).

Solvent extraction has successfully been used to recycle the zinc back to the operation (see Fig. 5.28). Zinc is extracted to greater than 95% (pH > 2) with

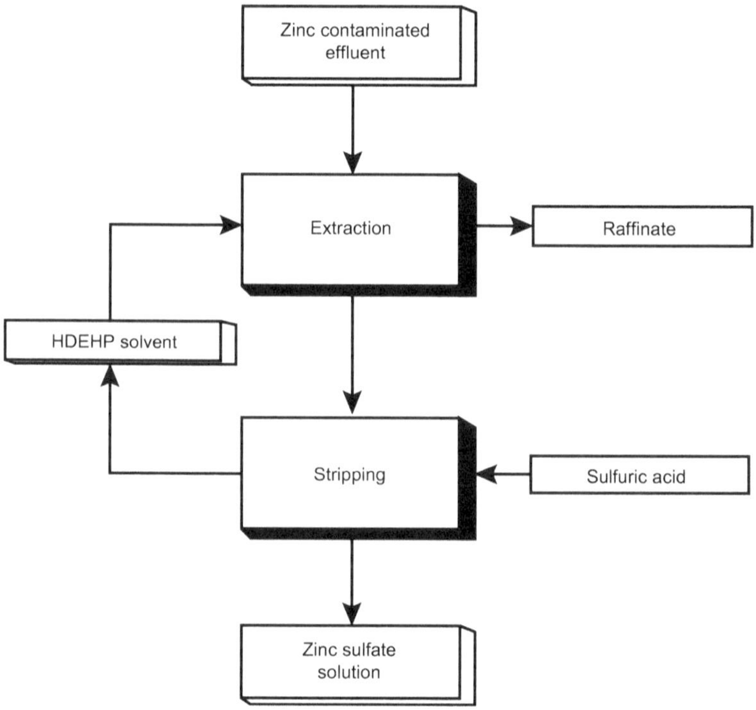

FIGURE 5.28 Extraction of zinc from weak acid effluents.

DEHPA (25%) in kerosene in two or three stages.[322,323] Stripping is performed using a sulfuric acid solution. By adjusting the net flow rate of the solution, the concentration of zinc in the strip solution may be increased to 50 g/dm^3 or more. Thus, the resulting zinc sulfate solution can be reused directly in the spinning bath.

The extraction of zinc is pH dependent. The lower the pH the lower the extraction of zinc. In addition, the extraction of each zinc ion (Zn^{2+}) liberates two hydrogen ions (H^+). Thus, the more zinc in solution, the less is extracted. This disadvantage can be overcomed by some neutralization (pH $>$ 2) to meet the environmental standards of today.

Extraction of Nickel from Cadmium Electrolytes

In the electrowinning of cadmium, nickel is an interfering element that has to be continuously removed from the weak acidic cadmium sulfate solution. Solvent extraction is used to remove nickel contamination.[324] The nickel extraction can be performed with DEHPA under defined conditions. As described earlier for zinc, the extraction of nickel is pH dependent. The D-value for nickel decreases drastically with decreasing pH below 3.5. The extraction of each nickel ion liberates two hydrogen ions from the extractant DEHPA, which means that only a very small amount of nickel can be transferred to the solvent before the extraction stops. Adding a neutralization reagent (NaOH) into the mixer simultaneously with nickel extraction, this disadvantage can be avoided. Three large mixers in parallel are used to allow for the necessary reaction time. The flow sheet is similar to the zinc block diagram with two extraction stages, containing three parallel mixers followed by a settler, and two stripping stages with normal mixer-settler arrangement. The pH is monitored and the acidity controlled by the addition of NaOH in the two first mixers. The extraction efficiency for nickel is better than 98%.

As alternative, for elimination of the pH dependency at extracting nickel the sodium form of the reagent (NaDEHP) was used as extractant.[319] From a weak sulfuric acid rinse water containing 1–2 g/dm^3 Ni, the metal could be effectively (>99%) removed in two extraction stages. By loading the solvent with nickel, the transfer of sodium to the strip solution was minimized. Nickel was stripped from the loaded solvent with dilute sulfuric acid and recovery from the strip solution was best performed by electrowinning. The main problem with this procedure was high loss of extractant in the NaOH extraction reconditioning wash. A considerable reduction of this loss could be achieved by increasing the amount of sodium sulfate in the wash solution.

Extraction of Zinc and Chromium(III) from Electroplating Baths

The use of Zn–Cr(III) alloy plating has almost replaced the use of Cr(VI) in the electroplating industry due to its excellent corrosion resistance and its lower toxicity. Recently, a solvent extraction procedure for separating and selectively recovering the two metals, zinc and chromium, from electroplating wastewaters has been demonstrated.[325]

Zinc is extracted and stripped with DEHPA in the conventional way as previously described. The major challenge lies in the extraction and stripping of Cr(III), which requires a higher pH than zinc. Chromium hydroxide precipitation places an upper limit on the pH that can be used. The extraction of Cr(III) with DEHPA is successful, but stripping of the organic solvent is difficult. Now, it was found that using $DEHPNH_4$ salt as extractant, all stripping problems were overcome. The suggested flow sheet includes extraction of Zn with DEHPA (0.1 mol/dm^3) followed by the extraction of Cr(III) with NH_4DEHP (0.1 mol/dm^3). The stripping of both metals was performed with dilute sulfuric acid. A process for the recovery of Zn and Cr(III) with good separation from each other has been demonstrated. The treated wastewater contained the metals at levels meeting stringent environmental standards.

Extraction of Copper and Zinc from Brass Mill Flue Dust

The solution contains copper, zinc, and little iron.[326] The solid material is leached with sulfuric acid to produce a weak acid metal sulfate solution. The operation can be regarded as two separate interconnected processes, one for

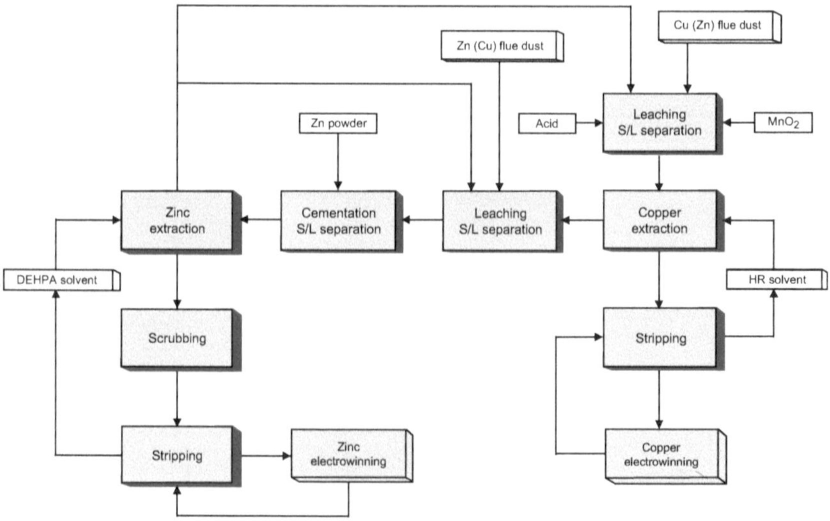

FIGURE 5.29 Extraction of copper and zinc from weak sulfuric acid leach solutions.

copper recovery and one for zinc (see Fig. 5.29). In the copper circuit, copper-rich material is leached in a pH-controlled leach at 60 °C to a final pH of about 2. Oxidizing conditions are maintained, by addition of MnO_2, to promote copper dissolution and to prevent cementation of metallic copper when the feed material contains metallic zinc (brass or iron). The leach solution, containing about 20 g/dm^3 zinc and 4 g/dm^3 copper, is filtered and fed to a solvent extraction circuit. Copper is extracted in four stages with a DEHPA (10%) in kerosene. Stripping is performed in three stages with a sulfuric acid.

The zinc circuit consists of a similar pH-controlled leach at 60 °C under oxidizing conditions. Zinc flue dust with low copper content is leached with the copper barren raffinate and with part of the zinc raffinate. The zinc leaching operation is maintained at pH about 2 for most of the leaching time and then slowly raised to a final pH of 4.5, reducing the iron level to below 10 ppm. The leach solution is filtered and cleaned from impurity metals such as Cu, Ni, and Cd by an ordinary cementation procedure, again filtered and finally fed to a solvent extraction circuit.

Zinc is extracted in three stages with DEHPA (50%) in kerosene solution. The amount of zinc transferred in the extraction procedure can be maintained as high as 20–25 g/dm^3 because the remaining zinc in the raffinate is recycled to leaching. Stripping of zinc is performed in two stages with sulfuric acid solution.

The process flow sheet was first tested for direct leaching of steel mill flue dust and production of zinc metal by electrowinning. The tests were performed in a continuously operating pilot plant, producing 10–20 kg/day zinc metal. The same pilot plant was then used for treating copper/zinc-rich brass mill flue

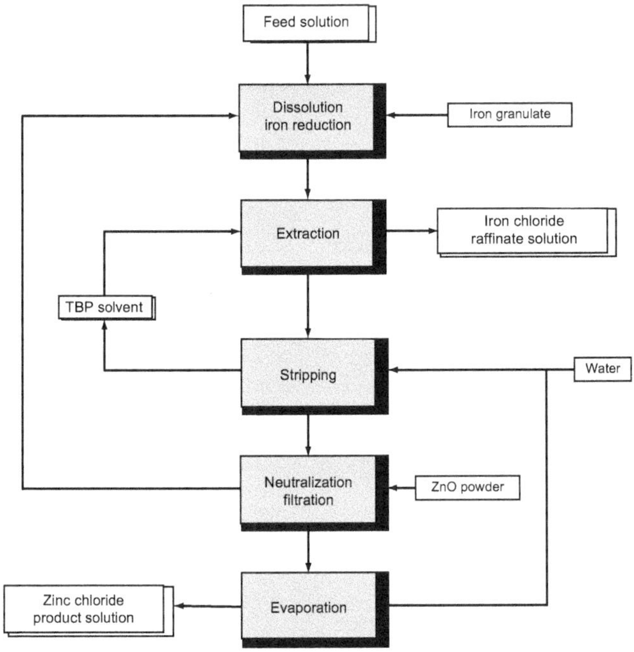

FIGURE 5.30 Extraction of zinc from spent hydrochloric acid pickling liquors.

dust in a closed loop operation, recycling all the zinc solvent extraction raffinate to the copper circuit leach section. In the zinc circuit leach section, only the amount of zinc-rich dust necessary for neutralization of the copper solvent extraction raffinate was used. The results obtained from the pilot plant tests indicated contamination problems within the solvent extraction loops. The estimation of economic data showed a weak return on the assets compared with the alkali route, and sensitivity toward the raw material price.

The authors[327] tried to use a similar flow sheet for the treatment of mine waters ($Fe = 3–5\ g/dm^3$, $Zn = 1\ g/dm^3$, and $Cu = 0.5–1\ g/dm^3$, $pH = 2–2.5$), but failed because of strong interference of iron, in the extraction of zinc. Preliminary removal of iron is costly and complicated, mainly because iron occurs in both its valences. To avoid coextraction of iron, the iron is initially reduced to its II-valent state. Then zinc is extracted from dilute hydrochloric acid solution by tributylphosphate (TBP). Example of the flow sheet of the zinc recovery from spent hydrochloric acid pickling liquors is presented in Fig. 5.30.

Extraction of Chromium(VI) from Surface Finishing Wastewater

Hexavalent chromate [Cr(VI)] is still used within the industry to meet critical high corrosion control and other metal surface finishing requirements. Cr(VI) is toxic and its control generates a hazardous, costly waste. Recently, a solvent extraction process[328] for the recovery of hexavalent chromium from surface finishing process water has been successfully demonstrated. The concentration level of Cr(VI) in the raffinate is low enough (0.1 mg/dm^3) for discharge to surface waters.

A tertiary amine (Alamine 336) is used as a selective extractant of Cr(VI). The extraction mechanism involves ion pairing between the amine and the dichromate anion, $HCrO_4^-$. The chrom is stripped with alkali to produce a strip liquor, containing 2% Cr(VI). This solution can be recycled. For removal and recovery of chromium (VI) from wastewaters of a galvanic process[329] a satisfactory performance of the pilot plant was achieved.

7.1.2. Solvent Extraction from Ammoniacal Feed Solutions

Recently, a new direction (AmMAR processes) in Hydrometallurgy has begun to develop intensively.[330] The main thread was the ammonia–ammonium chloride or carbonate leach solution and its extraordinary chemical flexibility. The general outline of an ammoniacal process for treatment of a material containing two or more of the metals: iron, chromium, copper, nickel, and zinc includes the following process steps:

1. Primary separation occurs in the leaching procedure, where the metals copper, zinc, and nickel are dissolved as metal ammonium chlorides or chloro-complexes, while iron and chromium remain in the solid residue as hydroxides.
2. Copper and nickel are subsequently removed from the leach liquor by solvent extraction. After selective stripping, nickel sulfate is produced by crystallization and copper metal by electrowinning.
3. Excess ammonia is evaporated from the resulting extraction raffinate and zinc is precipitated as carbonate by the remaining ammonium carbonate in the solution or by addition of carbon dioxide. The evaporated ammonia is absorbed in the filtrate and recycled to the leaching step.
4. To maintain the water balance, especially when a water-containing sludge is treated, an additional effluent treatment with lime, followed by ammonia evaporation, is necessary.

As an example of application of AmMAR concept to some waste recycling and consequent solvent extraction application is shortly described below.

Recovery of Nickel and Cadmium from Accumulator Waste

Nickel and cadmium are used in the production of NiFe rechargeable batteries. Using the AmMAR concept, the main leaching procedure to dissolve these valuable metals from spent accumulator scrap and production waste is performed in a two-step procedure, first with an ammonium carbonate solution and second with diluted sulfuric acid to obtain very high leaching efficiency. If the pH is kept about 3 the second leaching of iron is negligible. To reduce more iron

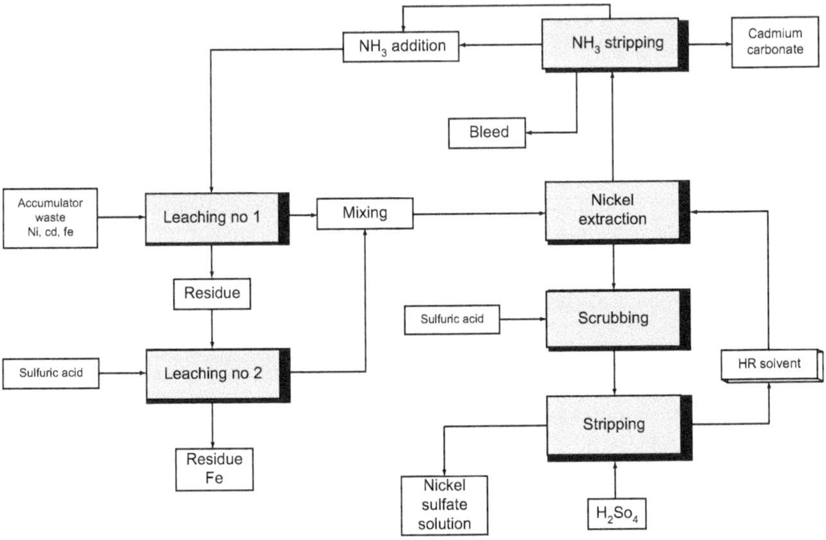

FIGURE 5.31 Recovery of nickel and cadmium from accumulator scrap.

in the leach solution, spent potassium hydroxide electrolyte from discarded batteries can be used for iron hydroxide precipitation (see Fig. 5.31).

Nickel is extracted from the leach solution using a hydroxyaryloxime (LIX 84) with significant amounts of coextracted ammonia and cadmium.[331] Scrubbing is therefore performed, first with an ammoniacal carbonate solution and then with very dilute sulfuric acid. Nickel is then stripped with strong sulfuric acid to produce a nickel sulfate solution, containing 90–100 g/dm^3 nickel. This solution is used directly in the production of new accumulators.

The procedure has several advantages: iron will remain in the leach residue; nickel and cadmium are almost completely leached and form complexes in solution; and, finally, the buffer capacity of the solution facilitates solvent extraction of metals with hydrogen ion exchange reagents.

The presence of carbonate in the ammoniacal system offers a possibility of controlling the conditions for the formation of cadmium carbonate.

After separation of nickel in the solvent extraction loop, cadmium is precipitated from the raffinate as carbonate. Thermal stripping reduces the ammonia concentration and some carbon dioxide will also be evaporated.

7.2. Recovery of Acids from Waste by Solvent Extraction

7.2.1. Recovery of Hydrofluoric and Nitric Acids from Stainless Steel Pickling Baths

A stainless steel pickling bath initially contains 2.2 mol/dm^3 HNO_3 and 1.6 mol/dm^3 HF. The bath is used until the iron concentration reaches 40–80 g/dm^3. At this stage, the bath contains about 50% unused acids in addition to the dissolved

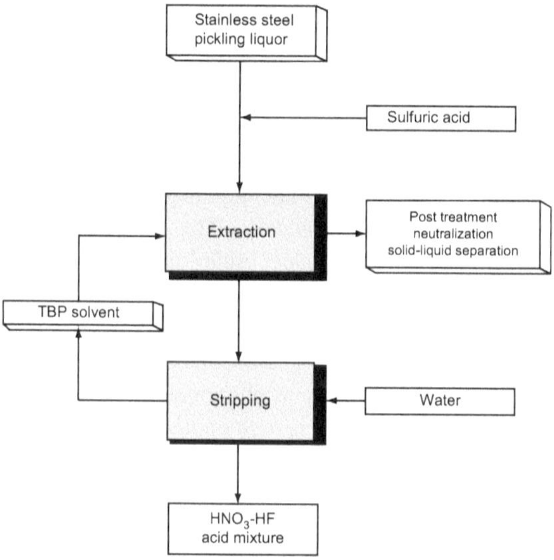

FIGURE 5.32 Recovery of acids from stainless steel pickling baths.

metals: iron, nickel, chromium, and molybdenum. In a used pickling bath, most of the acids are bound to the metals as complexes. By adding sulfuric acid to the pickling bath, sulfate will replace some of the nitrate and fluoride, leading to the formation of extractable, undissociated nitric and hydrofluoric acids.

The general procedure to remove monovalent acids from the aqueous solution by adding sulfuric acid and then extracting the released acids with an organic donor molecule, dissolved in an organic diluent, has been called acid exchange process.[332] The process consists of three main steps (see Fig. 5.32):

1. Addition of H_2SO_4 and extraction of HNO_3 and HF with an organic solution, containing 75%TBP in kerosene, in a pulsed column operation.
2. Precipitation of the metals (Cr, Fe, and Ni) left in the aqueous raffinate.
3. Stripping HNO_3 and HF from the organic solution with water in a second pulsed column operation and recycling the acids back to the pickling bath.

A plant for treating 600 dm^3/h spent pickling liquid (corresponding to an annual production of 25,000 t stainless steel) was in operation for some years in Sweden.

Similar extraction procedures[333] using a mixture of trialkylphosphine oxides (CYANEX 923) for extraction of the mineral acids from electroplating and stainless steel pickling baths are reported. The possibility of using nitric acid as an oxidizing agent in a leaching operation is often prohibitive because many reagents used in solvent extraction are not stable in an oxidizing solution. However, a nitric acid (nitrate) leach solution can be converted to a weak sulfuric acid (sulfate) solution, well suited for a solvent extraction treatment.

7.2.2. Purification of Green Phosphoric Acid (H_3PO_4) to Food Grade Quality

By reacting apatite with sulfuric acid, phosphoric acid and gypsum will result. Most phosphoric acid is produced in this way and is normally used as raw material (green acid) to produce fertilizers. A small amount, however, is further refined to phosphoric acid of food grade quality. The green phosphoric acid from a phosphoric acid plant contains 25–30% H_3PO_4. The acid is heavily entrained with impurity cations, among which arsenic, cadmium, and uranium are the most toxic. In addition, anions, like chloride, fluoride, and sulfate must be considered. Selective extraction of phosphoric acid[334] with TBP, according to the block diagram in Fig. 5.33, has been used to produce acid of food grade.

The purification operation starts with addition of chemicals to precipitate arsenic and sulfate in a pretreatment step. The acid is then cooled and left for an appropriate aging time. A flocculent is added and the solid content is allowed to flock (precipitate) in a clarifier. The clear phosphoric acid is fed to the solvent extraction operation, where about 60% of the acid is extracted countercurrently with TBP in kerosene. The resulting raffinate containing most of the contaminating substances, together with the slurry from the clarifier, is returned to the phosphoric acid plant. In this way, the impurities in the green acid will be

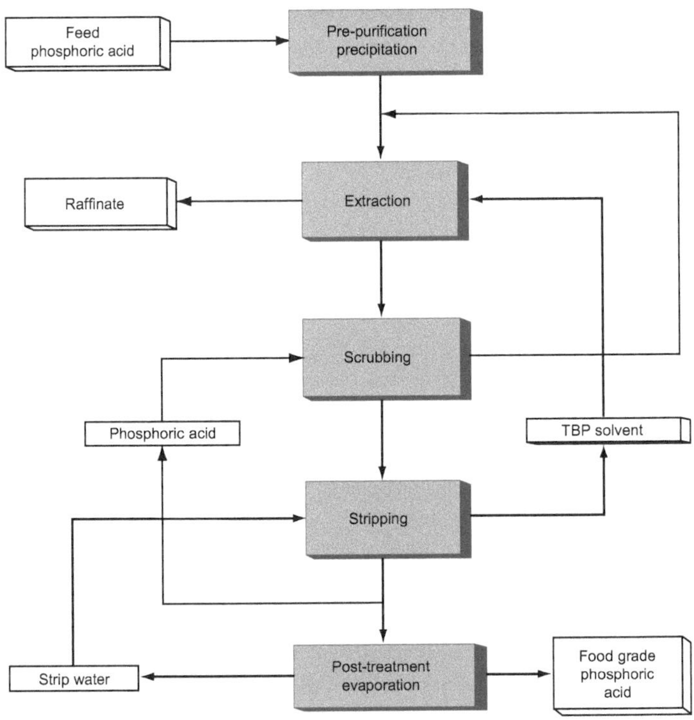

FIGURE 5.33 Purification of phosphoric acid to food grade quality.

returned to fertilizer production without creating additional environmental problems. After extraction, the organic solvent, loaded with phosphoric acid, is scrubbed with a phosphoric acid solution. In this way, coextracted contamination is removed. The scrub liquor is recycled to extraction.

After adjusting the temperature to about 40 °C, the organic solvent is treated with clean water to strip the phosphoric acid. This acid, containing about 30% H_3PO_4, is heated and water evaporated to produce concentrated (85%) phosphoric acid of food grade quality. In addition, a process for the production and purification of citric acid has been reported, utilizing a similar flow sheet.

7.2.3. Removal of Arsenic Acid (H_3AsO_4) from Copper Electrolytes

The use of low-grade ores with large amounts of harmful impurities results in increased accumulation of these impurities in the electrolyte. One of the most harmful is arsenic, which seriously affects the quality of the copper cathodes. Arsenic occurs in the sulfuric acid electrolyte as arsenic acid, and as such can be extracted with TBP. The extraction is favored by increasing acidity and low temperature. Arsenic control by solvent extraction using TBP has been demonstrated[335] in copper electrorefining plant practice and recently an arsenic removal process with undiluted TBP has been suggested. Centrifugal extractors were used, as the separation conditions (emulsions) in extraction and stripping were very slow. After filtration of the barren electrolyte, containing about 8 g/dm^3 As, the sulfuric acid concentration was raised to 180 g/dm^3. The electrolyte was fed to extraction, where about 50% of the arsenic was transferred in two stages to the organic solvent.

After a scrubbing stage with dilute sulfuric acid to minimize copper coextraction, arsenic was stripped with water at elevated temperature. The process was demonstrated on a semi-industrial scale-removal of 1 kg/h As. The operation of the extractors was reliable. The phase entrainment in each stage was less than 0.05% at an organic/aqueous phase = 1/1 ratio. TBP only extracts As(V) and therefore other extractants active for As(III), Sb, and Bi have been tested. Some years ago, a two-extractant procedure[336] was suggested and recently, the use of a mixture of alkylphosphine oxides (R_3PO) has been reported.[337,338]

7.2.4. Cyanide Recovery by Solvent Extraction

A solvent extraction process for the recovery of cyanide from acidified solutions using organophosphorus solvating extractants has recently been suggested.[339] The most favorable extractant for effective HCN loading is a mixture of trialkylphosphine oxides (R_3PO) (see Fig. 5.34). The distribution, D_{HCN} is well over 10, which indicates that a four-stage extraction would be sufficient to remove cyanide to less than 10 mg/dm^3 from a starting concentration of 200 mg/dm^3 at an aqueous/organic phase ratio = 5/1. The extraction kinetics are reported to be fast.

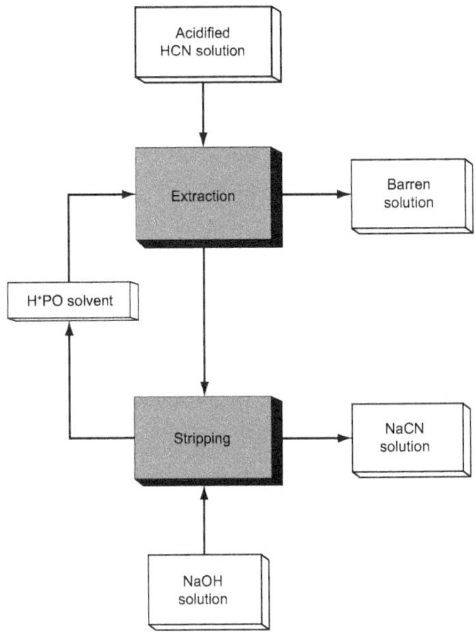

FIGURE 5.34 Recovery of cyanide as NaCN.

The stripping of the organic loaded solvent is accomplished by neutralization of the organic solvent with an alkali solution (e.g., NaOH) to reconstitute a strong cyanide solution for recycling.

7.3. Separation of Organic Pollutants by Solvent Extraction

Although this chapter concentrates on environmental problems associated with metals, the discharge of organic compounds is also subjected to stringent limits. The compounds that cause most concern are pesticides, herbicides, phenols, biphenyls, polychlorinated hydrocarbons (PCBs), and polyaromatic hydrocarbons (PAHs). Considerable amounts of oily sludge are generated during the storage, transportation, and refining of petroleum.

Solvent extraction has many features that make this separation technique applicable for the removal of organic pollutants from wastewaters.[340,341] The organic solutes can be recovered, and there is thus a potential for economic credit to the operation. In comparison with biological treatment, it is not subject to toxicity instabilities. The disadvantage is that, even with solvents having low solubilities in water, solvent losses can be substantial owing to high flow rates of wastewaters. In practice, solvent extraction, therefore, has to be combined with other separation methods, such as distillation, to fulfill the environmental restrictions. Complex process considerations have then to be taken into account to evaluate the economics of different alternatives.

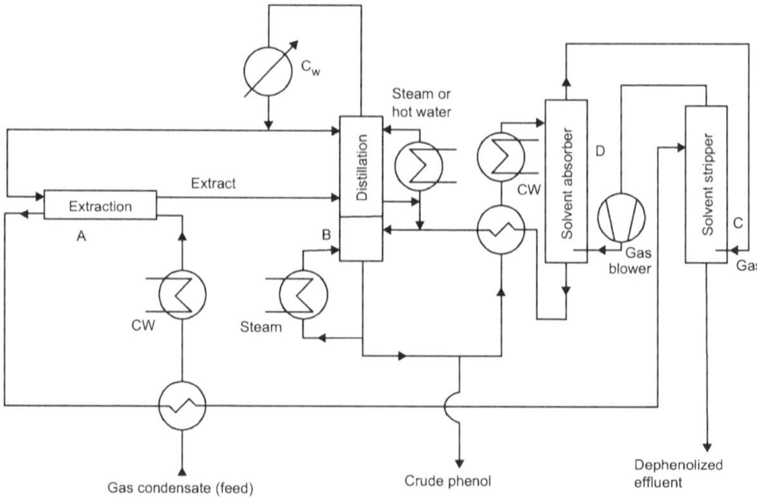

FIGURE 5.35 The Phenosolvan process.

7.3.1. Phenosolvan Process

In this process,[342] the phenolic effluent is contacted with the solvent in a multistage mixer-settler countercurrent extractor (see Fig. 5.35). The extract, containing phenol, is separated into phenol and solvent by distillation and solvent is recycled to the extractor. The aqueous raffinate phase is stripped from solvent with gas, and the solvent is recovered from the stripping gas by washing with crude phenol and passed to the extract distillation column. Initially the process utilized butyl acetate as a solvent, but more recently isopropyl ether has been used, although the latter has a much lower partition coefficient for phenol. The reason for this choice of solvent is that the separation of solvent and phenol by distillation is easier and less costly. In addition, isopropyl ether is not subject to hydrolysis, which can be a problem when using esters as solvents.

The Phenosolvan process is used in the treatment of wastewaters from plants involving phenol synthesis, coke ovens, coal gasification, low pressure carbonization, and plastic manufacturing. The residual phenol content after dephenolization is usually in the range 5–20 mg/dm^3 and plants treating 500 m^3 water per day are in operation.

REFERENCES

1. Rydberg J, Cox M, Musikas C, Choppin GR, editors. *Solvent extraction principles and practice*. New York, Boca Raton: Society of Chemical Industry; 2004.
2. Ritcey GM. Solvent extraction. *Principles and applications to process metallurgy*. Amsterdam: Elsevier; 2006. book 1.
3. Ritcey GM. Solvent extraction. *Principles and applications to process metallurgy*. Amsterdam: Elsevier; 2006. book 2.

4. Lo TC, Baird MHI, Hanson C. *Handbook of solvent extraction*. New York: Wiley-Interscience; 1983.
5. Dudley KA, Moroney AS, Braaksma MO, Readett DJ. Copper solvent extraction oxime selectivity: Operational experience at Girilambone Copper Company. In: *Proceedings of the 4th COPPER 99-COBRE 99 international conference*, Phoenix, 1999, vol. 4. p. 247–62.
6. Preston JS. Solvent extraction of base metals by mixtures of organophosphoric acids and non-chelating oximes. *Hydrometallurgy* 1983;**10**:187.
7. Preston JS. Solvent extraction of nickel and cobalt by mixtures of carboxylic acids and non-chelating oximes. *Hydrometallurgy* 1983;**11**:105.
8. Szymanowski J. *Hydroximes and copper hydrometallurgy*. Boca Raton, Florida: CRC Press; 1993.
9. Dalton RF, Seward GW. *Proceedings of reagents in the minerals industry*. London: Institute of Mining and Metallurgy; 1984. p. 106.
10. Dalton RF, Seward GW. The role of aryl aldoxime reagents in the technical advancement of copper production by solvent extraction. In: *Proceedings of ISEC'86*, 1986, vol. 2. p. 11. Frankfurt-am-Main: DECHEMA.
11. Murthy TKS, Koppiker KS, Gupta CK. Solvent extraction in extractive metallurgy. In: Li NN, Navratil JD, editors. *Recent developments in separation science*, vol. 8. Boca Raton, Florida: CRC Press; 1986. p. 1.
12. Dalton RF, Price R, Henmana E, Hoffmann B. *Separation processes in hydrometallurgy*. Chichester: Ellis-Horwood; 1987. p. 466.
13. Burkin R, editor. *Critical reviews in applied chemistry, Vol. 17; Extractive metallurgy of nickel*. Chichester: John Wiley & Sons; 1987.
14. Chapman TW. Extraction-metals processing. In: Rousseau RW, editor. *Handbook of separation process technology*. New York: John Wiley & Sons; 1987. p. 467. Chap. 8.
15. Sadeghi R, Mohebbi A, Baniasadi M. CFD modeling of the launder of settler of an industrial copper solvent extraction plant: A case study on Sarcheshmeh copper complex, Iran. *Int J Miner Process* 2010;**98**:55–65.
16. Barnes JE, Stetchfield JH, Williams GOR. J. Solvent extraction of Fe(III) by di-(2-ethylhexyl) phosphoric acid. *Inorg Nucl Chem* 1976;**38**:1065.
17. Schulz WW, Navratil JD, editors. *Science and technology of tributyl phosphate*, vols. 1 and 2. Boca Raton, Florida: CRC Press; 1987.
18. Nishimura S. *Extraction metallurgy '81*. London: Institute of Mining and Metallurgy; 1981. p. 404.
19. Flett DS. New reagents or new ways with old reagents. *J Chem Technol Biotechnol* 1999;**74**:99.
20. Mackenzie JMW, Virnig MJ, Boley BD, Wolfe GA. *ALTA 1998 nickel cobalt pressure leaching and hydrometallurgy forum*. Melbourne: ALTA Metallurgical Services; 1998.
21. Yamada H, Tanaka M. Solvent extraction of metal carboxylates. *Adv Inorg Chem Radiochem* 1985;**29**:143.
22. Osseo-Asare K, Miller JD, editors. *Hydrometallurgy; research, development and plant practice*. New York: Metallurgical Society AIME; 1982.
23. *Proceedings of extraction metallurgy conferences*. London: Institution of Mining and Metallurgy; 1989.
24. Lapidus GT, et al. Thermodynamic models for gold and silver extraction from copper-ammonia-thiosulfate solutions. In: Moyer BA, editor. *Proceedings of ISEC 2008 international solvent extraction conference. Fundamentals to industrial Applications,* Montreal, Canada: The Canadian Institute of Mining, Metallurgy and Petroleum; 2008. Vols. 2. p. 299–304.
25. Sole KC, Cole PM. Purification of nickel by solvent extraction, in ion exchange. In: Marcus Y, Sengupta AK, editors. *Ion exchange and solvent extraction*, vol. 15. New York: Marcel Dekker; 2002. p. 143.

26. Soldenhoff K, Wilkins D. *Solvent extraction for the 21st century*. In: *Proceedings of ISEC'99*, vol. 1. London: Society of Chemical Industry; 2001. p. 777.
27. Grant RA, Drake VA. *Proceedings of ISEC'2002*, Cape Town, 2002, 940, vol. 1. p. 123.
28. Baroncelli F, Carlini D, Gasparini GM, Simonetti E. *Proceedings ISEC '88*, vol. 4. Moscow: Moscow; Vernadsky Inst.; Academy of Sciences USSR; 1988. p. 236.
29. Demopoulos GP, Pouskouleli G, Ritcey GM. *Proceedings of ISEC '86*. Frankfurt-am-Main. Munich: DECHEMA; 1986. vol. 2. p. 581.
30. Villaescusa I, Miralles N, de Pablo J, Salvado V, Sastre AM. *Solv Extn Ion Exch* 1993;**11**:613.
31. Argiropoulos G, Cattrall RW, Hamilton IC, Paimin R. Value adding through solvent extraction. In: *Proceedings of ISEC'96*. University of Melbourne; 1996.
32. Kordosky GA, Sierokoski JM, Virnig MJ, Mattison PL. Gold solvent extraction from typical cyanide leach solutions. *Hydrometallurgy* 1992;**30**:291.
33. Virnig MJ, Kordosky GA. *Value adding through solvent extraction*. In: *Proceedings of ISEC'96*, vol. 1. University of Melbourne; 1996. p. 311.
34. Gamino-Arroyo Z, et al. Recovery of silver and gold from thiourea leaching Au-Ag ores liquors by D2EHDTPA extraction. In: Moyer BA, editor. *Proceedings of ISEC 2008 international solvent extraction conference*. Montreal, Canada: The Canadian Institute of Mining, Metallurgy and Petroleum; 2008. p. 293–8. in 2 vols.
35. Mansur MB, et al. Solvent extraction for metal and water recovery from industrial wastes and effluent. In: Moyer BA, editor. *Proceedings of ISEC 2008 international solvent extraction conference*. Montreal, Canada: The Canadian Institute of Mining, Metallurgy and Petroleum; 2008. p. 311–6. in 2 vols.
36. Zhao J, Wu Z, Chen J. *Value adding through solvent extraction*. In: *Proceedings of ISEC'96*, vol. 1. University of Melbourne; 1996. p. 629.
37. Capela RS, Paiva AP. In: Proceedings of ISEC 2002. Cape Town: 2002. p. 335.
38. Facon S, Rodriguez MAvila, Cote G, Bauer D. *Solvent extraction in the process industries*. In: *Proceedings of ISEC93*, vol. 1. Elsevier Applied Science; 1993. p. 557.
39. Woollam SF, Grant RA. In: Moyer BA, editor. *Proceedings of ISEC 2008 international solvent extraction conference*. Montreal, Canada: The Canadian Institute of Mining, Metallurgy and Petroleum; 2008. p. 281–6. in 2 vols.
40. S6nchez-Loredo MG, et al. In: Moyer BA, editor. *Proceedings of ISEC 2008 international solvent extraction conference*. Montreal, Canada: The Canadian Institute of Mining, Metallurgy and Petroleum; 2008. p. 1049–55. in 2 vols.
41. Tasker PA, et al. Anion selectivity in zwitterionic amide-functionalised metal salt extractants. In: Moyer BA, editor. *Proceedings of ISEC 2008 international solvent extraction conference*. Montreal, Canada: The Canadian Institute of Mining, Metallurgy and Petroleum; 2008. p. 1457–62. in 2 vols.
42. Chayama K, et al. In: Moyer BA, editor. *Proceedings of ISEC 2008 international solvent extraction conference*. Montreal, Canada: The Canadian Institute of Mining, Metallurgy and Petroleum; 2008. p. 1585–90. in 2 vols.
43. Grant RA, et al. In: Moyer BA, editor. *Proceedings of ISEC 2008 international solvent extraction conference*. Montreal, Canada: The Canadian Institute of Mining, Metallurgy and Petroleum; 2008. p. 415–20. in 2 vols.
44. Du Preez A, Preston JS. Proceedings of ISEC 2002. Cape Town: 2002. p. 896.
45. Pouskouleli G, Kelebek S, Demopoulos GP. In: Davies GA, editor. *Separation processes in hydrometallurgy*. Chichester: Ellis-Horwood; 1987. p. 175.
46. Narita H, et al. Separation of palladium in hydrochloric acid solution with sulfide-containing amide compound. In: Moyer BA, editor. *Proceedings of ISEC 2008 International Solvent*

Extraction Conference. Montreal, Canada: The Canadian Institute of Mining, Metallurgy and Petroleum; 2008. p. 1445–50. in 2 vols.

47. Wisniewski M, et al. Kinetics of zinc and iron ion extraction from acidic solutions with phosphonium ionic liquid. In: Moyer BA, editor. *Proceedings of ISEC 2008 International Solvent Extraction Conference*. Montreal, Canada: The Canadian Institute of Mining, Metallurgy and Petroleum; 2008. p. 1295–300. in 2 vols.
48. Grant RA, Murrer BA. Extraction of platinum(IV) from hydrochloric acid solutions by amine alamine 304 in xylene. Estimation of the interaction coefficient between PtCl62– and H+ European *Patent No. 0210004*, 1986.
49. Dhara SC. Solvent extraction with alkyl amines. In: Kudryk V, Corrigan DA, Laing WW, editors. *Proceedings of precious metals mining, extraction, and processing*. Littleton, Collorado: Metallurgical Society AIME; 1984. p. 199.
50. Dolgikh VI, Borikov PI, Borbat VF, Kouba EF, Gindin LM. Solvent extraction in hydrometallurgy. *Tsvet Metal* 1967;**40**:30.
51. Fedorenko NV, Ivanova TI. Solvent extraction of rhodium from chloride solutions. *Russ. J Inorg Chem* 1965;**10**:387.
52. Benguerel E, Demopoulos GP, Cote G, Bauer D. Speciation and separation of rhodium (III) from chloride solutions: a critical review. *Solv Extn Ion Exch* 1994;**12**:49.
53. Benguerel E, Demopoulos GP, Harris GB. Speciation and Separation of Rhodium (III) from Chloride Solution: A Critical Review. *Hydrometallurgy* 1996;**40**:135.
54. Brown CG, Sherrington LG. Neutral organophosphorus extractants. *J Chem Technol Biotechnol* 1979;**29**:193.
55. Preston JS, du Preez AC. *Mintek Report M378*. Randburg, South Africa: Council for Mineral Technology; 1988.
56. Li D. Solvent extraction for the 21st century. *Proceedings of ISEC'99,* 2000; vol. 2. p. 108.
57. Yuan C, Ma H, Pan C, Rickelton WA. Value adding through solvent extraction. *Proceedings of ISEC'96,* 1996; vol. 1. p. 733.
58. Vertis A, Nagy S, Klencsar Z. *Handbook of nuclear chemistry*. Amsterdam: Kluwer Academic Publishers; 2003. in 3 vols.
59. Loveland WD, Morrissey DJ, Seaborg GT. *Modern nuclear chemistry*. New York: John Wiley & Sons; 2006.
60. Choppin GR, Liljenzin J-O, Rydberg J. *Radiochemistry and nuclear chemistry*. 3rd ed. Woburn, Massachusetts: Butterworth-Heinemann; 2002.
61. Musikas C, Schulz WW, Liljenzin J- O. Solvent extraction in nuclear science and technology. In: Rydberg J, Cox M, Musikas C, Choppin GR, editors. *Solvent extraction principles and practice*. New York: CRC Press; 2004.
62. Encyclopedia of materials: science and technology. 2008. Ganguly G. Nuclear reactor fuel fabrication, p. 6324–39.
    Rodrigues P. Nuclear fuel cycles, p. 6300–03.
    Laidler JJ. Nuclear fuel reprocessing, p. 6303–09.
    Ramanujan A. PUREX and THOREX processes, p. 7918–24.
63. *Encyclopedia of physical science and technology*; 2004.
    Simnad MT. Nuclear reactor materials and fuels, p. 775–815.
    Knief RA. Nuclear fuel cycles, p. 655–70.
    Chenoweth WL, Pool TC. Uranium, p. 351–7.
    Forsberg CW. Radioactive wastes, p. 643–9.
64. Friedlander G. *Introduction to radiochemistry*; 2007.
65. Bryan JC. *Introduction to nuclear science*; 2008.

66. Neeb K- H. The radiochemistry of nuclear power plants with light water reactors. *Europe: Walter De Gruyter Inc*; 1997.
67. Ehmann WD, Vance DE. *Radiochemistry and nuclear methods of analysis*. New York: John Wiley & Sons; 1991.
68. Musikas C, Le Marois G, Fitoussi R, Cuillerdier C. Actinide separations. In: Navratil JD, Schultz WW, editors. *ACS symposium series 117*. Washington, D.C.: American Chemical Society; 1980. p. 131.
69. Coleman CF, Brown KB, Moore JG, Crouse D. *J Ind Eng Chem* 1958;**50**:1756.
70. Lloyd PJD. In: Lo TC, Baird MHI, Hanson C, editors. *Handbook of solvent extraction*. New York: John Wiley & Sons; 1983. p. 763.
71. Ritcey GM. 2006. *Solvent extraction, principles and applications to process metallurgy*, vol. 2. New York: Elsevier Scientific; 1979.
72. Singh H, Mishra SL, Vijayalakshmi R. Uranium recovery from phosphoric acid by solvent extraction using a synergistic mixture of di-nonyl phenyl phosphoric acid and tri- *n*-butyl phosphate. *Hydrometallurgy* 2004;**73**(1–2):63–70.
73. Jha RK, Gupta KK, Kulkarni PG, Gurba PB, Janardan P, Changarani RD, Dey PK, Pathak PN, Manchanda VK. Third phase formation studies in the extraction of Th(IV) and U(VI) by N, N-dialkyl aliphatic amides. *Desalination* 2008;**232**(1–3):225–33.
74. Nagasaki T, Shinkai S. The scavenging of uranyl from phosphate-containing media. *J Chem Soc Perkin Trans* 1991;**2**:1063.
75. Soldenhoff K, Wilkins D, Shamieh M, Manis A. Mini-plant testing of uranium solvent extraction process for acid leach liquors containing chloride. *Proceedings of ISEC 2005*; 2005.
76. Zuo Y, Chen J, Li D. Reversed micellar extraction and separation of thorium(IV) from rare earth(III) by primary amine N1923 in ionic liquid. *Sep Purif Technol* 2008;**63**(3): 684–90.
77. Meguro Y, Iso S, Yoshida Z, Tomioka O, Enokida Y, Yamamoto I. Decontamination of uranium oxides from solid wastes by supercritical CO_2 fluid leaching method using HNO_3–TBP complex as a reactant. *J Supercrit Fluids* 2004;**31**(2):141–7.
78. Kumar P, Pal A, Saxena MK, Ramakumar KL. Supercritical fluid extraction of uranium and thorium from solid matrices. *Desalination* 2008;**232**(1–3):71–9.
79. Trofimov TI, Samsonov MD, Kulyako YM, Myasoedov BF. Dissolution and extraction of actinide oxides in supercritical carbon dioxide containing the complex of tri- *n*-butylphosphate with nitric acid. *C R Chim* 2004;**7**(12):1209–13.
80. Miles JH. In: Schulz WW, Burger LL, Navratil JD, editors. *Science and technology of tributyl phosphate*, vol. 3. Boca Raton, Florida: CRC Press; 1990. Chap. 1, Part 2.
81. Sorel C, et al. In: Moyer BA, editor. *Proceedings of ISEC 2008 international solvent extraction conference*. Montreal, Canada: The Canadian Institute of Mining, Metallurgy and Petroleum; 2008. p. 715–20. in 2 vols.
82. Fermvik A, et al. Proceedings of ISEC 2008 international solvent extraction conference. In: Moyer BA, editor. Montreal, Canada: The Canadian Institute of Mining, Metallurgy and Petroleum; 2008. p. 551–6. in 2 vols.
83. Modolo G, et al. In: Moyer BA, editor. *Proceedings of ISEC 2008 International Solvent Extraction Conference*. Montreal, Canada: The Canadian Institute of Mining, Metallurgy and Petroleum; 2008. p. 521–6. in 2 vols.
84. Meyer M, Burgat R, Faure S, Batifol B, Hubinois J-C, Chollet H, Guilard R. Thermodynamic studies of actinide complexes. 1. A reappraisal of the solution equilibria between plutonium(IV) and ethylenediaminetetraacetic acid (EDTAH4) in nitric media. *C R Chim* 2007;**10**: 929–7.

85. Guo JX, Sun SX, Li MX, Yin ZL, Pang H, Wang ZL, Meng ZL. The effects of diluent on the extraction behavior of uranium (VI) with benzyloctadecyldimethyl ammonium chloride (BODMAC) from chloride solution. *Proceedings of ISEC 2005*; 2005.
86. Sano Y, Kamiya M, Shibata A, Koizumi T, Koyama T. Neptunium co-recovery with uranium and plutonium in next process. *Proceedings of ISEC 2005*; 2005.
87. Schulz WW, Navratil JD. In: Li NN, editor. *Recent developments in separations science*, vol. 7. Boca Raton, Florida: CRC Press; 1980. p. 31.
88. Ozawa M, Nemoto S, Toghashi A, Kawata T, Onishi A. *Solv Extr Ion Exch* 1992;**10**:829.
89. Law JD, Brewer KN, Herbst RS, Todd TA, Wood DJ. Development and demonstration of solvent extraction processes for the separation of radionuclides from acidic radioactive waste. *Waste Manag* 1999;**19**:27–37.
90. Herbst RS, Brewer KN, Law JD, Tranter TJ, Todd T. A separation of actinides and other elements from simulated sodium-bearing waste using the TRUEX process. In: Proceedings from SPECTRUM '94. Atlanta, Georgia: 1994.
91. Law JD, Brewer KN, Todd TA. Actinide partitioning from actual Idaho chemical processing plant acidic tank waste using centrifugal contactors. In: *Proceedings from GLOBAL '97*, Yokohama, Japan: 1997.
92. Horwitz EP, Dietz ML, Fisher DE. SREX: A new process for the extraction and recovery of strontium from acidic nuclear waste streams. *Solvent Extr Ion Exch* 1991;**9**(1):1.
93. Wood DJ, Law JD. Evaluation of the SREX solvent extraction process for the removal of ^{90}Sr and hazardous metals from acidic nuclear waste solutions containing high concentrations of interfering alkali metal ions. *Sep Sci Technol* 1997;**32**(1–4):241.
94. Myaesoedov BF, Chmutova MK, Smirnov IV, Shadrin AU. *Global'93: Future nuclear systems: Emerging fuels cycles and waste disposal options*. La Grange Park, Illinois: American Nuclear Society; 1993. p. 581.
95. Horwitz EP, Shulz WW. An improved extraction chromatographic resin for the separation of uranium from acidic nitrate media. In: Bond AH, Dietz ML, Rogers RD, editors. *Metal ion separation and preconcentration progress and opportunities*. Washington, D.C: American Chemical Society; 1998.
96. Babain VA, Shadrin AY, Lazarev LN, Kiseleva RN. Extraction of cesium and strontium by solutions of chlorinated cobalt dicarbollide in diethyleneglycol ditetrauoropropyl ether. *Radiokhimiya* 1993;**35**:81.
97. Madic C, Hudson MJ. *High level liquid waste partitioning by means of completely incinerable extractants*. European Commission, Nuclear Science and Technology, EUR18038 EN; 1998. p. 208.
98. Madic C, Hudson MJ, Liljenzin JO, Glatz JP, Nannicini R, Facchini A, Kolarik Z, Odoj R. *New partitioning techniques for minor actinides*. European Commission, Nuclear Science and Technology, EUR19149 EN; 2000. p. 286.
99. Grüner B, Kvíčalová M, Selucký P, Lučaníková M. Anionic alkyl diglycoldiamides with covalently bonded cobalt bis(dicarbollide)(1−) ions for lanthanide and actinide extractions. *J Organometallic Chem* 2010;**695**(9):1261–4.
100. Liljenzin JO, Rydberg J, Skarnemark G. Synthetic development of c3-symmetric triphenoxymethane based reagents for the selective recognition and sequestration of lanthanides and actinides. *Sep Sci Technol* 1980;**15**:799.
101. Cecille L, Dworschak H, Girardi E, Hunt BA, Mannone F, Mousty E. The TRUEX Process – A Process for the Extraction of the Transuranic Elements from Nitric Acid. Waste utilizining Modified PUREX Solvent Extraction. In: Navratil JD, Schulz WW, editors. *Actinides separations*. Washington, D.C: American Chemical Society; 1980. p. 427.

102. Liljenzin JO, Persson G, Svantesson I, Wingefors S. The CTH-process for HLLW treatment, Part I – General description and process design. *Radiochim Acta* 1984;**35**:155.
103. Persson G, Wingefors S, Liljenzin JO, Svantesson I. The CTH-Process for HLLW treatment, Part II – Hot test. *Radiochim Acta* 1984;**35**:163.
104. Persson G, Svantesson I, Wingefors S, Liljenzin JO. Hot test of a TALSPEAK procedure for separation of actinides and lanthanides using recirculating DTPA-lactic acid solution. *Solv Extr Ion Exch* 1984;**2**(1):89.
105. Dakshinamoorthy A, Dhami PS, Naik PW, Dudwadkar NL, Munshi SK, Dey PK, Venugopal V. Separation of palladium from high level liquid waste of PUREX origin by solvent extraction and precipitation methods using oximes. *Desalination* 2008;**232**(1–3):26–36.
106. Guillaume B, Germain M, Puyou M, Rouger H. *Experimental study of CO_2 Absorption in a hollow fiber membrane contactor. Proceedings of international conference on nuclear fuel reprocessing, RECORD 87*, vol. 1. Paris: Soc. Francaise d'Energie Nucleaire; 1987. p. 433.
107. Schulz WW. *Macroreticular Ion Exchange Resin Cleanup of Purex Process Solvent; U.S. Atomic Energy Commission Report ARH-SA-58*. Richland, Washington, D.C: Atlantic Richfield Hanford Co.; 1970.
108. Mailen JC. Secondary Purex solvent cleanup: laboratory development. *Nucl Technol* 1998;**83**:182.
109. Choppin G, Liljenzin JO, Rydberg J. *Radiochemistry and nuclear chemistry*. 3rd ed. Oxford: Butterworth-Heinemann; 2002. p. 593.
110. CURE. *Clean Use of Reactor Energy, U.S. Department of Energy Report WHC-EP-0268*. Richland, Washington, D.C: Westinghouse Hanford Co.; 1990.
111. Mukaiyama T, Kubota M, Takizuka T, Ogawa T, Mizumoto M, Yoshida Y. Partitioning and transmutation program OMEGA at JAERI. *Proceedings of the GLOBAL'95 conference, Versailles. France* 1995; vol. 1. p. 110.
112. Yoshizuka K, Pietzsch HJ, Seifert S, Stephan H. Novel computational chemistry for molecular design of radioactive metal complexes. In: Moyer BA, editor. *Proceedings of ISEC 2005 international solvent extraction conference*. Montreal, Canada: The Canadian Institute of Mining, Metallurgy and Petroleum; 2005. in 2 vols, p. 952–60.
113. Barre JY, Bouchard J. French R&D strategy for the back-end of the fuel cycle. In: *Proceedings of the Global '93*, Seattle, Washington, D.C., September 1993, p. 12–7.
114. Musikas C, Vitart X, Pasquiou JY, Hoel P. Diamides as actinides extractants for various wastes treatment. In: King CJ, Navratil JD, editors. *Chemical separations*, vol. 2. Denver: Litarvan Literature; 1986. p. 359.
115. Kolarik Z, Mullich U, Gassner F. Extraction of Am(III) and Eu(III) nitrates by 2,6-di(5,6-dipropyl-1,2,4-triazin-3-yl)pyridine. *Solv Ext Ion Exch* 1999;**17**(23):1155.
116. Modolo G, Odoj R. Influence of the purity and irradiation stability of Cyanex 301 on the separation of trivalent actinides from lanthanides by solvent extraction. *J. Radioanal Nucl Chem* 1998;**228**(1,2):83.
117. Charbonnel MC, Boubals N, Giroux S, Guillaumont D. Thermodynamic properties accounting for differences in affinity and selectivity of BTPs toward Eu(III) and Am(III) extraction. in 2 vols. In: Moyer BA, editor. *Proceedings of ISEC 2005 International Solvent Extraction Conference*. Montreal, Canada: The Canadian Institute of Mining, Metallurgy and Petroleum; 2005. p. 467–75.
118. Bosland L, Adnet JM, Berthon L, Pareau D, Stambouli M. Extraction kinetics of actinides and lanthanides trivalent nitrates by DMDOHEMA molecule. In: Moyer BA, editor. *Proceedings of ISEC 2005 International Solvent Extraction Conference*. Montreal, Canada: The Canadian Institute of Mining, Metallurgy and Petroleum; 2005. in 2 vols, p. 1036–43.

119. Weigl M, Geist A, Müllich U, Gompper K. Kinetics of americium extraction with 2,6-di(5,6-dipropyl-1,2,4-triazin-3-yl) pyridine. In: Moyer BA, editor. *Proceedings of ISEC 2005 international solvent extraction conference*. Montreal, Canada: The Canadian Institute of Mining, Metallurgy and Petroleum; 2005. in 2 vols, p. 671–9.
120. Hill MGB, Hudson C, Iveson MJ, Madic PB, Youngs C, Drew TGA. Solvent extraction and lanthanide complexation studies with new terdentate ligands containing two 1,3,5-triazine moieties. *Dalton Trans* 2004;**2**:244–51.
121. Maina TH, Chiotellis E. Synthesis, radiochemistry and biological evaluation of technetium-99m complexes with 1,8-diamine-3,6-dithiaoctane (DDO) ligands. *Int J Radiat Appl Instrum. Part B. Nucl Med Biol* May 1992;**19**(4):481–9.
122. Dileep CS, Jagasia P, Dhami PS, Achuthan PV, Dakshinamoorthy A, Tomar BS, Munshi SK, Dey PK. Distribution of technetium in PUREX process streams. *Desalination* 30 November 2008;**232**(1–3):157–65.
123. Pillai MRA, Kothari K, Banerjee S, Samuel G, Suresh M, Sarma HD, Jurisson S. Radiochemical studies of ^{99m}Tc complexes of modified cysteine ligands and bifunctional chelating agents. *Nucl Med Biol* July 1999;**26**(5):555–61.
124. Stary J. *The solvent extraction of metal chelates*. Oxford: Pergamon Press; 1964.
125. De AK, Khopkar SM, Chalmers RA. *Solvent extraction of metals*. London: Van Nostrand Reinhold; 1970.
126. Alders L. *Liquid–liquid extraction*. New York: Amsterdam; Elsevier; 1995.
127. Zolotov YA, Bodnya VA, Zagruzina AN. Liquid-liquid extraction in inorganic analysis: current status. *CRC Crit Rev Anal Chem* 1982;**14**:2.
128. Keller R, Mermet JM, Otto M, Widmer HM, editors. *Analytical chemistry*. Weinheim, Germany: Wiley-VCH, Verlag GmbH; 1998. p. 25.
129. Rydberg J, Cox M, Musikas C, Choppin GR. *Solvent extraction principles and practice*. 2nd ed. New York: Marcel Dekker Publishers Inc.; 2004.
130. Kislik VS, editor. *Liquid membrane. Principles and applications in chemical separations and wastewater treatment*. Amsterdam: Elsevier; 2010.
131. Valcarcel M. Liquid-liquid extraction in continuous flow analysis. In: Alegret S, editor. *Developments in solvent extraction*. West Sussex, UK: Ellis Horwood Series in Analytical Chemistry; 1988. p. 135.
132. Preconcentration of trace elements. In: Lobinsky L, editor. *Comprehensive analytical chemistry*, vol. XXV. Amsterdam: Elsevier; 1992.
133. Tobiszewski M, Mechlińska A, Zygmunt B, Namieśnik J. Green analytical chemistry in sample preparation for determination of trace organic pollutants. *TrAC Trends Anal Chem* 2009;**28**(8):943–51.
134. Persson B-A, Vessman J. The use of selectivity in analytical chemistry – some considerations. *TrAC Trends Anal Chem* 2001;**20**(10):526–32.
135. Tagushi S, Goto K. In: Alfassi ZB, Wai CM, editors. *Pre-concentration techniques for trace elements*. Boca Raton, Florida: CRC Press; 1989. p. 31.
136. Kumar AR, Riyazuddin P. Chemical interferences in hydride-generation atomic spectrometry. *TrAC Trends Anal Chem* 2010;**29**(2):166–76.
137. Luciano RM, Bedendo GC, Carletto JS, Carasek E. Isolation and preconcentration of Cd(II) from environmental samples using polypropylene porous membrane in a hollow fiber renewal liquid membrane extraction procedure and determination by FAAS. *J Hazard Mater* 2010;**177**(1–3):567–72.
138. Martinis EM, Berton P, Monasterio RP, Wuilloud RG. Emerging ionic liquid-based techniques for total-metal and metal-speciation analysis. *TrAC Trends Anal Chem. November* 2010;**29**(10):1184–201.

139. Cortina JL, Warshawsky A. Distribution studies of Zn(II), Cu(II) and Cd(II) with levextrel resins containing di(2,4,4-trimethylpentyl)phosphonic acid. In: Marinsky JA, Marcus Y, editors. *Ion exchange and solvent extraction*, vol. 13. New York: Marcel Dekker; 1996. p. 195.
140. Herrera-Herrera AV, Asensio-Ramos M, Hernández-Borges J, Rodríguez-Delgado MÁ. Dispersive liquid-liquid microextraction for determination of organic analytes. *TrAC Trends Anal Chem* 2010;**29**(7):728–51.
141. Sarafraz-Yazdi A, Amiri A. Liquid-phase microextraction. *TrAC Trends Anal Chem* 2010;**29**(1):1–14.
142. Mester Z, Sturgeon R, Pawliszyn J. Solid phase microextraction as a tool for trace element speciation. *Spectrochim Acta Part B: Atomic Spectrosc* 2001;**56**(3, 29):233–60.
143. Xiaoqi S, Yang J, Chen J, Jiutong MA. Solvent impregnated resin prepared using task-specific ionic liquids for rare earth separation. *J Rare Earths* 2009;**27**(6):932–6.
144. Beyer A, Biziuk M. Applications of sample preparation techniques in the analysis of pesticides and PCBs in food. *Food Chem* 2008;**108**(2):669–80.
145. Hu B, Song W, Xie L, Shao T. Determination of 33 pesticides in tea using accelerated solvent extraction/gel permeation chromatography and solid phase extraction/gas chromatography-mass spectrometry. *Chinese J Chromatogr* 2008;**26**(1):22–8.
146. Camel V. Microwave-assisted solvent extraction of environmental samples. *TrAC Trends Anal Chem* 2000;**19**(4):229–48.
147. Abrha Y, Raghavan D. Polychlorinated biphenyl (PCB) recovery from spiked organic matrix using accelerated solvent extraction (ASE) and Soxhlet extraction. *J Hazard Mater* 2000;**80**(1–3):147–57.
148. Dunford NT, Irmak S, Jonnala R. Pressurized solvent extraction of policosanol from wheat straw, germ and bran. *Food Chem* 2010;**119**(3):1246–9.
149. Haskins SD, Kelly DG, Weir RD. Novel pressurized solvent extraction vessels for the analysis of polychlorinated biphenyl congeners in avian whole blood. *Anal Chim Acta* 2010;**677**(1):19–23.
150. Cheng KL, Ueno K, Imamura T. *Handbook of organic analytical reagents*. Boca Raton, Florida: CRC Press; 1982.
151. Townshend A, Burns DT, Guibault GG, Lobinski R, Marczenko Z, Newman EJ, Onishi H, editors. *Dictionary of organic analytical reagents*. London: Chapman & Hall; 1993.
152. Kudo Y, Usami J, Katsuta S, Takeda Y. Solvent extraction of permanganates (Na, K) by 18-crown-6 ether from water into 1,2-dichloroethane: Elucidation of an extraction equilibrium based on component equilibria. *Talanta* 2003;**59**(6):1213–8.
153. Takeda Y, Yasui A, Morita M, Katsuta S. Extraction of sodium and potassium perchlorates with benzo-18-crown-6 into various organic solvents. Quantitative elucidation of anion effects on the extraction-ability and -selectivity for Na^+ and K^+. *Talanta* 2002;**56**(3):505–13.
154. Misra NL, Dhara S, Óvári M, Záray Gy, Aggarwal SK, Varga I. Determination of low atomic number elements at trace levels in uranium matrix using vacuum chamber total reflection X-ray fluorescence. *Spectrochim Acta Part B: Atomic Spectrosc* 2010;**65**(6):457–60.
155. Kagawa T, Ohno M, Seki T, Chikama K. Online determination of copper in aluminum alloy by microchip solvent extraction using isotope dilution ICP-MS method. *Talanta* 2009;**79**(4):1001–5.
156. Shigematsu T, Kudo K. Substoichiometric determination of trace impurities in zinc selenide. *J Radioanal Chem* 1978;**45**:373.
157. Tasaki T, Oshima T, Baba Y. Selective extraction and transport of copper(II) with new alkylated pyridinecarboxylic acid derivatives. *Talanta* 2007;**73**(2):387–93.
158. Fouad EA. Separation of copper from aqueous sulfate solutions by mixtures of Cyanex 301 and LIX® 984N. *J Hazard Mater* 2009;**166**(2–3):720–7.
159. Hoentsch E, Flachowsky J. Nuclear and radiochemical analysis. *Isotopenpraxis* 1977;**13**:432.

160. Iqbal M, Ejaz M, Chaudhri SA, Zamiruddin A. Extraction of arsenic(III) from chloride-iodide solutions. *J Radioanal Chem* 1978;**42**:335.
161. Beiraghi A, Babaee S. Separation and preconcentration of ultra trace amounts of beryllium in water samples using mixed micelle-mediated extraction and determination by inductively coupled plasma-atomic emission spectrometry. *Anal Chim Acta* 2008;**607**(2):183–90.
162. Hirayama N, Taga J, Oshima S, Honjo T. Sulfonamide-type di-Schiff base ligands as chelate extraction reagents for divalent metal cations. *Anal Chim Acta* 2002;**466**(2):295–301.
163. Oshima S, Hirayama N, Kubono K, Kokusen H, Honjo T. Ion-pair extraction behavior of divalent transition metal cations as charged complexes with *N, N′*-bis(2-pyridylmethylidene)-1,2-diiminoethane and its analogues. *Anal Chim Acta* 2001;**441**(1):157–64.
164. Shtefak M, Kirsh M, Rais I. Critical evaluation of stability constants for nucleotide complexes with protons and metal ions. *Zh Anal Khim* 1976;**32**:1364.
165. Takeda Y, Suzuki S, Ohyagi Y. A Comparison of the trace analysis of heavy metal aerosols with atomic absorption spectroscopy. *Chem Lett* 1978;**12**:1377.
166. Khajeh M. Optimization of microwave-assisted extraction procedure for zinc and copper determination in food samples by Box-Behnken design. *J Food Comp Anal* 2009;**22**(4):343–6.
167. Maurí-Aucejo AR, Arnandis-Chover T, Marín-Sáez R, Llobat-Estellés M. Application of pressurized fluid extraction to determine cadmium and zinc in plants. *Anal Chim Acta* 2007;**581**(1):78–82.
168. Olszanowski A. Solvent extraction of cadmium(II) from chloride solutions by pyridyl ketoximes. *Hydrometallurgy* 2011;**105**(3–4):284–9.
169. Anthemidis AN, Zachariadis GA, Farastelis CG, Stratis JA. On-line liquid–liquid extraction system using a new phase separator for flame atomic absorption spectrometric determination of ultra-trace cadmium in natural waters. *Talanta* 2004;**62**(3):437–43.
170. Leermakers M, Baeyens W, Quevauviller P, Horvat M. Mercury in environmental samples: Speciation, artifacts and validation. *TrAC Trends Anal Chem* 2005;**24**(5):383–93.
171. Li Z, Wei Q, Yuan R, Zhou X, Liu H, Shan H, Song Q. A new room temperature ionic liquid 1-butyl-3-trimethylsilylimidazolium hexafluorophosphate as a solvent for extraction and preconcentration of mercury with determination by cold vapor atomic absorption spectrometry. *Talanta* 2007;**71**(1):68–72.
172. Quevauviller P, Filippelli M, Horvat M. Method performance evaluation for methylmercury determination in fish and sediment. *TrAC Trends Anal Chem* 2000;**19**(2–3):157–66.
173. Argollo RM, Schilling JG. Determination of germanium by atomic absorption spectrometry following volatile hydride generation. *Anal Chem Acta* 1978;**96**:117.
174. Oshima S, Kubono K, Kokusen H, Honjo T. Selective extraction of gallium from aluminum and indium using tripod phenolic ligands. *Talanta* 2001;**53**(4):857–62.
175. Taher MA. Differential pulse polarography determination of indium after column preconcentration with [1-(2-pyridylazo)-2-naphthol]-naphthalene adsorbent or its complex on microcrystalline naphthalene. *Talanta* 2000;**52**(2):301–9.
176. Chung NH, Nishimoto J, Kato O, Tabata M. Selective extraction of thallium(III) in the presence of gallium(III), indium(III), bismuth(III) and antimony(III) by salting-out of an aqueous mixture of 2-propanol. *Anal Chim Acta* 2003;**477**(2):243–9.
177. Coedo AG, Dorado MT, Padilla I. Evaluation of different sample introduction approaches for the determination of boron in unalloyed steels by inductively coupled plasma mass spectrometry. *Spectrochim Acta Part B: Atomic Spectrosc* 10 January 2005;**60**(1):73–9.
178. Rao TP, Kala R. On-line and off-line preconcentration of trace and ultratrace amounts of lanthanides. *Talanta* 2004;**63**(4):949–59.
179. Kaiyanarman S, Khopkar SM. Solvent extraction separation of hafnium with 4-methyl-3-pentene-2. *Anal Chem* 1977;**49**:1192.

180. Charib A, Morris DFC. Determination of rhenium at ultratrace levels by selective laser. *Talanta* 1978;**25**:569.
181. Bayyari MA, Nazal MK, Khalili FI. The effect of ionic strength on the extraction of thorium(IV) from perchlorate solution by didodecylphosphoric acid (HDDPA). *Arab J Chem* 2010;**3**(2):115–9.
182. Madrakian T, Afkhami A, Mousavi A. Spectrophotometric determination of trace amounts of Uranium(VI) in water samples after mixed micelle-mediated extraction. *Talanta* 2007;**71**(2):610–14.
183. Fujino O, Umetani S, Ueno E, Shigeta K, Matsuda T. Determination of uranium and thorium in apatite minerals by inductively coupled plasma atomic emission spectrometry with solvent extraction separation into diisobutyl ketone. *Anal Chim Acta* 7 September 2000;**420**(1):65–71.
184. Shimomura S, Sakurai H, Morita I, Mino Y. Determination of germanium by atomic absorption spectrometry following volatile hydride generation. *Anal Chem Acta* 1978;**96**:69.
185. Steinnes E. Analytical applications of solvent extraction. *Radiochem Radioanal Lett* 1978;**33**:205.
186. Mane CP, Anuse MA. Extraction behaviour of 2-octylaminopyridine towards lead(II) from succinate media and its separation from other toxic metals. *J Hazard Mater* 2008;**152**(3):1146–54.
187. Aguilar JC, de San Miguel ER, de Gyves J, Bartsch RA, Kim M. Design, synthesis and evaluation of diazadibenzocrown ethers as Pb^{2+} extractants and carriers in plasticized cellulose triacetate membranes. *Talanta* 2001;**54**(6):1195–204.
188. Luconi MO, Olsina RA, Fernández LP, Silva MF. Determination of lead in human saliva by combined cloud point extraction–capillary zone electrophoresis with indirect UV detection. *J Hazard Mater* 2006;**128**(2–3):240–6.
189. Zhu Z, Zhang W, Cheng CY. A literature review of titanium solvent extraction in chloride media. *Hydrometallurgy* 2011;**105**(3–4):304–13.
190. Hasemann K, Bock R. Spectrochemical trace analysis for metals and metalloids. *Z Anal Chem* 1985;**274**:185.
191. Huffman EH, Beaufait LJ. The fractional separation of zirconium and hafnium by extraction with TTA. *J Am Chem Soc* 1949;**71**:3179.
192. Wadelin C, Mellon MG. Photometric determination of phosphate using metal oximates. *Anal Chem* 1953;**25**:1668.
193. Moreda-Piñeiro J, Alonso-Rodríguez E, López-Mahía P, Muniategui-Lorenzo S, Prada-Rodríguez D, Moreda-Piñeiro A, Bermejo-Barrera A, Bermejo-Barrera P. As, Cd, Cr, Ni and Pb pressurized liquid extraction with acetic acid from marine sediment and soil samples. *Spectrochim Acta Part B: Atomic Spectrosc* 2006;**61**(12):1304–9.
194. Shemirani F, Baghdadi M, Ramezani M. Preconcentration and determination of ultra trace amounts of arsenic(III) and arsenic(V) in tap water and total arsenic in biological samples by cloud point extraction and electrothermal atomic absorption spectrometry. *Talanta* 28 February 2005;**65**(4):882–7.
195. Garboś S, Bulska E, Hulanicki A, Fijałek Z, Sołtyk K. Determination of total antimony and antimony(V) by inductively coupled plasma mass spectrometry after selective separation of antimony(III) by solvent extraction with *N*-benzoyl- *N*-phenylhydroxylamine. *Spectrochim Acta Part B: Atomic Spectrosc* 2000;**55**(7):795–802.
196. Madrakian T, Bozorgzadeh E. Spectrophotometric determination of Sb(III) and Sb(V) in biological samples after micelle-mediated extraction. *J Hazard Mater* 2000;**170**(2–3):809–13.
197. Nakamura K, Ozawa T. Spectrophotometric determination of micro amounts of mercury, cadmium, and zinc by stepwise extraction with tribenzylamine. *Anal Chim Acta* 1976;**86**:147.

198. Li X, Wei C, Deng Z, Li M, Li C, Fan G. Selective solvent extraction of vanadium over iron from a stone coal/black shale acid leach solution by D2EHPA/TBP. *Hydrometallurgy* 2011;**105**(3–4):359–63.
199. Poedniok J, Buhl F. Speciation of vanadium in soil. *Talanta* 2003;**59**(1):1–8.
200. Lapenko LA, Gibalo IM. Solvent Extraction of niobium, tantalum and some other elements with tetraphenylarsonium chloride. *Zh Anal Khim* 1976;**31**:481.
201. Spivakov BY, Orlova YA, Malyutina TM, Kirilova TL. Extractant effect on the retention of the stationary phase in a rotating coiled column. *Zh Anal Khim* 1979;**34**:161.
202. Kamada T, Shiroishi T, Yamamoto Y. Use of transition elements to enhance sensitivity for selenium. *Talanta* 1978;**25**:15.
203. Sargar BM, Anuse MA. Liquid–liquid extraction study of Tellurium(IV) with *N*- *n*-octylaniline in halide medium and its separation from real samples. *Talanta* 2001;**55**(3):469–78.
204. Nan J, Yan X- P. On-line dynamic two-dimensional admicelles solvent extraction coupled to electrothermal atomic absorption spectrometry for determination of chromium(VI) in drinking water. *Anal Chim Acta* 2005;**536**(1–2):207–12.
205. Abdollahi H. Simultaneous spectrophotometric determination of chromium(VI) and iron(III) with chromogenic mixed reagents by H-point standard addition method and partial least squares regression. *Anal Chim Acta* 2001;**442**(2):327–36.
206. Shrivas K, Agrawal K, Harmukh N. Trace level determination of molybdenum in environmental and biological samples using surfactant-mediated liquid–liquid extraction. *J Hazard Mater* 15 January 2009;**161**(1):325–9.
207. Alkan M, Kharun M, Chmilenko F. Spectrophotometric determination of molybdenum with Alizarin Red S in the presence of poly(sulfonylpiperidinylmethylene hydroxide). *Talanta* 2003;**59**(3):605–11.
208. Filik H, Çengel T, Apak R. Selective cloud point extraction and graphite furnace atomic absorption spectrometric determination of molybdenum (VI) ion in seawater samples. *J Hazard Mater* 2009;**169**(1–3):766–71.
209. Musil J, Dolezal J. Preconcentration of trace amounts of tungsten on Amberlite XAD-7 for its spectrophotometric determination in hot spring water. *Anal Chim Acta* 1977;**92**:301.
210. Roberts RF. Quantitative analysis of major, minor and trace elements in fluid. *Anal Chem* 1977;**49**:1861.
211. Tribalat S, Beydon J. Isoelement du technetium. *Anal Chim Acta* 1953;**8**:22.
212. Fujinaga T, Lee HL. Polarographic study of solutions of cadmium(11) in acetylacetone. *Talanta* 1977;**24**:395.
213. Bermejo-Barrera P, Moreda-Piñeiro A, González-Iglesias R, Bermejo-Barrera A. Multivariate optimization of solvent extraction with 1,1,1-trifluoroacetylacetonates for the determination of total and labile Cu and Fe in river surface water by flame atomic absorption spectrometry. *Spectrochim Acta Part B: Atomic Spectrosc* 2002;**57**(12):1951–66.
214. Saçmacı S, Kartal S. Selective extraction, separation and speciation of iron in different samples using 4-acetyl-5-methyl-1-phenyl-1 *H*-pyrazole-3-carboxylic acid. *Anal Chim Acta* 2008;**623**(1):46–52.
215. Nayl AA. Extraction and separation of Co(II) and Ni(II) from acidic sulfate solutions using Aliquat 336. *J Hazard Mater* 2010;**173**(1–3):223–30.
216. Bidabadi MS, Dadfarnia S, Haji Shabani AM. Solidified floating organic drop microextraction (SFODME) for simultaneous separation/preconcentration and determination of cobalt and nickel by graphite furnace atomic absorption spectrometry (GFAAS). *J Hazard Mater* 2010;**166**(1):291–6.

217. Zolotov YA, Petrukhin OM, Shevehenko UN, Dunina UY, Rukhadze EG. Extraction of platinum metals from hydrochloric acid medium. *Anal Chim Acta* 1978;**100**:613.
218. Ougiyanagi J, Meguro Y, Yoshida Z, Imura H, Ohashi K. Solvent effect on distribution ratio of Pd(II) in supercritical carbon dioxide extraction and solvent extraction using 2-methyl-8-quinolinol. *Talanta* 2003;**59**(6):1189–98.
219. Bagher Gholivand M, Nozari N. Extraction and spectrophotometric determination of trace amount of Pd(II) with 2,2′-dithiodianiline. *Talanta* 2000;**52**(6):1055–60.
220. Shinotsuka K, Suzuki K. Simultaneous determination of platinum group elements and rhenium in rock samples using isotope dilution inductively coupled plasma mass spectrometry after cation exchange separation followed by solvent extraction. *Anal Chim Acta* 12 November 2007;**603**(2):129–39.
221. Lobinski R, Marczenko Z. UV-VIS Spectroscopy and its Applications. *CRC Crit Rev Anal Chem* 1992;**23**:55.
222. Armenta S, Garrigues S, de la Guardia M. Green analytical chemistry. *TrAC Trends Anal Chem* 2008;**27**(6):497–511.
223. Zorov NB, Gocci TF, Winefordner JD. Comparison of figures of merit for fluorimetric analysis of rhodamine complexes of gallium and indium by excitation with pulsed laser and by xenon arc sources. *Talanta* 1976;**23**:455.
224. Zolotov YA, Petrukhin OM, Nagy VY, Volodarskii LB. Application of Stable Rradical-Containing Complexing Reagents for the Esr Determination of Metals. *Anal Chim Acta* 1980;**115**:1.
225. Zolotov YA, Kuzmin NM, Petrukhin OM, Spivakov BY. Liquid-liquid extraction in inorganic analysis: Current status and prospects. *Anal Chim Acta* 1986;**180**:137.
226. Matsuoka S, Yoshimura K, Waki H. Sorption-spectroscopic and test methods for the determination of metal ions on the solid-phase of ion-exchange materials. *Mem Fac Sci Kyushu Univ Ser C* 1991;**18**:55–62.
227. Compaño R, Ferrer R, Guiteras J, Prat MD. Spectrofluorimetric detection of trace elements with 2,6-diacetylpyridine functionalized amberlite XAD-4 by flow injection and atomic spectroscopy polymeric matrix. *Analyst* 1994;**119**:1225.
228. Kirkbright GF, Narayanaswamy R, Welti NA. Optical fibres in chemical sensing – a review. *Analyst* 1984;**109**:15.
229. Narayanaswamy R, Sevilla F. Reversible sensor for High Alkalinity Measurement. *Anal Chim Acta* 1986;**189**:365.
230. Scamerhorn JF, Harwell JH. *Surfactant-based separations processes*. New York: Marcel Dekker; 1989. p. 139.
231. Hinze WL, Pramauro E. Use of surfactant – mediated phase separation (cloud point extraction) with affinity ligands. *Crit Rev Anal Chem* 1993;**24**:133.
232. Carabias-Martínez R, Rodríguez-Gonzalo E, Moreno-Cordero B, Peŕrez-Pavoí n JL, Garcia-Pinto C, Fernaí ndez Laespada E. Surfactant cloud point extraction and preconcentration of organic compounds prior to chromatography and capillary electrophoresis. *J Chromatogr A* 2000;**902**:195.
233. Quina FH, Hinze WL. Surfactant-mediated cloud point extractions: an environmentally benign alternative separation approach. *Ind Eng Chem Res* 1999;**38**:4150.
234. Huddleston JG, Willauer HD, Griffin ST, Rogers RD*Metal Ion Separations in Aqueous Biphasic Systems and Using Aqueous Biphasic Extraction Chromatography. Ind Eng Chem Res* 1999;**38**:2523.
235. Pinto G, Peírez Pavoín C, Moreno Cordero JL, Romero Beato B, García Saí nchez E, Anal SJ. Micelle mediated extraction of metal ions from water samples. *Atomic Spectrom* 1996;**11**:37.

236. Akita S, Rovira M, Sastre AM, Takeuchi H. Cloud-Point Extraction of Gold(III) with Nonionic Surfactant—Fundamental Studies and Application to Gold Recovery from Printed Substrate. *Sep Sci Technol* 1998;**33**:2159.
237. Horvath WJ, Huie CW. Screening of urinary coproporphyrin using cloud point extraction and chemiluminescence detection. *Talanta* 1993;**9**:1385.
238. Rezende HC, Nascentes CC, Coelho NMM. Cloud point extraction for determination of cadmium in soft drinks by thermospray flame furnace atomic absorption spectrometry. *Microchem J* 2011;**97**(2):118–21.
239. García Pinto C, Peírez Pavoín JL, Moreno Cordero B. Cloud point extraction coupled with microwave or ultrasonic assisted back extraction as a preconcentration step prior to gas chromatography. *Anal Chem* 1995;**67**:2606.
240. Stangl G, Niessner R. Cloud point extraction of napropamide and thiabendazole from water and soil. *Mikrochim Acta* 1994;**113**:1.
241. Ferrer R, Beltran JL, Guiteras J. Use of cloud point extraction methodology for the determination of PAHs priority pollutants in water samples by high-performance liquid chromatography with fluorescence detection and wavelength programming. *Anal Chim Acta* 1996;**330**:199.
242. Eiguren A, Sosa Z, Santana JJ. Application of cloud point methodology to the determination of polychlorinated dibenzofurans in sea water by high performance liquid chromatography. *J Analyst* 1999;**124**:487.
243. Calvo Seronero L, Fernandez Laespada ME, Peírez Pavoí n JL, Moreno Cordero B. *J Chromatogr A* 2000;**897**:171.
244. Eiguren A, Sosa Z, Santana J. Application of cloud-point methodology to the determination of polychlorinated dibenzofurans in sea water by high-performance liquid chromatography. *J Anal Chim Acta* 1998;**358**:145.
245. Moreno B, Perez JL, García C, Fernandez ME. Cloud Point Methodology: A New Approach for Preconcentration and Separation in Hydrody- namic Systems of Analysis. *Talanta* 1993;**40**:1703.
246. Carabias-Martínez R, Rodríguez-Gonzalo E, Domínguez-Alvarez J, Hernaíndez-Meíndez. Cloud point extraction as a preconcentration step prior to capillary electrophoresis. *J Anal Chem* 1999;**771**:2468.
247. Janata J, Josowitz M, de Vaney DM. Chemical sensors. *Anal Chem* 1994;**66**:207R.
248. Camman K, Lemke U, Rohen A, Sander J, Wilken H, Winter B. Chemical sensors – from molecules, complex mixtures to cells – supramolecular imprinting strategies. *Angew Chem* 1991;**109**:519.
249. Szczepaniac W, Oleksy J. Ion-selective electrodes in analytical chemistry. *J Anal Chim Acta* 1986;**189**:237.
250. Pretsch E, Buchi R, Amman D, Simon W. *Essays on analytical chemistry*. New York: Pergamon Press; 1977. p. 321.
251. Xie RY, Christian GD. Flow injection study of lithium selectivity. *Analyst* 1987;**112**:61.
252. Pioda LAR, Stankova Y, Simon W. Ion selective polymeric membranes as chemically selective coulometric electrodes. *Anal Lett* 1969;**2**:665.
253. Scholer RP, Simon W. The crystal and molecular structure of bis(cyclopentadienyl)(-allyl)(carbon disulphide)niobium. *Chimia* 1970;**24**:372.
254. Hobby PC, Moody GJ, Thomas JDR. Studies of calcium ion-selective electrodes in the presence of anionic surfactants. *Analyst* 1983;**108**:581.
255. Geissler M, Kunze R. Ion-selective electrodes based on the phosphonium salt. *Anal Chim Acta* 1986;**189**:245.

256. Peters DG, Hayes JM, Hieftje GM. *Modern chemical analysis*. Philadelphia: W.B. Saunders; 1976.
257. Hamilton RJ, Sewei PA. *Introduction to high performance liquid chromatography*. London: Chapman & Hall; 1977.
258. Wakui Y, Matsunaga H, Suzuki TM. Distribution of Rare Earth Elements between (2-Ethylhexyl Hydrogen 2-Ethylhexylphosphonate)-Impregnated Resin and Acid Aqueous Solution. *Anal Sci* 1988;**4**:325.
259. Muscatello AC, Yarbro SL, Marsh SF. In: Cecille L, Casarci M, Pietrelli L, editors. *New separation chemistry techniques for radioactive waste and other specific applications*. Amsterdam: Elsevier Science Publishing; 1991. p. 88.
260. Dietz ML, Horwitz EP, Nelson DM, Walgren M. Extraction Chromatography: Progress and Opportunities. *Health Phys* 1991;**61**:871.
261. Horwitz EP, Dietz ML, Nelson DM, La Rosa JJ, Fairman W. Concentration and separation of actinides from urine using a supported bifunctional organophosphorus extractant. *Anal Chim Acta* 1990;**238**:263.
262. Snyder LR, Kirkland JJ. *Modern liquid chromatography*. New York: John Wiley & Sons; 1974.
263. Parris NH. *Instrumental liquid chromatography*. Amsterdam: Elsevier; 1976.
264. Lawrence JF, Frei RW. *Chemical derivitization in liquid chromatography*. Amsterdam: Elsevier; 1976.
265. Horvath C, editor. *High-performance liquid chromatography*, vols. 1–3. New York: Academic Press; 1980–1983.
266. Mulier W. Precoating. *Kontakte (Merck)* 1986;**3**:3.
267. Braun R, Ghersini G. *Extraction chromatography*. Amsterdam: Elsevier; 1975.
268. Veuthy JL, Caude M, Rosset R. Metal speciation by supercritical fluid extraction with on-line analysis. *Analysis* 1990;**18**:103.
269. Hawthorne SB. Analytical scale supercritical fluid extraction. *Anal Chem* 1990;**62**:633.
270. Vannoort RW, Chervet JP, Lingeman H, De Jong GJ, Brikman AUTh. Supercritical fluid chromatography and extraction in surface coatings analysis — review of representative applications. *J Chromatogr* 1990;**505**:45.
271. Sugiyama K, Saito M, Hondo T, Seda M. New double-stage analysis method. *J Chromatogr* 1985;**332**:107.
272. Anderson IGM. Supercritical fluid extraction coupled to packed column supercritical fluid chromatography. In: Westwood SA, editor. *Supercritical fluid extraction and its use in chromatographic sample preparation*. UK: Blackie Academic & Professional, Chapman & Hall; 1993. p. 112.
273. Foreman JK, Sockweli PB. *Topics in automatic chemical analysis*, vol .1. Chichester, UK: Ellis Horwood; 1979.
274. Valcarecel M, Luque de Castro MD. *Flow-injection analysis*. Chichester, UK: Ellis Horwood; 1987.
275. Furman WB. *Continuous flow analysis: Theory and practice*. New York: Dekker; 1976.
276. Coakley WA. Handbook of automatic analysis. *Continuous flow techniques*; 1982. New York: Dekker.
277. Ruzicka J, Hansen EH. *Flow-injection analysis*. New York: Wiley; 1981.
278. Valcarcel M, Gallego M. Separation techniques. In: Bruguera JL, editor. *Flow injection atomic spectroscopy*. New York: Marcel Dekker; 1989.
279. Luque de Castro MD, Valcarcel M. Integration of separation and detection in continuous-flow systems. *Trends Anal Chem* 1991;**10**:114.

280. MacLaurin P, Andrew KN, Worsfold PJ. Flow injection analysis. In: McLennan F, Kowalski BR, editors. *Process analytical chemistry*. Glasgow: Blackie Academic & Professional; 1995. p. 159.
281. Valcarcel M, Luque de Castro M. *Non chromatographic continuous separation techniques*. Cambridge, UK: Royal Society of Chemistry; 1990.
282. Shelly DC, Rossi TM, Warner IM. Multiple solvent extraction system with flow technology. *Anal Chem* 1982;**54**:87.
283. Gallego M, Valcarcel M. Indirect continuous automatic determination of pharmaceuticals by atomic absorption spectroscopy. *Anal Chim Acta* 1985;**169**:161.
284. Rao TP, Metilda P, Gladis JM. Preconcentration techniques for uranium(VI) and thorium(IV) prior to analytical determination—an overview. *Talanta* 2006;**68**(4):1047–64.
285. Gopalkrishnan M, Radhakrishnan K, Dhami PS, Kulkarni VT, Joshi MV, Patwardhan AB, Ramanujam A, Mathur JN. Determination of trace impurities in uranium, thorium and plutonium matrices by solvent extraction and inductively coupled plasma atomic emission spectrometry. *Talanta* 1997;**44**(2):169–76.
286. May CC, Worsfold PJ, Keith-Roach MJ. Analytical techniques for speciation analysis of aqueous long-lived radionuclides in environmental matrices. *TrAC Trends Anal Chem* 2008;**27**(2):160–8.
287. Desideri D, Feduzi L, Meli MA, Roselli C. Sequential determination of Am, Cm, Pu, Np and U by extraction chromatography. *Microchem J* 2011;**97**(2):264–8.
288. Hou X, Roos P. Critical comparison of radiometric and mass spectrometric methods for the determination of radionuclides in environmental, biological and nuclear waste samples. *Anal Chim Acta* 2008;**608**(2):105–39.
289. Zamani A-A, Yaftian MR. Solvent extraction of thorium, lanthanum and europium ions by bis(2-ethylhexyl)phosphoric acid using 2-nitrobenzo-18-crown-6 as ion size selective masking agent. *Sep Purif Tech* 1 December 2004;**40**(2):115–21.
290. Dybczyriski R. Role of ion-exchange and extraction chromatography in neutron activation analysis. *J Chromatogr* 1992;**600**:17–36.
291. Li Z, Conti PS. Radiopharmaceutical chemistry for positron emission tomography. *Adv Drug Deliv Rev* 2010;**62**(11):1031–51.
292. Qiao J, Hou X, Miró M, Roos P. Determination of plutonium isotopes in waters and environmental solids: a review. *Anal Chim Acta* 2009;**652**(1–2):66–84.
293. Keller C. *Radiochemistry. English edition published by Ellis Horwood*. Chichester, UK: Ellis Horwood Series in Physical Chemistry; 1988.
294. Wennersten R. The extraction of malic acid from fermentation broth using a solution of tertiary amine. *J Chem Tech Biotechnol* 1983;**33**:85.
295. Lloyd-Jones E. Recent advances in liquid-liquid extraction. *Chem Ind* 1967:1590.
296. Kertes AS, King C. Extraction chemistry of fermentation product carboxylic acids. *J Biotechnol Bioeng* 1986;**28**:269–82.
297. Canari R, Hazan B, Bloch R, Eyal AM. Value adding through solvent extraction. In: Shallcross DC, Paimin R, Prvcic LM, editors. *Proceedings of ISEC'96*, vol. 2. University of Melbourne Press; 1996. p. 1517.
298. Yagodin GA, Yurtov E, Golubkov AS. *Separation and concentration of amino acids using liquid emulsion membranes. Proceedings of ISEC'86*, vol. 3. Munich: DECHEMA; 1986. p. 677.
299. Lukhezo M, Kelly NA, Reuben BG, Dunne LJ, Verrall MS. Value adding through solvent extraction. In: Shallcross DC, Paimin R, Prvcic LM, editors. *Proceedings of ISEC'96*, vol. 1. University of Melbourne Press; 1996. p. 87.

300. Hano T, Matsumoto M, Ohtake T, Sasaki K, Hori F, Kawano Y. Separation of lactic acid from aqueous solution. *J. Chem Eng Jpn* 1990;**23**(6):734.
301. Schugerl K. *Solvent extraction in biotechnology.* Berlin: Springer-Verlag; 1994. 113–25.
302. Verrall MS. In: Thornton JD, editor. *Science and practice in solvent extraction. Oxford*, vol. 2. Oxford Science Publications; 1992. Chap. 3.
303. *Kirk-Othmer encyclopedia of chemical technology.* 4th ed., vol. 10. New York: John Wiley & Sons; 1993. p. 125.
304. Reuben BG, Sjoberg K. *Chem technology* 1981;**11**:315.
305. Reschke M, Schurgel K. Distribution equilibrium of penicillin G between water and organic phase. *Chem Eng J* 1994;**28**:B1–29.
306. Somekh GS. *Liquid-liquid extraction processes & process engineering. Proceedings of ISEC 71*, vol. 1. London: Lyon: Society of Chemical Industry; 1971. p. 323.
307. Kane L, Romanow-Garcia S, Nakamura D. Hydrocarbon gas processing. *Hydrocarb Process* 1995;**74**:36.
308. Bailes PJ. Oxide solid solutions as catalysts. *Chem Ind*; 1977:69.
309. Bender ML, Komiyama M. *Cyclodextrin chemistry.* Berlin: Springer-Verlag; 1978.
310. Uemasu I. Value adding through solvent extraction. In: Shallcross DC, Paimin R, Prvcic LM, editors. *Proceedings of ISEC'96*, vol. 2. University of Melbourne Press; 1966. p. 1635.
311. Pratt MWT. In: Lo TC, Baird MHI, Hanson C, editors. *Handbook of solvent extraction.* New York: John Wiley & Sons; 1983. p. 605.
312. Rogers PL, Jeon J, Lee KJ, Lawford HG. Zymomonas mobilis forfuel ethanol and higher value products. *Adv Biochem Eng Biotechnol* 2007;**108**:263–88.
313. Fischer CR, Klein-Marcuschamer D, Stephanopoulos G. Selection and optimization of microbial hosts for biofuels production. *Metab Eng* 2008;**10**:295–304.
314. Alper H, Moxley J, Nevoigt E, Fink GR, Stephanopoulos G. Engineering yeast transcription machinery for improved ethanol tolerance and production. *Science* 2006;**314**:1565–8.
315. Zhao J, Yang S-T, Jiang W, Yang S. Evolution of solvent-producing Clostridium beijerinckii towards high butanol tolerance. Presented at 238th ACS national meeting, 2009.
316. Boles E. Genetic engineering of yeast strains for the production of bioalcohols from C6- and C5-sugars. Presented at 27th International Specialized Symposium on Yeasts, 2009.
317. Steen EJ, Chan R, Prasad N, Myers S, Petzold CJ, Redding A, et al. Metabolic engineering of Saccharomyces cerevisiae for the production of n-butanol. *Microb Cell Fact* 2008;**7**:36.
318. Simoni LD, Chapeaux A, Brennecke JF, Stadtherr MA. Extraction of biofuels and bio-feedstocks from aqueous solutions using ionic liquids. *Comp Chem Eng* 2010;**34**:1406–12.
319. Flett DS, Pearson D. Recovery of the metal values from effluents. *Chem Ind* 1975:639.
320. Martin D, Diaz G, Garcia MA, Saínchez F. Solvent extraction in the primary and secondary processing of zinc. In: *Proceedings of ISEC 2002*, Cape Town: South Africa; 2002, p. 1045.
321. Sole KC. Zinc solvent extraction in the process industries, *Proceedings of ISEC 2002*. South Africa: Cape Town; 2002. p. 1033.
322. Reinhardt H. Zinc solvent extraction in the process industries. In: *Proceedings of the international waste treatment and utilization conference*, Waterloo, Canada, 1978, p. 83.
323. Reinhardt H. *Proceedings of Hydrometallurgy '81.* London: Society of Chemical Industry; 1981. Section F1.
324. MEAB Report. Extraction of nickel with simultaneous neutralization, 1989. www.meab-mx.se.
325. Sze YKP, Xue L. *Sep Sci Technol* 2003;**38**:405.

326. Andersson SOS, Reinhardt H. Hydrometallurgical metals and acids recovery (MAR) recovery process. Review. In: *Proc. Int. Solvent Ext. Conf., ISEC 77*, CIM, Special vol. 21, The Canadian Institute of Mining and Metallurgy, 1979. p. 798–804.
327. Ottertun H, Strandell E. *Extraction of vanadium from athabasca tar sands fly ash. Proceedings of ISEC'77*, vol. 1. Toronto: Canadian Institute of Mining, Metallurgy and Petroleum; 1979. p. 501.
328. Monzyk B, Conkle HN, Rose JK, Chauhan SP. Solvent extraction in hydrometallurgy: present and future. In: *Proceedings of ISEC 2002*, Cape Town: South Africa; 2002, p. 755.
329. Alonso AI, Galan B, Gonzalez M, Ortiz I. Experimental and theoretical analysis of a nondispersive solvent extraction pilot plant for the removal of Cr(VI) from a galvanic process wastewaters. *Ind Eng Chem Res* 1999;**38**(4):1666–75.
330. Reinhardt H. *Proceedings of ISEC '93*, vol. 3. York, UK; London: Society of Chemical Industry; 1993. p. 1625.
331. Ritcey GM. *Proceedings of ISEC '93*, vol. 1. York, UK; London: Society of Chemical Industry; 1993. p. 189.
332. Martin D, Garcia MA, Diaz G, Falgueras J. *Neutralization sludge from the plating industries flue dust from steel and brass mills. Proceedings of ISEC '99*, vol. 1. Barcelona, Spain; London: Society of Chemical Industry; 2001. p. 201.
333. Demster JH, Bjoerklund P. Biological and toxicological effects of solvent extraction, *Proceedings of 4th annual meeting on hydrometallurgy section of the metallurgical society*. Toronto: CIM; 1994. p. 68.
334. Dias G, Martin D, Frias C, Peírez O. *Emerging applications of ZINCEX and PLACID technologies. Proceedings of ISEC '99*, vol. 2. Barcelona, Spain; London: Society of Chemical Industry; 2001. p. 1449.
335. Kalujta VV, Kuznetsov GI, Kravchenko AN, Lanin VP, Miraevshy GP, Pushkov AA, Travkin VF, Shklyar LI. In: *Proceedings of ISEC '88*, Moscow, 1988, vol. 1, p. 252.
336. Dreisinger DB, Leong BJY, Balint BJ, Beyad MH. *The solvent extraction of As, Sb and Bi from copper refining electrolytes using organophosphorus reagents. Proceedings of ISEC '93*, vol. 3. York, UK, London: Society of Chemical Industry; 1993. p. 1271.
337. Wang C, Jiang K, Liu D, Wang H. NMR Spectroscopy in the liquid and gas phase. In: *Proceedings of ISEC 2002*. Cape Town: South Africa; 2002, p. 1039.
338. Cox M, Flett DS, Velea T, Vasiliu C. The extraction of copper, zinc, cadmium and lead from waste streams in the zinc-lead industry. In: *Proceedings of ISEC 2002*. Cape Town: South Africa; 2002, p. 995.
339. Dreisinger D, Wassink B, Ship K, King J, Hames M, Hackl R. Cyanide recovery by solvent extraction, *Proceedings of ISEC 2002*. South Africa: Cape Town; 2002. p. 798.
340. Hutchins D. (Nova Chemicals Ltd., Canada) Continuous solvent extraction process for the separation of aromatic hydrocarbons from a mixed feed stock. Can Pat Appl CA 1997-2214305 (1999), p. 31.
341. Zubaidy EAH, Abouelnasr DM. Fuel recovery from waste oily sludge using solvent extraction. *Process Safety Environ Prot* 2010;**88**(5):318–26.
342. Wohler E. *Removal and recovery of phenol and ammonia from gas liquor*. Frankfurt: Lurgi Kohle und Mineraloltechnik GmbH; 1978.

Part II

Novel Competitive Complexation/Solvation Theory (CCST) of Solvent Extraction: Principles and Practice

1. INTRODUCTION

Complex chemistry, which occurs in extraction processes, is still not completely understood. The complicated behavior of extraction systems conflicts, in some cases, with the classic stoichiometric ion-exchange models. Firstly, we have met with difficulties explaining the phenomenon "Dual maximum behavior" observed at titanium(IV) and many other transition metals extraction: two regions of effective extraction were observed, one at low acidity and the other at high acidity of aqueous solutions (for details and citations see Chapter 3). This phenomenon was found for most families of extractants: acidic (organic acids), basic (amines or their mixtures), and neutral complexants. Ad hoc arguments using acid–base and coordination chemistry concepts were not so convincing.

The experimental data fitted to basic equations of the classic theories describe well the extraction by strong extractants, forming strong complexes but are far to be satisfied for the extractants forming weak complexes with the solute. In all cases, these equations are reliable only for the separation factors in the certain limits, namely $0.2 \leq Z \leq 0.8$. These conditions are rarely respected in practice. In addition, during studied extractive reactions the intermediates formed become extractants also, strongly complicate interaction chemistry, and cast doubt on the reliability of equilibrium (extraction) constants obtained by basic equations.

The difference in the extraction trends must be attributed also to solvation (coordination) effects. Solvents are known to have numerous different effects on the reactivity: general effects stemming from polarity and polarizability and specific effects originating from π- or n-donor ability, hydrogen-bond forming capacity, etc. The solvents range from the very inert through non-polar but polarizable to the polar, polarizable and hydrogen-bonding (as water). In order to analyze solvation effects, identification and, most importantly, quantification of solute–solvent interactions must be accomplished. Complexity of solvent structure makes this task extremely difficult.

Modeling extraction systems researchers use only basic equations of an interaction reaction for a given solute ion and extractant. But a strong, sometimes critical influence of solvation effects on the extraction efficiency is also discussed in the same works. The polar, polarizable, hydrogen-bonding solvents change the extraction order dramatically. Solvent effects on UV-VIS absorbancy or NMR chemical shifts leads to anomalous relationship between these experimental data and extraction constants.

It is well known that the stoichiometry of the extracted species changes as loading increases. This phenomenon was described through aggregation mechanisms. Aggregation was observed at extraction of acids, metal ions, and salts. The formation of linear, cyclic aggregates and reversed micelles were proved using different experimental and analytical techniques. Classic models do not handle this very important problem.

The methodology of coordination chemistry in solution has been applied extensively to the determination of the equilibrium constants for solvent replacement in the coordination spheres of ions. Coordination models, explaining qualitatively some mechanisms of metal solvent extraction, meet difficulties when modeling anion exchangers, acid solvent extraction, and, especially, at the attempts of quantification of the models.

The physicochemical modeling approach, preliminarily developed for the acid–amine extraction systems (see Chapters 7 and 9), was introduced to explain the mechanisms of solvent extraction by different acidic, basic, or mixed extractants. This theory is based on the modified competitive preferential solvation (COPS) theory, the Lewis acid–base concept, the concept of amphoteric properties of extractants, and the concept of different aggregation structures formation at increasing extractant loading. Based on these well-known concepts a novel model for solvent extraction processes was developed. It interprets well both strong and weak ion–molecular interactions and its equations describe the processes in all concentration range. This model explains some anomalies which required previously ad hoc arguments.

In the following chapters principles of the novel approach to solvent extraction and its applications in acid and metal extraction are presented and interpreted on the base of this model. The data available from the literature and our experiments are used for interpretation. Mathematical description and suitable experimental techniques for verification of the model are presented.

Chapter 6

Backgrounds of the Competitive Complexation/Solvation Theory of Solvent Extraction

Chapter Outline

1. INTRODUCTION

Novel competitive complexation/solvation theory (CCST) of solvent extraction is based on the concepts well known in chemistry. These are H-bonding and proton transfer, competitive preferential solvation theory, electronic acid-base theory, and amphoterity and aggregation. These concepts presented in short in this Chapter with the literature references for the purpose to make the CCST material easier to understand.

2. COMPLEXATION THROUGH THE H-BONDING AND PROTON TRANSFER

During the last decades the nature of H-bond, H-bonded complexes, and proton transfer have been intensively investigated by different methods.[1–13] However, interpretation of the experimental data is very difficult. The simplest and most clearly defined H-bonded species are carboxylic acid dimers. Strong evidence may be mentioned for the presence of higher aggregates.[6,7]

Reactions can be divided into single, competing, and coupling. In a process described by a single chemical equation, equilibrium constant are determined

Solvent Extraction. DOI: 10.1016/B978-0-444-53778-2.10006-8

by the initial concentrations of all reactants. This is what many experimentators do in solvent extraction (SX) and what is far from real processes because the SX are not single reactions in most cases. Two or more reactions are competing if common solute reacts with different solvents (extractant, enhancer, diluent, etc.). Two reactions are coupled when a product of one reaction is a reactant for the next. This reactant is usually nominated as an intermediate or adduct. The equilibrium concentrations of the species participating in one reaction depend on the initial concentrations of all chemicals (including adducts) involved in all reactions. Competing and coupling play an important role in hydrogen bond and proton transfer reactions and SX processes.

Proton transfer in general and specifically in organic solvents has been investigated using many techniques. The highly known organic compounds are proton acceptors and/or proton donors, and may be converted into highly reactive intermediates by proton transfer:

$$\mathrm{SH + E \leftrightarrow S{-}H\ldots E} \tag{6.1}$$

$$\mathrm{SH + E \leftrightarrow S^{-}\ldots H{-}E^{+}} \tag{6.2}$$

$$\mathrm{S^{-}\ldots H{-}E \leftrightarrow E^{+}{-}H + S^{-}} \tag{6.3}$$

$$\mathrm{S{-}H + A \leftrightarrow S{-}H\ldots A} \tag{6.4}$$

$$\mathrm{S{-}H\ldots A \leftrightarrow S^{-} + H^{+}A} \tag{6.5}$$

Where S, H, E, and A are solute, proton, extractant, and active solvent, respectively. As a rule reactions (6.4) and (6.5) in extraction experiments are omitted to simplify consideration.

These reactions are typical examples of competing and coupled reactions. In most cases, to explain experimental data of intermolecular proton transfer, three-step mechanism is applied:

$$\mathrm{E + S{-}H \leftrightarrow E\ldots SH \leftrightarrow E^{+}H\ldots S^{-} \leftrightarrow E^{+}H + S^{-}} \tag{6.6}$$

At excess of solute or extractant is added, and the following interactions are found:

$$\mathrm{E\ldots H{-}S + SH \leftrightarrow E\ldots H{-}S\ldots SH \leftrightarrow E^{+}{-}H\ldots S^{-}\ldots SH \leftrightarrow E^{+}{-}H + (SHS)^{-}} \tag{6.7}$$

or

$$\mathrm{E^{+}{-}H\ldots S + E \leftrightarrow (EHE)^{+}S^{-} \leftrightarrow (EHE)^{+} + S^{-}} \tag{6.8}$$

In many cases, especially when aprotic solvents with low dielectric permittivity complexes are formed with dipole moment $\mu \geq 5$ D, these complexes interact to form aggregates.[2,5–7] Clusters of molecules are formed by H-bonding in solution and their structures vary strongly with solvent (its polarity, proton acceptor or proton–donor properties), stoichiometry, concentration, temperature, and impurity content.[2] The lack of definitive and general techniques for proving the product structures, neutrals, ions, is a serious limitation in studying acid–base interactions.

Several physical properties of H-bonded complexes, e.g., excess of dipole moment ($\Delta\mu$), change of proton, carbon-13, nitrogen-15 chemical shifts ($\Delta\delta$), frequency shift ($\Delta\nu$), when plotted against ΔpK_a, pK_a, pH, PA (proton affinity) or ΔH, give titration curve relations. As a rule, these techniques do not distinguish between intermediate reactions, especially rapid, and show summarized results.

Comparison of spectral properties (IR, Raman) of isostructural systems with hydrogen and coordination bonds, their capacity to form complexes, ionic pairs, or ions is shown to be similar.[3] H-Bonded intermediates of VI–VIII group transition metal derivatives with different strength acids were investigated.[4] Proton transfer through hydrogen or coordinate-bonded intermediates leads to the formation of a classical ion exchange complex.

Acid–base reactions are distinguished by strength of the hydrogen bond formation: from weak (nonspecific solvation) to specific and strong ionic or ion-pair complex formation. It is also possible to distinguish between intermediates formed with fast or slow proton migration and proton localization.

Different aspects of the nature of H-bonded complexes and proton transfer are covered in the referenced monographs and reviews.[2,5–13]

3. DISTRIBUTION ISOTHERM

The shape of most general extraction isotherm consists of four stages (regions) at loading of organic phase by extracted solute (see Fig. 6.1).

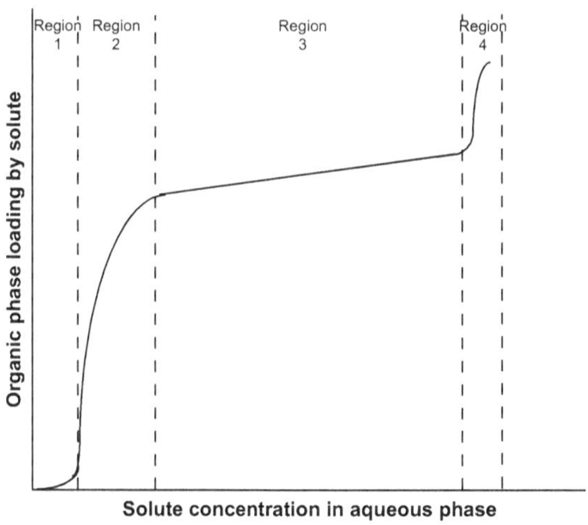

FIGURE 6.1 General scheme of organic phase loading as a function of solute concentration in the aqueous phase.

Region 1: Comparatively low extraction at low solute concentration in the aqueous phase.
Region 2: Drastic rise of distribution curve where a small increase of the solute concentration in the aqueous phase has a strong effect on the organic phase concentration.
Region 3: The distribution curve approaches saturation and levels off to nearly a plateau.
Region 4: Above stoichiometric extraction is observed at high solute concentrations with massive formation of the third phase. The distribution curve rises once more.

Division of the distribution curve into the regions depends on many factors such as properties and concentrations of solute extracted, the properties of extractants and solvents (modifiers, synergistic agents) used, temperature, acidity of the treated (feed) aqueous solutions and the nature of initial compounds, the nature of the organic compounds formed, their solubility in aqueous phase, aggregation, etc. Some examples of carboxylic and mineral acid extraction by amines[14–20] are presented in Figs. 6.2 and 6.3.

Analysis of the extraction (distribution) isotherm in Fig. 6.1 is based on the following observations, reported in a large number of experimental studies.

At extraction of "hard"[5,21–23] mineral acids by "hard" organic bases ($pK_B >> pK_A$) region 1 is not observed (see in Fig. 6.2 extraction isotherm of HCl) or may be observed only at extremely low acid concentrations in the aqueous phase. Region 2 is very steep and distribution ratio reaches stoichiometric saturation of the organic phase at relatively low aqueous phase concentrations. The type of bonding in the complex is predominantly ionic. Region 3 is wide and the slope of the curve is slight. Over-stoichiometric loading, at $Z > 1$, in region 4 (where Z denoted as the saturation factor, is the acid to base equivalent ratio in the organic phase) is observed (if at all) at very high solute concentrations in the aqueous phase.

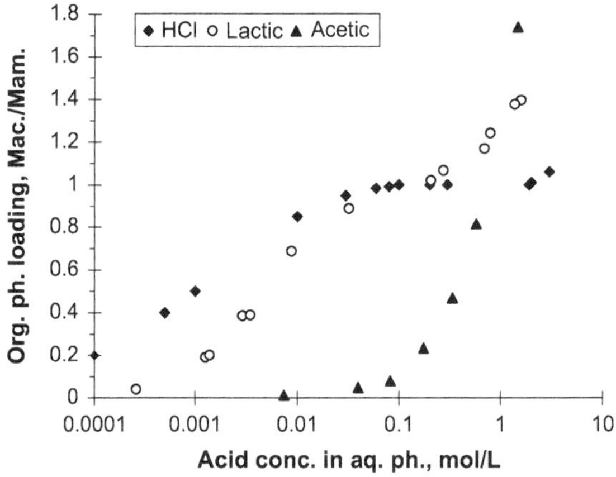

FIGURE 6.2 Extraction isotherms of different monobasic acids by Alamine-336 (0.29–0.5 mol/L in different diluents). *(Source: Data from Refs. 14,15,18–20,25,26) Notes: Concentrations of Alamine-336 and diluents used for extraction of (a) HCl: 0.5 mol/L in kerosene*[14,15]*; (b) lactic acid: 0.29 mol/L in chloroform*[18–20,25,26]*; (c) acetic acid: 0.29 mol/kg in 15vol% chloroform and n-heptan.*[25,26]

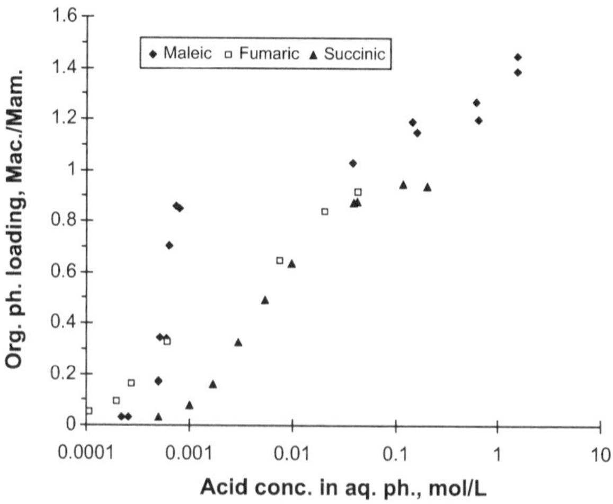

FIGURE 6.3 Extraction isotherms of different dicarboxylic acids by 0.29 mol/L Alamine-336 in chloroform. *(Source: Data from Refs. 15,25,26 with permission.)*

At "softening" of the extracted acids and the same "hardness" of the base, region 1 and region 2 are broadened in the range of the acid concentrations in the aqueous phase. The distribution ratio (slopes of the curves) depends on the apparent affinity (strength) of the constituents of extraction systems to form complexes, on miscibility of the complexes in the aqueous phase (hydrophilicity), on pH conditions. At extraction of weak hydrophobic carboxylic acids region 2 is not detectable. On the other hand, region 3 is narrowed, and transition to region 4 occurs at lower acid concentrations in the aqueous phase. The slope in region 3 depends strongly on the affinity of the complexes, already formed, to hydrogen bonds with molecules of the acid and different solvents, present in the extraction system. In region 4, the *Z*-values may approach 4 and even more at high solute concentrations.

When both acid and base are weakened the same trend but much stronger is continued. At weak acid and base (the value of pK_A close to pK_B) region 2 is not detectable. Regions 1 and 3 are weakly pronounced and only region 4 may occur at high concentrations of the solute.

Many authors,[14–16,24–26] modeling extraction systems, use the strength order of solute–extractant interactions for a given solute and extractant and do not take into account effect of interactions with other solvents. But in the same works, a strong, sometimes even critical influence of solvents, as synergistic agents (enhancers), modifiers, diluents, on the extraction is also discussed. The polar, polarizable, H-bonding solvents change the order dramatically. The overall solvent effect is composed of a general effect stemming from polarity and polarizability and of specific effect originated by π- or n-donor ability.

Appropriate way to represent these effects quantitatively is to use empirical or semiempirical parameters. For example, donating specific effect may be determined by ionization potential (eV) of the donor molecule or by charge (proton) transfer energy of the complexes.

Various models and equations, using different parameters, have been developed with the aim to describe and quantify solvation effects on the solute–extractant interaction strength in organic compositions. Some authors used the Hildebrand solubility parameter[8,27–29] as a measure of solvation, some used linear free-energy relationships,[30,31] and some combined "physical and chemical modeling" calculating the activity coefficients for the reactants and complexes.[32,33] Headley with coworkers,[34–36] studying solvation effects on the basicity of amines, introduced many parameters, such as hydrogen bond donor and acceptor abilities, solvent's dipolarity/polarizability, and dipole moment. All models developed are very interesting from the scientific point of view but they require a considerable number of adjustable parameters and development of calculation techniques to be useful for the technological practice.

Stability constants can generally be estimated using classical theories for solvates of definite geometry and solvation (coordination) number, n: the equilibrium constants for the replacement of some solvent (in our case, water molecules, W) in the solvate $S(W)^{Z+}$ by another solvent (extractant molecules, E) to form the solvated ion $S(E)^{Z+}$.[8,27–29] This replacement reaction is generally studied in mixed solvents, but the values may be extrapolated to apply to the pure solvents. Marcus,[27–29] Grunwald,[37] Covington,[38,39] and Cox[40] derived equations relating the equilibrium constants for solvent replacement to standard molar Gibbs free energy of solute (S) transfer from a solvent (W) to its mixture with another solvent (W + E).

According to Cox:

$$\Delta G_t^{\infty}(\mathrm{S, W \rightarrow E}) = -n\,\mathrm{RT}\ln \overline{K}_{\mathrm{SE}} \tag{6.9}$$

where the $\overline{K}_{\mathrm{SE}}$ is the average equilibrium constant for W→E solvent replacement and n is the solvation (coordination) number.

According to Covington[38] rather complicated relationship between chemical shift NMR (δ) and solvent composition was proposed:

$$\frac{\delta - \delta_{\mathrm{SW}}}{\delta_{\mathrm{SE}} - \delta_{\mathrm{SW}}} = \frac{K_1Y + 2K_1K_2Y^2 + 3K_1K_2K_3Y^3 + 4K_1K_2K_3K_4Y^4}{4(1 + K_1Y + K_1K_2Y^2 + K_1K_2K_3Y^3 + K_1K_2K_3K_4Y^4)} \tag{6.10}$$

Where $n = 4$ and $Y = a_{\mathrm{SE}}/a_{\mathrm{SW}}$ (a_{i} – activity of solute species).

At $\overline{K} = (K_1K_2K_3K_4)^{1/4}$ average constant Eqn (6.2) may be simplified and presented in the linear form:

$$\frac{1}{\delta - \delta_{\mathrm{SW}}} = \frac{1}{\delta_{\mathrm{SE}} - \delta_{\mathrm{SW}}}\left(1 + \frac{1}{\overline{K}Y}\right) \tag{6.11}$$

According to Scatchard–Deranleau equation[41–43] measurements for UV-VIS (ε) (if all solute species obey Beer's law):

$$\frac{\varepsilon_{SW} - \varepsilon}{C_E} = \overline{K}_{SE}(\varepsilon_{SW} - \varepsilon_E) - \overline{K}_{SE}(\varepsilon_{SW} - \varepsilon) \quad (6.12)$$

$$Z = \frac{\varepsilon_{SW} - \varepsilon}{\varepsilon_{SW} - \varepsilon_{SE}} = \frac{C_{SE}}{C_S^0} = \frac{\overline{K}_{SE} C_E}{1 + \overline{K}_{SE} C_E} \quad \text{at } 0 < Z < 1 \quad (6.13)$$

It has been shown[38–43] that data obtained by these equations are reliable only between certain limits of saturation, Z, namely $0.15–0.2 < Z < 0.7–0.8$ or namely for the 1 and 2 regions of the general extraction isotherm (see Fig. 6.1).

This condition is rarely respected in practice. While for strong complexes it would be easy to fulfill it, weak complexes require relatively high solute concentrations to reach, if ever, the appropriated degree of saturation. In this case the solute becomes one of the constituents of the reaction medium. Such transformation of complexing solute into complexing and solvating molecules casts doubt on the reliability of equilibrium constants of weak complexes. According to the theory of charge transfer complexes, the absorption intensity should increase with increasing components' interaction. However, the opposite behavior is observed frequently with $K = 0$ and $\varepsilon = \infty$ as a limiting case. This means that slope and intercept of above equations are both equal to 0 and determination of K and ε impossible. In other cases negative intercepts and positive slopes were found. This implies negative association constants and that is thermodynamically unacceptable.

These phenomenological anomalies stimulated intensive research activity. Several *ad hoc* theories have been proposed.[44–47] But all of them are reliable only in specific cases. Competitive preferential solvation, COPS, theory of weak molecular interactions,[48–51] modified for extraction systems, is verified and shows reliability in the whole concentration range. It can be applied to any kind of physico-chemical techniques (e.g., chemical kinetics, NMR, UV-V, fluorescence, relaxation time measurements). COPS theory modified for extraction systems[52–54] is used in the presented work to describe and quantify correlations between complexation and solvation effects.

4. MODIFIED COMPETITIVE PREFERENTIAL SOLVATION THEORY

The following postulates (A–C) of the modified COPS theory are presented.

A. Solute in liquid phases is surrounded by a solvation shell. This means that a solute interacts with all the constituents of the environment, and therefore, "free" solute, free extractant, or free solute–extractant complex do not exist. The molecules in the solvation shell continuously and fastly relax between complexing and solvating states. Therefore, coefficient, k_i, represents the total, complexing, and solvating affinity constant.

This statement was proved in many experimental investigations. For example, while studying the mechanisms of the extraction of carboxylic acids with tri-*n*-octylamine[26] the authors found that the equilibrium between the ion-pair (complexing) form and H-bonded (solvating) form does not exist. In order to explain these results, the authors had to introduce a "molecular resonance theory" with a parameter "overall equilibrium constant," which is similar to the above statement of the COPS theory.

It is necessary to clarify the terms "complexation" and "solvation" used in this article. According to Covington,[39] strong, "long range interactions" which are electrostatic in origin and are normally treated by Born theory lead to complexation. Complexation (including ion pairing) occurs when the interacting partners meet each other with well-defined orientations, allowing the favorable overlap of their orbitals or charge (proton, electron) transfer. Any other situation leads to solvation. The latter are the so-called weak or "short-range interactions."[39]

B. Composition of the solvation shell in a mixture of solute, S, extractant, E, solute–extractant complex formed, SE, solvents, *A* (including water, W), and diluents, D, depends on affinity constant k_i and on the number of potentially available solvent (including extractant) molecules present in the mixture, i.e., on the actual concentration of every component.[48,52,53]

Really, we have to use here and below the activity of a solvent, which plays the same role as the concentration in ideal solutions (obeying the gas laws) and is equal to concentration at infinite dilution.[27,28] Provided that the concentration of solute is small relative to solvents in the mixture, the ratio of the activity coefficients of two solvents participating in the replacement equilibrium can be approximated by the ratio of their molar concentrations or mole fractions.[29]

The greater the factor $k_{SE}C_E$ (or $k_{SW}C_W$, or $k_{SA}C_A$, or $k_{SD}C_D$), the better the component solvates the solute (preferential solvation). The mole fraction scale is useful for the entire solute composition range. For practical and theoretical purposes, involving statistical thermodynamics, it is advisable to deal with the molar volume concentrations (C, in M units or mol/dm^3). In this case, it is essential that the densities of the pure components, as well as the densities of composed solutions, be known.

The solute is considered as if it is partitioned among all components of the system. Neglecting the solvent–solvent interactions (at this stage of considerations) the system is described as

$$C_{SO} = C_{Saq} + C_{Sorg} = C_{Saq} + C_{SE} + C_{SW} + C_{SA} + C_{SD} \qquad (6.14)$$

where C_{SO}, is the initial (or total) concentration of the solute; C_{Saq} is the concentration of the solute in the aqueous phase; C_{Sorg} is the concentration of the solute in the organic phase; C_{SE}, C_{SW}, C_{SA}, and C_{SD} are the molar fractions of the solute in the mixed solvation shell of the organic phase: in extractant, water, active solvent, and diluent, respectively, at equilibrium.

Water in the solvation shell, C_{Worg}, is considered to be one of the active solvents. So, we are speaking here of water coextraction with solute and assume that all water, determined in the organic phase after its separation and centrifugation, is present in the solvation shell. Situations in which water is extracted by extractants or solvents, such as at $C_{Sorg} << C_{Worg}$, will be considered for definite extraction systems.

Since the composition of the medium is expressed in the molar volume fractions, the sum of all partitioning factors is equal to unity:

$$\sum_{1}^{n} P_{Si} = 1 \tag{6.15}$$

where P_{Si} is the generalized partitioning factor of the solute in homogenous media; extracting components are $n = 4$ (E, extractant; A, active solvent; W, water (in organic); D, inert diluent).

Partitioning factor P_{Si} is

$$P_{Si} = \frac{k_{Si} C_i}{\sum_1^n k_{Si} C_i} \tag{6.16}$$

where C_i are concentrations of extracting components in the bulk organic phase (C_E, C_A, C_{Worg}, C_D) and k_{Si} are affinity constants of solute to every component.

This means that the mixed solvation shell medium is considered to be equivalent to a weighed mixture of solutions in pure solvent components.

The concentrations of the solute, partitioned between the components of the solvation shell, are[52,53]

$$C_{Si} = C_{Sorg} P_{Si} \tag{6.17}$$

or

$$C_{SE} = C_{Sorg} \frac{k_{SE} C_E}{k_{SE} C_E + k_{SW} C_{Worg} + k_{SA} C_A + k_{SD} C_D} \tag{6.18}$$

$$C_{SW} = C_{Sorg} \frac{k_{SW} C_{Worg}}{k_{SE} C_E + k_{SW} C_{Worg} + k_{SA} C_A + k_{SD} C_D} \tag{6.19}$$

$$C_{SA} = C_{Sorg} \frac{k_{SA} C_A}{k_{SE} C_E + k_{SW} C_{Worg} + k_{SA} C_A + k_{SD} C_D} \tag{6.20}$$

$$C_{SD} = C_{Sorg} \frac{k_{SD} C_D}{k_{SE} C_E + k_{SW} C_{Worg} + k_{SA} C_A + k_{SD} C_D} \tag{6.21}$$

A different approach is needed for solvents that are soluble in both the organic and aqueous phases.[52,53] For solvent A', partially soluble in aqueous and organic phases, the concentration is

$$C_{SA'} = C_{Sorg} \frac{k_{SA'} C_{A'org}}{k_{SE} C_E + k_{SW} C_{Worg} + k_{SA'} C_{A'org} + k_{SD} C_D} \tag{6.22}$$

where $C_{A'org} = C_{A'aq}P_{A'}$. $P_{A'}$ is the partition factor of solvent A' between the organic and aqueous phases, determined experimentally in the E–A–W system (at the absence of solute).

Solute in the solvation shell is present in both the complexed and solvated forms. Therefore, partitioning factor P_{SE} can be divided into two:[52,53]

$$P_{SE} = P_{SEc} + P_{SEs} \tag{6.23}$$

where P_{SEc} and P_{SEs} represent partitioning factors of the solute in the complexed and solvated forms, respectively. So, solute concentrations in the complexed, C_{SEc}, and solvated, C_{SEs}, forms in the solvation shell are determined by equations

$$C_{SEc} = C_{Sorg}\frac{k_{SEc}C_E}{k_{SEc}C_E + k_{SEs}C_E + k_{SW}C_{Worg} + k_{SA}C_A + k_{SD}C_D} \tag{6.24}$$

$$C_{SEs} = C_{S\,org}\frac{k_{SEs}C_E}{k_{SEc}C_E + k_{SEs}C_E + k_{SW}C_{Worg} + k_{SA}C_A + k_{SD}C_D} \tag{6.25}$$

where k_{SEc} and k_{SEs} are affinity constants of the solute toward extractant in the complexed and in the solvated forms, respectively.

Partitioning factors P_{SE}, P_{SW}, P_{SA}, P_{SD} are proportional to the 1/SAF (solvent attenuation factor) of the Headley solvent effects theory.[34–36] However, the last one is more specific and needs its reaction parameters determined in the gas phase, using ion cyclotron resonance spectrometry. "Critical micelle concentration" (cmc) parameter of the Fendler micelle formation theory[55] is also proportional to the partitioning factor of the COPS theory.

C. The actual value of the physicochemical property X (e.g., chemical shift, δ, in NMR, molar extinction coefficient, ε, in UV-VIS, changes in free energy, ΔG, in potentiometric titration), measured in mixed solvents, is considered to be the weighed sum of this same property measured in pure solvent constituents.[47,51,52] The effects of solvents on this property X are additive:

$$X = X_{SE}P_{SE} + X_{SW}P_{SW} + X_{SA}P_{SA} + X_{SD}P_{SD} \tag{6.26}$$

Extraction system with a monobasic extractant E, with a relatively dilute solution of monobasic solute S, with water in organic phase W and a diluent D (excluded on this stage of consideration) may be described:

$$P_{SW} + P_{SE} = 1 \tag{6.27}$$

$$C_{Worg} = \frac{1 - C_E\nu_E}{\nu_W} \tag{6.28}$$

where ν_E and ν_W are partial molar volumes of extractant and water.

Using Eqn (6.25) we obtain the linear equation for the measured property X:

$$\frac{X_{\mathrm{SW}} - X}{C_{\mathrm{E}}} = \frac{k_{\mathrm{SE}}}{k_{\mathrm{SW}}} \nu_{\mathrm{W}} (X_{\mathrm{SW}} - X_{\mathrm{SE}}) - \left[\frac{k_{\mathrm{SE}}}{k_{\mathrm{SW}}} \nu_{\mathrm{W}} - \nu_{\mathrm{E}} \right] (X_{\mathrm{SW}} - X) \qquad (6.29)$$

It can be seen that the treatment by COPS theory reached simple linear mathematical forms directly but classical treatment of Covington, Scatchard, Deranleau, etc., comes to the simple forms through many complicated steps and simplifications. And COPS linear equations always has correct negative slope even in cases which show anomalies.

5. ELECTRONIC ACID–BASE THEORY AND AMPHOTERITY

The coordination process, interpreted as an acid–base reaction,[5,8,22,23] distinguishes between low electronegative but high polarizable (soft) bases and high electronegative but low polarizable (hard) bases. As an extension of complex formation reactions the metal ions are classed in general as acids. Metal ions that form their most stable complexes with hard bases, containing as a rule, fluorine, oxygen, nitrogen, are described as hard acids, while those that form their most stable complexes with soft bases, containing the heavier elements of these groups, are soft acids. Hard bases prefer hard acids and soft bases prefer soft acids. The type of bonding between the acid and the base is of minor importance: it is predominantly ionic in the case of hard acids, and predominantly electron sharing in the case of soft acids. Some ligands, containing either monodentate and polydentate groups, form both kinds of bonds (with electron donors of carboxyl-, phosphoryl-, sulfuryl-, amino-groups) with chelate formation.

There are solvents in which, though a slight self-ionization is observed, this cannot longer serve as a foundation for the acid–base reactions.[55,56] They are mainly organic solvents, particularly those containing nitrogen, phosphorus–oxygen, sulfur–oxygen functional groups. In these solvents the acid–base act is initiated not by self-ionization but by solvation through H-bonding of the dissolved compounds. The comprehensive theory of Usanowitsch[57] stated that all coordinately unsaturated atomic groupings are acids and all coordinately saturated atomic groupings are bases.

The dual maximum behavior (two regions of effective metal extraction: one at low acidity and the other at high acidity of the aqueous solutions) of Ti(IV)[58,59] and many other transition metals[60] was found for most families of metal extractants: acidic (organic acids), basic (amines or their mixtures), and neutral complexants. Examples with references,[61–82] which show the universality of the dual behavior at metals' extraction, are presented in Table 6.1.

Pearson[21,22] classifies many substances as both bases and acids: any specie with pK_a value higher than the given one may be conjugate base to it. According to the COPS theory[44] ions (including metal ions) in solutions cannot

TABLE 6.1 Dual Maximum Acidity Dependence of Metal ions at their Extraction with Different Types of Extractants

No Periodic Table	Metal ions	Acidity / Refs.	M/l 10-8, 10-6, 10-4, 10-2; pH 8, 6, 4, 2	2, 4, 6, 8, 10, 12; −0.30, −0.60, −0.78, −0.90, −1.0, −1.08
VIII A	Fe(III)	(61)	M,D	D; M
VII A	Re (VII)	(62,63) (61,64)	18C6; TOMAC	18C6; TOMAC
VI A	Mo (VI)	(65,66) (67,68)	TOMAC; Neutr.	TOMAC; Neutr.
VI B	Te (II)	(69)	18C6	18C6
V A	Nb (V) V (V)	(61,70) (71, 72)	Neutr.; TOMAC	Neutr.; TOMAC
V B	Sb(V)	(71, 72) (73, 74) (75)	Neutr.; TDA	M,D; D; M; Neutr.; TDA
IV A	Ti (IV) Zr (IV) Hf (IV)	(58,59) (77,78) (72–74)	TOMAK, TOA; M,D; 18C6	M,D,TOA; 18C6
IV B	Sn (IV)		Alcoh.	M,D; Alcoh.; D; M
III B	In (III)	(61,62) (79–81) (82,83)	M,D; TDA; Neutr.	D; M; TDA; Neutr.

☐ Acidity Borders

be considered as free but solvated (coordinated) with charged ligands and/or neutral molecules, depending on their oxidation state and solvation (coordination) number. These species may be classed both as acids or bases and as amphoteric depending on the properties and concentrations of groups and molecules bonded to ions. An empirical quantum chemical method's approach, applied for the treatment of ion solvation (coordination) in the bulk solvent, employs the concepts of donor–acceptor interactions.[5,46,47] This approach recognizes the coordinate bonds that may be formed, i.e., allows for some overlap of the electron orbitals of the ion and the solvent in the solvated ion. So, the dual maximum behavior of metal extraction data (Table 6.1) may be explained by amphoteric properties of the solute and extractant species in solutions at different acidities.

6. AGGREGATION

Aggregation is a well-known process in liquid–liquid extraction. At extraction of metals or acids into immiscible organic compounds, the solvent stoichiometry of extracting species changes as loading increases. This process is described as aggregation. The simplest and most clearly defined aggregates are carboxylic acid dimers. Strong evidence may be mentioned for the presence of higher aggregates.[6,7] Aggregation was observed at the extraction of metal ions by acidic extractnts[73,83–89] of metal ions and acids by basic and neutral extractans.[14–16,18,19,27,90–93]

Many investigators correspond the molecular aggregates to reversed micellar structures.[94–96] Some of them propose the formation of linear aggregates,[84,89] which at a critical concentration of solute in organic phase is structurally reorganized into reversed micelles[84–87,94,95] or cluster-like aggregates[52–54] in nonpolar aprotic solvents. Three regimes of aggregation behavior are considered as a function of metal ions M^{2+} concentration in organic phase of acidic organophosphorus extractants:[84–89] the tetrametric species of type ML_2*2HL → linear aggregates → reversed micelles. The authors proved the formation of linear, cyclic aggregates and reversed micelles using many different experimental and analytical techniques. Formation of three-dimensions cluster-like aggregates is expected also.[52–54] The linear and subsequently three-dimensional aggregation was observed with other organophosphorus extractants, namely, phosphonic and phosphinic acid esters.[86]

REFERENCES

1. Lo TC, Hanson C, Baird MHI, editors. *Handbook of solvent extraction*. New York: Wiley; 1983.
2. Szafran M. Recent aspects of the proton transfer reaction in H-bonded complexes. *J Mol Struct* 1996;**381**:39–64.
3. Epstein LM, Shubina ES, Kavtsov DN. Hydrogen and coordination bonds in isostrutural systems. Are hydrogen bonds unique? *J Mol Struct* 1988;**177**:327–38.
4. Gutsul EI, Belkova NV, Epstein LM, Shubina ES, Peruzzini M, Bianchini C, Zanolini F, Ledos AL, Maresca O. Hydrogen bonded intermediates and proton transfer to PP_3MH. In: *Proceedings of XVth International Conference on Horizons in Hydrogen Bond Research*. Germany: Free University of Berlin; 2003. p. 3P06.
5. Huyskens PL, Luck WAP, Zeegers-Huyskens T, editors. *Intermolecular forces*. Berlin: Springer-Verlag; 1991.
6. Douglas B, et al., editors. *Concepts and models of inorganic chemistry*. 3rd ed. JohnWiley & Sons; 1994. p. 309–49.
7. Dotre JC, Teixeira J, editors. *Hydrogen-bonded liquids*. The Netherlands: Kluwer; 1991.
8. Marcus Y. *Ion solvation*. London: Wiley-Interscience; 1985.
9. Schuster P, Zundel G, Sandorfy C, editors. *The hydrogen bond. Recent developments in theory and experiments*, vols. 1–3. Amsterdam: Holland; 1976.
10. Ratajczak H, Orville-Thomas WJ, editors. *Molecular interactions*, vols. 1–3. Chichester: Willey; 1980. p. 1981–2.
11. Mayer JM, Rhile IJ. Thermodynamics and kinetics of proton-coupled electron transfer: stepwise vs. concerted pathways. *Biochim Biophys Acta Bioenerg* 2004;**1655**:51–8.
12. Scheiner S. Relationship between strength of hydrogen bond and barrier to proton transfer. *J Mol Struct* 1994;**307**:65–71.
13. Maechal Y. What mechanisms for transfers of protons through H-bonds? *J Mol Liq* 1991;**48**(2–4):233–51.

14. Tamada J, King C. *Extraction of Carboxylic Acids by Amine Extractants*. Ph.D. Thesis. Lawrence Berkeley Laboratory, University of California, Applied Science Division; 1989.
15. Tamada J, Kertes A, King C. Extraction of carboxylic acids with amine extractant. 1. Equilibria and law of mass action modeling. *Ind Eng Chem Res* 1990;**29**:1319–26.
16. Eyal AM, Canari R. PH dependence of carboxylic and mineral acid extraction by amine-based extractants: effects of pKa, amine basicity, and diluent properties. *Ind Eng Chem Res* 1995;**34**(5):1789–98.
17. Yang S, White S, Hsu S. Extraction of carboxylic acids with tertiary and quaternary amines: effect of pH. *Eng Chem Res* 1991;**30**:1335–48.
18. Tamada J, Kertes A, King C. Extraction of carboxylic acids with Amine extractant. 2. Chemical interaction and interpretation of data. *Ind Eng Chem Res* 1990;**29**:1327–32.
19. Tamada J, Kertes A, King C. Extraction of carboxylic acids with Amine extractant. 3. Effect of temperature, Water coextraction and Process considerations. *Ind Eng Chem Res* 1990;**29**:1333–41.
20. Canari R, Hazan B, Eyal A. Carboxylic acids extraction by basic extractants: the effects of pH, ion concentration and acid-base property of the extractant and extracted acid. *Proceedings of ASEC'96*, vol. 2. 1996. p. 1517–22.
21. Pearson RG, Vogelsong DC. Developmentsin crystal field theory. *J Amer Chem Soc* 1958;**80**:1038.
22. Pearson RG. Hard and soft acids and bases. *J Amer Chem Soc* 1963;**85**:3533.
23. Luder WF, Zuffanti S. *The electronic theory of acids and bases*. New York: Wiley; 1946.
24. Vieux AS. Extraction of some dicarboxylic acids by tri-isooctylamine. *Anal Chim Acta* 1974;**68**:415–24.
25. Wardel JM, King CJ. Solvent equilibria for extraction of carboxylic acids from water. *J Chem Eng Data* 1979;**23**(2):144–8.
26. Mei Wen, Yiyan Yang, Youyuan Dai. Studies on the mechanism of extraction of carboxylic acids with tri-n-octylamine by chemical complexation. *Proceedings ISEC'96*, vol. 1. 1996. p. 195–200.
27. Marcus Y. Preferential solvation of ions in mixed solvents. *Rev Anal Chem* 1980;**53**(5):127–35.
28. Marcus YA. quasi-lattice quasi-chemical theory of preferential solvation of ions in mixed solvents. *Austr J Chem* 1983;**36**:1719–31.
29. Marcus Y. *The properties of solvents*. Wiley-Interscience; 1998.
30. Spala EE, Ricker NL. Thermodynamic model for solvating solutions with physical interactions. *Ind Eng Chem Process Des Dev* 1982;**21**:409–15.
31. Taft RW, Abboud JLM, Kamlet MJ, Abraham MN. Linear solvation energy relations. *J Solution Chem* 1985;**14**(3):153–86.
32. Shorter J. *Correlation analysis in chemistry*. Oxford: Clarendon Press; 1973.
33. Nagata I. Ternary liquid-liquid equilibria: the associated solution theory. *Fluid Phase Equilib* 1977/1978;**1**:267–75.
34. Headley AD. Solvent effects on the basicity of alkyl-substituted dimethylamines. *J Org Chem* 1988;**53**:312–4.
35. Headley AD, McMyrry ME, Starnes SD. Effects of substituents on the Acidity of Acetic acid. *J Org Chem* 1994;**59**:1863–6.
36. Headley AD. Quantitative analysis of solvation effects and influence of alkyl substituents on basicity of amines. *J Org Chem* 1991;**56**:3688–91.
37. Grunwald E, Baughman G, Kohnstam G. The solvation of electrolytes in dioxan-water mixtures, as deduced from the effect of solvent change on the standard partial molar free energy. *J Am Chem Soc* 1960;**82**:5801–11.

38. Covington AK, Newman KE, Lilley TN. *J Chem Soc Faraday Trans 1* 1973;**69**:963.
39. Covington AK, Newman KE. Thermodynamics of preferential solvation of electrolytes in binary solvent mixtures. *Adv Chem Ser* 1976;**155**:153–96.
40. Cox BG, Parker AJ, Waghorne WE. Coordination and ionic solvation. *J Phys Chem* 1974;**78**:1731–40.
41. Scatchard G. The attraction of proteines for small molecules and ions. *Ann NY Acad Sci* 1949;**51**:660–7.
42. Deranleau DA. Theory of the measurement of weak molecular complexes. I. General considerations. *J Amer Chem Soc* 1969;**91**:4044–9.
43. Deranleau DA. Theory of the measurement of weak molecular complexes. II. Consequence of multiple equilibria. *J Amer Chem Soc* 1969;**91**:4050–62.
44. Orgel LE, Mulliken RS. Molecular Complexes and Their Spectra. Contact Charge-Transfer Spectra. *J Amer Chem Soc* 1957;**79**:4839–43.
45. Carter S, Murrell JN, Rosch EJ. Donor-Acceptor properties. *J Chem Soc* 1965:2048–51.
46. Kuntz Jr ID, Gasparo FP, Johnston Jr MD, Taylor RP. Molecular Interactions and the Benesi-Hildebrand Equation. *J Amer Chem Soc* 1968;**90**:4748–54.
47. Homer J, Everdell MH, Jackson CJ, Whitney PM. Molecular complexes. Part 7.—Some comments on the processing of data obtained from nuclear magnetic resonance studies. *J Chem Soc Faraday II* 1972;**68**:874–82.
48. Nagy OB, Muanda H, Nagy JB. Competitive preferential solvation theory of weak molecular Interactions. *J Chem Soc Faraday Trans I* 1978;**74**:2210–29.
49. Parbhoo B, Nagy OB. Application of competitive preferential solvation theory to ion-molecular interactions. *J Chem Soc Faraday Trans I* 1986;**82**:1789–93.
50. Szpakowska M, Nagy OB. Application of the preferential solvation theory to coordinative solute-solvent interactions. *J Chem Soc Faraday Trans I* 1989;**85**(9):2891–9.
51. Nagy OB, Muanda M, Nagy JB. Quantititive evaluation of solute-solvent interactions. Environmental effects on the n-butylaminolysis of tetrachloro-n-butylphtalamide. *UJ Phys Chem* 1979;**83**:1961–79.
52. Kislik V. Competitive complexation/solvation theory of solvent extraction. II. Solvent extraction of metals by acidic extractants. *Sep Sci Technol* 2002;**37**(11):2623–57.
53. Kislik V. Application of preferential solvation concept for Interpretation of the mechanisms of acids solvent extraction by amines. *Sep Sci Technol* 2001;**36**(13):2899–926.
54. Kislik V. Application of preferential solvation model for interpretation of the mechanisms of solvent extraction. Proceedings of the XVth International Symposium on Physico-Chemical Methods of Mixtures Separation, "ARS SEPARATORIA 2000", Poland, 2000, p. 34–40.
55. Fendler EJ. Proton magnetic resonance investigation of alkylammonium carboxylate micelles in nonaqueous solvents. *J Phys Chem* 1973;**77**(11):1432–6.
56. Devis MM. *Acid-base behavior in aprotic organic solvents. National Bureau of Standards.* Washington, D.C. Monograph 105; 1968.
57. Usanowitsch M. Theory of acids and bases. *J Allg Chem (USSR)* 1939;**9**:182–98.
58. Kislik V, Eyal A. Acidity dependence of Ti(IV) extraction: a critical analysis. *Solvent Extr Ion Exch* 1993;**11**(2):258–66.
59. Kislik V, Eyal A. Extraction of titanium(IV) by mixtures of mono- and d-(2-ethylhexyl) phosphoric acids esters. *Solvent Extr Ion Exch* 1993;**11**(2):285–10.
60. Tsuboi I, Kasai S, Yamamoto T, Komasawa I, Kunugita E. Recovery of rare metals from coal fly ash. Part B. In: *Proceedings of ISEC-90.* Amsterdam: Elsevier; 1992. p. 1199–204.
61. Levin IS, Azarenko TG. Extraction of rare metals by alkylphosphorus acids. *Zh Anal Khim* 1968;**23**(7):1013–7.

62. Jewad SK, Al-Murab SR, Jalhoom MG. Studies on the extraction of perrhnate by use of crown ethers. Part A. In: *Proceedings of ISEC-90*. Amsterdam: Elsevier; 1992. p. 907–11.
63. Zagorodnyaya AN, Ponomareva EI, Abisheva ZS, Tarasova EG. Rhenium extraction by tri-n-octylamine hydrosulfate. Part A. In: *Proceedings of ISEC-90*. Amsterdam: Elsevier; 1992. p. 955–60.
64. Klimov IT. Extraction of Fe(III) with mixtures of cyclohexanone with isoamyl acetate, $CHCl_3$, CCl_4 and bromobenzene. *Gidrokhim Materialy* 1964;**37**:114.
65. Hirai T, Komasawa I. Separation and purification of vanadium and molibdenium by solvent extraction and subsequent reductive stripping. *Proceedings of ISEC-90*. Amsterdam: Elsevier; 1992. p. 1015–20.
66. Jalhoom MG. Extraction of pertechnetate ions by use of crown ethers from sulfuric acid solutions. *Radiochimica Acta* 1986;**39**:195.
67. Jalhoom MG. Distribution of tetra-n-alkylammonium salts between water and organic solvents. *J Radioanal Nucl Chem Lett* 1986;**104**:131.
68. Haesebroek G, Vliegen J, De Schepper A. Solvent extraction separation of niobium and tantalum at MHO: potentialities of Cyanex 923 versus MIBK. *Proceedings of ISEC-90*. Amsterdam: Elsevier; 1992. Part, B1115–20.
69. Levin IS. Concentration and separation of indium, gallium, thallium, antimony, and bismuth by extraction with alkylphosphoric acids. *Zh Prikl Khim* 1962;**35**:2368–79.
70. Levin IS, Shatalova AA. Separation of Sb(III) by extraction with alkylphosphoric acids. *USSR Talanta* 1967;**14**(7):801.
71. Dakar GM, Iofa BZ. Extraction of complex acids by O-containing solvents. V. Mechanism of extraction of pentavalent Sb. *Radiokhimia* 1963;**5**(4):428–34.
72. Stronski I. Radiochemical studies on the extraction and stabilization of metal chelates and on the separation of metal salts by extractive chromatography. *Zh Physik Khim* 1966;**3**:329–36.
73. Goto H, Kakita Y. Extraction of iodide complexes of Cu, Pb, Bi, Sb, Cd and Te with methyl isobutyl ketone. *Nippon Kagaku Zasshi* 1966;**87**(9):962–71.
74. Lyle SJ, Shendrikar AD. Some aspects of N-benzoylphenylhydrxylamine and cupferron reactions with tin, antimony and other selected elements. *Anal Chim Acta* 1966;**36**(3):286–97.
75. Iofa BZ, Mitrofanov KP. Investigation of the extraction mechanism of Sb from the HCl solutions by dibutyl ether. *Radiokhimia* 1964;**6**:419–28.
76. Deorkar NV, Khopkar SM. Solvent extraction separation of thorium, uranium, zirconium and hafnium with crown ethers. In: *Proceedings of ISEC-90*. Amsterdam: Elsevier; 1992. p. 467–72. Part A.
77. Lomako VL, Starobinets GL. Effect of a solvent on the kinetics of draining extraction of zinc thiocyonate complexes. *Vesti Akad Nauk Belorus Ser Khim Nauk* 1971;**6**:634–42.
78. Watanabe H, Akatsuka K. Solvent extraction of metal ions from sulfuric acid solutions with a mixed extractant-long-chain primary amine and bis (2-ethylhexyl) phosphate. *Nippon Kagaku Zasshi* 1968;**89**(3):280.
79. Good ML, Holland FF. Extraction of indium(III) and gallium(III) from aqueous chloride media by long-chain alkyl amines and quaternary salts. *J Inorg Nucl Chem* 1961;**26**(2):321–7.
80. Kleiner KE, Markova LV. Extraction of indium(III) from water and sulfuric acid solutions by MIBK. *Zh Anal Khim* 1969;**8**:279–84.
81. Hasegawa Yu. Behavior of indium(III) complexes in several liquid-liquid distribution systems. *Bull Chem Soc Japan* 1970;**43**(9):2665–9.
82. Golinski M. *Cerium removal from lanthanide nitrate solutions. Proceedings of ISEC-71*. Amsterdam: Elsevier; 1971. p. 603–10.

83. Neuman RD, Nai-Fu-Zhou, Wu J, Jons MA, Gaonkar AG. General model for aggregationof metal-extractant complexes in acidic organophosphorus solvent extraction systems. *Sep Sci Technol* 1990;**25**(13–15):1655–74.
84. Neuman RD, Jons MA, Zhou NF. Photo-correlation spectroscopy applied to hydrometallurgical solvent-extraction systems. *Colloid Surf* 1990;**46**(1):45–61.
85. Wilkinson G, Gillard RD, MaccLeverty JA, editors. *Comprehensive coordination chemistry.* John Wiley & Sons; 1987. p. 2. pp. 43–76.
86. Gaonkar AG, Garver TM, Neuman RD. H-1-NMR spectroscopic investigation of reversed micellization in metal organophosphorus surfactant system. *Colloid Surf* 1988;**30**(3–4): 265–73.
87. Yu ZJ, Ibrahim TH, Neuman RD. Aggregation behavior of cobalt(II), nickel(II) and copper(II) bis(2-ethylhexyl) phosphate complexes in n-heptane. *Solv Extr Ion Exch* 1998;**16**(6):1437–63.
88. Neuman RD, Ibrahim TH. Novel structural model of reversed micelles: the open water-channel model. *Langmuir* 1999;**15**:10–2.
89. Evers A, Hancock R. A study of aggregation in the organic phase of solvent extraction systems and its influence on the extraction process. *Report Natl Inst Metall Repub S Afr* 1977;**34**:24–38.
90. Hogfeldt E, Fredlund F. On aggregation of the complex between nitric acid and trilaurylamine dissolved in octan and m-xylene. *Acta Chem Scandinavica* 1964;**18**:543–7.
91. Hogfeldt E, Fredlund F, Odberg L, Merenge G. Extraction of water, nitric and perchloric acids by nitrobenzene. *Acta Chem Scandinavica* 1973;**18**(5):1781–94.
92. Grinstead RR. Base strengths of amine in liquid-liquid extraction systems. *Proceedings of ASEC'67* 1968:427–36.
93. Davis MM. Acid-Base Behavior in Aprotic Organic Solvents. In: Lagowski, editor. *The chemistry of nonaqueous solvents*, vol. 3. New York: Academic Press; 1970. p. 1–32.
94. Dega-Szafran Z, Szafran M. *Heterocycles* 1994;**37**:627–34.
95. Tani Hirofumi, Kamidate Tamio, Watanabe Hiroto. Micelle-mediated extraction. *J Chrom* 1997;**780**(1–2):229–41.
96. Anticó E, Masana A, Hidalgo M, Salvadó V, Iglesias M, Valiente M. Solvent extraction of yttrium from chloride media by di(2-ethylhexyl)phosphoric acid in kerosene. Speciation studies and gel formation. *Anal Chim Acta* 1996;**327**(3):267–76.

Chapter 7

Competitive Complexation/ Solvation Theory of Solvent Extraction: General Mechanisms and Kinetics

Chapter Outline

1. BASIC STATEMENTS

Competitive complexation/solvation theory of solvent extraction is based on four statements; each of them are very well-known facts in the chemistry that is described in the previous chapter. The statements are as follows:

I. Extraction act is initiated by solvation of the solute compounds in organic solvents with formation of hydrogen-(coordinate-)bonded intermediates (adducts). Kinetics of proton, electron transfer, and formation of final complexes are stepwise and depends on kinetics of step reactions, competition, coupling, strength of H-bonding and many other factors, as mentioned in Chapter 6. The kinetics of the step reactions may be fast, so, final products are formed during extraction process, or may be slow, so, extracted species are some of intermediates and the latter process can be called as a pseudo-equilibrium extraction.

Solvent Extraction. DOI: 10.1016/B978-0-444-53778-2.10007-X

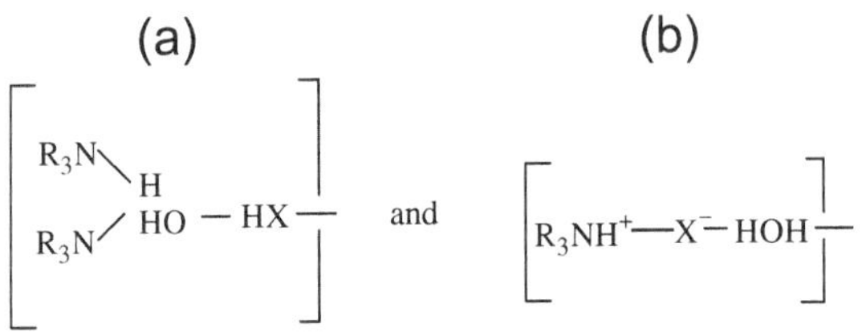

FIGURE 7.1 Example of extreme structures of nuclei aggregates with (a) only H-bond (pure region 1, where $P_{SEc} \rightarrow 0$) and (b) only ion-pair bond (pure region 2, where $P_{SEs} \rightarrow 0$) between acid and amine.

II. Strong (ion-exchange, chemical, electrostatic), weak (solvating, physical, intermolecular, H–bond, coordinate) interactions and aggregation are the basic interactions which have to be taken into consideration in all range of extracted solute concentrations. Competition between these interactions forms the four region shape (Chapter 6, Fig. 6.1) of the general extraction isotherm.

III. Solutes, extractants, active solvents, and their interacting species are considered amphoteric and may perform as acids (electron acceptors) or bases (electron donors), depending on the structure of their functional groups and composition of the organic phase, and on the structure of the solute species and composition of the aqueous phase.

IV. Aggregation mechanisms and species formed are changing with increasing loading of organic phase by solute, and should be considered and analyzed in all regions of the general extraction isotherm (Chapter 6, Fig. 6.1).

As can be seen from Chapter 6, all statements of the presented theory separately are well-known concepts in chemistry, but taken together for consideration in a specific way they permit to introduce a novel solvent extraction modeling approach. This approach is an attempt to describe the nature of solvent extraction with quantitative evaluation of different compounds, formed in the organic phase at loading from low to high concentrations of the solute. The following considerations of general extraction isotherm (Chapter 6, Fig. 6.1) from the presented theory point of view explain the essence of the theory.

2. EXTRACTION SYSTEMS WITH LOW SOLUTE CONCENTRATIONS IN ORGANIC PHASE

At low concentrations of the solute and high concentrations of the other components in the organic phase the extraction process is mainly characterized by the formation of single solute-extractant molecules, solvated by additional molecules of extractant and/or other solvents (solvation shell). This intermediate specie is denoted as a nucleus aggregate. The nucleus aggregate is open to bulk solvents and is characterized by the fast exchange with the bulk solvents.

Here, we have to distinguish between three nominations of the same physico-chemical reality: solvation shell, nucleus aggregate, and coordination complex. According to the competitive preferential solvation (COPS) theory (see Chapter 6, Section 4), the first nomination considered the solute as interacting and microscopically partitioning between different solvents (in the case that solvent–solvent interactions may be neglected) to form some kind of a virtual physical unit, called the solvation shell. It is a statistical thermodynamics' denomination, which is convenient for mathematical description. The second nomination, nucleus aggregate, describes the same but real physical unit with stereospecific bonds and orientation in the bulk organic solution. The reader can conclude that the "coordination complex" and "nucleus aggregate" nominations describe the same unit, also in case of metals with the coordination sphere of definite number and geometry. In the presented theory the nomination of "nucleus aggregate" is mainly used as universal, not only for metals, but also sometimes two others are used.

Formation of the nuclei aggregates (single solvation shells) is typical mainly in the extraction regions 1 and 2 (see Chapter 6, Fig. 6.1). Competition between strong, complexing, and weak, solvating interactions (see Fig. 7.1) influences mainly the shape of the distribution isotherm between two limiting shapes: pure solvating (Fig. 7.1a) as a region 1 and pure complexing (Fig. 7.1b) as a region 2.

Region 1 is characterized by solute/extractant equivalent (Meq/Meq) concentration ratios, $C_{\mathrm{Sorg}}/C_{\mathrm{SE}} < 1$.[1–4] The structure of the nuclei aggregates in this region is formed mainly through the weak interactions between solute and solvents (including extractant), $P_{\mathrm{SE_S}} > P_{\mathrm{SE_C}}$. These are hydrogen-bond, coordination-bond, polarity, or polarizability stabilization intermediates formed. The contribution of strong proton, electron transfer, ion pair solute–extractant interactions may also be considered, but with less probability. The pH dependence in this region is not pronounced,[5–8] but stability pH limitations of the aggregates formed are observed.

For simplicity, let us consider in the three component S–E–W system that only one molecule of solute S participates in formation of single solvation shell. Therefore, from Eqn (6.15) in Chapter 6 we obtain that $C_{\mathrm{SE}} = P_{\mathrm{SE}}$ and $C_{\mathrm{SW}} = P_{\mathrm{SW}}$. Figure 7.2 presents an example of nuclei aggregate structures formation in region 1 at extraction of monobasic acids by monobasic amines: five possible structures at two possible compositions of the solvation shell and one possible $C_{\mathrm{Sorg}}/C_{\mathrm{SE}} = 1/2$ (Meq/Meq) ratio. We can expect all five nuclei aggregate associations at $Z = 1/2$, shown in Fig. 7.2, but with different probability. Examples of affinity constant ratios data in region 1 for acetic acid (see Chapter 6, Fig. 6.2), obtained, using experimental data from the Ref. 9 are presented in Table 7.1.

Region 2. Formation of nuclei aggregates in this region is driven mainly by strong (ionic, electrostatic, ion pair) interactions: $P_{\mathrm{SE_S}} < P_{\mathrm{SE_C}}$ and $C_{\mathrm{Sorg}}/C_{\mathrm{SE}} \approx 1$. Kinetics of proton, electron transfer in the H-bonded intermediates is very fast. The solute in the nuclei aggregates of region 2 forms mainly strong complexes

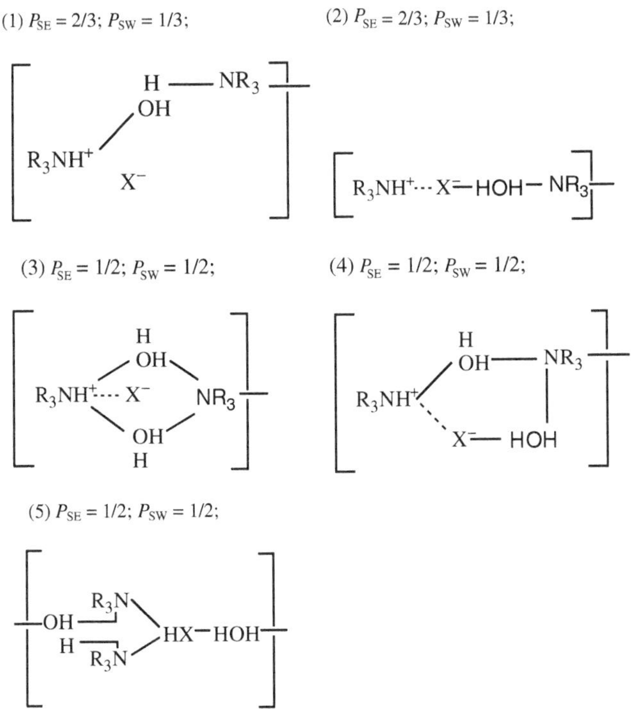

FIGURE 7.2 Possible structural schemes of nuclei aggregates formation in the region 1 at the same molar equivalent ratios $Z = C_{Sorg}/C_{SE} = 1/2$ in the solvation shell. The ionic (electrostatic) bonds are represented by dotted lines while the hydrogen bonds are marked by continuous lines.

with extractant, but, in some extent, weak solute–solvent intermediates (H-bonding, coordinate) formed are stable during the extraction operating time. Interactions in this region are strongly dependent on pH or anion concentration in the aqueous phase. Divergence of the distribution curve slope from the 1–1 equivalent of the S–E complex is explained by weak, solvating interactions.

As an example, four possible structures of nuclei aggregates in the region 2 with the same molar equivalent ratios $Z = C_{Sorg}/C_{SE}(\text{Meq}/\text{Meq}) \approx 1$ at two possible compositions of the solvation shell are presented in Figure 7.3. Examples of affinity constant ratios data in the region 2 for lactic and citric acids, obtained, using experimental data from the references,[9,10] are presented in Table 7.1.

It can be seen that the difference between the regions 1 and 2 is mainly in the magnitude of the affinity constants ratios:

$$\frac{\dfrac{k_{SEc}}{k_{SW}}}{\dfrac{k_{SEs}}{k_{SW}}} = \frac{k_{SEc}}{k_{SEs}} \tag{7.1}$$

TABLE 7.1 Affinity Constant Ratios and Averaged Equilibrium Constants of Acetic, Lactic Acid Extraction by Alamine 336 in Chloroform (Experimental Data from the Ref. 9), and Citric Acid Extraction by Alamine 304 in Kerosene (Experimental Data from Ref. 10) Calculated for Different Regions of the Extraction Isotherm Using Equations of the Presented Theory

Extraction systems	Regions in the extraction isotherm	Affinity constant ratios k_{SE}/k_{SW}	Averaged equilibrium constants $\overline{K}_{Si}$	Hypothetical Z^* (intercept) in pure E at $Q = 0$	Suggested aggregate formed at competitive complexation/ solvation interactions
Acetic acid-Alamine 336 in Chloroform-water	Formation of nuclei aggregates, region 1	0.51	0.25	−0.11	See Fig. 7.1a or 7.2(4)
	Formation of three-dimensional aggregates, region 4	93.3	45.7	−39.3	
Lactic acid-Alamine 336 in Chloroform-water	Formation of nuclei aggregates, region 2	69.4	34.0	−0. 62	See Fig. 7.1b or 7.3(3), or 7.3(4)
	Formation of linear or cyclic aggregates, region 3	1.02	0.50	0.83	See Fig. 7.6(5)
Citric acid-Alamine 304 in kerosene-water	Formation of nuclei aggregates, region 2	69.5	49.4	0.01	See Fig. 7.3(1), 7.3(2), 7.3(5) or Fig. 7.1b
	Formation of linear or cyclic aggregates, region 3	0.52	0.37	0.55	See Fig. 7.6(1), 7.6(2), 7.6(5)

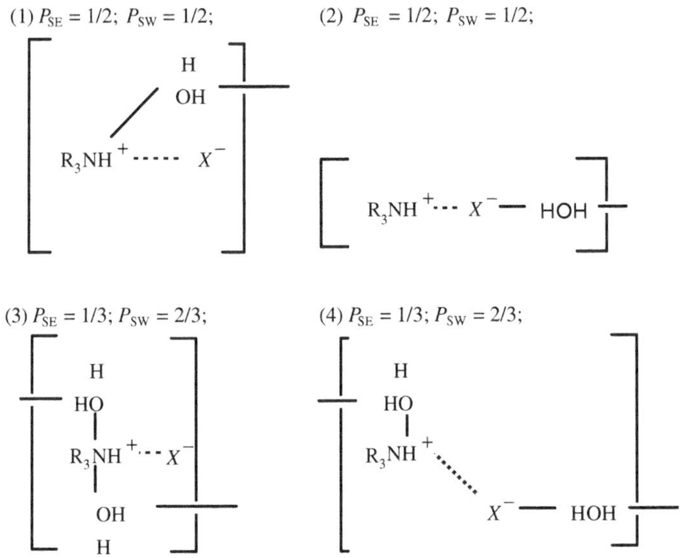

FIGURE 7.3 Possible structural schemes of nuclei aggregates' formation in the region 2 at the same molar equivalent ratios $Z = C_{Sorg}/C_{SE} = 1$ in the solvation shell.

at the same concentrations of extractant, water in organic phase (at $C_{SW} = P_{SW}$, see above) and affinity constant of solute in pure water. So, all structure schemes of the nuclei aggregates for region 1 (see Figs. 7.1c and 7.2, except the 2–5) may belong to region 2 (see Figs. 7.1c and 7.3) but at $P_{SEc} > P_{SEs}$. It follows that the slopes of the distribution curve (see Chapter 6, Fig. 7.1) for regions 1 and 2 may have all magnitudes between the two extreme structures: $Z = 1/2$ where $P_{SEc} \rightarrow 0$ and $Z = 1$ where $P_{SEs} \rightarrow 0$ (see Fig. 7.1a and b), respectively.

For strong mineral acids extraction by strong amines the contribution of H-bonding is negligible[9–12,15] and nucleus aggregate structure is close to b) in Fig. 7.1; for weak carboxylic or amino acids extraction by weak amines the contribution of the ion-pair bonding is negligible (Fig. 7.1a).[9–12,15,16] The reader will find detailed descriptions in Chapter 8 of the book.

Now, using the Eqn (6.25) from the modified COPS theory (Chapter 6, Section 4), we can develop expressions for different measured solute properties X in organic phase.

For NMR, where $X = \delta$ is the chemical shift:

$$\frac{\delta_{SW} - \delta}{C_E} = \frac{k_{SE}\nu_{Worg}}{k_{SW}}(\delta_{SW} - \delta_{SE}) - \left[\frac{k_{SE}}{k_{SW}}\nu_{Worg} - \nu_E\right](\delta_{SW} - \delta) \qquad (7.2)$$

For UV–VIS, where $X = \varepsilon$ is the molar extinction coefficient:

$$\frac{\varepsilon_{SW} - \varepsilon}{C_E} = \frac{k_{SE}\nu_W}{k_{SW}}(\varepsilon_{SW} - \varepsilon_{SW}) - \left[\frac{k_{SW}}{k_{SW}}\nu_W - \nu_E\right](\varepsilon_{SW} - \varepsilon) \quad (7.3)$$

where

$\frac{A}{C_{Sorg}} = \varepsilon \frac{A_{SE}}{C_{Sorg}} = \varepsilon_{SE} \frac{A_{SW}}{C_{Sorg}} = \varepsilon_{SW}$ and A is a measured absorbency;

For free energy (or enthalpy) gradient, where $X = \Delta G$ (at constant temperature and pressure):

$$\frac{\Delta G}{RT} = -\left(\frac{k_{SE}}{k_{SW}} \times \frac{C_E \nu_E}{C_{Worg}\nu_W}\right) \quad (7.4)$$

From the slope of the Eqn (6.25) (Chapter 6, Section 4) plots $(X_{SW} - X)/C_E$ vs $(X_{SW} - X)$ one can obtain the affinity constant ratio k_{SE}/k_{SW}:

$$\text{Slope} = \frac{k_{SE}}{k_{SW}}\nu_W - \nu_E \quad (7.5)$$

The intercepts give the hypothetical differences between the property (e.g., chemical shifts, molar extinction coefficients, etc.), measured in pure extractant and in pure water:

$$\text{Intercept} = \frac{k_{SE}}{k_{SW}}\nu_W(X_{SW} - X_{SE}) \quad (7.6)$$

Thus, the agreement between the direct experimental determination of the measured property in pure solvents and their graphically obtained values of $X_{SW} - X_{SE}$ may be examined.

Saturation factor Z, derived from Eqn (6.25) (Chapter 6, Section 3):

$$Z = \frac{k_{SE}\nu_W C_E}{k_{SW} + (k_{SE}\nu_W - k_{SW}\nu_E)C_E} \quad (7.7)$$

or, using Eqns (6.23) and (6.24) from the Chapter 6, Section 3:

$$Z = \frac{C_E\nu_E}{1 - \left(\nu_W - \frac{k_{SW}}{k_{SE}}\nu_E\right)C_{Worg}} = \frac{P_{SE}}{P_{SE} + \frac{k_{SW}}{k_{SE}}\nu_E C_{Worg}} \quad (7.8)$$

and

$$\frac{1}{Z} = 1 + \frac{k_{SW}C_{Worg}}{k_{SE}P_{SE}\nu_E} = 1 + \frac{k_{SW}C_{Worg}}{k_{SE}C_E} \quad (7.9)$$

Experimental determination of the solvation shell composition, C_{SE} and C_{SW}, and affinity constants ratio, k_{SE}/k_{SW}, permits us to find out the most convenient, in this case, nucleus aggregate. Quantitative, or even semi-quantitative, spectrometric determination of C_{SEc} and C_{SEs} (composition of the solute in the complexed and solvated forms; see Chapter 8, Section 3-2) permits us to set up an exact form of the nucleus aggregate.

Considering nuclei aggregates formation at metal extraction we reinvestigated the data of Ti(IV) extraction by DEHPA in our previous works.[17–19] At aqueous phase acidity 0.01 – 0.5 M (pH = 2.0–0.3) (see Figs. 7.4a and 7.5a the nucleus aggregate formed is: $[Ti{*}(DEHP)4{*}2H_2O]_{org}$. Here, four anionic ligands of DEHP- are bonded to titanium central atom. Two H-bonded water molecules occupy Ti(IV) ion coordination sites up to saturation in the first (inner) coordination sphere. It was intermediate colorless specie with slow kinetics of final product formation: after aging of the separated organic phase during 2–3 months, its color changed to yellow and additional water separated from organic phase.

At extraction of titanium from the strong acidic aqueous phase, 7.0–8.0 M HCl (see Figs. 7.4b and 7.5b), the nucleus aggregate formed is

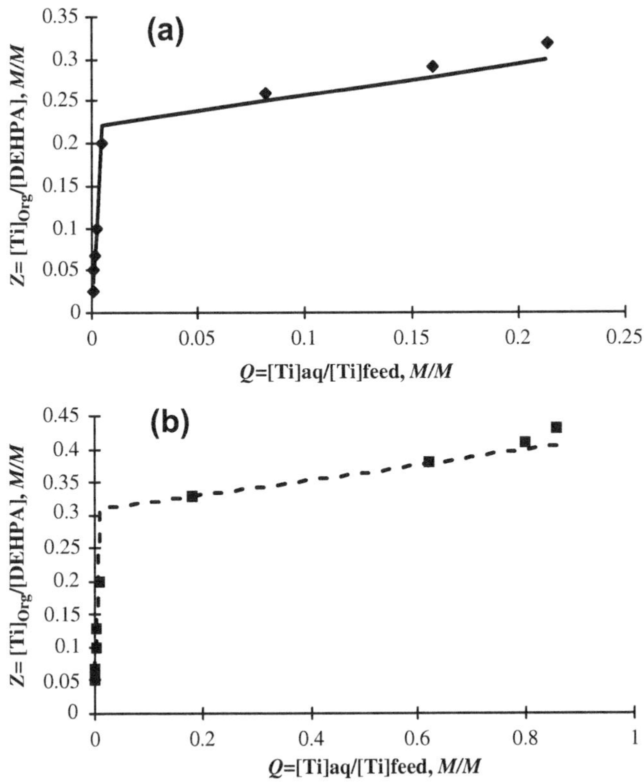

FIGURE 7.4 Comparison of experimentally obtained (points) data [17] and calculated according to the presented theory data (lines) of extraction of titanium(IV) chlorides by DEHPA in benzene from the aqueous solutions at initial acidities (HCl): (a) 0.1 mol/kg and (b) 7.6 mol/kg.

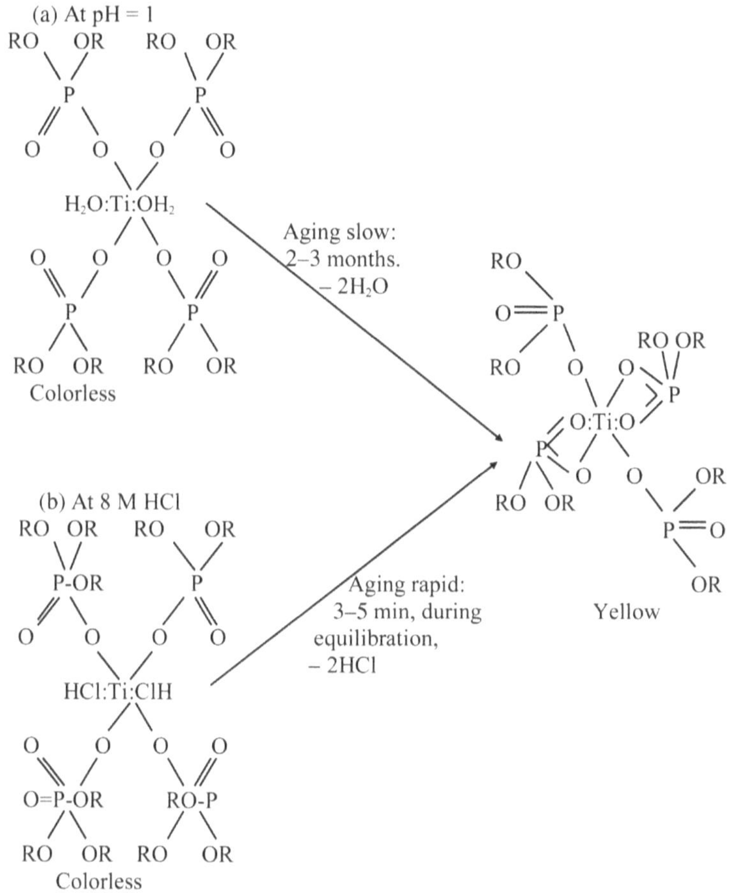

FIGURE 7.5 Structural scheme examples of nuclei aggregates formation at equilibration of titanuim chloride aqueous solutions with DEHPA in kerosene and transformation of structures after aging of the separated organic phase: at (a) pH = 1 and (b) 8 M HCl.

$[TiCl(4-x)*(DEHP)x*(4-x)HCl*(4-x)DEHPA]_{org}$, where x, equal 2, is the number of DEHP-anion ligands, bonded to titanium central atom in the first coordination sphere. It was also an intermediate specie, but kinetics of final product formation here is very fast: during equilibration, first 3–5 min the organic phase was also colorless, but then, it yellowed and did not change after aging. So, in comparison with dehydration, the rate of HCl separation and transformation of monodentate complex to polydentate was very high and finished during extraction process.[17]

More details of structure formation and transformation at extraction of some transition metals (cobalt, nickel, copper, zinc, etc.)[20] the reader will find in Chapter 10 of the book.

The same relations (3–9) may be used in analysis of metal extraction systems, using NMR, UV-VIS, or ΔG measured properties. For standard molar Gibbs free energy (or enthalpy) gradient, $X = \Delta G$, at definite metal atom coordination number n and metal ions, NM = 1, participating in a single reaction act, relation will be

$$\Delta G = -n\, RT\, \ln\left(\frac{k_{\mathrm{ME}}}{k_{\mathrm{MW}}} \times \frac{N_{\mathrm{E}}}{n - N_{\mathrm{E}}}\right) \tag{7.10}$$

N_{E} represents the number of molecules of extractant E = HL, bonded both as the charged anion L- (complexed, charge-transfer bonds) and as solvated (hydrogen, coordination bonds) neutral molecules HL. The metal central ion is coordinated by z-charged ligands L-, by $N_{\mathrm{E}} - z$ neutral HL molecules and by NW water molecules, filling metal ion coordination sites up to saturation.

Partitioning factors in this case are

$$P_{\mathrm{ME}} = N_{\mathrm{E}}/n + 1 \tag{7.11}$$

and

$$P_{\mathrm{MW}} = N_{\mathrm{W}}/n + 1 \quad \text{or} \quad P_{\mathrm{MW}} = (n - N_{\mathrm{E}})/n + 1 \tag{7.12}$$

Partitioning factors P_{MEc} and P_{MEs} may be determined separately:

$$P_{\mathrm{MEc}} = \frac{z}{n+1} \tag{7.13}$$

$$P_{\mathrm{MEs}} = \frac{(N_{\mathrm{E}} - z)}{n+1} \tag{7.14}$$

Some experimental techniques are able to detect separately the strong (complexed, charge transfer) and the weak (H-bonded, coordinate) bonds or to distinguish between them (e.g., UV–VIS spectrometry for copper), while others may monitor only the average complexed and solvated forms together (e.g., NMR spectroscopy, potentiometric titration). For example, for Cu–DEHPA–water system, measuring UV–VIS spectral shift, we can determine $k_{\mathrm{MEc}}/k_{\mathrm{MW}}$ and $k_{\mathrm{MEs}}/k_{\mathrm{MW}}$ separately. Measuring absorbency A at a given wavelengths we can distinguish between A_{c} as an absorbency of Cu in the complexed form

$$A_{\mathrm{c}} = \varepsilon_{\mathrm{MEc}} C_{\mathrm{MEc}} \tag{7.15}$$

And A_{s} as an absorbency of Cu in the solvated form

$$A_{\mathrm{s}} = \varepsilon_{\mathrm{MEs}} C_{\mathrm{MEs}} \tag{7.16}$$

Introducing $\varepsilon^{\mathrm{f}}_{\mathrm{SEc}}$ (or $\varepsilon^{\mathrm{f}}_{\mathrm{SEs}}$) as a hypothetical molar extinction coefficient of the complexing (or solvating) interaction:

$$\varepsilon^{\mathrm{f}}_{\mathrm{MEc}}(\text{or } \varepsilon^{\mathrm{f}}_{\mathrm{MEs}}) = \varepsilon_{\mathrm{MEc}}(\text{or } \varepsilon_{\mathrm{MEs}}) \frac{k_{\mathrm{MEc}}(\text{or } k_{\mathrm{MEs}})}{k_{\mathrm{ME}}} \tag{7.17}$$

we obtain

$$\frac{C_{\mathrm{Morg}}}{A_{\mathrm{c}}(\text{or } A_{\mathrm{s}})} = \left[1 - \frac{\nu_{\mathrm{E}} k_{\mathrm{MW}}}{\nu_{\mathrm{W}} k_{\mathrm{ME}}}\right] \frac{1}{\varepsilon^{\mathrm{f}}_{\mathrm{MEc}}(\text{or } \varepsilon^{\mathrm{f}}_{\mathrm{MEs}})} + \frac{k_{\mathrm{MW}}}{k_{\mathrm{ME}} \nu_{\mathrm{W}} C_{\mathrm{E}} \varepsilon^{\mathrm{f}}_{\mathrm{MEc}}(\text{or } \varepsilon^{\mathrm{f}}_{\mathrm{MEs}})} \tag{7.18}$$

or

$$\frac{C_{\mathrm{Morg}} C_{\mathrm{E}}}{A_{\mathrm{c}}(\text{or } A_{\mathrm{s}})} = \frac{k_{\mathrm{MW}}}{k_{\mathrm{ME}} \nu_{\mathrm{W}} \varepsilon^{\mathrm{f}}_{\mathrm{MEc}}(\text{or } \varepsilon^{\mathrm{f}}_{\mathrm{MEs}})} + \left[1 - \frac{\nu_{\mathrm{E}} k_{\mathrm{MW}}}{\nu_{\mathrm{W}} k_{\mathrm{ME}}}\right] \frac{C_{\mathrm{E}}}{\varepsilon^{\mathrm{f}}_{\mathrm{MEc}}(\text{or } \varepsilon^{\mathrm{f}}_{\mathrm{MEs}})} \tag{7.19}$$

3. SYSTEMS WITH MEDIUM CONCENTRATIONS OF THE SOLUTE

With increasing concentration of solute the nuclei aggregates interact, grow in size via a step-wise aggregation, and form linear or cyclic aggregates. This is region 3, in which the equivalent of solute to the equivalent of extractant ratio is $C_{\mathrm{Sorg}}/C_{\mathrm{SE}} \geq 1$.

Region 3 is characterized mainly by bridging of the nuclei aggregates through the solute, extractant, or active solvent (including water) molecules. Competition between these components influences the slope value of the curve. The interactions are driven mainly by intermolecular H-bonds. All components of the linear aggregates are also open to the bulk phase, and exchange between the aggregate and the bulk solvents continue to be fast.

An example of three types of linear aggregates, formed at extraction of acids by amines via aggregation of nuclei aggregates, are shown in Fig. 7.6: nuclei aggregates bridged by molecules of water (Schemes 1 and 2) and/or of acid (Schemes 4 and 5) and/or of amine (Scheme 3).

From the published experimental data it may be inferred as a general rule that the aggregation at the strong amine–acid interactions (ionic, ion pair) is enhanced by the same factors as those which influence the polarity of amine salts: by the strength of the acid and by the basicity of the amines. For the mineral acids the order found is $HCl < HNO_3 < HBr < HClO_4 < HHSO_4$, which is nearly exactly the order of polarity of the amine-acid salt increasing.[12,16] Most results show that aggregation comes to a stop at the stage of low oligomers, with degree within 2–6. This fact is established for the salts of secondary and tertiary amines.

For example, when hydrochloric acid is extracted from diluted aqueous solutions, where activity of water in the system is high, water is extracted to organic phase in an equimolar ratio, i.e., one water molecule to each molecule of the amine-acid ion pair formed.[12,16,21,22] At extraction from concentrated acid solutions coextraction of water decreases to very small concentrations in the organic phase. Authors[16] suggested and proved it by infra-red spectra

(1) $P_{SE} = \geq 2/5$; $P_{SW} \leq 3/5$; $C_{Sorg}/C_{SE} = 1$.

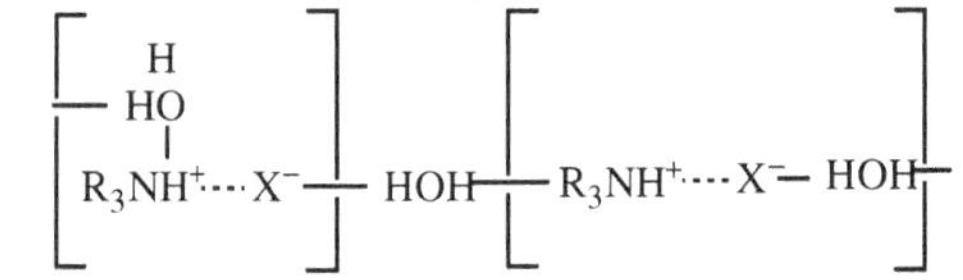

(2) $P_{SE} = \geq 4/7$; $P_{SW} = \leq 3/7$; $C_{Sorg}/C_{SE} = 1/2$.

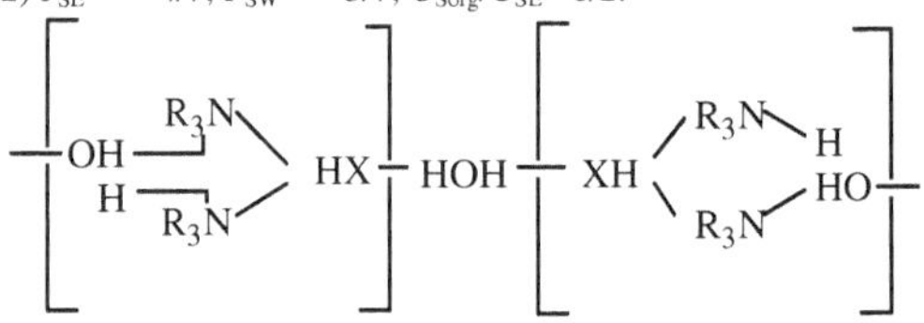

(3) $P_{SE} = \leq 2/5$; $P_{SW} \geq 3/5$; $C_{Sorg}/C_{SE} = 2/3$.

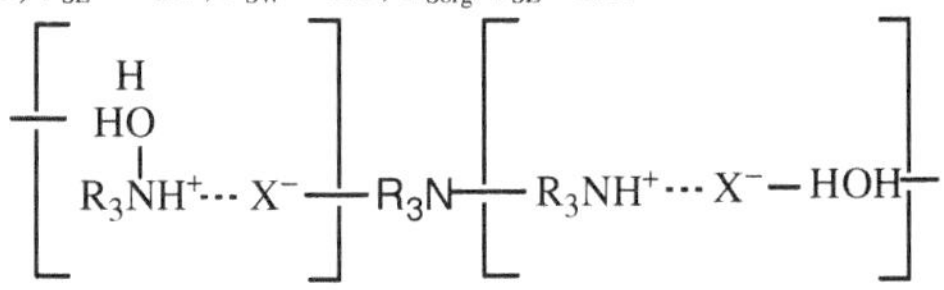

(4) $P_{SE} => 1/2$; $P_{SW} = \geq 4/7$; $C_{Sorg}/C_{SE} = 3/4$.

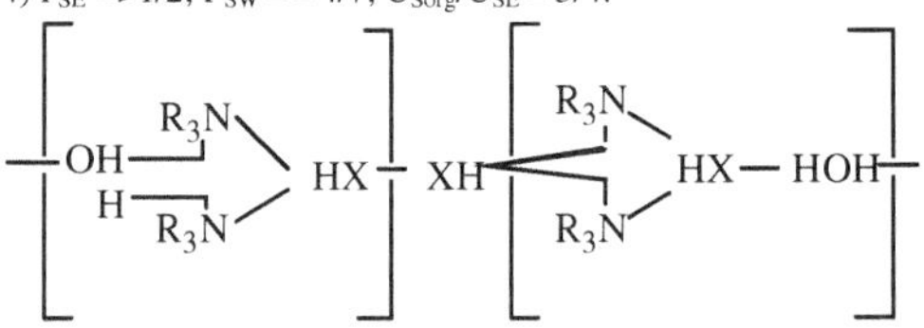

(5) $P_{SE} = \leq 3/4$; $P_{SW} \geq 1/4$; $C_{Sorg}/C_{SE} = 3/2$.

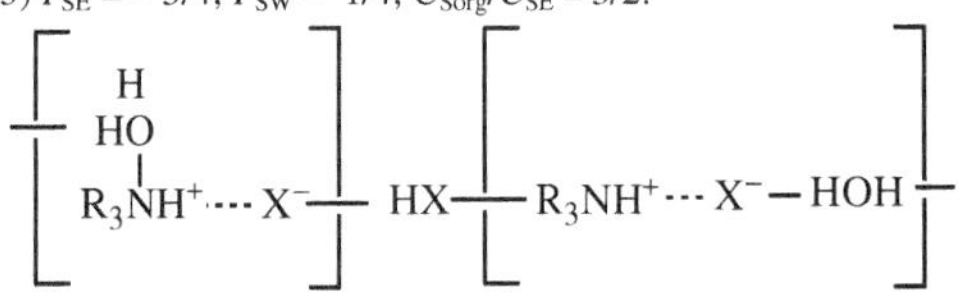

FIGURE 7.6 Some examples of the structural schemes of linear aggregates' formation in the region 3 at different molar equivalent ratios in the solvation shell.

studies that six-membered rings, high stability benzene-like cycles, are formed in all cases, no matter what is its hydration degree.

Examples of two external six-membered ring structures are shown in Fig. 7.7. Here it can be seen that from two to three nuclei aggregates of type in Fig. 7.1b are participating in the formation of the cycle, loosing molecules of water, partly up to all, at increasing acid concentration in the organic phase and decreasing activity of water. But in all (extreme and intermediate) cases the ratio $C_{Sorg}/C_{SE} = 1$.

Relatively weak anion X^- can be bonded to the N–H^+ cation by electrostatic forces as well as by hydrogen bonds through the water molecules. So, water molecules or anions X^- are closing the ring at low and middle acid concentrations, respectively.

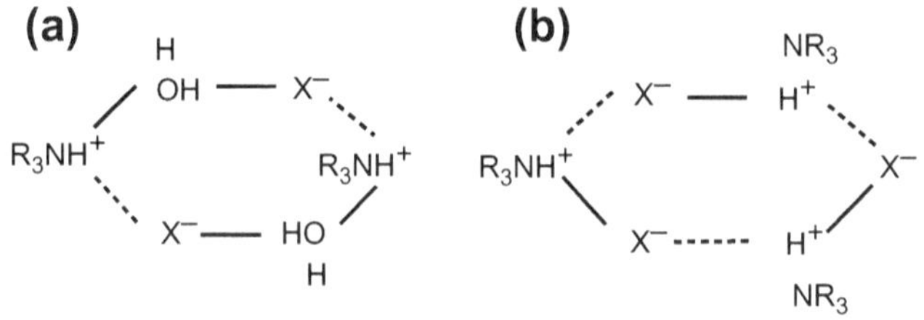

FIGURE 7.7 Example of two extreme six-membered ring structures formation at extraction of strong (mineral) acids by strong amine: (a) two nuclei aggregates bridged through water molecules, three nuclei aggregates lost water molecules and bridged through the acid anion. For both the ratio $C_{Sorg}/C_{SE} = 1$.

At extraction from high acid concentration aqueous solutions excess of acid amounts, $C_{Sorg}/C_{SE} \geq 1$, increases in parallel with water concentration in the organic phase,[8,28] especially for weaker acids (stronger base is the anion X^-). The second O–H bond of water molecule remains free, since the NH^+ group does not dispose of free electron pairs. The stronger base is the anion X^- (weaker acid), the stronger it will perturb the O–H bond in water, the stronger will be the polarization. At this stage it may be supposed that acid molecules are transferred into or formed in the organic phase the structure of $H_3O^+X^-$ ion pairs.[28] This means that the association of aggregates is going through these ion pair bridges. These ion pairs bond to the acid anion X^- of a hydrated ring and to the anionic poles of a second ring by means of its free OH group, and the association degree of linear or planar aggregate increases. It links cycles into larger units. If it is true, in the IR spectrum there should appear a new band in the region of low frequencies belonging to the hydrogen-bonded H_3O^+ cations.

For some of monocarboxylic acids the strength order is trichloroacetic acid (0.70)>monochloroacetic acid (2.86) > gluconic acid (3.70)>formic acid (3.75)>lactic acid (3.86)>acetic acid (4.76)>butanoic acid (4.85)>propionic acid (4.87), where the magnitudes of pK_a are in the parentheses.[13,23] The strength of basicity and polarization orders of the anions X^- are inversive. So, at extraction by trioctylamine, we can expect here the association of all types of nuclei aggregates through the bridges by water molecule (stronger acids) or/and by $H_3O^+X^-$ ion pairs (middle-strength acids) or/and by HX molecule (weak acids) into cyclic aggregates. More than one equivalent of acid to one equivalent of extractant, $C_{Sorg}/C_{SE} \geq 1$, is typical for these systems.

Probability of nuclei aggregates bridging through the amine molecule, as it is shown in Fig. 7.6(3), is very low. It may happen at extraction of weak acids by weak amines, where nuclei aggregates of structure in Fig. 7.2(a) may be bridged through amine and water molecules by nonspecific hydrogen bonds.[24–26]

Examples of affinity constant ratios data in the region 3 for lactic and citric acids, obtained using experimental data,[9,19] are presented in Table 7.1. Primary and secondary amines show considerably greater aggregation than tertiary amines.[21]

Similar mechanisms may be considered in region 3 describing metal extraction systems. Bridging of nuclei aggregates may occur by metal ion (or its salt), or by molecules of extractant, or by active solvent (including water). Competition between these components influences the slope value of the curve in the region 3. Experimental data, presented in Fig. 7.4 for Ti(IV) extraction by DEHPA, demonstrate correctness of the above suggestions. At extraction of titanium from the low acidity (Fig. 7.4a) aqueous phase titanium ion (or molecule) serves as a bridging component, interacting with DEHPA molecules of two (or three) nuclei aggregates. The slope or the value of the affinity constant ratio $k_{TiE}/k_{TiW} = 0.37$. At extraction from the high-acidity aqueous phase (Fig. 7.4b) bridging of nuclei aggregates is suggested mainly through the HCl molecules with formation of stable cyclic aggregates. In this case, the value of affinity constant ratio ($k_{TiE}/k_{TiW} = 0.11$) is much less. This mechanism may be compared with one, considered at extraction of strong mineral acids by tertiary amines (see Fig. 7.7b).

4. SYSTEMS WITH HIGH SOLUTE CONCENTRATIONS

At high or very high solute concentrations, upon reaching a critical size, the structural reorganization of the linear (or cyclic) aggregates occurs and supramolecular structures, reversed micelle-like, or cross-linked cluster-like, are formed.[16,27–29] Three-dimensional aggregates are typical for region 4.

Region 4 is characterized by $P_{SEs} < P_{SEc}$ and $C_{Sorg}/C_{SE} > 1$ (2–4 and may be more). Above-stoichiometric loading and massive third-phase formation are typical for this region. For example, at extraction of titanium by DEHPA[17] and citric acid by Alamine 304[11] from high concentration aqueous solutions massive formation of the third phase was observed. After aging, the separated organic phase, loaded by titanuim, was transformed into gel, loaded by citric acid, into solid polymeric substance. Interactions in this region are driven by the three-dimensional structure formation and kinetics laws (cross-linking, micellation, gelation, polymerization). As a rule, the exchange rate between the components between the solvation shell and bulk solution cannot be described by CCST equations, presented above. It depends on the orientation of the polar groups: inside or outside of the aggregate relative to the bulk phase. Components of the voluminous aggregates are mainly closed to the bulk solvents and the exchange between aggregate components and bulk solvents is controlled by diffusion kinetics.

Host–guest interaction models can be used for analysis of these shapes; guest molecules are confined by different type of interactions, in the cavities of the host system. High polarity and well-defined structures with known number of sites are very suited for the extraction and reaction purpose. Structure can change from a globular reverse micellar arrangement to a cylindrical amphoteric shape. As a result of this, the polar interior is directed toward the aqueous or organic phases.

For example, extraction by trioctylamine of monocarboxylic acids at high acid concentration aqueous solutions excess of acid amounts, $C_{Sorg}/C_{SE} > 1$, increases in parallel with water concentration in the organic phase[13,16] especially for weaker acids (stronger base is the anion X^-). As was suggested in Section 3, at this stage the acid molecules are transferred into or formed in the organic phase the structure of $H_3O^+X^-$ ion pairs.[16] This means that association of aggregates is going through these ion pair bridges. These ion pairs bond to the acid anion X^- of a hydrated ring and to the anionic poles of a second ring by means of its free OH group and the association degree increases.

Presence of the internal aqueous phase in the micelle-like aggregates has to show the strong increase of both the acid and the water (free, not associated) in organic phase. This phenomenon was proved by indirect experiment (see Fig. 7.8): samples of organic phase (Alamine 304 (trilauryl amine)) in kerosene, loaded with citric acid up to 0.5, 0.7, and 0.9 mol acid/mol amine were titrated by 0.1 M NaOH immediately after separation of aqueous phase and dilution of the samples with iso-propanol. The sudden increase of the sample acidity in the

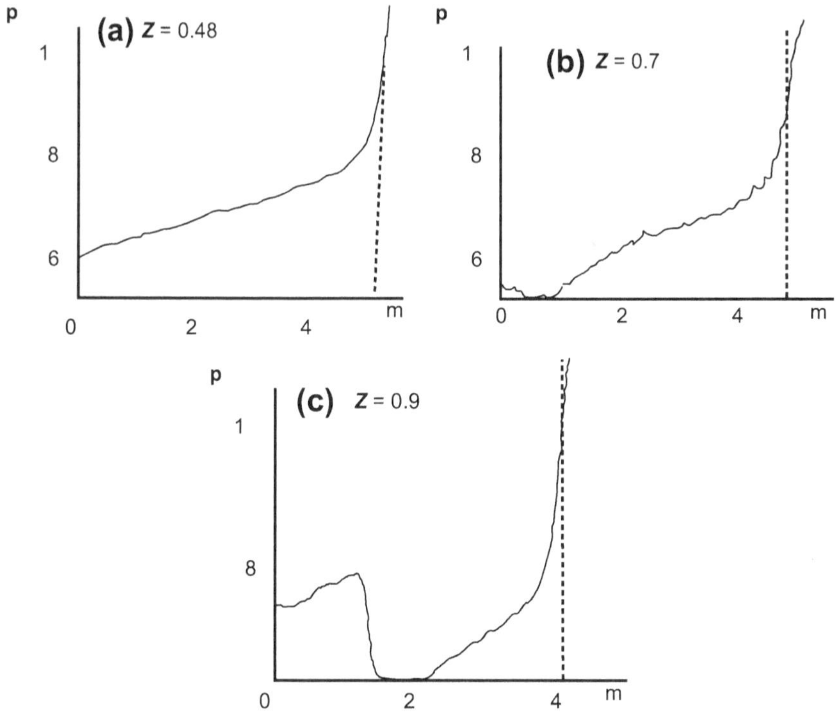

FIGURE 7.8 Titration by 0.1 M NaOH of loaded with citric acid organic phase samples, immediately after their dilution with iso-propanol, using autotitrator Titralab 90. Extractant: 1 mol/kg alamine 304 (trilauryl amine) in kerosene (isopar K). (a) Z (Cit. ac./A 304) = 0.48 M/M, initial pH = 5.838; (b) Z = 0.7 (M/M), initial pH = 6.327; (c) Z = 0.9 (M/M), initial pH = 6.953.

middle of titration of high loaded (0.9 M/M) samples witnesses destraction of the closed cavities (closed reversed micelles), filled with aqueous solution.

In the case of cluster-like aggregates formation, the concentration of water constituent in the solvation shells may increase, be the same or even decrease compared with region 3.

The rates of the three-dimensional aggregates formation and destruction may differ very much, especially in the case of closed micelle-like aggregates formed. So, the researchers, which investigate the extraction systems by dilution of very high loaded by solute organic phase,[12] have to be sure that they come to the equilibrium state.

5. COMPARISON OF CONVENTIONAL AND PRESENTED THEORIES DESCRIPTION

Comparing Eqn (7.10) of the CCST with Eqns (6.1)–(6.3) in Chapter 6 we obtain

$$\overline{K}_{SE} = \frac{k_{SE}}{k_{SW}} \times \frac{N_E}{n - N_E} \tag{7.20}$$

where N_E is a number of extractant molecules participating in a single reaction act.

Comparing Eqns (7.2) and (7.3) of the presented theory with Scatchard–Deranleau Eqns (7.11) and (7.12), presented in Chapter 6 we obtain

$$\overline{K}_{SE} = \frac{k_{SE}}{k_{SW}} \nu_W - \nu_E \tag{7.21}$$

As can be seen, the right parts of Eqns (7.5) and (7.21) are identical. So, the slope is identical to the average equilibrium constant of the classical theories.

One can see that the classical equilibrium constant parameter $\overline{K}_{SE}$ represents, in fact, an affinity constant ratio k_{SE}/k_{SW} when the active solvent (here, water in organic phase) is included in the thermodynamic treatment.

Using the classical theories, it is impossible to explain why equilibrium constant $K_{ex} = 0$ or negative without employing some *ad hoc* arguments for each individual case. The presented theory handles this problem in a straightforward way. According to Eqns (7.5) and (7.21), the slope can be negative (normal case), positive, or equal zero depending on the relative magnitude of components $\frac{k_{Si}}{k_{Sj}} \nu_j$ and ν_i.

The presented theory explains the dual maximum behavior (see Chapter 6) using coordination model[8,30] and the acid-base concept[1,2] with amphoteric properties of solvents.

The presented theory considers metal ions in solutions as bonded with charged ligands and neutral molecules in the solvation shells, depending on their oxidation state and solvation (coordination) number. These cationic, neutral, or anionic species may be classed as both conjugate acids and bases, depending on properties and concentrations of ion groups and molecules bonded.

Metal salts, dissolved in water, hydrolyzed, and coordinated by water molecules (hydrated), or dissolved in strong acids, coordinated by acid molecules (solvated), can be conjugate base or acid, respectively, to the same extractant.[5,13,21,31] Example with Ti(IV) was presented above, in Section 3.

These considerations are valid for most of the metals and extractants (acidic, basic, neutral) presented in Table 6.1 of Chapter 6. They explain the universality of dual behavior of different extractants and prove their amphoterity, depending on the composition of the solute in the aqueous phase and its acidity.

6. SUMMARIZING REMARKS FOR THE CCST

In the chemical modeling approach,[9,10,12–14,16,21,23,32,33] the authors consider the system as a static, where one of the mechanisms (ion pair, anion exchange, H-bonding) is a dominating one and is not changed at changing concentrations of solute. They have developed the mathematical models in which hypothetical complexes of a specific stoichiometry are formed. This approach is the useful tool to describe the data quantitatively, if complexation is strong, as in the region 2. The presented theory overcomes some limitations of chemical modeling approach. Here, different regions are analyzing separately and separate mathematical descriptions are developed for different regions of the extraction system. It shows that the contradictions between King,[13–15,34] Yang,[32] and Eyal[12,23] groups in the interpretation of experimental data are only seeming ones: really, they are analyzing different regions, in which the interaction mechanisms are different.

The theory introduces an active solvent (including water) as a quantitative parameter, which is participating and influencing the formation of different compositions of the aggregates (solvation shells) at changing of solute concentration in the organic phase.

Only relative values (affinity constant ratios) can be measured in solution because of the ubiquitous nature of molecular interactions. Affinity constant ratio's value of unity suggests the same values of solvation effects of the solute with extractant in the organic phase and with water in the aqueous phase; a large value of k_{SE}/k_{SW} means strong or very strong complexation effects with extractant. A positive sign of a slope means that the solvation of the reactant molecules is stronger than the product molecules. The negative slope shows that the solvation of product molecule is stronger than the reactant's.

The linearity of the Eqn (6.19) plots (see Chapter 6) over the whole concentration range means that the same affinity constant ratio is valid at low and high solute concentrations and does not depend on its concentration. Extraction of some carboxylic acids by alcohols may serve as examples of such systems (e.g., the reader is referred to Nagy,[35–38] Laub,[39] and Purnel.[40] Interchanges in the linearity of Eqn (6.19, Chapter 6), or the different slopes in the loading curve over the solute concentration range mean that the affinity constant ratios and/or concentrations of the components in the solvation shell are different. It means, also, that the different charge-transfer complexation mechanisms, and consequently the different values of k_{SEc}, or the different solvation mechanisms and consequently different values of k_{SEs}, take place at different solute concentration. So, we have to determine either k_{SEc}/k_{SEs}.

Regions 1–4 in the general distribution curve (Chapter 6, Fig. 6.1) with different values of the slopes testify different compositions of the aggregates formed and different influence of the constituents of extraction system on the interaction mechanisms over the solute concentration range. It is evident that in this case we will obtain different values of the hypothetical chemical shifts in pure solvents (water). This analysis is useful when studying polybasic solutes or extractants, and especially when studying mixtures of extractants, such as acid–amine or amine–amine mixtures.

REFERENCES

1. Kislik V. Competitive complexation/solvation theory of solvent extraction. II. Solvent extraction of metals by acidic extractants. *Sep Sci Technol* 2002;**37**(11):2623–57.
2. Kislik V. Application of preferential solvation concept for interpretation of the mechanisms of acids solvent extraction by amines. *Sep Sci Technol* 2001;**36**(13):2899–926.
3. Kislik V, Eyal A. Acidity dependence of Ti(IV) extraction: a critical analysis. *Solvent Extr Ion Exch* 1993;**11**(2):258.
4. Kislik V. Application of Preferential solvation model for Interpretation of the Mechanisms of Solvent Extraction. In: *Proceedings of the XVth International Symposium on Physico-chemical Methods of Mixtures Separation*. Poland: ARS SEPARATORIA 2000; 2000. p. 34–40.
5. Marcus Y, Kertes AS. *Ion exchange and solvent extration of metal complexes*. London: Wiley-Interscience; 1969.
6. Komasawa I, Otake T, Hattori I. The effects of diluent in the liquid-liquid extraction of copper and nickel using 2-hydroxy-5-nonylbenzophenone oxime. *J Chem Eng Japan* 1983; **16**(5):377–88.
7. Kolarik Z, Grimm R. Acidic organophosphorus extractants. XXIII. Distribution of mono-2-ethylhexyl phosphoric acid in two liquid phase systems. *J Inorg Nucl Chem* 1976; **38**(5):1761–3.
8. Sekine T, Hasegawa Y. *Solvent extraction chemistry*. New York: Marcel Dekker Inc; 1977.
9. Tamada J, King C. *Extraction of Carboxylic Acids by Amine Extractants*. Ph.D. Thesis. Lawrence Berkeley Laboratory, University of California, Applied Science Division; 1989.
10. Vieux AS. Extraction of some dicarboxylic acids by tri-isooctylamine. *Anal Chim Acta* 1974;**68**:415–24.
11. Kislik V, Eyal A. unpublished experimental data, 1998.
12. Eyal AM, Canari R. pH dependence of carboxylic and mineral acid extraction by amine-based extractants: effects of pKa, amine basicity, and diluent properties. *Ind Eng Chem Res* 1995;**34**(5):1789–98.

13. Tamada J, Kertes A, King C. Extraction of carboxylic acids with amine extractant. 3. Effect of temperature, water coextraction and process considerations. *Ind Eng Chem Res* 1990;**29**:1333–41.
14. Tamada J, Kertes A, King C. Extraction of carboxylic acids with amine extractant. 1. Equilibria and law of mass action modeling. *Ind Eng Chem Res* 1990;**29**:1319–26.
15. Tamada J, Kertes A, King C. Extraction of carboxylic acids with amine extractant. 2. Chemical interaction and interpretation of data. *Ind Eng Chem Res* 1990;**29**:1327–32.
16. Wardel JM, King CJ. Solvent equilibria for extraction of carboxylic acids from water. *J Chem Eng Data* 1979;**23**(2):144–8.
17. Kislik V, Eyal A. Extraction of titanium(IV) by mixtures of mono- and d-(2-ethylhexyl) phosphoric acids esters. *Solvent Extr Ion Exch* 1993;**11**(2):285–310.
18. Kislik V, Eyal A. Mechanisms of titanium (IV) and some other transition metals extraction by organophosphorus acids. *Proceedings of ISEC'93* 1993;vol. 2:1353–60.
19. Kislik V, Eyal A. Hybrid liquid membrane (HLM) and supported liquid membrane (SLM) based transport of titanium (IV). *J Mem Sci* 1996;**111**:273–81.
20. Neuman RD, Jons MA, Zhou NF. Photo-correlation spectroscopy applied to hydrometallurgical solvent-extraction systems. *Colloids and Surfaces* 1990;**46**(1):45–61.
21. Grinstead RR. Base strengths of amine in liquid-liquid extraction systems. *Proceedings of ASEC'67* 1968;427–36.
22. Dotre JC, Teixeira J, editors. *Hydrogen-bonded liquids*. The Netherlands: Kluwer; 1991.
23. Canari R, Hazan B, Eyal A. Carboxylic acids extraction by basic extractants: the effects of pH, ion concentration and acid-base property of the extractant and extracted acid. *Proceedings of ASEC'96* 1996;**2**:1517–22.
24. Headley AD. Solvent effects on the basicity of alkyl-substituted dimethylamines. *J Org Chem* 1988;**53**:312–4.
25. Headley AD, McMyrry ME, Starnes SD. Effects of substituents on the acidity of acetic acid. *J Org Chem* 1994;**59**:1863–6.
26. Headley AD. Quantitative analysis of solvation effects and influence of alkyl substituents on basicity of amines. *J Org Chem* 1991;**56**:3688–91.
27. Spala EE, Ricker NL. Thermodynamic model for solvating solutions with physical interactions. *Ind Eng Chem Process Des Dev* 1982;**21**:409–15.
28. Fendler EJ. Proton magnetic resonance investigation of alkylammonium carboxylate micelles in nonaqueous solvents. *J Phys Chem* 1973;**77**(11):1432–6.
29. Toda F, Hyoda S, Okada K, Hirtsu R. Isolation of anhydrous hydrasine as stable inclusion complexes with hydroquinone and p-methoxyphenol and their solide state reaction with esters which gives pure hydrolyzides. *J Chem Soc Chem Commun* 1995;1531–2.
30. Kislik V, Eyal A. Heavy metals removal from wastewaters of phosphoric acid production: a comparison of hybrid liquid membrane (HLM) and aqueous hybrid liquid membrane (AHLM) technologies. In: *Proceedings of the conference on membranes in drinking and industrial*, vol. 1. Paris, France: Water Production; 2000. p. 503–514.
31. Deranleau DA. Theory of the measurement of weak molecular complexes. I. General considerations. *J Amer Chem Soc* 1969;**91**:4044–9.
32. Yang S, White S, Hsu S. Extraction of carboxylic acids with tertiary and quaternary amines: Effect of pH. *Eng Chem Res* 1991;**30**:1335–48.
33. Mei Wen, Yiyan Yang, Youyuan Dai. Studies on the mechanism of extraction of carboxylic acids with tri-N-octylamine by chemical complexation. *Proceedings ISEC'96* 1996;**1**: 195–200.
34. Hogfeldt E, Fredlund F, Odberg L, Merenge G. Extraction of water, nitric and perchloric acids by nitrobenzene. *Acta Chemica Scand* 1973;**18**(5):1781–94.

35. Nagy OB, Muanda H, Nagy JB. Competitive preferential solvation theory of weak molecular interactions. *J Chem Soc Faradey Trans1* 1978;**74**:2210–29.
36. Szpakowska M, Nagy OB. Application of the preferential solvation theory to coordinative solute-solvent interactions. *J Chem Soc, Faradey Trans1* 1989;**85**(9):2891–9.
37. Nagy OB, Muanda M, Nagy JB. Quantitative evaluation of solute-solvent interactions. Environmental effects on the n-butylaminolysis of tetrachloro-n-n-butylphthalimide. *J Phys Chem* 1979;**83**(15):1961–79.
38. Parbhoo B, Nagy O. Application of the competitive preferential solvation theory to ion-molecule interactions. *J Chem Soc Faradey Trans 1* 1986;**82**:1789–93.
39. Laub RJ, Purnell JH. Solution and complexing Studies. III. Futher evidence for a microscopic partitioning theory of solution. *J Amer Chem Soc* 1976;**98**(1):30–5.
40. Purnell JH, Vargas de Andrade. Solution and complexing studies. I.Gas-liquid chromatografic investigation of supposed complexing systems. II. Comparison and correlation of nuclear magnetic resonance and gas-liquid chromatografic data. *J Amer Chem Soc* 1975;**97**:3585–90.

Chapter 8

CCST in Engineering Design, Procedures, and Calculations

Chapter Outline

1. INTRODUCTION

When the classical theories are experimentally tested, water in the initial aqueous solution of solute-in-water, S, is replaced by extractant E, to arrive, finally, to the solution of solute-in-pure E. The corresponding variation of the physicochemical property of S constitutes a substitution profile, the linearization of which makes it possible to determine the thermodynamic and spectroscopic properties of S–E–W interactions and distributions. As a rule, these theories[1–5] do not distinguish quantitatively between complexation and solvation in the S–E–W interactions. Also, they do not predict quantitatively the influence of different active solvents (enhancers, modifiers, synergistic agents, mixed extractants) and diluents.

The presented approach takes into account the influence of different solvents (including water extracted) and gives a key for quantitative analysis and preliminary prediction of suitable extraction systems.

Solvent Extraction. DOI: 10.1016/B978-0-444-53778-2.10008-1

2. ENGINEERING CONSIDERATIONS IN EXPERIMENTAL INVESTIGATION OF CCST

As mentioned in the previous chapters, a composition of the solvation shell in a mixture depends on affinity constant k_i and on the number of potentially available solvent (extractant, active solvent, water) molecules present in the mixture, i.e., on the actual concentration of every component. The greater the factor $k_{SE}C_E$ (or $k_{SW}C_W$, or $k_{SA}C_A$) the better the component solvates solute. Competition between extractant, active solvent, and water molecules to interact with the solute molecules in the solvation shell determines the distribution curve slope. According to the theory, the solute–solvent interactions are different for every solvent and result in different affinity constant ratios k_{SE}/k_{SW}, k_{SE}/k_{SA}, k_{SA}/k_{SW}, k_{EA}/k_{AW}, and concentration ratios C_{SE}/C_{SW}, C_{SA}/C_{SW}, C_{SE}/C_{SA}, C_{EA}/C_{AW}, which have to be determined. Preliminary data of the extraction compositions can be calculated using the results of three component systems. Hence, the experimental studies simplify to the series of measurements for the S–E–A, S–A–W, and E–A–W systems, analogically to S–E–W system (for details see Chapter 7).

Let us consider a S–E–W extraction system in which solute ion, S, N_S-molecules–Extractant, E, N_E-molecules–Water, W, N_W-molecules participate in a single reaction act. Here, z is the oxidation state of solute ion.

We are repeating the basic equations of the CCST presented in the previous chapters:

$$\frac{X_{SW} - X}{C_E} = \frac{k_{SE}}{k_{SW}}\nu_W(X_{SW} - X_{SE}) - \left[\frac{k_{SE}}{k_{SW}}\nu_W - \nu_E\right](X_{SW} - X) \tag{8.1}$$

where X is the value of the measured property.

Saturation factor Z:

$$Z = \frac{C_E\nu_E}{1 - \left(\nu_W - \dfrac{k_{SW}}{k_{SE}}\nu_E\right)C_{Worg}} = \frac{P_{SE}}{P_{SE} + \dfrac{k_{SW}}{k_{SE}}\nu_E C_{Worg}} \tag{8.2}$$

and

$$\frac{1}{Z} = 1 + \frac{k_{SW}C_{Worg}}{k_{SE}P_{SE}\nu_E} = 1 + \frac{k_{SW}C_{Worg}}{k_{SE}C_E} \tag{8.3}$$

Affinity constant ratio k_{SE}/k_{SW} is determined from the slope of Eqn (8.1) plots: $(X_{SW} - X)/C_{Aorg}$ versus $(X_{SW} - X)$. The intercept gives the hypothetical difference between the property, measured in pure extractant and in pure active solvent. Determination of the affinity constant ratio k_{SE}/k_{SW} may be realized by conventional equilibrium extraction experiment series. Several physical properties of complexes, e.g., excess of dipole moment ($\Delta\mu$) change of proton, carbon-13, nitrogen-15 chemical shifts ($\Delta\delta$), frequency shift ($\Delta\nu$), when plotted against ΔpK_a, pK_a, pH, PA (proton affinity), or ΔH, give titration curve relations. As a rule, these techniques do not distinguish between intermediate reactions, especially rapid, and show summarized results.

For NMR, where $X = \delta$ is the chemical shift:

$$\frac{\delta_{SW} - \delta}{C_E} = \frac{k_{SE} v_{Worg}}{k_{SW}}(\delta_{SW} - \delta_{SE}) - \left[\frac{k_{SE}}{k_{SW}} v_{Worg} - v_E\right](\delta_{SW} - \delta) \quad (8.4)$$

For UV–VIS, where $X = \varepsilon$ is the molar extinction coefficient:

$$\frac{\varepsilon_{SW} - \varepsilon}{C_E} = \frac{k_{SE} v_W}{k_{SW}}(\varepsilon_{SW} - \varepsilon_{SE}) - \left[\frac{k_{SE}}{k_{SW}} v_W - v_E\right](\varepsilon_{SW} - \varepsilon) \quad (8.5)$$

where

$\frac{A}{C_{Sorg}} = \varepsilon \frac{A_{SE}}{C_{Sorg}} = \varepsilon_{SE} \frac{A_{SW}}{C_{Sorg}} = \varepsilon_{SW}$ and A is a measured absorbency;

For free energy (or enthalpy) gradient, where $X = \Delta G$ (at constant temperature and pressure):

$$\frac{\Delta G}{RT} = -\ln\left(\frac{k_{SE}}{k_{SW}} \times \frac{C_E v_E}{C_{Worg} v_W}\right) \quad (8.6)$$

or for the metals:

$$\Delta G = -n\, RT \ln\left(\frac{k_{ME}}{k_{MW}} \times \frac{N_E}{n - N_E}\right) \quad (8.7)$$

where N_E represents the number of molecules of extractant E = HL, bonded both as the charged anion L^- (complexed, charge-transfer bonds) and as the solvated (hydrogen, coordination bonds) neutral molecules HL. The metal central ion is coordinated by z-charged ligands L^-, by $N_E - z$ neutral HL molecules, and by N_W water molecules, filling metal ion coordination sites up to saturation.

Partitioning factors in this case are

$$P_{ME} = \frac{N_E}{n + 1} \quad (8.8)$$

and

$$P_{MW} = \frac{N_W}{n + 1} \quad \text{or} \quad P_{MW} = \frac{(n - N_E)}{n + 1} \quad (8.9)$$

Partitioning factors P_{MEc} and P_{MEs} may be determined separately:

$$P_{MEc} = \frac{z}{n + 1} \quad (8.10)$$

$$P_{MEs} = \frac{N_E - z}{n + 1} \quad (8.11)$$

From the slope of the Eqn (8.1) (Chapter 6, Section 4) plots $(X_{SW} - X)/C_E$ versus $(X_{SW} - X)$ one can obtain the affinity constant ratio k_{SE}/k_{SW}:

$$\text{Slope} = \frac{k_{SE}}{k_{SW}} \nu_W - \nu_E \tag{8.12}$$

The intercepts give the hypothetical differences between the property (e.g., chemical shifts, molar extinction coefficients), measured in pure extractant and in pure water:

$$\text{Intercept} = \frac{k_{SE}}{k_{SW}} \nu_W (X_{SW} - X_{SE}) \tag{8.13}$$

Thus, the agreement between the direct experimental determination of the measured property in pure solvents and their graphically obtained values of $X_{SW} - X_{SE}$ may be examined.

3. EXPERIMENTAL TECHNIQUES FOR THE PRESENTED THEORY

Experimental determination of the solvation shell composition, C_{SE} and C_{SW}, and affinity constants ratio, k_{SE}/k_{SW}, permit us to find out the most convenient, in this case, nucleus aggregate. Quantitative, or even semi-quantitative, spectrometric determination of C_{SEc} and C_{SEs} (composition of the solute in the complexed and solvated forms (see Chapter 6, Section 4)) permits us to set up an exact form of the nucleus aggregate.

In many cases, studying interaction structures, chemical relaxation techniques,[6] dynamic NMR spectroscopy, ultrasonic methods,[7] IR, UV, dielectric methods are useful. Mass spectroscopic methods permit precise determination of proton affinities and equilibrium constants. As a rule, some techniques monitor only the average complexed and solvated forms together (e.g., NMR spectroscopy, potentiometric titration).

Some experimental techniques are able to detect separately the strong (complexed, charge transfer) and the weak (H-bonded, coordinate) bonds or to distinguish between them. For example, for Cu–DEHPA–water system, measuring UV-VIS spectral shift, we can determine k_{MEc}/k_{MW} and k_{MEs}/k_{MW} by Eqn (8.3) (for details see Chapter 7). Measuring absorbency A at a given wavelength we can distinguish between the complexed and the solvated forms.

For the complexed form at the absorbencies A_c

$$A_c = \varepsilon_{MEc} C_{MEc} \tag{8.14}$$

For the solvated form at the absorbencies A_s

$$A_s = \varepsilon_{MEs} C_{MEs} \tag{8.15}$$

Introducing ε^{f}_{SEc} (or ε^{f}_{SEs}) as a hypothetical molar extinction coefficient of the complexing (or solvating) interaction:

$$\varepsilon^{f}_{MEc}(\text{or }\varepsilon^{f}_{MEs}) = \varepsilon_{MEc}(\text{or }\varepsilon_{MEs})\frac{k_{MEc}(\text{or }k_{MEs})}{k_{ME}} \tag{8.16}$$

we obtain

$$\frac{C_{Morg}}{A_c(\text{or }A_s)} = \left[1 - \frac{\nu_E k_{MW}}{\nu_W k_{ME}}\right]\frac{1}{\varepsilon^{f}_{MEc}(\text{or }\varepsilon^{f}_{MEs})} + \frac{k_{MW}}{k_{ME}\nu_W C_E \varepsilon^{f}_{MEc}(\text{or }\varepsilon^{f}_{MEs})} \tag{8.17}$$

or

$$\frac{C_{Morg}C_E}{A_c(\text{or }A_s)} = \frac{k_{MW}}{k_{ME}\nu_W \varepsilon^{f}_{MEc}(\text{or }\varepsilon^{f}_{MEs})} + \left[1 - \frac{\nu_E k_{MW}}{\nu_W k_{ME}}\right]\frac{C_E}{\varepsilon^{f}_{MEc}(\text{or }\varepsilon^{f}_{MEs})} \tag{8.18}$$

The molar extinction coefficient (or chemical shift, or free energy change) of the solute in pure solvents (in this case water), ε_{SW}, and the variation of the apparent molar extinction coefficient, ε, are measured as the water molecules are stripped off stepwise from the solvation shell of the solute and replaced by molecules of the extractant E, throughout the whole concentration range, until reaching the corresponding value in pure extractant, ε_{SE}. The affinity constant ratios are determined by the slope of the plot ($\varepsilon_{SW} - \varepsilon$) versus ($\varepsilon_{SW} - \varepsilon_{SE}$)/$C_E$. The intercept of the plot gives the hypothetical molar extinction coefficient (or chemical shift in NMR measurements) of the solute in pure water or in pure extractant. Thus, the agreement between the direct experimental determination of the chemical shifts in pure solvents and their graphically obtained values at $\varepsilon_{SW} - \varepsilon = 0$ or $\varepsilon_{SE} - \varepsilon = 0$ can be compared.

The changes in the linearity of the plots (intersection points), or the different slopes in the loading curve (Fig. 6.1 in Chapter 6) over the concentration range, mean that different complexation mechanisms, or different solvation mechanisms, take place at different solute concentrations. It is evident that in this case we will obtain different values of the hypothetical chemical shifts in pure water. It means that we have different compositions of the solvation shell, depending on solute concentration. In this case we have to use one more experimentally measured property, such as UV-VIS or quantitative IR measurements, to determine either k_{SEc} or k_{SEs}, or both (see Chapter 9). This analysis is useful when studying polyvalent metals or polybasic extractants, and especially when studying mixtures of extractants.

A serious limitation for studying acid–base interactions is a lack of definitive and general techniques for proving the structures of the product species. Very often the species are postulated in order to fit the experimental data with a given model. Most transition metals, post-transition and some main group elements, have well-defined solvation, or coordination numbers in solution: the metal ion is coordinated by a definite number, *n*, of charged ligands and neutral molecules with a well-defined geometry. However, in the mixture of solvents the competition between constituents for coordination site takes place according to their affinity constants and concentrations; thus, the contribution of each of them in a definite coordination number cannot be well defined. Many other cations, such as alkali metals, most organic cations, and especially anions, form diffusive, ill-defined solvation shells. And problems arise with mixed extractants, diluents, impurities (e.g., coextracted water), ions, which are generally neglected in spectroscopic investigations.

Any necessary chemical or physicochemical property of the solute in pure solvents (extractant, water, synergistic agent) and their mixtures are measured directly. For example, preliminary data, obtained for the acids and metal extraction systems, used in our experiments (see Chapters 9 and 10) are presented in Tables 8.1 and 8.2.

TABLE 8.1 Some Physico-Chemical Properties of the Initial Constituents of Citric Acid-Alamine 304-Water-Active Solvent (Alcohols) Extraction Systems

Components	Molecular weight of solvent (g/mol)	Density (g/cm^3)	Molar volume, v (dm^3/mol)
Alamine 304	541		
1.0 mol/kg Alamine 304 in kerosene		0.760	0.711
1-Octanol	130.23		
1.0 mol/kg Alamine 304 + 0.40 mol/kg 1-Octanol		0.778	0.167
1.0 mol/kg Alamine 304 + 0.77 mol/kg 1-Octanol		0.792	0.164
Citric acid	192.14		
0.78 mol/kg citric acid aqueous solution		1.070	0.180
1.56 mol/kg citric acid aqueous solution		1.122	0.171
1-Propanol	60.1		
1.56 mol/kg citric acid + 0.95 mol/kg 1-Propanol		1.109	0.054

TABLE 8.2 Some Physico-Chemical Properties of Initial Constituents of Acid–Amine–Water and Acid–Amine–Active Solvent–Water Extraction Systems

Components	Molecular weight of solvent (g/mol)	Density (g/cm^3)	Molar volume, v (dm^3/mol)
Alamine 304	541	0.851	0.637
Alamine 336	392	0.800	0.490
Acetic acid	60.05	1.049	0.057
Lactic acid	90.08	1.249	0.072
Citric acid	192.14	1.542	0.125
Chloroform	119.37	1.498	0.080
Nitrobenzene	123.12	1.199	0.103
1-Octanol	130.23	0.825	0.158
1-Propanol	60.1	0.804	0.075
1.0 mol/kg Alamine 304 in kerosene		0.760	0.711
1.0 mol/kg Alamine 304 + 0.40 mol/kg 1-Octanol		0.778	0.167
1.0 mol/kg Alamine 304 + 0.77 mol/kg 1-Octanol		0.792	0.164
0.78 mol/kg citric acid aqueous solution		1.070	0.180
1.56 mol/kg citric acid aqueous solution		1.122	0.171
1.56 mol/kg citric acid + 0.95 mol/kg 1-Propanol		1.109	0.154

4. DETERMINATION OF EXTRACTION CONSTANT AND ITS COMPARISON WITH CCST AFFINITY CONSTANT RATIOS

Most transition metals, post-transition and some main group elements, have well-defined solvation, or coordination numbers in solution (see Chapters 6 and 7): the metal ion is coordinated by a definite number, n, of charged ligands and neutral molecules with a well-defined geometry. However, in the mixture of solvents (i.e., extractant and water) the competition between constituents for coordination site takes place according to their affinity constants and concentrations; thus, the contribution of each of them in a definite coordination number cannot be well defined. Many other cations, such as alkali metal cations and most organic cations form diffusive, ill-defined solvation shells.

If solvates of definite geometry and coordination number, n, are formed, their stability constants can generally be estimated, i.e., the equilibrium constants for the replacement of some solvent (in our case, water molecules, W) in the solvate $M(W)^{Z+}$ by another solvent (extractant molecules, E) to form the solvated ion $M(E)^{Z+}$.[8] This replacement reaction is generally studied in mixed solvents, but the values may be extrapolated to apply to the pure solvents. Grunwald,[9] Covington,[10] Cox,[11] and Marcus[12,13] derived equations relating the equilibrium constants for solvent replacement to standard molar Gibbs free energy of ion M transfer from a solvent (W) to its mixture with another solvent (W + E). According to Cox:[11]

$$\Delta G_t^{\infty}(\mathrm{M, W \rightarrow E}) = -n\,RT\ln \overline{K}_{\mathrm{ME}} \tag{8.19}$$

where the $\overline{K}_{\mathrm{ME}}$ is the average equilibrium constant for W → E solvents replacement and n is the solvation (coordination) number.

According to Scatchard–Deranleau equation[14–16] for UV-VIS (ε) (if all metal species obey Beer's law) or NMR (δ) measurements we obtain equations for extraction constant:

$$\frac{\varepsilon_{\mathrm{ME}} - \varepsilon}{C_{\mathrm{E}}} = \overline{K}_{\mathrm{ME}}(\varepsilon_{\mathrm{ME}} - \varepsilon_{\mathrm{MW}}) - \overline{K}_{\mathrm{ME}}(\varepsilon_{\mathrm{ME}} - \varepsilon) \tag{8.20}$$

or

$$\frac{\varepsilon_{\mathrm{ME}} - \varepsilon}{\varepsilon_{\mathrm{ME}} - \varepsilon_{\mathrm{MW}}} = \frac{C_{\mathrm{ME}}}{C_{\mathrm{M}}^{0}} = \frac{\overline{K}_{\mathrm{ME}} C_{\mathrm{E}}}{1 + \overline{K}_{\mathrm{ME}} C_{\mathrm{E}}} \quad \text{at } 0 < z < 1 \tag{8.21}$$

Comparing the Eqns (8.20) and (8.21) for extraction constant of classical theories with the Eqns (8.4) and (8.5) for affinity constant ratios of the presented theory we obtain

$$\overline{K}_{\mathrm{ME}} = \frac{k_{\mathrm{ME}}}{k_{\mathrm{MW}}} \times \frac{N_{\mathrm{E}}}{n - N_{\mathrm{E}}} \tag{8.22}$$

$$\overline{K}_{\mathrm{MEc}} = \frac{k_{\mathrm{MEc}}}{k_{\mathrm{MW}}} \times \frac{z}{n - N_{\mathrm{E}}} \tag{8.23}$$

$$\overline{K}_{\mathrm{MEs}} = \frac{k_{\mathrm{MEs}}}{k_{\mathrm{MW}}} \times \frac{N_{\mathrm{E}} - Z}{n - N_{\mathrm{E}}} \tag{8.24}$$

where

$$\overline{K}_{\mathrm{ME}} = \overline{K}_{\mathrm{MEc}} \times \overline{K}_{\mathrm{MEs}} \tag{8.25}$$

The correctness of these relations in Eqns (8.22)–(8.24) between the averaged extraction parameter $\overline{K}_{ME}$ of the classical theories and affinity constant ratios, k_{ME}/k_{MW}, of the CCST, obtained by the measurements of the same property: chemical shifts, extinction coefficients, potentiometric titrations, etc., is one of the main tasks at the experimental verification of the presented model.

Using the classical theories, it is impossible to explain why equilibrium constant K equals zero or negative, without employing some *ad hoc* arguments for each individual case. The presented theory handles this problem in a straightforward way. According to Eqn (8.12), the slope can be negative (normal case), positive, or zero depending on the relative magnitude of components $\frac{k_{Si}}{k_{Sj}}\nu_j$ and ν_i.

So, experimenting with the CCST we have to determine for any extraction system:

1. Affinity constant ratios, k_{SE}/k_{SW}, k_{SE}/k_{SA}, k_{SA}/k_{SE}, etc.
2. Averaged equilibrium constants, $\overline{K}_{Si}$.
3. Hypothetical Z^* (intercept) in pure E at $C_{\text{tot. Aq}} = 0$.
4. Suggested aggregates formed at the CCST interpretation.

5. ANALYTICAL METHODS USED FOR THE CCST VERIFICATION

The concentration of citric acid in the aqueous solutions was determined by titration with a standard 0.1 M solution of NaOH. Phenolphthalein was used as an indicator (all three protons of the citric acid are titrated). Accuracy of the analytic method was checked by titration of standard aqueous citric acid solutions.

Citric acid concentration in the loaded organic phase was determined by titration with 0.1 M NaOH using isopropanol as a cosolvent, and phenolphthalein as an indicator.

The concentrations of propanol in the aqueous solutions were determined by HPLC using OA KC column, 0.01 M H_2SO_4 eluent, and RI detection (retention time 32.9 min). In these tests the citric acid analysis was reconfirmed (retention time 10.8 min).

The propanol concentration in the organic phase was determined by extracting the latter into water. Three consequent contacts of each organic phase sample were conducted at org/aq weight ratio = 1/3 and ambient temperature. The three separated aqueous phases were combined and analyzed by HPLC technique.

Water concentration in the organic phase should be determined (e.g., using Karl Fisher techniques) in all experiments, in order to check its difference in the solvation shells of regions 1, 2, and 3 (see Fig. 6.1 in Chapter 6).

6. TRANSFERABILITY OF THE VALUES OF AFFINITY CONSTANT RATIOS

One of the biggest advantages of the CCST is independence and transferability of the values of affinity constant ratios from one system to the other. The theory of partitioning in the homogenous media implies that the components act independently in their interaction with a given solute. In other words, the values of k_{SE} are independent on the values of k_{SW} or k_{SA} and are transferable from one system to the other. As a result, predictions of the influence of different solvents on the effectiveness and selectivity of the extraction process are possible once some solvent parameters are ascertained. Transferability is attained by different independent experiments with three component systems, proving relations:

$$\frac{k_{S_1E}}{k_{S_2E}} = \frac{k_{(S_1E)_1}}{k_{(S_1W)_1}} \times \frac{k_{(S_2W)_2}}{k_{(S_2E)_2}} = \frac{k_{(S_1E)_1}}{k_{(S_1A)_1}} \times \frac{k_{(S_2A)_2}}{k_{(S_2E)_2}} \tag{8.26}$$

$$\frac{k_{SE_1}}{k_{SE_2}} = \frac{k_{(SE_1)_1}}{k_{(SW)_1}} \times \frac{k_{(SW)_2}}{k_{(SE_2)_2}} = \frac{k_{(SE_1)_1}}{k_{(SA)_1}} \times \frac{k_{(SA)_2}}{k_{(SE_2)_2}} \tag{8.27}$$

$$\frac{k_{SA_1}}{k_{SA_2}} = \frac{k_{(SA_1)_1}}{k_{(SW)_1}} \times \frac{k_{(SW)_2}}{k_{(SA_2)_2}} = \frac{k_{(SA_1)_1}}{k_{(SE)_1}} \times \frac{k_{(SE)_2}}{k_{(SA_2)_2}} \tag{8.28}$$

where the systems (all in the same inert diluent) experimentally verified are S_1–E–W, S_1–E–A, and S_2–E–W, S_2–E–A, respectively, for Eqn (8.26); S–E_1–W, S–E_1–A, and S–E_2–W, S–E_2–A, respectively, for Eqn (8.27); S–A_1–E, S–A_1–W, and S–A_2–E, S–A_2–W, respectively, for Eqn (8.28).

The basic idea of the homogenous partitioning, independence, and transferability of affinity constants was developed by Purnell with coworkers.[17,18] They have experimentally proved it for about 180 organic systems, by chromatographic techniques. Nagy with coworkers[19–22] developed the mathematical analysis of homogenous partitioning and experimentally proved it, by NMR, UV-VIS, potentiometric titrations, and kinetic measurement techniques.

Validity of the presented approach, verified experimentally, gives a key for quantitative analysis and preliminary prediction of suitable extraction systems for different metals' separation.

REFERENCES

1. Lo TC, Hanson C, Baird MHI, editors. *Handbook of solvent extraction*. New York: Wiley; 1983.
2. Rydberg J, Cox M, Musikas C, Choppin GR, editors. *Principles and practices of solvent extraction*. 2nd ed. New York: Marcel Dekker. Inc; 2004.
3. Ritcey GM. *Solvent extraction. Principles and applications to process metallury*. New York: CRC; 2006.

4. Ionescu G, Hritcu D, Ionescu I. Theoretical and experimental aspects regarding the distribution equilibrium in the aqueous phase-phenylacetic acid (AFA)—extraction solvent system. *Rev Roumaine Chim* 1993;**38**(8):927–33.
5. Takeda Y, Takagi C. Solvent effects on extraction of sodium picrate with 15-crown-5. The determination of ion-pair formation constant of crown ether-metal salt 1:1:1 complex in water. *Bull Chem Soc Japan* 1994;**67**(1):56–60.
6. Strehlow H, Knohe W. *Fundamental methods of chemical relaxation*. Wenheim: Verlag Chemie; 1977.
7. Limbach HH. Dynamic NMR Spectroscopy in the Presence of Kinetic Hydrogen/Deuterium Isotope Effects. In: Huyskens PL, Luck WAP, Zeegers-Huyskens T, editors. *Intermolecular forces*. Berlin: Springer-Verlag; 1991.
8. Marcus Y. *Ion solvation*. London: Wiley-Interscience; 1985. p. 194–216.
9. Grunwald E, Baughman G, Kohnstam G. The solvation of electrolytes in dioxan-water mixtures, as deduced from the effectof solvent change on the standard partial molar free energy. *J Am Chem Soc* 1960;**82**:5801–11.
10. Covington AK, Newman KE. Thermodynamics of preferential solvation of electrolytes in binary solvent mixtures. *Adv Chem Ser* 1976;**155**:153–96.
11. Cox BG, Parker AJ, Waghorne WE. Coordination and ionic solvation. *J Phys Chem* 1974;**78**:1731–40.
12. Marcus Y. A quasi-lattice quasi-chemical theory of preferential solvation of ions in mixed solvents. *Austr J Chem* 1983;**36**:1719–31.
13. Marcus Y. Preferential solvation of ions in mixed solvents. *Rev Anal Chem* 1980;**5**:267–82.
14. Scatchard G. The attraction of proteines for small molecules and ions. *Ann NY Acad Sci* 1949;**51**:660.
15. Deranleau DA. Theory of the measurement of weak molecular complexes. I. General considerations. *J Am Chem Soc* 1969;**91**:4044–9.
16. Deranlea DA. Theory of the measurement of weak molecular complexes. II. Consequence of multiple equilibria. *J Am Chem Soc* 1969;**91**:4050–62.
17. Laub RJ, Purnell JH. Solution and complexing Studies. III. Further evidence for a microscopic partitioning theory of solution. *J Am Chem Soc* 1976;**98**(1):30–5.
18. Purnell JH, Vargas de Andrade. Solution and complexing studies. I.Gas-liquid chromatografic investigation of supposed complexing systems. II. Comparison and correlation of nuclear magnetic resonance and gas-liquid chromatografic data. *J Am Chem Soc* 1975;**97**:3585–90.
19. Nagy OB, Muanda H, Nagy JB. Competitive preferential solvation theory of weak molecular interactions. *J Chem Soc Faraday Trans 1* 1978;**74**:2210–29.
20. Szpakowska M, Nagy OB. Application of the Preferential solvation theory to Coordinative Solute-Solvent interactions. *J Chem Soc Faraday Trans 1* 1989;**85**(9):2891–9.
21. Nagy OB, Muanda M, Nagy JB. Quantitative evaluation of solute-solvent interactions. Environmental effects on the n-butylaminolysis of tetrachloro- *N-n*-butylphthalimide. *J Phys Chem* 1979;**83**(15):1961–79.
22. Parbhoo B, Nagy O. Application of the competitive preferential solvation theory to ion-molecule interactions. *J Chem Soc Faraday Trans 1* 1986;**82**:1789–93.

Chapter 9

The CCST in Solvent Extraction of Acids by Amine-Based Extractants

Chapter Outline

1. INTRODUCTION

1.1. Interaction Mechanisms

The most common approach to investigate an acid extraction behavior is to perform equilibrium distribution studies and fit an extraction equation to the data in order to determine species stoichiometry using Nernst's distribution

Solvent Extraction. DOI: 10.1016/B978-0-444-53778-2.10009-3

law with sets of mass action law equilibria equations.[1] According to the type of distribution reactions, extractants for carboxylic acids were divided into three main categories[1,2]: (1) carbon-bonded oxygen-donor extractants (inert aliphatic, aromatic, and substituted hydrocarbons are included in this type); (2) phosphorus-bonded oxygen-donor extractants, and (3) high-molecular-weight aliphatic amine extractants. The first two are classified as solvation extractants, based on the strength and specificity of the solvation bonds; the third is based on charge transfer (ion-pair formation and/or proton transfer) reactions with amines. Amine-based extractants are used for consideration and analysis of new modeling approach, while the carbon- and phosphorus-bonded oxygen-donor extractants are considered here as active solvent additives.

Using the order of acid–base strength for a given acid and amine interactions the authors[3–14] have developed models in which hypothetical acid–base complexes of a specific stoichiometry are formed. As a rule, they consider the system where one of the mechanisms is dominating and is not changed with changing acid concentrations. In addition, over-stoichiometric complexation at high acid loading is described through hydrogen bonding,[7,11] aggregation[5,13] and even polymerization.[15] But generalization that can be drawn from these studies is that the stoichiometry (and maybe nature) of the solute–extractant complexes are changing with increasing acid concentrations.

1.2. Acids' Extraction Modeling by CCST

Complicated behavior of extraction systems conflicts, in many cases, with the stoichiometric ion-exchange models. Coordination models, explaining some mechanisms of metal solvent extraction, meet difficulties when modeling acid extraction, and, especially, when quantification attempts are made.[16] The competitive complexation/solvation (CCS) theory strongly simplifies this task.

CCS modeling approach is introduced here to explain the mechanisms of solvent extraction. For the following reading convenience the four basic statements of the CCST approach are repeated here:

I. Extraction act is initiated by solvation of the solute compounds in organic solvents with formation of hydrogen- (coordinate-) bonded intermediates (adducts). Kinetics of proton, electron transfer, and formation of final complexes is stepwise and depends on kinetics of step reactions, competition, coupling, strength of H-bonding, and many other factors. Interacting mechanisms and final compounds formed are changing with increasing extracted solute concentration in organic phase (see Chapter 6, Fig. 6.1).

II. Strong (ion-exchange, chemical, electrostatic) and weak (solvating, physical, intermolecular, H-bond, coordinate) interactions and aggregations are the basic interactions that have to be taken into consideration in all ranges of extracted solute concentrations. Competition between these interactions forms the four region shape (Chapter 6, Fig. 6.1) of the general extraction isotherm.

III. Solutes, extractants, active solvents, and their interacting species are considered amphoteric and may perform as acids (electron acceptors) or bases (electron donors), depending on the structure of their functional groups and composition of the organic phase, and on the structure of the solute species and composition of the aqueous phase.

IV. Aggregation mechanisms and species formed are changing with increasing loading of organic phase by solute, and should be considered and analyzed in all regions of the general extraction isotherm (Chapter 6, Fig. 6.1).

CCS theory constitutes a general framework for interpretation of ion–molecular interaction data in the solvent extraction systems. Mathematical description for the process simulation presents the possibility of analysis of extraction systems of interest and behavior predictions.

Some extraction systems, e.g., acid–amine–water, were analyzed in Chapter 7. In this chapter the CCST approach is described using some more complicated acid–base systems. Influence of active solvents on the acid solvent extraction by amine-based extractants is interpreted on the basis of the presented theory. Active solvent is introduced as an additional parameter to the quantitative consideration of the extraction systems. Data, available from the literature[1,6–14] and our experiments,[17,18] are used for interpretation.

1.3. Active Solvents

The difference in the extraction trends must be attributed to solvation effects. Strong, sometimes even critical, influence of active solvents, used as synergistic (enhancing) agents, modifiers, and/or diluents, on the distribution of acids into organic phase is discussed in many research publications.[2,6,9–11,13,15–21] Polar, polarizable, hydrogen-bonding solvents change the extraction order dramatically. In order to analyze solvation effects, identification, and most importantly, quantification of solute–solvent interactions must be accomplished.

Solvents can be classified into different ways.[20] For our purpose, classification of solvents by their influence on the extraction systems' effectiveness is useful. Division into inert (nonpolar) and active (polar) solvents is made here.

Inert solvents. When used in extraction systems as diluents for amines, solvents such as alkanes, benzene, alkyl-substituted aromatics, and kerosene result in relatively low distribution of acids into organic phase.[6–9] These diluents do not affect extraction mechanisms and are added in order to improve physical properties of the extraction system, e.g., reduction of viscosity and enhance separation of the phases. Alkanes, being nonpolar, provide very little solvation. Aromatic diluents provide slightly higher solvation values, which have been rationalized as solvation due to interaction of the aromatic

π electrons with the complexes.[2,16,20] All these diluents may serve as modifiers (if at all), stabilizing to varying degree by polarizability effect of the amine–acid complexes and their aggregates. The influence of the inert solvents on the extracting parameters of the system is not taken into account at this stage.

Active solvents (e.g., alcohols, esters, ethers, chlorinated hydrocarbons, etc.) have functional groups that enable much stronger interactions with the components of the extraction system. For each solvent, the extent of these solute–solvent interactions is different. Identification and quantification of solute–solvent interactions are very difficult tasks due to the complexity of solvent structure.

Coextraction of water. There are many publications about coextraction of water and its influence on extraction process viability.[2,4–13,17–23] Water molecule is self-complementary as regards optimal hydrogen bonding since it involves two donors (H atoms) and two acceptors (electron pairs) with tetrahedral coordination.[7,14] Despite the complete saturation of H-bonds valences, water is not fully satisfied, since the O(H)O hydrogen bonds cannot attain lengths short enough to prevent porous supertetrahedral structures. Hence, water participates in the solvation as one of the active solvents.

Alcohol solvents, added to amine–acid extraction systems, give unusually high formation constants, higher than would be expected from polarity arguments alone.[24–27] Hydroxyl group can donate one H atom to a hydrogen bond and possesses two oxygen electron pairs as acceptors.[28] Amino group involves two (primary), one (secondary), or no (tertiary) donor H atoms and one acceptor electron pair. Therefore, alcohols and amines can form only one hydrogen bond per hydroxyl and amino groups, but have a potential to form additional hydrogen bonds at specific compositions and conditions. So, hydroxyl groups in alcohols and amino groups in amines are complementary as regards hydrogen bond donors (HBDs) and acceptors both stoichiometrically and geometrically. This is expected to lead to molecular recognition among alcohols and amines toward formation of 1-1 alcohol–amine complexes, driven by some increase of the number of hydrogen bonds as compared with the uncomplexed constituents. It is known[28–30] that alcohol–amine complexes usually interact and form zigzag, helical, or cyclic geometrical patterns. Comparing hydrogen bonds of the type O(H)O, N(H)N, and N(H)O, authors[28,29] come to conclusion of complete preference of N(H)O type. The sizes and shapes of the residues R (amine) and R′ (alcohol) also may affect the specific intermolecular interactions. Linear aliphatic side chains appear to fit well into H-bond architecture.[30] Three-dimensional assemblies adopt right- and left-handed helical shapes. One NH_2 group is bonded to three OH groups as two acceptors and one donor, indicating that each hydrazine group is slightly fixed by six hydrogen bonds.[30]

Chlorinated aromatics promote about a fivefold greater equilibrium formation constant than benzene.[8] In case of ketone solvents formation constants are 10-fold larger. Nitrobenzene and chlorinated hydrocarbons-containing amine extractants have distribution ratios of 100-fold greater than benzene-containing ones.

For some solvents, e.g., phosphorus-bonded oxygen-donor solvents, equilibrium constants, higher than would be expected from polarity considerations, are explained by specific hydrogen bonding between proton of the solvent and acid–amine complex.[8,19]

Many models were developed to quantify the effect of solvents on the acid–amine complexation.[3,5–7,9,11,15,16,18–21,23–26,30] Various parameters and equations have been developed to describe solvation properties and their effects on the acid–base strength of organic compositions. A simplified approach is to classify solvents as protic or aprotic. Some authors used the Hildebrand solubility parameter, δ, as a measure of solvation,[31] linear free-energy relationships,[24] and combined "physical and chemical modeling", calculating the activity coefficients for the reactants and complexes.[32–34] Headley and coworkers[16,35,36] studied solvation effects of active solvents on the basicity of amines, and introduced many parameters, such as hydrogen bond donor (HBD) and hydrogen bond acceptor (HBA) abilities, solvent's dipolarity/polarzability, dipole moment, etc., in order to quantify the solvation effects. These models are qualitative, or successfully fit the data for some solvents but do not follow for the others. All models, developed in this field, require a considerable number of adjustable parameters and development of calculation techniques to be useful for the technological practice. The CCS theory strongly simplifies this task, establishes the correlation between complexation and solvation, and permits for preliminary quantification of complex extraction systems.

1.4. Influence of Temperature

Effect of temperature on extraction of acids was studied experimentally.[1,2,7,21,23,25] There is a slight effect, if any, of temperature in the range of 20–90 °C on the distribution of propanoic, lactic, and succinic acids into alcohols, ketones, diethyl carbinol, and ethers.[1] Significant effect of temperature on acid extraction was observed for alkylphosphates and amines as extractants.

As a rule, the distribution of the acid into the organic phase (a variety of amines in a variety of diluents with modifiers) decreases with increasing temperature. For some acids the concentrating effect in a product stream may be achieved using temperature-swing process.[36] Studying the effect of temperature on extraction of citric acid by extractants, composed of tridecylamine with alcohol modifiers and xylene or nitrobenzene[5–7,36] or by Alamine 336,[27] or by tributylphosphate[21,26,27] in a variety of diluents the authors have found a sharp decrease of the acid distribution into organic phase with increasing temperature. At extraction of lactic, tartaric, succinic, and citric acids by trilaurylamine in xylene[1,5–7] increasing temperature decreases the distribution ratios by factors of 2–10, depending on acid and its concentration.

The temperature-swing regeneration includes forward extraction at a low temperature, obtaining acid's loaded organic phase and an aqueous raffinate, containing feed components that have to be separated, and back-extraction at a higher temperature, obtaining acid's loaded aqueous product stream and regenerated organic phase. Back-extraction of some acids may be achieved with fresh water without dilution of the acid.

2. ACID–AMINE SYSTEMS IN THE CCST INTERPRETATION

CCS theory constitutes a framework for interpretation of ion (or ion-pair)–molecular interaction data in the solvent extraction systems. Composition and behavior predictions of an extraction system of interest can be analyzed. For details of CCST, see Chapter 7.

Solute in the phases is surrounded by a solvation shell.[17,18,37,38] This means that it interacts with all the constituents of the environment and, therefore, "free" solute, free extractant, or free solute–extractant complex do not exist. Components of the solvation shell compete to form a complex according to their electronic-geometric affinity k_i (interaction strength), which is constant at a given temperature and pressure. Coefficient, k_i, represents the total, complexing and solvating affinity constant. Competition between the constituents of the solvation shell takes place according to their affinity constants k_i and the number of potentially available molecules of every component, i.e., its actual concentration. A set of equations were developed for calculation of the generalized partitioning factors of the solute, P_{Si}, and its concentrations, C_{Si}, in the homogenous media of the solvation shell at equilibrium (see Chapters 6 and 7).

Interaction mechanisms and species formed in the organic phase change with increasing acid concentration (Statement I). The general extraction isotherm consists of four regions (Chapter 6, Fig. 6.1). The influencing effect of solvents may be different from one region to the other. Many authors, considering extraction systems of interest, do not distinguish between different regions in which different interaction mechanisms dominate. Ignoring this fact introduces contradictions in the interpretations of the experimental data (for details see Chapter 6, 7).

Consider a system containing an acid, S, amine, E, water, W, active solvent, A, and inert solvent (diluent), D. Some preliminary assumptions are as follows:

1. All components of the system are monomers in a pure, initial state.
2. The volume change on solvent mixing is neglected.
3. The diluent D is inert enough and do not participate in solvation.
4. Water in the organic phase after equilibration, separation, and centrifugation is present as bonded only in the solvation shell, i.e., any solubility of water in the bulk organic solvent mixture is excluded. Situations in which water is extracted by extractants or solvents, such as at $C_{Sorg} << C_{Worg}$, will be considered for definite extraction systems.

From the point of view of CCS theory:

$$C_S^0 = C_{Saq} + C_{Sorg} = C_{Saq} + C_{SE} + C_{SW} + C_{SA} + C_{SD} \quad (9.1)$$

where C_S^0 is the initial (or total) concentration of the acid; C_{Saq} is the concentration of the acid in the aqueous phase; C_{Sorg} is the concentration of the acid in the organic phase; C_{SE}, C_{SW}, C_{SA}, and C_{SD} are the concentrations of the acid in the organic phase, partitioning between amine, water, active, and inert solvents, respectively, at equilibrium. At this stage, the inert solvent, D, is not considered participating in solvation.

The value of a physicochemical property X (e.g., chemical shift δ in NMR, molar extinction coefficient ε in UV–VIS, changes in free energy ΔG, etc.), measured in mixed solvents, is considered to be the weighed sum of this same property measured in pure solvent constituents. The contribution of each solvent on the shift of this property, X, is additive. Considering three-component system, i.e., solute–extractant–water, we obtain, after some algebra, the linear equation for the measured property X:

$$\frac{X_{SW} - X}{C_E} = \frac{k_{SE}}{k_{SW}} V_W (X_{SW} - X_{SE}) - \left[\frac{k_{SE}}{k_{SW}} V_W - V_E \right] (X_{SW} - X) \quad (9.2)$$

and the saturation factor Z at $C_E > C_S^0$:

$$Z = \frac{C_{Sorg}}{C_S^0 - C_{Sorg}} = \frac{X_{SW} - X}{X_{SW} - X_{SE}} = \frac{k_{SE} V_W C_E}{k_{SW} + (k_{SE} V_W - k_{SW} V_E) C_E} \quad (9.3)$$

where k_{SE}/k_{SW} is the affinity constant ratio; C_S^0 is the initial (or total) concentration of the solute; C_{Saq}, C_{Sorg}, C_{SE}, C_{SW} are the solute concentrations in the aqueous phase, in the organic phase, and in the mixed solvation shell of the organic phase (extractant, water, active, or inert solvent), respectively, at equilibrium; v_E and v_W are partial molar volumes of extractant and water.

The linearity of Eqn (9.2) plots over the whole concentration range means that the same affinity constant ratio is valid at low and high acid concentrations and does not depend on concentration. Extraction of acetic acid by Alamine 336 with 15% v/v chloroform in *n*-heptane, presented in Fig. 9.1, may be an example of such a system. Interchanges in the linearity of Eqn (9.2), or the different slopes in the loading curve over the acid concentration range mean that the affinity constant ratios and/or concentrations of the components in the solvation shell are changing. For examples, see Figs. 9.2–9.6. It means, also, that the different charge-transfer (chemical) complexation mechanisms, and consequently the different values of k_{SEc}, or the different solvation mechanisms and consequently different values of k_{SEs}, take place at different solute concentration (k_{SEc} and k_{SEs} are affinity constants of the solute toward extractant in the complexed and solvated forms, respectively). Therefore, we have to use one more experimentally measured property, such as UV–VIS or quantitative IR spectroscopy measurements, to determine either k_{SEc}/k_{SEs}.

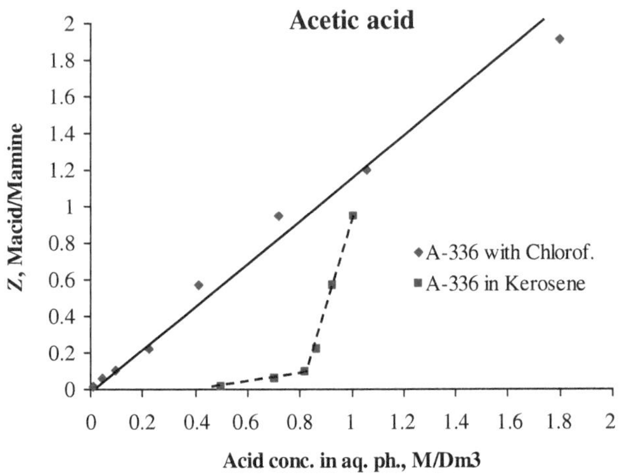

FIGURE 9.1 Comparison of experimental and calculated data for acetic acid extraction by Alamine 336 in kerosene and by Alamine 336 with 15% v/v chloroform in *n*-heptane. *(Source: Data from Ref. 1 with permission)*

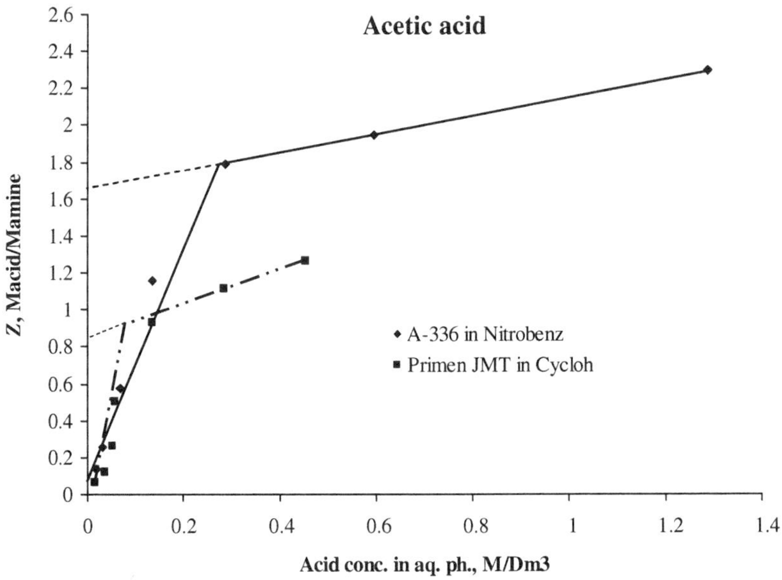

FIGURE 9.2 Comparison of experimental and calculated data for the acetic acid extraction by Alamine 336 in nitrobenzene and by Primene JMT in cyclohexane. *(Source: Data from Ref. 1 with permission)*

Regions 1–4 (Chapter 6, Fig. 6.1) in the distribution curve with different slope values testify different compositions of the aggregates formed and different influence of the extraction system constituents on the interaction mechanisms over acid concentration range. Different values of the hypothetical chemical

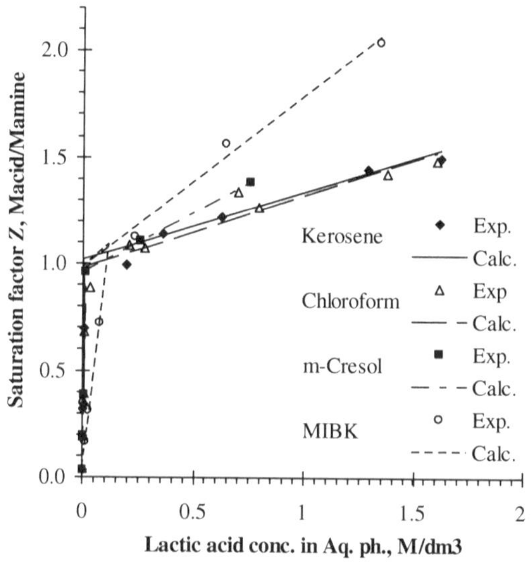

FIGURE 9.3 Comparison of experimental and calculated data for lactic acid extraction by Alamine 336 in different solvents: kerosene, chloroform, *n*-Cresol, and MIBK.[1,17] *(Source: Data from Ref. 1 with permission)*

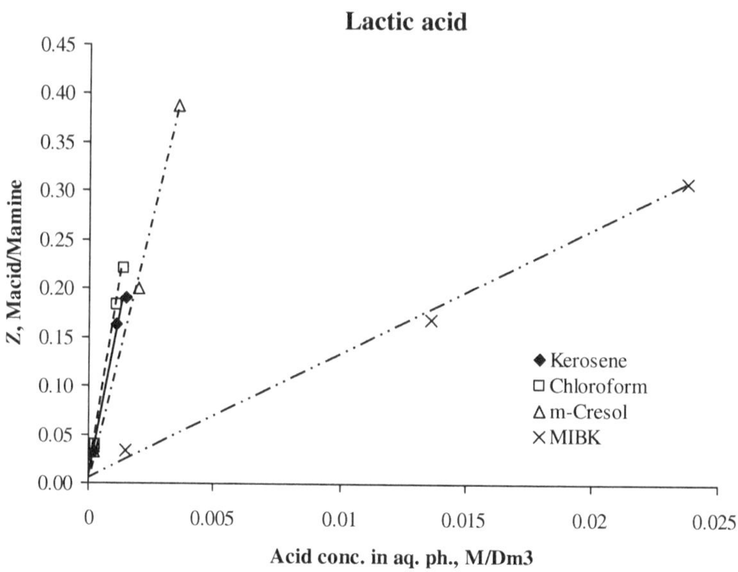

FIGURE 9.4 Comparison of experimental and calculated data for lactic acid extraction by Alamine 336 in different solvents: kerosene, chloroform, *n*-Cresol, and MIBK[1,17]; presented in Region 2. *(Source: Data from Ref. 1 with permission)*

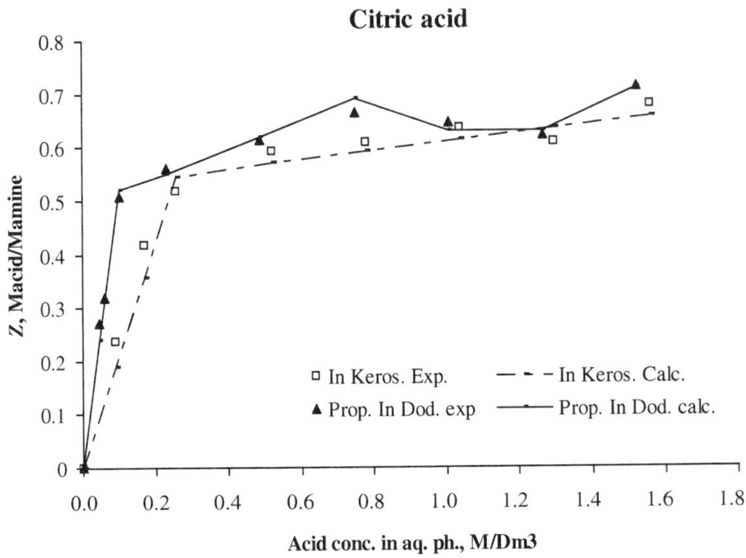

FIGURE 9.5 Comparison of experimental and calculated data for citric acid extraction by Alamine 304 in kerosene and by Alamine 304 with 1 M/kg propanol in *n*-dodecane.[17]

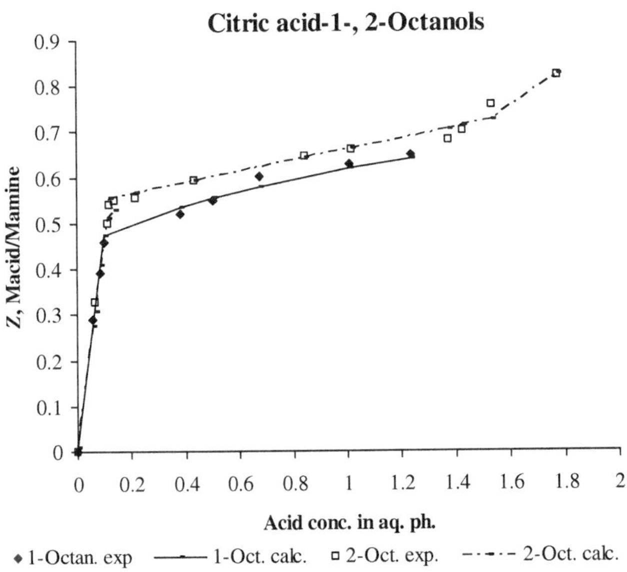

FIGURE 9.6 Comparison of experimental and calculated data for citric acid extraction by Alamine 304 with 1 M/Kg 1- Octanol in kerosene and by Alamine 304 with 1 M/kg 2-Octanol in kerosene.[17]

shifts in pure water are evident. This analysis is a useful tool when studying polybasic solutes or extractants, and especially when studying mixtures of extractants, such as acid–amine or amine–amine mixtures.

In the CCS theory, different regions are analyzed separately and the mathematical descriptions developed for different interactions between the components of the extraction system.

Some extraction systems are analyzed in the following sections using the statements of the CCS theory.

2.1. The Acid–Amine (in Inert Diluent)–Water (S–E–W) System

Let us consider a relatively simple system containing monobasic amine, E, monobasic acid, S, water, W, and inert organic diluent, D. Four preliminary assumptions, mentioned above, are valid in this case also.

Using different measured properties, X, we obtain relations, derived from Eqn (9.2);
for NMR, where $X = \delta$ is chemical shift:

$$\frac{\delta_{\mathrm{SW}} - \delta}{C_{\mathrm{E}}} = \frac{k_{\mathrm{SE}} V_{\mathrm{Worg}}}{k_{\mathrm{SW}}}(\delta_{\mathrm{SW}} - \delta_{\mathrm{SE}}) - \left[\frac{k_{\mathrm{SE}}}{k_{\mathrm{SW}}} V_{\mathrm{Worg}} - V_{\mathrm{E}}\right](\delta_{\mathrm{SW}} - \delta) \qquad (9.4)$$

for UV–VIS, where $X = \varepsilon$ is molar extinction coefficient:

$$\frac{\varepsilon_{\mathrm{SW}} - \varepsilon}{C_{\mathrm{E}}} = \frac{k_{\mathrm{SE}} V_{\mathrm{W}}}{k_{\mathrm{SW}}}(\varepsilon_{\mathrm{SW}} - \varepsilon_{\mathrm{SE}}) - \left[\frac{k_{\mathrm{SE}}}{k_{\mathrm{SW}}} V_{\mathrm{W}} - V_{\mathrm{E}}\right](\varepsilon_{\mathrm{SW}} - \varepsilon) \qquad (9.5)$$

where
$\dfrac{A}{C_{\mathrm{Sorg}}} = \varepsilon$, $\dfrac{A_{\mathrm{SE}}}{C_{\mathrm{Sorg}}} = \varepsilon_{\mathrm{SE}}$, $\dfrac{A_{\mathrm{SW}}}{C_{\mathrm{Sorg}}} = \varepsilon_{\mathrm{SW}}$ and A is a measured absorbency;
for free energy (or enthalpy) gradient, where $X = \Delta G$ (at constant temperature and pressure):

$$\frac{\Delta G}{RT} = -\ln\left(\frac{k_{\mathrm{SE}}}{k_{\mathrm{SW}}} \times \frac{C_{\mathrm{E}} V_{\mathrm{E}}}{C_{\mathrm{Worg}} V_{\mathrm{W}}}\right) \qquad (9.6)$$

In Figs. 9.1–9.6 the experimentally obtained isotherms for acetic and lactic acids, equilibrated with Alamine 336 (from Ref. 1,6) and for citric acid, equilibrated with Alamine 304 (from Ref. 17,24), are compared with the calculated, according to the CCST data. The difference between experimental and calculated data may be explained by the influence of chloroform, which was used as an active solvent in the case of acetic (Fig. 9.1) and lactic (Fig. 9.3) acids and of n-octanol in the case of citric acid (Fig. 9.6). The data for calculating the chloroform effect were not available. Effect of octanol is presented in Fig. 9.6 and discussed in Section 4.3. But, nevertheless, the figures show that the presented theoretical approach satisfactorily quantitatively describes the studied extraction systems and, even, indirectly shows the influence of an active solvent.

Considering the interactions between the acid and amine from the classical theories point of view, equilibrium constant, K_{SE}, can be determined using the equations, developed by Scatchard[39]; e.g., the NMR:

$$\frac{\delta_{SW}-\delta}{C_E} = K_{SE}(\delta_{SW}-\delta_{SE}) - K_{SE}(\delta_{SW}-\delta) \tag{9.7}$$

The comparison of Eqns (9.4) and (9.7) leads to the following relation:

$$K_{SE} = \frac{k_{SE}}{k_{SW}}V_W - V_E \tag{9.8}$$

Using the classical theories, it is impossible to explain why equilibrium constant K equals zero or negative, without employing some *ad hoc* arguments for each individual case. The presented theory handles this problem in a straightforward way. According to Eqn (9.8), the slope can be negative (normal case), positive, or zero depending on the relative magnitude of components $\frac{k_{Si}}{k_{Sj}}V_j$ and V_i.

2.2. The System with Mixture of Two Acids; Selectivity

Let us consider now expected interactions and associations formed at extraction of two monobasic acids, S_1 and S_2, where $pK_{aS1} > pK_{aS2}$, according to the classical theories [6–9,11,12] or

$$\frac{k_{S_1E}C_{S_1org}}{k_{S_1W}C_{S_1aq}} > \frac{k_{S_2E}C_{S_2org}}{k_{S_2W}C_{S_2aq}} \tag{9.9}$$

according to the presented theory (see Statement II). Preliminary assumptions, introduced in Section 2.1, are correct for both acids.

Comparing these two relationships the reader immediately can see the difference: classical theories do not take into account the concentrations of acids S_1 and S_2 in the organic and aqueous phases. The last relation includes also information about weak interactions of acid molecules not only with extractant, but also with other components of both the phases.

Let us consider expected interactions and associations formed in the mixture of extractant with two monobasic acids.

Case 1: One of the acids, S_1, is water immiscible (acid–base-coupled (ABC) extractants[40,41]).

Concentration of the water-immiscible acid in the aqueous phase $C_{S_1aq} \rightarrow 0$, therefore, relationship $\frac{C_{S_1org}}{C_{S_1aq}} >> \frac{C_{S_2org}}{C_{S_2aq}}$ will exist at about all the range of concentrations of the acid S_2. This means that the interaction of acid S_2 with extractant (i.e., its concentration in the solvation shells) will be very

small even at $pK_{aS1} > pK_{aS2}$ or $\frac{k_{S_2E}}{k_{S_2W}} < \frac{k_{S_1E}}{k_{S_1W}}$. Only at very high S_2 concentrations in the aqueous phase the extraction of this acid may be detectable. It will be interactions and aggregates formation typical for Region 4.
The presented theory shows without any experiments that ABC may be useful extractants only for strong mineral or carboxylic acids. Choosing the ABC extractants for the separation of relatively weak acids (some carboxylic or amino acids) and using statements of classical theories, the researchers[40–42] could not take into account the concentration ratios in two phases.

Case 2: Extraction of two monobasic acids, miscible both in the organic and in the aqueous phases and differ mainly in their acidity, $pK_{aS1} > pK_{aS2}$.

A. Initial concentrations of both acids in the aqueous phase are relatively low (both acids/amine <1/1) and $C^0_{1aq} \geq C^0_{2aq}$.
Expected interactions at these concentrations correspond to Regions 1 and 2 of the general extraction isotherm with nuclei aggregates formed (see Chapter 7). According to Statements I and III of the CCST electrostatic, ion-pair formation mechanisms (Region 2) will be dominating for the stronger acid, S_1, and H-bonding (Region 1) will dominate for the weaker acid, S_2. According to Eqn (9.9) the weaker acid plays a synergistic, enhancing role for the distribution of a stronger acid into organic phase. At $\frac{k_{S_1E}}{k_{S_1W}}$ and $\frac{k_{S_2E}}{k_{S_2W}}$ constant, what is true for every region, the difference between magnitudes of concentration ratios (and selectivity) at equilibrium, $\frac{C_{S_1 org}}{C_{S_1 aq}}$ and $\frac{C_{S_2 org}}{C_{S_2 aq}}$, will increase with increasing of initial concentrations of both acids. The best selectivity results may be obtained for the acids S_1 and S_2, which will form nuclei aggregates with extremal structures, shown in Fig. 7.1a and 7.1b in Chapter 7, respectively.

B. Initial concentrations of both acids in the aqueous phase are high (reaching or prevailing stoichiometry 1/1 for stronger acid) and $C^0_{1aq} \geq C^0_{2aq}$.

At these concentrations of acids the expected interactions correspond to Region 3 of the general extraction isotherm with linear (planar) aggregates formed preferentially through the H-bond bridging of the nuclei aggregates by the weaker acid.

In this case, at $\frac{k_{S_1E}}{k_{S_1W}}$ and $\frac{k_{S_2E}}{k_{S_2W}}$ constant (but may have different values than in Regions 1 and 2), the difference between magnitudes of concentration ratios $\frac{C_{S_1 org}}{C_{S_1 aq}}$ and $\frac{C_{S_2 org}}{C_{S_2 aq}}$ (and selectivity) at equilibrium will pass through a maximum with increasing of initial concentrations of both acids. The maximum value depends on kinds of both acids: at reaching stoichiometry 1/1 saturation by a stronger S_1 acid, the selectivity will be maximal; then it will decrease because of the competition between the weaker S_2 and stronger S_1 acids for the H-bonding. According to the presented theory (see Statements I and III), for the same amine and every two acids used the maximal selectivity will be at different amine–stronger acid compositions. Preliminary data for these compositions may be calculated, using relations, derived from Eqns (9.2) and (9.3):

$$\frac{(X_{SW}-X)_{max}}{C_E}=\frac{k_{SE}}{k_{SW}}V_W(X_{SW}-X_{SE})-\left[\frac{k_{SE}}{k_{SW}}V_W-V_E\right](X_{SW}-X)_{max} \tag{9.10}$$

$$\frac{1}{Z_{max}}=1+\frac{k_{SW}}{k_{SE}}\times\frac{[C_{W\,org}]_{max}}{C_E} \tag{9.11}$$

Determining experimentally $[C_{Worg}]_{max}$ for both, S_1 and S_2 acids, at already obtained affinity coefficients' ratios $\frac{k_{S_1E}}{k_{S_1W}}$ and $\frac{k_{S_2E}}{k_{S_2W}}$, we can determine water concentration limits for selective separation of two acids. It is especially important at using amine–acid mixtures as extractants for the weak acids.

Many researchers use amine–acid mixtures as extractants (ABC) with equal (M/M) initial concentrations. This may be true composition only at application of strong mineral acids in a mixture. But as a rule, the optimal composition will differ from 1–1 mixture and depend on the properties of both S_1 and S_2 acids.

It is important that the presented theory predicts the synergistic, enhancing properties of the weak acid for the distribution of a stronger acid into organic phase.

Water-swing separation of acids. From the above considerations it follows that the selective separation of two acids may be achieved at back-extraction by water. This process is well known, especially for the ABC extractants.[40] The presented theory permits the preliminary evaluation of optimal separation parameters for the mixture of two acids, using the data, obtained at experiments with simple systems $E–S_1–W$ and $E–S_2–W$. It is one of the advantages of the theory.

2.3. The System with the Mixture of Two Amines

Let us consider expected interactions and associations formed at extraction of monobasic acid S by a mixture of two monobasic amines, at

$$\frac{k_{SE_1}}{k_{SW}}\times\frac{C_{Sorg}}{C_{Saq}}>\frac{k_{SE_2}}{k_{SW}}\times\frac{C_{Sorg}}{C_{Saq}} \tag{9.12}$$

According to Statement III of the presented theory, the weaker amine E_2 performs in this system as a weak acid and all considerations, presented in Section 2, are valid here, of course, at $pK_{bE2}>pK_{aS}$ or $\frac{k_{SE_2}}{k_{SW}}<\frac{k_{SE_1}}{k_{SW}}$:

1. At low concentrations of the extracted acid the weaker amine plays a synergistic, enhancing effect on the distribution of the acid into organic phase.
2. At high concentrations of the acid (reaching or prevailing its stoichiometry 1/1), the synergistic, enhancing effect of the weak amine will reach

a maximum in the distribution of the acid into organic phase; then the distribution will decrease because of competition between the acid S and weak amine E_2 for the H-bonding. For the every weak amine E_2 and every acid extracted the maximal enhancing effect will be at different amine E_2–acid compositions.

It is evident from the theory that the initial concentration of the weak amine E_2 has to be lower than amine E_1 and acid S. If the amine E_2 is immiscible in the aqueous phase, above considerations for two acids, where one is immiscible, are valid in this case also at $C_{SE_2aq} \rightarrow 0$ and

$$\frac{C_{E_2org}}{C_{E_2aq}} >> \frac{C_{Sorg}}{C_{Saq}}$$

Preliminary data for these compositions may be calculated using independent experiments.

3. THE SYSTEMS WITH ACTIVE SOLVENT AS AN ADDITIVE

Many authors do not distinguish between different regions of the general extraction isotherm in which different interaction mechanisms dominate. The influencing effect of solvents may be different from one region to the other. Ignoring this fact introduces many contradictions in the interpretations of the experimental data obtained.

Consider a system that contains a solute (acid) of interest, S, a water-immiscible amine, E, in an inert also water-immiscible diluent, D, and active solvent, A, which may be water-immiscible, partly or completely soluble in both organic and aqueous phases. Above assumptions are valid in this case also.

As it was mentioned above, a composition of the solvation shell in a mixture depends on affinity constant k_i and on the number of potentially available solvent (amine, active solvent, and water) molecules present in the mixture, i.e., on the actual concentration of every component. The greater the factor $k_{SE}C_E$ (or $k_{SW}C_W$, or $k_{SA}C_A$), the better the component solvates acid. Competition between amine, active solvent, and water molecules to interact with the acid molecules in the solvation shell determines the distribution curve slope. According to the theory, the solute–solvent interactions are different for every solvent and result in different affinity constant ratios k_{SE}/k_{SW}, k_{SE}/k_{SA}, k_{SA}/k_{SW}, k_{EA}/k_{AW} and concentration ratios C_{SE}/C_{SW}, C_{SA}/C_{SW}, C_{SE}/C_{SA}, C_{EA}/C_{AW}, respectively, which have to be determined. Preliminary data of the extraction compositions can be calculated using the results of three-component systems. Hence, the experimental studies simplify to the series of measurements for the S–E–A, S–A–W, and E–A–W systems, analogically to S–E–W system.[17] Below, relations for the S–E–A system are presented:

$$P_{SE} + P_{SA} = 1 \tag{9.13}$$

$$C_A = \frac{1 - C_E \overline{V}_E}{V_A} \tag{9.14}$$

$$\frac{X_{SA} - X}{C_E} = \frac{k_{SE}}{k_{SA}} V_A (X_{SA} - X_{SE}) - \left[\frac{k_{SE}}{k_{SA}} V_A - V_E\right] (X_{SA} - X) \tag{9.15}$$

Affinity constant ratio k_{SE}/k_{SA} is determined from the slope of Eqn (9.15) plots: $(X_{SA} - X)/C_{Aorg}$ versus $(X_{SA} - X)$. The intercept gives the hypothetical difference between the properties, measured in pure extractant and pure active solvent. Thus, the agreement between the direct experimental determination of the measured property in pure solvents and their graphically obtained values of $X_{SA} - X_{SE}$ may be examined.

Saturation factor Z:

$$Z = \frac{k_{SE} V_A C_E}{k_{SA} + (k_{SE} V_A - k_{SA} V_E) C_E} \tag{9.16}$$

or

$$\frac{1}{Z} = 1 + \frac{k_{SA}}{k_{SE}} \times \frac{C_A}{C_E} \tag{9.17}$$

Using the Scatchard[39] Eqn (9.7) for equilibrium constant K_{SE}, developed from the classical theories point of view, and comparing it with the Eqn (9.5) of the presented theory, we obtain relation:

$$K_{SE} = \frac{k_{SE}}{k_{SA}} V_A - V_E \tag{9.18}$$

Affinity constant ratio, determined by free energy gradient measurements, $X = \Delta G$:

$$\frac{k_{SE}}{k_{SA}} = \frac{C_A V_A}{C_E V_E} \exp\left[\frac{\Delta G_{SE}}{RT}\right] \tag{9.19}$$

Analogical relations may be written for the S–A–W and E–A–W systems also.

All needed parameters for the S–E–A–W system development and predictions can be obtained by these measurements.

In the case of water coextraction by free extractant, active or inert solvents, and of active solvent, soluble in water, concentrations of C_{Worg}, C_{Aorg}, and C_{Aaq} in the bulk organic and aqueous phases have to be measured in the absence of acid, S. Partitioning factor for the adduct A′, soluble in both phases is determined according to Statement III of the theory at

$$\frac{k_{SE} C_{Sorg}}{k_{SW} C_{Saq}} > \frac{k_{AE} C_{Aorg}}{k_{AW} C_{Aaq}} \tag{9.20}$$

An active solvent perform in this system as a weak acid and all considerations, presented above for the extraction system with two acids, are valid here. The reader can see the difference between classical and presented theories: the first do not take into account the concentration ratio of an active solvent in the organic and aqueous phases.

Let us consider expected interactions and associations formed in a system containing an amine, active solvent, and monobasic acid. Active solvent is miscible in both organic and aqueous phases. Several important cases may be considered.

3.1. Case 1: $\frac{k_{SE}}{k_{SW}} > \frac{k_{AE}}{k_{AW}}$ and $\frac{C_{Sorg}}{C_{Saq}} > \frac{C_{Aorg}}{C_{Aaq}}$

At low acid concentration (lesser its stoichiometry 1/1 with the amine) expected interactions correspond to Regions 1 and/or 2 of the general extraction isotherm with nuclei aggregates formed. According to the statement of amphoterity (Statement III), an active solvent performs as a weak acid. In this case, an active solvent, A, has a synergistic, enhancing effect on the distribution of the acid into the organic phase. This effect is explained by replacing one of the hydrogen-bonded molecules of the amine by a molecule of the active solvent in the solvation shell, or nucleus aggregate. Therefore, apparent concentration of amine increases for acid–amine interactions. Headley and coworkers [16,35,36] denominate this phenomenon as a "relative increase of amine basicity."

The difference between magnitudes of concentration ratios, $\frac{C_{Sorg}}{C_{Saq}}$ and $\frac{C_{Aorg}}{C_{Aaq}}$, at equilibrium, increases with increasing acid and active solvent concentrations.

3.2. Case 2: $\frac{k_{SE}}{k_{SW}} > \frac{k_{AE}}{k_{AW}}$ but $\frac{C_{S\ org}}{C_{S\ aq}} < \frac{C_{Aorg}}{C_{Aaq}}$

The synergistic effect of the active solvent decreases with increasing acid concentration down to a minimum, where $\frac{C_{Sorg}}{C_{Saq}} = \frac{C_{Aorg}}{C_{Aaq}}$. Then, the system comes to the first case. The extremal situation will be with the water-immiscible active solvent, when $\frac{C_{Sorg}}{C_{Saq}} << \frac{C_{Aorg}}{C_{Aaq}}$: only at very high concentrations of acid the synergistic effect of the active solvent can be detectable (if at all).

3.3. High Acid Concentrations

At concentrations of the acid, reaching or prevailing its stoichiometry with the amine, or at transition from Regions 1 and 2 to Regions 3 and 4 the synergistic, enhancing effect of the active solvent reaches a maximum with increasing concentration, then stabilizes or decreases. This may be explained by competition between acid and active solvent for nuclei aggregates bridging. The maximum value may be evaluated, using relations, analogical to Eqns (9.10 and 9.11):

$$\frac{(X_{SA} - X)_{max}}{C_E} = \frac{k_{SE}}{k_{SA}} V_A (X_{SA} - X_{SE}) - \left[\frac{k_{SE}}{k_{SA}} V_A - V_E\right] (X_{SA} - X)_{max} \quad (9.21)$$

$$\frac{1}{Z_{\max}} = 1 + \frac{k_{\mathrm{SA}}}{k_{\mathrm{SE}}} \times \frac{[C_{\mathrm{A}}]_{\max}}{C_{\mathrm{E}}} \tag{9.22}$$

For every solvent and every acid extracted the maximal enhancing effect will be at different solvent–acid compositions. But it is evident that the concentration of the solvent in the organic phase, C_{Aorg}, has to be lower than amine C_{E} and acid C_{sorg}.

If the additive A is immiscible in the aqueous phase, above considerations for two acids, where weak one is immiscible, are valid in this case also.

3.4. Case 4: $\frac{k_{\mathrm{SE}}}{k_{\mathrm{SW}}} < \frac{k_{\mathrm{AE}}}{k_{\mathrm{AW}}}$ and $\frac{C_{\mathrm{S\ org}}}{C_{\mathrm{S\ aq}}} < \frac{C_{\mathrm{Aorg}}}{C_{\mathrm{Aaq}}}$

The active solvent will play an antagonistic effect on the distribution of the acid into organic phase.

Comparing values of the relations, $\frac{k_{\mathrm{SE}}}{k_{\mathrm{SW}}} \times \frac{C_{\mathrm{Sorg}}}{C_{\mathrm{Saq}}}, \frac{k_{\mathrm{SE}}}{k_{\mathrm{SA}}} \times \frac{C_{\mathrm{E}}}{C_{\mathrm{A}}}$ and $\frac{k_{\mathrm{AE}}}{k_{\mathrm{AW}}} \times \frac{C_{\mathrm{Aorg}}}{C_{\mathrm{Aaq}}}$, obtained at experiments with S–E–W, S–E–A and A–E–W systems, respectively, one can calculate $[C_{\mathrm{Aorg}}]_{\max}$ and $[C_{\mathrm{Aaq}}]^0$ compositions for optimal extraction of the acid of interest at its different concentrations.

These are general considerations about enhancing effect of the active additives in the extraction system, arising from the presented theory.

It has to be stressed that the presented theory predicts the synergistic or antagonistic effect of neutral protic, polar/polarizable solvents when added to the basic amine extractant on extraction of acids. Classical theories do not provide direct knowledge about this effect.

4. EXPERIMENTAL APPLICATION OF THE CCST IN CARBOXYLIC ACIDS EXTRACTION

Some experimental results of extraction are presented below with an aim to illustrate applicability of the CCST.

4.1. Acetic Acid Extraction

Original data for acetic acid extraction by Alamine 336 and Primene JMT with different solvents and diluents such as kerosene, nitrobenzene, chloroform, and cyclohexane were taken from Ref. 1,4–9 and rearranged according to the CCST. Results are presented in Table 9.1 and Figs. 9.1 and 9.2.

Distribution of acetic acid into organic phase of Alamine 336 in kerosene (see Fig. 9.1) did not detect up to 0.5 M/L concentration in the aqueous phase; then extraction begins but with very low effect as it is in Region 1. And only at acid concentration of >0.8 M/L in the aqueous phase visible extraction begins. High values of affinity constant ratios at high aqueous phase acid concentrations permit us to suppose that in this range of the extraction curve some three-dimensional cyclic- or cluster-like types of acid–amine aggregates are forming.

TABLE 9.1 Affinity Constant Ratios and Averaged Equilibrium Constants of Acetic Acid Extraction by Alamine 336 with Different Active Solvents, Calculated for Different Regions of the Extraction Isotherm Using Equations of the CCST

Extraction systems	Affinity constant ratios, k_{SE}/k_{SW}	Averaged equilibrium constants, $\overline{K}_{Si}$	Hypothetical Z^*(intercept) in pure E at $Q = 0$	Regions in the general extraction isotherm	Suggested aggregate formed (see Chapter 4)
Acetic acid–Alamine 336 in kerosene–water	0.25		−0.11	Formation of nuclei aggregates, Region 1	See Fig. 5.4 or 5 or Fig. 7a
	45.7		−39.3	Formation of three-dimensional aggregates, Region 4	
Acetic acid–Alamine 336 15% v/v Chloroform in *n*-heptane–water	1.07	0.67	0.04	Formation of aggregate stable through Regions 1–3	See Fig. 5.4 or 5 or Fig. 7a
Acetic acid–Alamine 336 in nitrobenzene–water	8.66		0.019	Formation of nuclei aggregates, Regions 1–2	See Fig. 5.4, 5.5, or Fig. 7a
	0.51		1.64	Formation of linear or cyclic aggregates, Region 3	
Acetic acid–Primen JMT in cyclohexane–water	9.11	0.67	−0.14	Formation of nuclei aggregates, Regions 1–2	
	1.06		0.81	Formation of linear or cyclic aggregates, Region 3	

Source: Experimental data from Ref 1,17 with permission

So, we can suggest transition of extraction curve from Region 1 directly to Region 4: nucleus aggregates, forming in the beginning of the experiment serve as nucleation centers of formation of three-dimensional, voluminous aggregates or even as centers of polymerization.

We can see a quite different situation at extraction of acetic acid by Alamine 336 with 15% v/v chloroform in *n*-heptane (see Fig. 9.1). Acid is distributed to organic phase with formation of 1/1 M_{ac}/M_{am} stable compound (nuclei aggregates) in the range from 0 to 2.0 M/L aqueous phase acid concentrations, or, in the range of 1–3 regions of the CCST. This is an example of the linearity of Eqn (9.2) plots (see Chapters 6 and 7) over the whole concentration range when the same affinity constant ratio is valid at low and high solute concentrations and does not depend on its concentration.

Organic solution of Alamine 336 in nitrobenzene shows new type of extractant for acetic acid (see Table 9.1 and Fig. 9.2). The shape of the first range of the curve witnessed mainly strong interactions of Region 2. Two types of nucleus aggregates are mainly supposed to form: complexes with $M_{ac}/M_{am} > 1/1$, perhaps, with two molecules of acid to one molecule of extractant, SHEHS = 2/1 (see details in Chapter 6) and regular complex with 1/1 M_{ac}/M_{am} (EHS, see Chapter 6). Then, the shape is turning to the extraction mechanisms appropriate to Region 3. Linear (or cyclic) aggregates are formed with bridging partly through acid or solvent molecules.

Example with extraction of acetic acid by Primene JMT in cyclohexane (Table 9.1, Fig. 9.2) shows the same trend, but with different types of complexes formed. Region 2 of nuclei aggregates consists mainly of 1/1 M_{ac}/M_{am} and 1/2 M_{ac}/M_{am} (EHSE, see Chapter 6) complexes. Linear or cyclic aggregates, formed in Region 3, are bridged by acid molecules only.

More exact structures cannot be suggested because of lack of data required. The reader have to remember that these experimental data[1,6] were obtained with classical approach.

4.2. Lactic Acid Extraction

Data for lactic acid extraction by Alamine 336 in kerosene, *n*-Cresol, MIBK, and with chloroform in *n*-heptane solvents and diluents were rearranged from[1,6] and/or obtained in our laboratory.[17,23] Results are presented in Table 9.2 and Figs. 9.3 and 9.4.

Shapes of the extraction curves show the same mechanism of the strong complexes formation, appropriate to Region 2; all of them with 1/1 M_{ac}/M_{am} ratio, but with different acid/amine affinity ratios: A-336 in m-Cresol >> A-336 in kerosene > A-336 with chloroform in *n*-heptane >> A-336 in MIBK. Here, we can clearly see the influence of active solvents and/or diluents on extraction effectiveness. Region 2 transfers to Region 3 with formation of linear (cyclic) aggregates with bridges bonded through competed acid and solvent (or water)

TABLE 9.2 Affinity Constant Ratios and Averaged Equilibrium Constants of Lactic Acid Extraction by Alamine 336 with Different Active Solvents, Calculated for Different Regions of the Extraction Isotherm Using Equations of the CCST

Extraction systems	Affinity constant ratios, k_{SE}/k_{SW}	Averaged equilibrium constants, $\overline{K}_{Si}$	Hypothetical Z^*(intercept) in pure E at $Q = 0$	Regions in the general extraction isotherm	Suggested aggregate formed (see Chapter 4)
Lactic acid–Alamine 336 in kerosene–water	82.9	76.8	0.098	Formation of nuclei aggregates, Region 2	See Figs. 6.3, 6.4
	0.34		0.98	Formation of linear or cyclic aggregates, Region 3	See Fig. 8.5
Lactic acid–Alamine 336 in *m*-Cresol–water	233.0	164.0	−0.04	Formation of nuclei aggregates, Regions 1–2	See Figs 6.1, 6.2, 6.5, or Fig. 7b
	0.56		0.96	Formation of linear aggregates, Region 3	See Fig. 8
Lactic acid–Alamine 336 with 15% v/v Chloroform in *n*-Heptane–water	70.0	112.1	0.012	Formation of nuclei aggregates, Region 2	See Figs 6.3, 6.4
	0.34		0.98	Formation of linear or cyclic aggregates, Region 3	See Figs 5.4, 5.5, or Fig. 7a
Lactic acid–Alamine 336 in MIBK	9.1		0.05	Formation of nuclei aggregates, Region 2	See Fig. 8.5
	0.80		0.99	Formation of linear or cyclic aggregates, Region 3	

Source: Experimental data from Ref. 1,17 with permission molecules at acid/solvent (or water) molecules ratio: A-336 in MIBK > A-336 in m-Cresol > A-336 in kerosene = A-336 with chloroform in *n*-heptane.

The same may be stated that more exact structures cannot be suggested because of lack of data required.

4.3. Citric Acid Extraction

Some results, obtained for citric acid–Alamine 304 in kerosene–water (S–E–W) and citric acid–Alamine 304 in kerosene or dodecane as diluents-water-alcohol (1-(or 2-)octanol, 1-propanol) systems S–E–W–A, obtained in our laboratory,[17,23] are rearranged and presented in Table 9.3 and Figs 9.5 and 9.6 to illustrate the presented theory.

Shapes of the extraction curves show mechanism of the complexes formation, appropriate to Region 2. All of them are with 1/2 M_{ac}/M_{am} ratio (type of ESHE, see Chapter 6). It is evident from the acid/amine affinity constant ratios data, A-304 with 1-propanol, $k_{SEA}/k_{SAW} = 5.11$ (S–E–W–A system) > A-304 with 1-octanol, $k_{SEA}/k_{SAW} = 4.57$ > A-304 with 2-octanol, $k_{SEA}/k_{SAW} = 3.89$ > A-304 in kerosene, $k_{SE}/k_{SW} = 2.1$ (S–E–W system), that alcohols show a small synergistic (enhancing) effect. Differences are not high, but alcohols play here a strong modification (stabilization) role, preventing the system from the formation of the third phase.

Analysis of water content in the organic phase of the S–E–W system samples shows two water molecules on every molecule of acid in Region 2 (nuclei aggregates). Suggested structure examples of nuclei aggregates are presented in Fig. 9.7a. Analysis of the water and 1-propanol in the organic phase of the S–E–W–PrOH system samples shows the same two water molecules and one 1-propanol molecule content on every molecule of acid in Region 2. It seems that 1-propanol molecule is bound to the amine in the nucleus aggregate through the water molecule. Examples of suggested nuclei aggregate structures are presented in Fig. 9.7b. In the systems with octanols the water in the organic phase is not analyzed.

It should be noted, the absence of water in the organic phase after mixing of Alamine 304 in kerosene (E–W system) and Alamine 304 with 1-propanol in kerosene (E–A–W system) with water and separation of the aqueous phase: content of water was on the detect level.

At transition of the extraction systems to Region 3 the alcohols influence differently. Considering the S–E–W system at formation of linear and/or cyclic aggregates in Region 3, each nucleus aggregate is loosing one of water molecules on increasing aqueous phase acid concentration. Approaching to Region 4, water content decreases to slightly more than one (≈ 1.15) water molecules to one acid molecule. The nuclei aggregates are bridging through competition among second molecule of water, the neutral molecule of amine, or additional molecule of acid (see Fig. 9.8a). The linear aggregate is closed to the cyclic form perhaps through the alcohol or acid molecule. According to

TABLE 9.3 Affinity Constant Ratios and Averaged Equilibrium Constants for Citric Acid Extraction by Alamine 304 in Kerosene with Alcohols (1-or 2-Octanol, 1-Propanol) as Active Solvents, Calculated for Different Regions of the Extraction Isotherm Using Equations of the CCST

Extraction systems	Affinity constant ratios, k_{Si}/k_{Sj}	Averaged extraction constants, $\overline{K}_{Si}$	Hypothetical Z^a(intercept) in pure E at $Q = 0$	Regions in the general extraction isotherm	Suggested aggregate formed (see Chapter 4)
Citric acid–Alamine 304 in kerosene–water	2.1	1.93	0.00	Formation of nuclei aggregates, Region 2	See Fig. 3a
	0.082	0.37	0.55	Formation of linear or cyclic aggregates, Region 3	See Fig. 4a
Citric acid–Alamine 304 1-Propanol–Water	5.11	3.58	0.01	Formation of nuclei aggregates, Region 2	See Fig. 3a
	0.27	0.19	0.49	Formation of linear or cyclic aggregates, Region 3	See Fig. $4b_1$ See Fig. $4b_4$
	−0.05	−0.090	0.69	Change of aggregates composition, Region 3	
	0.34	0.190	0.19	Transition to Regions 4	
Citric acid–Alamine 304 1-Octanol–water	4.57	3.13	0.01	Formation of nuclei aggregates, Region 2	See Fig. 3b
	0.15	−0.108	0.48	Formation of linear or cyclic aggregates, Region 3	See Fig. $4b_1$

(Continued)

TABLE 9.3 Affinity Constant Ratios and Averaged Equilibrium Constants for Citric Acid Extraction by Alamine 304 in Kerosene with Alcohols (1-or 2-Octanol, 1-Propanol) as Active Solvents, Calculated for Different Regions of the Extraction Isotherm Using Equations of the CCST—cont'd

Extraction systems	Affinity constant ratios, k_{Si}/k_{Sj}	Averaged extraction constants, $\overline{K}_{Si}$	Hypothetical Z^a(intercept) in pure E at $Q=0$	Regions in the general extraction isotherm	Suggested aggregate formed (see Chapter 4)
Citric acid–Alamine 304 2-Octanol–water	3.89	2.87	0.008	Formation of nuclei aggregates, Region 2	See Fig. 3b
	0.12	−0.080	0.54	Formation of linear or cyclic aggregates, Region 3	See Fig. $4b_1$
	0.35	0.084	0.20	Transition of aggregate compositions to Region 4	

[a]*Hypothetical saturation factor, Z, calculated from the intercept of the graphically obtained hypothetical molar extinction coefficient* ε^f_{SE} *or* ε^f_{SEA} *of acid in pure amine or in the amine–alcohol mixture in the absence of aqueous phase.*
Source: Experimental data from the Ref. 17

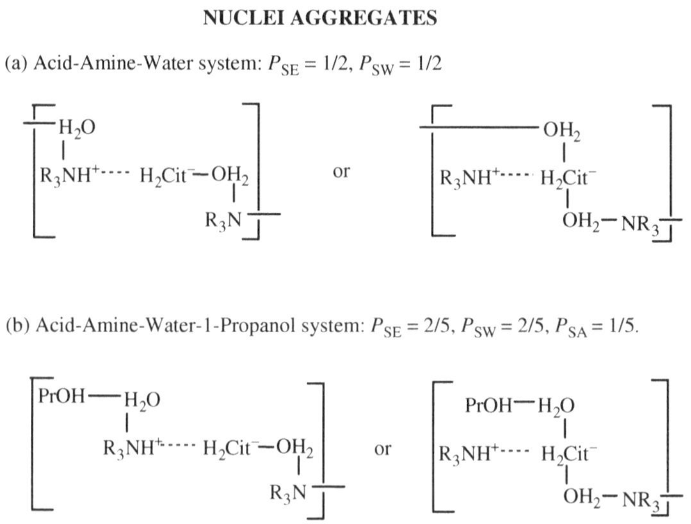

FIGURE 9.7 Suggested nuclei aggregate structures (Region 2), formed on extraction of citric acid by (a) Alamine 304 in kerosene (S–E–W system) and (b) Alamine 304 with 1-propanol in kerosene (S–E–W–A system).

the water concentration data we can suggest that the cycle contains 5–6 molecules of acid. Additional quantitative, or even semiquantitative, spectrometric determination of citric acid in the complexed (carboxylate) and solvated (neutral) forms (C_{SEc} and C_{SEs}) will indicate the correct structure of the cycle.[17]

Analysis of the water and 1-propanol in organic phase of the S–E–W–PrOH system samples shows the same two water molecules and one 1-propanol molecule content on every molecule of acid in Region 2 (see Fig. 9.7b). It seems that the 1-propanol molecule is bound to the amine in the nucleus aggregate through the water molecule. Examples of suggested nuclei aggregate structures are presented in Fig. 9.8b.

On formation of linear aggregates (Region 3) in the S–E–W–PrOH system approximately the same composition of the aggregate (solvation shell) is observed up to Z maximum in the extraction isotherm. This means that the linear aggregate is formed by bridging of nuclei aggregates mainly through the molecules of acid or amine as shown in Fig. 9.8, b_1 and b_2. After the Z maximum, a small decrease of water concentration and a small increase of 1-propanol concentration are observed with content of 1.75–1.85 water molecules and of 1.15–1.25 1-propanol molecules on every acid molecule at approaching to Region 4. It seems that here the linear aggregates are closed to the cyclic form through the additional molecule of 1-propanol, loosing one molecule of water. The cycle contains 5–6 nuclei aggregates.

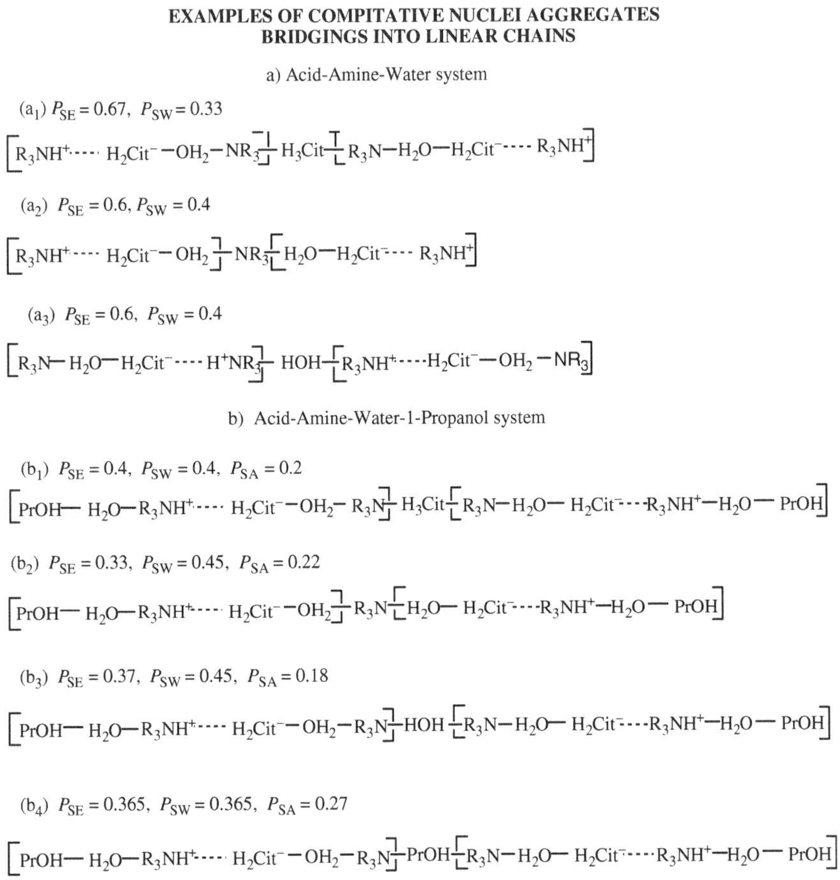

FIGURE 9.8 Suggested linear aggregate structures (Region 3), formed on extraction of citric acid by (a) Alamine 304 (S–E–W system) and (b) Alamine 304 with 1-propanol (S–E–W–A system).

Water content in the organic phase of the S–E–W–OctOH system samples was not analyzed. According to the data of affinity coefficient ratios and octanol concentrations in the organic phase samples with different acid loadings we can suggest that the interaction mechanisms and structure of nuclei and cyclic aggregates formed are similar to those of S–E–W–PrOH system. But that still needs to be proved experimentally.

The basic statement of the presented theory, such as cosolvent independence and transferability of affinity constant ratios from one system to the other, provides the key for inverse calculation of affinity constant ratios: k_{SE}/k_{SA}, k_{SA}/k_{SW}, k_{EA}/k_{AW} from the known values of k_{SE}/k_{SW}, k_{SEA}/k_{SAW} and respective concentrations of C_{Eorg}, C_{Sorg}, C_{Aorg}, C_{Worg}, C_{Saq}, C_{Aaq} at equilibrium.

Presence of the internal aqueous phase in the micelle-like aggregates (Region 4) is expected to show a strong increase of both acid and water concentrations in the organic phase. It was qualitatively observed on titration of the citric acid in the highly loaded ($Z \geq 0.7$) organic phase samples by NaOH (see Chapter 7, Fig. 7.8): the titration curves dramatically decreased and had minima at destruction of the micelles.

It is well known that micelle destruction by dilution of the solutions is sometimes a slow kinetic process. This means that kinetics of formation and destruction of the micelle-like aggregates may differ very much. Some researchers establish extraction experiments by equilibrating the highly loaded by acid organic samples with water or low acid concentration aqueous solutions[4,9]. Using this technique the equilibrium state has to be checked obligatory because of possible slow kinetics. In these experiments the equilibrium state has to be controlled very strongly.

4.4. Active Solvent-Swing Regeneration Technology

The active solvent-swing regeneration technology was first described by Baniel et al.[42] The composition of the acid-loaded organic phase at equilibrium can be changed by changing active solvent concentration. The forward extraction step is conducted with extractant composition containing high proportions of active solvent for efficient extraction. Prior to back-extraction, active solvent may be removed from the extract or another solvent may be added for efficient acid back-extraction to the aqueous phase.

The presented theory predicts the solvent-swing regeneration and permits its preliminary quantification. The values of $[C_{\mathrm{Worg}}]_{\mathrm{max}}$, $[C_{\mathrm{Aorg}}]_{\mathrm{max}}$, $\frac{[C_{\mathrm{Aorg}}]_{\mathrm{max}}}{C_{\mathrm{E}}}$, $\frac{[C_{\mathrm{Aorg}}]_{\mathrm{max}}}{C_{\mathrm{Aaq}}}$, and $[C_{\mathrm{Aaq}}]^0$ obtained by experiments with the three-component systems provide for evaluating the limits for optimal swing regeneration in the S–E–A–W system of interest.

4.5. Modifying Effect of Active Solvents

Active solvent with (or without) synergistic or antagonistic effects may serve as a modifier in an extraction system, stabilizing polar species in the organic phase by its dielectric properties. In general, the stabilizing effect is achieved by replacing water molecules and even acid molecules in case of a weak acid, by molecules of the active solvent in the solvation shell.

Below the reader can find some experimental results, obtained in our laboratory,[23] on investigation of systems with four components: citric acid–Alamine 304–water–active solvents (S–E–W–A) using conventional extraction procedures. The data were reinvestigated with an aim to explain and to prove the presented theory.

At low acid concentrations, in Regions 1 and 2, the active solvent molecules, replacing water, form stable hydrophobic nuclei aggregates. This leads to better separation of the organic and aqueous phases.

On extraction from concentrated acid solutions, or on transition to Region 3, bridging of nuclei aggregates takes place through the H-bonded solvent molecules with preferential formation of the stable low membered ring structures (see Fig. 9.8a) where water molecules are replaced partly or even completely by active solvent molecules. The coalescence of the cyclic aggregates into three-dimensional micelle-like or cluster-like aggregates (Region 4) and formation of the "third phase" takes place at much higher acid critical concentrations. Alcohols' modifying effect on the extraction systems is a good example. The tendency of alcohols to close aggregates into highly stable six-membered benzene-like cycles is a well-known fact.[26–29] On extraction of naphthenic acid by amine, photo correlation spectroscopy (PCS) data[43,44] showed that addition of 2-octanol into diluent moves the formation of reversed micelles to a much higher acid concentrations in the aqueous phase. Analogical results were obtained in our laboratory[23] on extraction of citric acid by trilauryl amine (A304) using 1-, or 2-octanol, or 1-propanol as modifying (and enhancing) solvents.

4.6. Temperature-Swing Regeneration

The effect of temperature on extraction of carboxylic acids was studied by many researchers.[9,23,41,42] As a rule, the distribution of acid into the organic phase (a variety of tertiary amines in a variety of diluents with modifiers) decreases with increasing temperature, sufficiently to allow back-extraction of some acids (e.g., citric, succinic, and lactic) with a fresh aqueous phase without dilution of the acid. The distribution ratios' decrease with temperature increase depends upon the type and concentration of acid. The temperature-swing regeneration includes forward extraction at a low temperature, obtaining acid-loaded organic phase and an aqueous raffinate, containing feed components that have to be separated, and back-extraction at a higher temperature, obtaining acid-loaded aqueous product stream and regenerated organic phase. For some acids, the acid concentrating effect in a product stream may be achieved using temperature-swing process.

The presented theory permits to achieve preliminary quantification of the temperature effect for the temperature-swing regeneration for the S–E–A–W extraction system using the data, obtained at experiments with the simple three-component S–E–W, S–A–W, S–E–A, and E–A–W systems.

From the Eqn (9.18) it follows:

$$\ln\left(\frac{k_{SE}}{k_{SW}} \times \frac{C_E V_E}{C_{Worg} V_W}\right) = \frac{-\Delta H_{SEW}}{RT} + \frac{\Delta S_{SEW}}{R} \tag{9.23}$$

$$\ln\left(\frac{k_{SE}}{k_{SA}} \times \frac{C_E V_E}{C_{Aorg} V_A}\right) = \frac{-\Delta H_{SEA}}{RT} + \frac{\Delta S_{SEA}}{R} \tag{9.24}$$

$$\ln\left(\frac{k_{EA}}{k_{AW}} \times \frac{C_{Aorg}V_A}{C_{Worg}V_W}\right) = \frac{-\Delta H_{EAW}}{RT} + \frac{\Delta S_{EAW}}{R} \tag{9.25}$$

$$\ln\left(\frac{k_{SA}}{k_{SW}} \times \frac{C_{Aorg}V_A}{C_{Worg}V_W}\right) = \frac{-\Delta H_{S\ AW}}{RT} + \frac{\Delta S_{SAW}}{R} \tag{9.26}$$

If the enthalpy ($-\Delta H$) and entropy (ΔS) of reaction are assumed to be constant over the temperature range, the Eqns (9.21–9.26) indicate that a plots of ln(…) versus 1/T give straight lines. The slopes are proportional to the enthalpy of reactions and the intercepts are proportional to the entropy.

Formation of complexes and aggregates in the organic phase makes the system more ordered and thus decreases the entropy.

Affinity coefficients ratios, k_{SE}/k_{SW}, k_{SE}/k_{SA}, k_{SA}/k_{SW}, k_{EA}/k_{AW} and different concentration ratios, C_{SE}/C_{SW}, C_{SE}/C_{SA}, C_{EA}/C_{AW}, in the solvation shell, are measured, experimenting with simple systems at various temperatures. Temperature dependence of water coextraction by free extractant and active or inert solvents in the bulk organic and aqueous phases is measured in the absence of acid. Temperature dependence of the solubility factors of active solvent FA $=$ f(T) and water FW $=$ f(T), and concentrations of active solvent $C_{Aorg} = f(T)$ and water $C_{Worg} = f(T)$ in organic phase are obtained.

According to the theory, preliminary data for the temperature-swing regeneration of S–E–A–W system of interest may be calculated using the above independent experiments' data for the simple systems. So, the experimental studies simplify to the series of measurements for the three-component, S–E–W, S–A–W, S–E–A, and E–A–W, systems.

Validity of the presented approach was verified experimentally in our laboratory[23] and has shown a good result between experimental and calculated results. This gives a key for quantitative prediction of suitable extraction systems at different acids separation.

5. SUMMARIZING REMARKS

The presented CCS modeling approach constitutes a general framework for interpretation of strong (ion exchange and electrostatic) and weak (H-bond, solvating, and coordinate) interactions data in the acid–solvent extraction systems. It is an attempt to describe the mechanisms of solvent extraction with quantitative evaluation of different compounds, formed in the organic phase at different concentrations of the acid and active solvent.

It is stressed once again that the presented theory predicts the synergistic, enhancing effect of weak amines, weak acids, or neutral protic, polar/polarizable solvents as additives to the basic amine extractant on extraction of acids. Classical theories do not indicate the direct knowledge about this effect.

Two more parameters, water and active adduct concentrations in the organic phase, are introduced to the quantitative consideration of extraction system in every region of the general extraction systems' isotherm. Mathematical description for the process simulation is proposed.

Preliminary predictions may be made before experimentation; e.g., for strong acid and amine Region 1 of the general extraction systems' isotherm may be excluded. Region 3 is a plateau with small magnitude of the slope. The slope depends on the influence of water and active solvent in the formation of linear (planar) aggregates. Region 4 occurs (if at all) at the very high concentrations of acid.

For weak acid and base Region 2 may be excluded and Region 1 is transferred to Region 3 without visible shape changes on the distribution curve. The slope of the curve shows strong dependency on the weak molecular interactions of undissociated acid, water, and active diluent, competing at nuclei, then forms linear aggregates.

Experimental determination of the affinity constant ratios and solvation shell compositions, quantitative, or even semiquantitative, spectrometric determination of the acid in the complexed and solvated forms permits to set up exact form of the aggregates formed and predict the behavior of the extraction system of interest and influence of its different components.

The presented theory predicts the synergistic or antagonistic effect of active solvents on distribution to organic phase in basic amine extractants. Classical theories do not indicate the direct knowledge about this effect.

Affinity constant ratios are easily acceptable through the independent measurements conducted with simple three-component systems. Experimental determination of the affinity constant ratios and the aggregate structures formed permits to analyze and predict the extraction system of interest. Relation (7) between equilibrium constant k_{Si} of the classical theories and affinity constant ratio, k_{Si}/k_{Sj}, obtained by the measurements of the same property, permits the use of available equilibrium constant data for calculations in the presented theory.

Active solvent parameters as additional data are introduced into the quantitative consideration of extraction system in every region of the general extraction isotherm. Mathematical description for the process simulation is proposed.

The concept of partitioning in the homogenous media implies that the extraction system components act independently in their interaction with a given solute: the values of k_{SE}/k_{SW} and k_{SA}/k_{SW} are independent and transferable from one system to the other. Cosolvent independence, established experimentally, allows to compare directly the complexing-solvating power (competition order) of various extractants, solvents, and diluents. Once determined and tabulated, the data can be used in any system, containing these components. Independence and transferability of affinity constants have been proved for many organic systems. Nevertheless, it has to be proved for different extraction systems of interest.

Although the theory has many simplifications and limitations, it is a starting point for the quantification of acid solvent extraction processes with an aim to predict suitable extraction systems for separation of different solutes.

REFERENCES

1. Tamada J, King C. *Extraction of carboxylic acids by amine extractants*. Ph.D. Thesis. Lawrence Berkeley Laboratory, University of California, Applied Science Division; 1989.
2. Komasawa I, Otake T, Hattori I. The effects of diluent in the liquid-liquid extraction of copper and nickel using 2-hydroxy-5-nonylbenzophenone oxime. *J Chem Eng Japan* 1983; **16**(5):377–88.
3. Hogfeldt E, Fredlund F, Odberg L, Merenge G. Extraction of water, nitric and perchloric acids by nitrobenzene. *Acta Chem Scandinavica* 1973;**18**(5):1781–94.
4. Eyal AM, Canari R. pH Dependence of carboxylic and mineral acid extraction by amine-based extractants: effects of pKa, amine basicity, and diluent properties. *Ind Eng Chem Res* 1995;**34**(5):1789.
5. Tamada J, Kertes A, King C. Extraction of carboxylic acids with amine extractant. 1. Equilibria and law of mass action modeling. *Ind Eng Chem Res* 1990;**29**:1319–26.
6. Tamada J, Kertes A, King C. Extraction of carboxylic acids with amine extractant. 2. Chemical interaction and interpretation of data. *Ind Eng Chem Res* 1990;**29**:1327–32.
7. Tamada J, Kertes A, King C. Extraction of carboxylic acids with amine extractant. 3. Effect of temperature, water coextraction and process considerations. *Ind Eng Chem Res* 1990;**29**:1333–6.
8. Yang S, White S, Hsu S. Extraction of carboxylic acids with tertiary and quaternary amines: effect of pH. *Eng Chem Res* 1991;**30**:1335.
9. Canari R, et al. Carboxylic acids extraction by basic extractants: the effects of pH, Ion concentration and acid-base property of the extractant and extracted acid. In: *In the Proceedings of the ASEC'96*, 1996, p. 1517–22.
10. Grinstead RR. Base strengths of amine in liquid-liquid extraction systems. In: *In the Proceedings of the ASEC'67*, 1968, p. 427–36.
11. Grinstead R, Davis J. Base strength of amine-amine hydrochloride systems in toluene. *J Phys Chem* 1968;**72**(5):1630.
12. Mrnka M, Kyrsova V, Celeda J. Mechanism of iron separation by extraction with tri-n-octylamine from chloride solutions. *Proceedings of ISEC'74* 1975;**1**:49–62.
13. Vieux AS, et al. Extraction of some dicarboxylic acids by tri-isooctylamine. *Anal Chim Acta* 1974;**68**:415–24.
14. Wardel JM, King CJ. Solvent equilibria for extraction of carboxylic acids from water. *J Chem Eng Data* 1979;**23**(2):144–8.
15. Kolarik Z, Grimm R. Acidic organophosphorus extractants. XXIII. Distribution of mono-2-ethylhexyl phosphoric acid in two liquid phase systems. *J Inorg Nucl Chem* 1976;**38**(5):1761–3.
16. Headley AD. Quantitative analysis of solvation effects and influence of alkyl substituents on basicity of amines. *J Org Chem* 1991;**56**:3688–91.
17. Kislik V, Eyal A, Hazan B. Competitive complexation/Solvation theory of solvent extraction. III. Influence of active solvents on acid solvent extraction by amine based extractants. *Sep Sci Technol* 2003;**38**(8):1681–704.
18. Kislik V. Application of preferential solvation concept for interpretation of the mechanisms of acid solvent extraction by amines. *Sep Sci Technol* 2001;**36**(13):2899–926.
19. Lo TC, et al., editors. *Handbook of solvent extraction*. New York: Wiley; 1983.
20. Marcus Y. *Ion solvation*. London: Wiley-Interscience; 1985.
21. Kertes AS, King CJ. Extraction chemistry of fermentation product carboxylic acids. *Biotech Bioeng* 1986;**28**:269–82.

22. Kislik V, Eyal A. Extraction of titanium (IV) by mixtures of mono- and d-(2-ethylhexyl) phosphoric acids esters. *Solvent Extr Ion Exch* 1993;**11**(2):285–310.
23. Eyal A, Kislik V, Hasan B. Experimental research project for private company, 1996–1998, unpublished data.
24. Taft RW, Abboud JLM, Kamlet MJ, Abraham MN. Linear solvation energy relations. *J Solution Chem* 1985;**14**(3):153–86.
25. Shorter J. *Correlation analysis in chemistry.* Oxford: Clarendon Press; 1973.
26. Fendler EJ, et al. Proton magnetic resonance investigation of alkylammonium carboxylate micelles in nonaqueous solvents. *J Phys Chem* 1973;**77**(11):1432–6.
27. Mei Wen, Yiyan Yang, Youyuan Dai. Studies on the mechanism of extraction of carboxylic acids with Tri-N-Octylamine by chemical complexation. In: *In the Proceedings of the ISEC'96*, 1996. vol. 1, p. 195–200.
28. Ermer O, Eling A. Molecular recognition among alcohols and amines: Super-tetrahedral crystal architectures of linear diphenol-diamine complexes and aminophenols. *J Chem Soc Perkin Trans* 1994;**2**:925–44.
29. Loehlin JH, Etter MC, Gendreau C, Cervasio E. Hydrogen-bond patterns in several 2:1 amine-phenol cocrystals. *Chem Mater* 1994;**6**:1218–21.
30. Fumio Toda, Hyoda S, Okada K, Hirtsu R. Isolation of anhydrous hydrasine as stable inclusion complexes with hydroquinone and p-methoxyphenol and their solide state reaction with esters which gives pure hydrolyzides. *J Chem Soc Chem Commun* 1995;1531–2.
31. Devis MM. *Acid-base behavior in aprotic organic solvents.* Washington, DC: National Bureau of Standards, Monograph 105; 1968.
32. Anet FAL, Carter P, Wenstein S. Effects of steric compression on coupling constants. *J Chem Soc* 1965;**67**:2048–57.
33. Spala EE, Ricker NL. Thermodynamic model for solvating solutions with physical interactions. *Ind Eng Chem Process Des Dev* 1982;**21**:409–15.
34. Nagata I, et al. Ternary liquid-liquid equilibria: the associated solution theory. *Fluid Phase Equilib* 1977/1978;**1**:267–75.
35. Headley AD. Solvent effects on the basicity of alkyl-substituted dimethylamines. *J Org Chem* 1988;**53**:312–4.
36. Headley AD, McMyrry ME, Starnes SD. Effects of substituents on the acidity of acetic acid. *J Org Chem* 1994;**59**:1863–6.
37. Nagy OB, Muanda H, Nagy JB. Competitive preferential solvation theory of weak molecular interactions. *J Chem Soc Faraday Trans 1* 1978;**74**:2210–29.
38. Nagy OB, et al. Quantitative evaluation of solute-solvent interactions. Environmental effects on the n-butylaminolysis of tetrachloro-N-n-butylphthalimide. *J Phys Chem* 1979;**83**:15.
39. Scatchard G. The attraction of proteins for small molecules and ions. *Ann NY Acad Sci* 1949;**51**:660.
40. Eyal AM. Acid extraction by acid-base-coupled extractants. In: Marinsky JA, Marcus Y, editors. *Ion exchange and solvent extraction*, vol. 13. New York: Marcel Dekker Inc.; 1997. p. 31–93.
41. Cohn-Syzov N, Eyal AM. Extraction and interactions in two-phase, multi-acid, multi-base, systems. In: In the *Proceedings of the ISEC'99*, 1999, p. 363.
42. Baniel AM, et al. Recovery of acids from aqueous solutions. *U.S. Patent 4,275,234*, 1981.
43. Parbhoo B, Nagy O. Application of the competitive preferential solvation theory to ion-molecule interactions. *J Chem Soc Faraday Trans 1* 1986;**82**:1789–93.
44. Laub RJ, Purnell JH. Solution and complexing Studies. III. Further evidence for a microscopic partitioning theory of solution. *J Am Chem Soc* 1976;**98**(1):30–5.

Chapter 10

Competitive Complexation/ Solvation Theory in Metal Solvent Extraction

Chapter Outline

1. INTRODUCTION

The complicated behavior of metal extraction systems conflicts in some cases with the classic stoichiometric ion-exchange models. The dual maximum behavior of titanium(IV) and many other transition metals (see Chapter 6,

Solvent Extraction. DOI: 10.1016/B978-0-444-53778-2.10010-X

Table 6.1), found in most families of extractants such as acidic (organic acids), basic (amines or their mixtures), and neutral complexants, is difficult to explain from classical theories point of view. The dual behavior phenomenon is easily explained by presenting competitive complexation/solvation theory (CCST) approach.

The difference in the extraction of metals must be attributed to solvation (coordination) effects. Many authors[1–6] modeling extraction systems use the order of an interaction reaction for a given metal ion and extractant only. But in the same works, a strong, sometimes even critical influence of solvation (hydration) effects on the extraction is also discussed: the polar, polarizable, hydrogen-bonding solvents change the extraction order dramatically.

The methodology of coordination chemistry in solution has been applied extensively to the determination of the equilibrium constants for solvent replacement in the coordination spheres of metal ions. Coordination models, explaining qualitatively some mechanisms of metal solvent extraction, including the dual extraction behavior for different extractants, meet difficulties when modeling anion exchangers, and, especially, at the attempts of quantification of the models. The competitive complexation/solvation modeling approach (see Chapter 7) is introduced to explain the mechanisms of solvent extraction of metal ions by different acidic, basic, or mixed extractants.

2. CCST BASIC STATEMENTS

CCST of solvent extraction is based on four statements, which are well-known facts in the chemistry. These statements, modified for metal solvent extraction, are explained in the following subsections.

2.1. Stepwise Solvation–Complexation Interactions

Extraction act is initiated by solvation of the metal compounds in organic solvents with the formation of hydrogen- (coordinate-) bonded intermediates (adducts). Kinetics of proton, electron transfer, and formation of final compounds is stepwise and depends on kinetics of step reactions, competition, coupling, strength of hydrogen bonding, and many other factors (mentioned in Chapters 6 and 7). The kinetics may be fast and the final products formed during extraction process; may be slow, so, extracted species are some of intermediates and the latter process can be called as a pseudo-equilibrium extraction.

Metal ions, as well as all other species, in the phases are surrounded by a solvation shell.[7] This means that they interact with all the constituents of the environment and, therefore, "free" metal ions, free extractant, or free metal–extractant complex do not exist. Components of the solvation shell compete to form a complex according to their electronic-geometric affinity k_i (interaction strength), which is constant at a given temperature and pressure. The molecules in the solvation shell continuously and rapidly relax between complexing and solvating states.

Metal salts dissolved in water are hydrolyzed and coordinated by water molecules (hydrated), whereas those dissolved in strong acid are coordinated by acid molecules (solvated).[1,2,8,9] The cationic, neutral, or anionic species formed strongly depend on pH (acidity) of the solution. Example of the approximate distribution of titanium(IV) chloride species as a function of pH (acidity) is presented in Fig. 10.1. The diagram was summarized using reviewed literature and experimental data.[1,2,10,11] Depending on the ligand complexing order, the similar picture of titanium compounds in other mineral acids were obtained.

The presented approach suggests that acid–base act is initiated not by self-ionization in aqueous or organic phases, as it stated in classical theories, but by solvation of the species by organic environment. They are mainly organic solvents, particularly those containing nitrogen, phosphorus-oxygen, and sulfur-oxygen functional groups.

This is the first step of the interactions in extraction process with formation of intermediates (adducts).

In our investigation of titanium(IV) extraction with di(2-ethylhexyl)phosphoric acid (DEHPA),[2,10,11] the following reactions were suggested between Ti species and extractant in the regions IIa and IIb in Fig. 10.1. For the Region IIa, at aqueous phase acidity is 0.01–0.5 M (pH = 2.0–0.3):

$$[\mathrm{Ti(OH)_2 * 4H_2O}]^{2+}_{\mathrm{aq}} + 4\mathrm{DEHPA_{org}} \leftrightarrow [\mathrm{Ti(OH)_2 * (DEHP)_2 * 2H_2O} * 2\mathrm{DEHPA}]_{\mathrm{org}} + 2\mathrm{H^+aq} + 2\mathrm{H_2O} \quad (10.1)$$

Here, acid–base act is initiated by solvation of the dissolving compounds. Then, ion-transfer reactions take place: two DEHPA molecules lose protons to form complex with $\mathrm{Ti(OH)_2^{2+}}$ cation. So, four anionic ligands and two water molecules are bonded to titanium central atom to occupy titanium(IV) ion coordination sites up to saturation in the first (inner) coordination sphere. Two neutral, DEHPA molecules are bonded by hydrogen bonding to the water dipoles in the second (outer) coordination sphere.[2,10,11] Experimental data, presented in Fig. 10.2a, proves these suggestions.

This colorless aggregate is not stable and slowly (during several weeks of organic phase aging) loses its water molecules and partly DEHPA, forming yellow complex with bidentate bonds of titanium atom with oxygens of DEHP phosphoryl group:

$$[\mathrm{Ti(OH)_2 * (DEHP)_2 * 2H_2O * 2DEHPA}]_{\mathrm{org}} \xrightarrow{\text{aging}} [\mathrm{TiO * (DEHP)}]_{\mathrm{org}} + 3\mathrm{DEHPA_{org}} + 3\mathrm{H_2O} \quad (10.2)$$

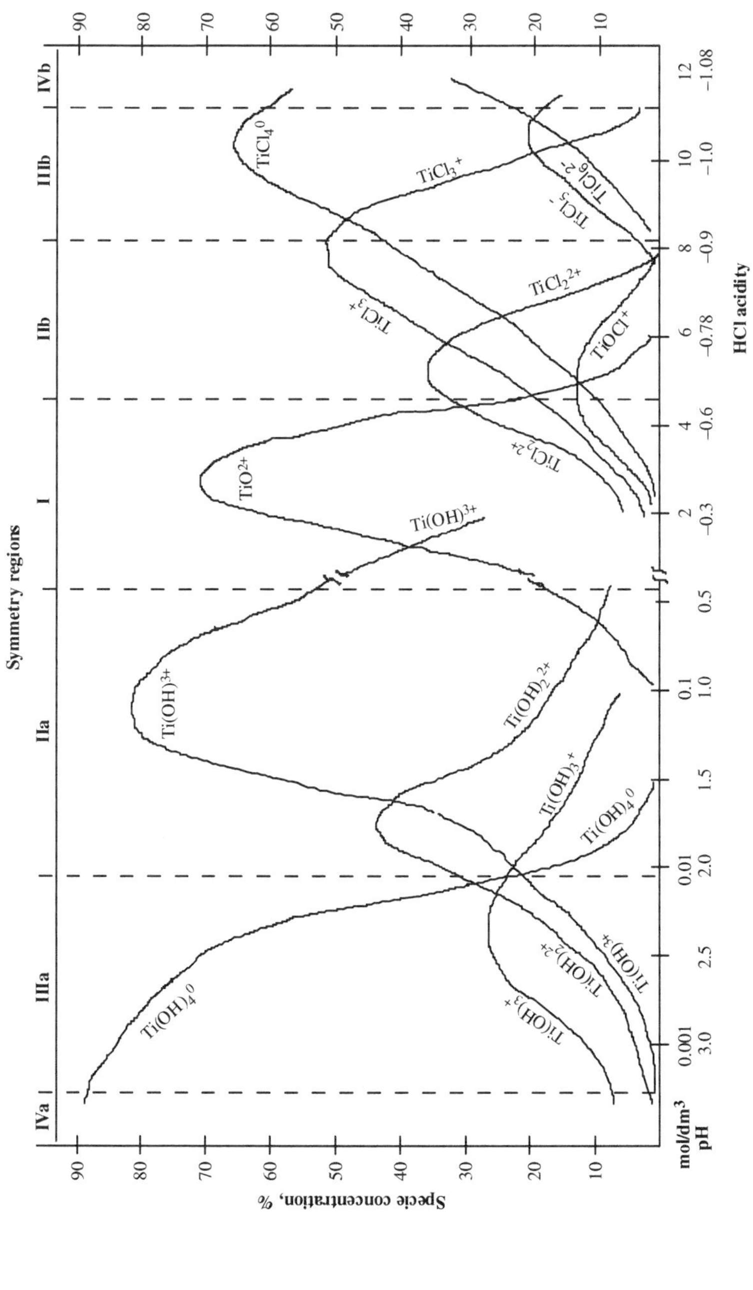

FIGURE 10.1 Approximate composition of titanium(IV) species in hydrochloric acid aqueous solutions (summarized, reviewed, and experimentally obtained data[1,2,8,9]).

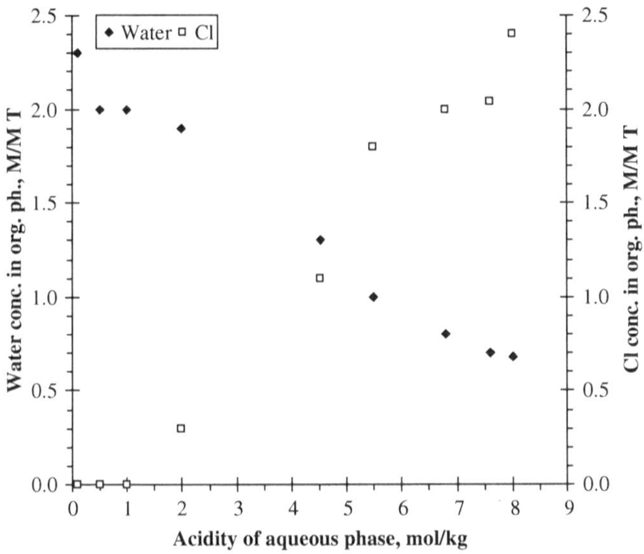

FIGURE 10.2 Dependence of water and chlorine concentration in organic phase on acidity of the aqueous phase at extraction of titanium(IV) by DEHPA.

Analogically, the structure of the nucleus aggregate formation at extraction of titanium from the strongly acidic aqueous phase, 6.0–8.0 M HCl, may be described as:

$$[TiCl_2 * 2HCl * 2H_2O]^{2+}_{aq} + 4DEHPA_{org} \rightarrow [TiCl_2 * (DEHP)_2 * 2HCl * 2DEHPA]_{org} + 2H^{+}aq \quad (10.3)$$

This aggregate is not stable, but dehydration and desolvation with bidentate bond formation is fast and is completed within short time of the phases equilibration contact:[10,11]

$$[TiCl_2 * (DEHP)_2 * 2HCl * 2DEHPA]_{org} \rightarrow [TiO * (DEHP)]_{org} + 3DEHPA_{org} + 2HCl_{aq} + H_2O \quad (10.4)$$

Concentration of water and chlorine anion in organic phase of titanium(IV), after equilibration with DEHPA, was analyzed, which is presented in Fig. 10.2. These data prove the above considerations.

There are many publications about coextraction of water and its influence upon extraction process viability.[8–19] As a rule, water coextraction depends on the nature (structure) of extractant (or mixtures of extractants), metal ions extracted and solvents used. Some researchers studied interactions between metal cations and water molecules using different methods such as perturbation, molecular dynamic simulations, and quantum chemical calculations.[18] Water molecules in the close vicinity of cations are strongly structured and present a constrained orientation toward the ion. Water molecules only are free to rotate at cation–water oxygen distance of 2 Å. At larger distances, the ion–water interactions are mainly electrostatic. Interaction energy vary from 17 kcal/mol for Rb to 57 kcal/mol for Mg, and 98 kcal/mol for Be cations, interacting with a single water molecule of the first hydration shell. Polarization effects are arised from ion–water interactions.

Alcohols compete with water for coordination sites around cations.[8c,20,21] In this approach, coextracted water plays a very active role forming H-bonded intermediates with extractant and other components of organic phase.

Low extraction of titanium by DEHPA from the aqueous phases with medium acidities, ≈2 M_{HCl}/kg (see Fig. 10.1, Region I) showed that interaction through solvation begins practically only at relatively high concentrations of titanium (see Fig. 10.3) in the aqueous phase. Affinity constant ratio $k_{TiDEHPA}/k_{TiW} = 0.036$ (see Table 10.1) is very small for detection of titanium transfer from the aqueous to organic phase at low concentrations.

Only at relatively high titanium concentrations in the aqueous phase its visual transfer into organic phase was observed. May be in this case, the interactions with extractants begin through solvation with aggregation. Affinity ratio $k_{TiDEHPA}/k_{TiW} = 0.3$ testifies the interaction mainly through solvation.

At extraction of divalent transition metals (i.e., cobalt, nickel, and copper) with acidic organophosphorus extractants, e.g., DEHPA,[22–25] nucleus

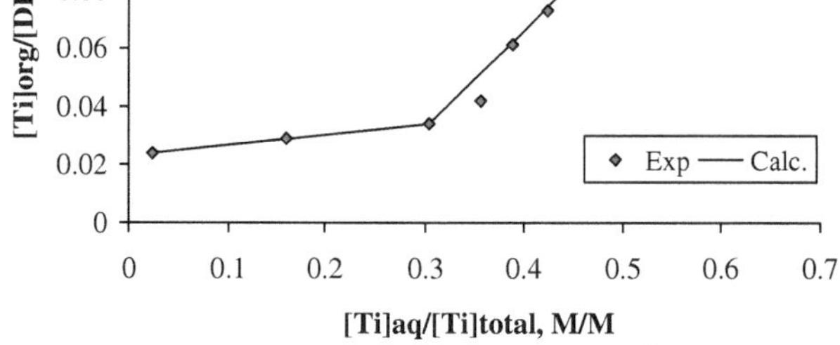

FIGURE 10.3 Comparison of experimentally obtained (points) and calculated (lines) data of extraction of titanium(IV) chlorides by DEHPA in benzene from the aqueous solutions at initial acidities (HCl): 2.0 mol/kg.

TABLE 10.1 Some Physicochemical Properties of the Initial Constituents of titanium(IV)–DEHPA–Water Extraction System

Components	Initial concentration (mol/kg)	Acidity H^+ (M/dm^3)	Density (g/cm^3)	Molar volume, v (dm^3/mol)
Extractant, E: DEHPA in benzene	1.0	1.04	0.89	0.36
Titanium chloride in water	0.1	0.1	1.0	0.018
Titanium chloride in 2.0 mol/kg HCl	0.1	1.9	1.06	0.046
Titanium chloride in 8.0 mol/kg HCl	0.1	6.7	1.21	0.040

metal-extractant aggregate formed is typically tetraligated species, e.g., $Cu(DEHP)_2 \cdot 2DEHPA$ or $Ni(DEHP)_2 \cdot 2H_2O \cdot 2DEHPA$. In the case of Ni, similar to that with titanium, two neutral, DEHPA molecules are bonded by hydrogen bonding to the water dipoles in the second (outer) coordination sphere. The authors stated that this coordination complex was stable. Interesting data were obtained at extraction of copper with different extractants:[11,24] with DEHPA, vinylsulfonic acid (VSA) and Cyanex C302. Analysis, immediately after separation of phases, showed: (1) the complex with DEHPA had no water; (2) the complex with VSA has two molecules of water but loses it during an hour; the complex with Cyanex C302 has water, much more stable and loses it during 24–48 h. The sequence of interaction steps and formation of intermediates and final product may be shown in example of extraction with VSA:

$$Cu * 4VSA * 2H_2O \rightarrow Cu(VS)_2 * 2H_2O * 2VSA \rightarrow Cu(VS)_2 * 2VSA \tag{10.5}$$

In all cases, the final product was yellow tetrahedral complex (nucleus aggregate). These facts permit us to conclude that with all mentioned extractants copper forms intermediate with water, but the transformation kinetics to the final products goes very fast with DEHPA, relatively fast with VSA and relatively slow with Cyanex C302. I think the same happens with Ni: reaction of losing water and formation of final product goes very slow and the extraction process here may be qualified as quasi-equilibrium process. It was the same at extraction of titanium(IV) from low-acidity aqueous phase (see Fig. 10.1 and sections below).

2.2. Competition Between Basic Interactions at Extraction

Strong (ion-exchange, chemical, electrostatic) and weak (solvating, physical, intermolecular, H-bond, and coordinate) interactions and aggregations are the basic interactions that have to be taken into consideration in all range of extracted metal concentrations. Competition between these interactions forms the four region shapes (Chapter 6, Fig. 6.1) of the general extraction isotherm.

The division of the distribution curve between the regions depends on many factors, such as properties and concentrations of metal extracted, the properties of the extractants and solvents (modifiers, synergistic agents) used, temperature, acidity of the treated (feed) aqueous solutions and the structure of initial compounds, the structure of the organic compounds formed, and aggregation.

The field of extraction in literature is full of a strong, sometimes even critical influence of solvation (hydration) effects on the extraction process. The polar, polarizable, hydrogen-bonding solvents change the extraction order dramatically. Solvents are known to have numerous different effects on the reactivity.[8] As a rule, two types of solvent effects are distinguished: general effects stemming from polarity and polarizability and specific effects originating from π- or n-donor ability, hydrogen bond forming capacity, etc. The solvents range from the very inert through nonpolar but polarizable to the polar, polarizable and hydrogen-bonding (as water) solvents. The selectivity and position of a given solvent is ion-specific and depends on its (solvent) basicity, but also on its bulk concentration, size of the ion, mono-poly-dentate nature, and other factors. In general, electron-withdrawing groups in the solvent molecules increase acidity and decrease basicity; electron-donating groups act in the opposite way. An increased extraction of metals may be achieved by the use of oxygen-containing solvents, polar in nature, due to their ability to replace water molecules. Such a cooperative effect is essentially a synergistic extraction. In order to analyze solvation effects, identification and, most importantly, quantification of solute–solvent interactions must be accomplished. Complexity of solvent structure makes this task extremely difficult.

Coordination models, explaining qualitatively some mechanisms of metal solvent extraction meets difficulties when modeling anion exchangers, acid solvent extraction and, especially, at the attempts of quantification of the models. The physicochemical modeling approach, competitive preferential solvation (COPS) approach, developed preliminary for the weak molecular interactions[26] and also applicable to strong–weak, ion–molecular interactions[7] and acid–base systems[5] was modified for metal extraction systems.[9,10] Below the reader will find the mathematical description used for the CCST model in metal extraction.

Metal ions, as well as all other species, in the phases are surrounded by a solvation shell.[26] This means that they interact with all the constituents of the environment and, therefore, "free" metal ions, free extractant, or free metal–extractant complex do not exist. Components of the solvation shell compete to form a complex according to their electronic-geometric affinity k_i (interaction strength), which is constant at a given temperature and pressure. The molecules in the solvation shell continuously and rapidly (10–11 s for alkali metal halides[27]) relax between complexing and solvating states. Therefore, coefficient, k_i, represents the total, complexing and solvating affinity constant.[26] In all cases, the competition between the constituents of the solvation shell for coordination sites takes place according to their affinity constants and concentrations; thus, the contribution of each one of them in a definite coordination number cannot be well defined.

To proceed further, it is necessary to clarify the terms "complexation" and "solvation" used in this work. Strong, "long-range interactions,"[27] which are electrostatic in origin and are normally treated by Born theory, lead to complexation. Complexation (including ion-exchange) occurs when the interacting partners meet each other with well-defined orientations, allowing the favorable overlap of their orbitals or charge (electron) transfer. Any other interactions lead to solvation. The last are so-called weak or "short-range interactions".[27]

Many transition, post-transition, and some main group elements have well-defined solvation shells, or coordination spheres in solution.[4,5,7–9,26–29] This is not the case, however, for many other cations, the solvation shells of which are diffuse and ill-defined.[8a] Any given arrangement, or architecture of the solvation shell persists for only a very short time. So, as a rule, measured property, e.g., NMR chemical shift, or UV spectral shift, can be formulated as the weighted average of the property, arising from all possible arrangements.

A composition of the solvation shell (or coordination complex) in a mixture of metal, M, extractant, E, metal–extractant complex formed, ME, solvents, A (including water, W), and diluents, D, depends on affinity constant k_i and on the number of potentially available molecules of every component, present in the organic phase, i.e., on the actual concentration[1] of every component.[9,26] The metal ions are considered as if they are partitioned among all constituents of the solvation shell medium and are equivalent to a weighed mixture[2] of solutions in pure solvent components.[30]

As ions form definite solvates when placed in a pure solvent, they may be expected also to form them in a mixture of solvents.[8] The two solvents, A_1 and A_2, are considered as competing ligands. Here, the question of partitioning (and

1. Really we have to use here and below the activity of a solvent, which plays the same role as the concentration in ideal solutions (obeying the gas laws) and is equal to concentration at infinite dilution. Provided that the concentration of solute (acid) is small relative to solvents in the mixture, the ratio of the activity coefficients of two solvents participating in the replacement equilibrium can be approximated by the ratio of their molar concentrations or mole fractions. For simplicity of the following considerations we apply a simplification where activity of every component of the system is considered equal to its concentration.

2. For the present purpose the volume change on mixing of the solvents is neglected.[8b]

selectivity) arises. Stepwise solvent replacement generally takes place as the concentration of A_2 in the bulk mixture increases, until the last A_1 molecule is displaced (pure A_2). The solvation numbers are not definite at the replacement reactions and depend on their bulk concentrations, mono-poly-dentate nature, sizes, etc.

2.3. Acid–Base Interaction Model and Amphoterism

Metal ions, extractants, active solvents, and their species formed are considered amphoteric and may perform as acids (electron acceptors) or bases (electron donors), depending on the structure of their functional groups and composition of the organic phase, and on the structure of the metal species and composition of the aqueous phase.

According to the Lewis[4,5,31] and Pearson[32,33] classification, many substances can be both bases and acids. These species have amphoteric properties, depending on the ligand. Any specie with pK_a value higher than the given one may be conjugate base to it. Classification of metal ions as acids (see Section 1) is not constant, but depends on its oxidation state and influenced by groups and/or molecules bound to it.[5,31,34]

There are solvents in which a slight self-ionization is observed, and this cannot serve as a foundation for the acid–base reactions. They are mainly organic solvents, particularly those containing nitrogen, phosphorus-oxygen, and sulfur-oxygen functional groups. In these solvents, the acid–base act is initiated not by self-ionization, but by solvation of the dissolved compounds. An empirical quantum chemical methods' approach is applied for the treatment of metal ion solvation (coordination) in the bulk solvent, which employs the concept of donor–acceptor interactions. The comprehensive theory of Usanowitsch[35] stated that all coordinately unsaturated atomic groupings are acids and all coordinately saturated atomic groupings are bases. Coordination approach recognizes the coordinate bond that may be formed, i.e., allows some overlap of the electron orbitals of the ion in the solvated ion.

As stated in Section 2.1 titanium(IV), e.g., forms very different species, depending on acidity of aqueous solution. This is evident from Fig. 10.1, which presents titanium(IV) chloride species as a function of pH (acidity). The diagram shows the evident symmetry of complexing properties of titanium(IV) in acidic aqueous solutions. Region II can be divided into two symmetrical parts: IIa in the range of 0.5–0.01 M HCl (pH $\approx$ 0.3–2.0) and IIb in the range of 6.0–8.0 M HCl. The $[TiO]^{2+}$ ion typical in Region I becomes unstable and changes to $[Ti(OH)_2 \cdot 4H_2O]^{2+}$ in the IIa and to $[TiCl_2 \cdot 2HCl \cdot 2H_2O]^{2+}$ in the IIb, with coordination number of 6. The same, Regions III and IV are divided to IIIa (0.01–0.005 M HCl or pH = 2.0–3.3), IVa (<0.0005 M HCl or pH > 3.3) and IIIb (8.0–11.0 M HCl), IVb (>11.0 M HCl). In the Region III, electrolytic activity of titanium(IV) species decreases and a considerable part of them are electroneutral: $Ti(OH)_4 \cdot 2H_2O$ in the IIIa and $TiCl_4 \cdot 2HCl$ in the IIIb. In the

Region IV, the anionic complexes of titanium(IV), may be with coordination number of 8 are supposed to exist in the aqueous solutions. Similar to pictures in Fig. 10.1 for titanium compounds with other mineral acids were obtained analyzing literature data.

Dealing with the above symmetry classification and the Statement 2 of the model, extraction of titanium hydroxide ion, coordinated (hydrated) by water molecules, with acidic organic extractants is considered as an acid–base interaction in which titanium(IV) complexed ion at pH $\geq$ 2 of the aqueous phase behaves as a conjugate base (electron donor) to extractant. Extractant in an inert organic diluent (e.g., DEHPA in kerosene[1,6]) behaves in this case as an electron acceptor (acid). The complexed titanium atom in organic phase is coordinated via phosphoryl (sulfuryl, carboxyl) oxygen of organic acid and water molecules. Suggested reaction at extraction of titanium(IV) at pH = 1 in the aqueous phase is presented in Eqn (10.1).

At pH ≤ -0.9 ($\geq$8 mol/kg HCl) of the aqueous solutions, titanium ion, coordinated (solvated) by HCl molecules, behaves as a conjugate acid (electron acceptor) to the same acidic organic extractants. In this case, extractant behaves as an electron donor (base). The complexed titanium atom in organic phase is coordinated via chlorine anions of HCl and phosphoryl (carbonyl) oxygen atoms of organic acid molecules (see Eqn (10.3)).

These considerations are valid for most metals and extractants: acidic, basic, and neutral, which are presented in Table 6.1 of Chapter 6. They explain the universality of the dual behavior of different extractants and prove their amphoterity, depending on the composition of the solute in the aqueous phase and its acidity. The type of bonding here is predominantly ionic in the case of hard acids and predominantly electron sharing in the case of soft acids. Some extractants, containing either monodentate and polydentate groups, form both kinds of bonds (with electron donors of carboxyl-, phosphoryl-, sulfuryl-, amino-groups) with chelate formation. For the majority of chelating systems it seems to be a rule that whenever the coordination number of the metal equals double its oxidation (or ionic charge) number, the chelate formed satisfies the coordination requirements of the metal and the metal is readily extractable into both polar and nonpolar solvents.[6,36] On the other hand, when a metal ion has a coordination number higher than twice its ionic charge, the metal retains one or more water (or acid) molecules in its first coordination sphere in order to satisfy its coordination requirements.

2.4. Aggregation

Aggregation mechanisms and species formed are changing with increasing loading of organic phase by metal specie, and should be considered and analyzed in all regions of the general extraction isotherm (Chapter 6, Fig. 6.1).

It is well known that the stoichiometry of the extracted species changes as loading increases. This phenomenon was described by some authors through aggregation mechanisms.[2,9,10,22,23,26,37–41] Aggregation was observed at extraction of metal ions by acidic extractants,[2,10,22,23,39–41] of metal ions and acids by basic and neutral extractans.[9,15,26,37,38]

According to CCST at low concentrations of the metal in the organic phase, metal–extractant complex, surrounded by its solvation shell (or coordination sphere) forms a geometric structure, which is denoted as a nucleus aggregate. The solvation shell in the nucleus aggregate is open to bulk solvents and is characterized by the fast exchange with the bulk solvents. These are Regions 1 and 2 on the shape of general extraction isotherm (Chapter 6, Fig. 6.1) in which the ratio $C_{Morg}/C_{ME} \leq 1$.

Once again we have to repeat an important statement: What are the solvation shell, the nucleus aggregate, and the coordination complex? These are three nominations of the same physicochemical reality. The first nomination considers a metal ion as interacting and microscopically partitioning between different solvents (in case that solvent–solvent interactions may be neglected) forming some kind of a virtual physical entity, called solvation shell. It is a statistical thermodynamics' nomination, which is convenient for mathematical description. The second description, nucleus aggregate, describes the same but real physical entity with stereo-specific bonds and orientation in the bulk organic solution. "Coordination complex" and "nucleus aggregate" nominations are equal for metals with the coordination sphere of definite number and geometry. In the presented theory, the nomination of "nucleus aggregate" is used as universal, not only for metals.

Region 1 (Chapter 6, Fig. 6.1) is characterized by metal–extractant equivalent concentration ratios, $C_{Morg}/C_{ME} < 1$ (more than one equivalent of extractant to one equivalent of metal) and $P_{MEs} \geq P_{MEc}$. The structures of the nuclei aggregates in this region are formed mainly through the weak interactions (hydrogen, coordination bonding), but, to some extent, through the strong electrostatic, charge transfer interactions of metal ion with extractant. The pH dependence in this region is not pronounced,[2,23,27,37,41] but stability pH limitations of the formed complexes are observed.

Formation of nuclei aggregates in Region 2 is driven mainly by strong (electrostatic, ion-exchange, ion-pair) interactions: $P_{MEs} < P_{MEc}$ and $C_{Morg}/C_{ME} \approx 1$. The structures of the nuclei aggregates in this region are formed mainly through the cation exchange of metal ions with proton of organic acid or with another cation, but, to some extent, through the weak H-bond or coordination interactions with molecule of extractant and/or other components. Divergence of the distribution curve slope from the 1/1 equivalents of the M–E complex is explained by weak interactions. The interactions in this region are strongly dependent on pH or anion concentration in the aqueous phase.

According to the presented theory (see Chapter 7), the difference between the Regions 1 and 2 is mainly in the magnitude of the affinity constants ratios k_{MEc}/k_{Mes} at the same concentrations of extractant, water in organic phase, and affinity coefficient of metal ion in pure water. So, all coordination structure schemes of the nuclei aggregates in Region 1 may belong to Region 2, but at $k_{MEc} > k_{MEs}$. It follows that the slopes of the distribution curve in Chapter 6, Fig. 6.1 for Regions 1 and 2 may have all magnitudes between two extremal structures: at $k_{MEc} \rightarrow 0$ and at $k_{MEs} \rightarrow 0$.

When metal concentration is increased the nuclei aggregates interact and form linear, or ringed (cyclic) aggregates.[16,20–23,25,41] This is Region 3, in which the equivalent of metal to the equivalent of extractant ratio is: $C_{Morg}/C_{ME} > 1$. The Region 3 is characterized by the bridging of nuclei aggregates with metal ion (or its salt), molecules of extractant, or active adduct (including water). Competition between these components influences the slope value of the curve in the Region 3. All components of the linear (or cyclic) aggregates are also open to the bulk solvents and exchange between the aggregate components and the bulk solvents continue to be fast.

At high concentrations of metal ion, upon reaching a critical size, the structural reorganization of the linear (or cyclic) aggregates occurs and three-dimensional supramolecular structures, reversed micelle-like, or cross-linked cluster-like, are formed.[20–23,25,41,42] This is Region IV. Above-stoichiometric loading and massive third-phase formation are typical for this region.

Region IV is characterized by $P_{MEs} < P_{MEc}$ and $C_{Morg}/C_{ME} >> 1$. Interactions in this region are driven by the three-dimensional structures formation laws (cross-linking, micellation, gelation, and polymerization). Host–guest interaction models can be used for analysis of these shapes; guest molecules are confined by different types of interactions, in the cavities of the host system. High polarity and well-defined structures with known number of sites are very suited for extraction and reaction purpose. Structure can change from a globular reverse micellar arrangement to a cylindrical amphoteric shape. As a result of this, the polar interior is directed toward the aqueous or organic phases. For details, see supramolecular compounds formation and application in solvent extraction, presented in the chapters of Part III.

The exchange rate (and therefore, preferential competition) between the components in the solvation shell and the bulk phase depends on the orientation of the polar groups relative to the bulk phase, inside or outside the aggregate cavity. Components of the voluminous aggregates are mainly closed to the bulk solvents and the exchange between aggregate components and bulk solvents are controlled by diffusion kinetics.

The rates of three-dimensional aggregate formation and destruction may differ very strongly, especially in the case of micelle-like aggregates. So, equilibrium state has to be controlled, especially at stripping experiments.

At extraction of titanium from its high concentration aqueous solutions[6] massive formation of the third phase was observed. After aging of the separated organic phase it was transformed into gel.

Spectrometric investigation of titanium(IV) loaded in DEHPA at contact with aqueous phases of different acidities (ranged from 0.1 to 7.6 M/Kg) proved all these considerations. Broad peaks (characterized aggregates) of absorbance at concentration of [Ti] = 30 ppm in organic phase (see Fig. 10.4a) really consist of two (or more) peaks: one was distinguished at 278–290 nm (I) and the other at 298–310 nm (II).

On increasing of titanium(IV) concentration in organic phase from 5 to 30 ppm peak (I) had blue shift: from 278 to 290 nm but peak (II) conversely red shift: from 310 to 300 nm. Dilution of 30 ppm titanium concentration organic sample down to 6 ppm by adding benzene destructs the aggregates to the single (nucleus) state and only peak (I) was retained what is seen in Fig. 10.4b.

On prolonging mixing time of the aqueous–organic phases contact, much more broader absorbance peak appears and peak (II) moves to 310 nm (see Fig. 10.4d).

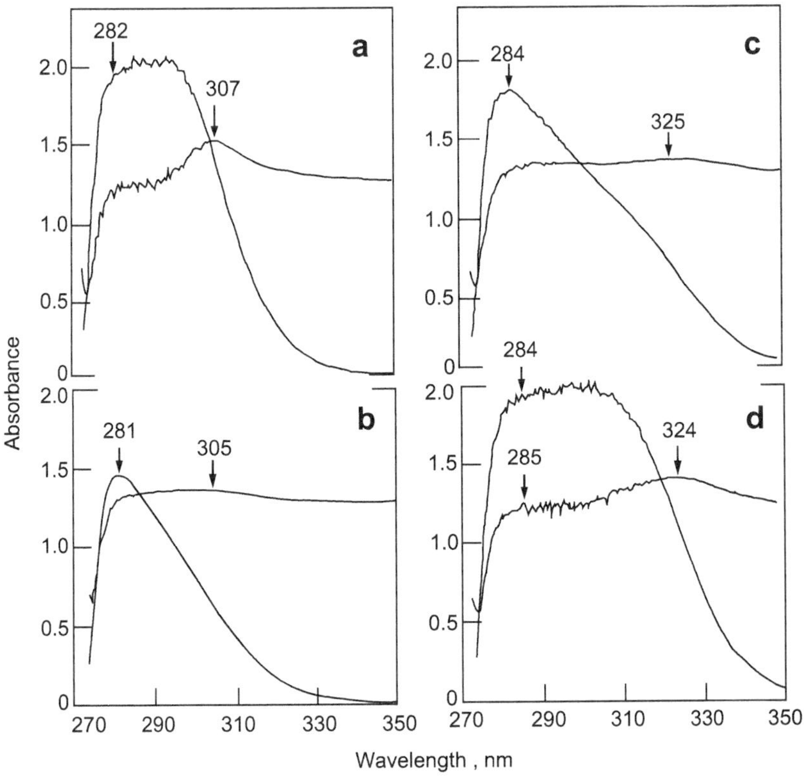

FIGURE 10.4 Absorption spectra and their derivatives of titanium(IV), loaded into 6.0 mmol/kg DEHPA in benzene organic solution, obtained on contacting with 0.62 mmol/kg titanium(IV), 0.1 M/kg HCl aqueous feed solutions. (a) 30 ppm titanium(IV) in organic phase (colorless); (b) the same sample (a), diluted by benzene to 6 ppm Ti (colorless); (c) the same sample (a) after aging during 2 months (yellow). *(Source: Data from Ref. 6)*

UV spectra were studied during the aging of organic samples, measured after 3 days, a week, a month, and 2 months. The results are as follows:

1. For the samples after extraction from 7.6 M/kg HCl aqueous feed absorbance intensity of peak (I) decreases and that of peak (II) increases. Both move after the first 3 days to lower wavelengths: peak (I) to 279–281 nm and peak (II) to 295–302 nm, and then UV spectra became stable. New peak at 323–325 nm begins to grow (see Fig. 10.5).
2. For the samples after extraction from 0.1 M/kg HCl aqueous feed absorbance intensity of peak (I) increases during the first 3 days and that of peak (II) decreases. Then, after a week, an opposite behavior was observed: peak (II) increases and moves to the lower wavelengths. After a month, the UV spectra of the samples became stable and were about the same as that of the samples, obtained after extraction from 7.6 M/kg HCl feed. Peak at 325 nm grows and has blue shift.

These results show a complex interaction and aggregate formation during extraction process and afterwards in the loaded organic phase. Suggested nuclei aggregates are formed through complexation (peak (I)) and H-bonds (peak (II)). Dehydration (or desolvation), indicated by the shifts of the peaks, witnessed about linear or cyclic aggregate formation. Appearance of a new peak at 330 nm witnesses, to our suggestion, beginning of voluminous aggregate formation with bidentate bonds.

At aging of organic phase samples during 2 months the solution color was changed from colorless to yellow and the peak at 325–335 nm appears and its intensity grows. Appearance of this peak witnesses the voluminous aggregate forming. The growing intensity and its red shift from 335 to 325 nm indicate the growth of the three-dimensional aggregates (Fig.10.5). For details, see Refs. 2,6.

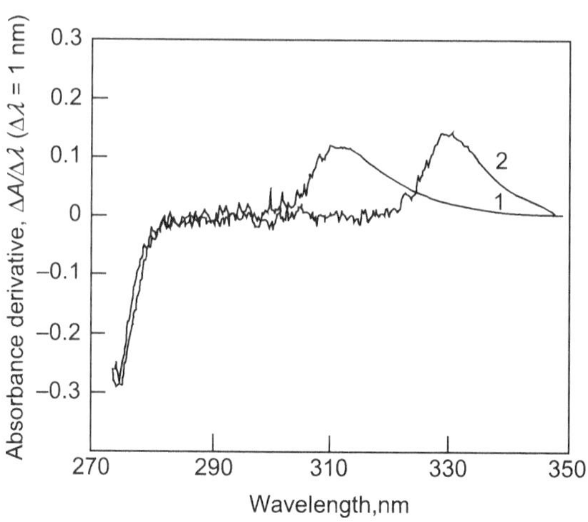

FIGURE 10.5 UV absorption spectra derivatives of titanium(IV) in 1.0 M/kg DEHPA in benzene, obtained after contacting with 0.1 M/kg Ti in 4.5 M/kg HCl aqueous solution. (1) After phases mixing 5 min (organic phase colorless); (2) after phase mixing 30 h (organic phase yellow). *(Source: Data From Ref. 6)*

The UV data results are proved statement of the CCST that in some extraction processes the reliable contact time brings the system only to pseudo-equilibrium state and kinetics of final product formation is slow, or very slow.

Similar UV absorbance data were obtained at investigation of copper(II) extraction by DEHPA, VSA, and Cyanex C302 in kerosene as diluent. UV–Vis study of VSA–copper complexes in organic solution are presented in Fig. 10.6.[43] Absorbance spectra of Cu(II) in aqueous solution show two main peaks at 800 and 230 nm. In VSA organic solution at low copper concentrations these peaks are shifted: peak 800 to 780 nm (red shift) and peak 230–280 nm (blue shift). At increasing copper concentration in organic solution magnitude of peaks increases, they are broadening and shift: the first peak shifts to 750–760 nm (continue red shift) and second to 285–310 nm (continue blue shift).

At $CuCl_2$–VSA molar ratio of 1/2, both peaks are strongly broad and new peak at 386 nm appears. Three regimes of aggregation behavior are considered a function of metal ions M^{2+} concentration in organic phase of acidic

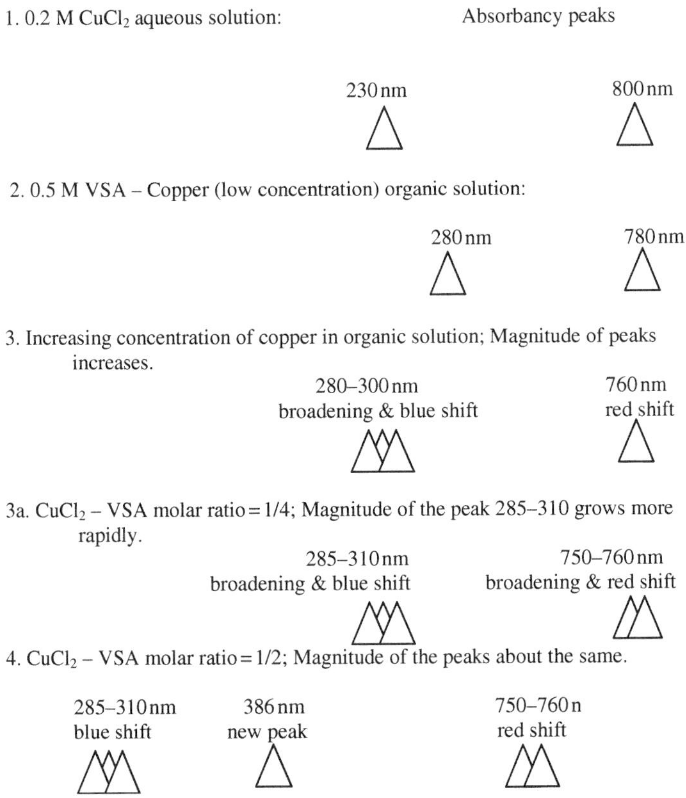

FIGURE 10.6 UV–Vis study of vinylsulfonic acid (VSA)–copper complexes in organic solution.

extractants:[11,22,23] the tetrametric species of type $ML_2 \cdot 2HL$ —> linear aggregates —> reversed micelles.

One more experimental proof of closed reversed micelles formation is presented in Chapter 7, Fig. 7.8, where titration of the high-loaded (0.9 M/M) organic samples was followed by destruction of the closed reversed micelles, filled with aqueous solution, what was indicated by the sudden increase of the sample acidity in the middle of titration of samples.

3. MATHEMATICAL SUPPLEMENT FOR THE CCST IN METAL EXTRACTION

3.1. General Considerations

Let us consider an extraction system in which E is a monobasic organic acid, M is a monovalent metal ion, A is an active solvent, W is water (in organic), D is an inert diluent, C_{i} are concentrations of extracting components in the bulk organic phase (C_{E}, C_{A}, C_{Worg}, C_{D}), and k_{Mi} are affinity constants of metal ion to every component. A commonly used concentration scale is the mole fraction.[32] Solvent–solvent interactions are neglected.

A set of equations may be obtained for a relatively dilute solution of metal at the equal volumes of the phases:

$$C_{\mathrm{M}}^{0} = C_{\mathrm{Maq}} + C_{\mathrm{Morg}} = C_{\mathrm{Maq}} + C_{\mathrm{ME}} + C_{\mathrm{MW}} + C_{\mathrm{MA}} + C_{\mathrm{MD}} \qquad (10.6)$$

where C_{M}^{0}, is the initial (or total) concentration of the metal ion; C_{Maq} is the concentration of the metal in the aqueous phase; C_{Morg} is the concentration of the metal in the organic phase; C_{ME}, C_{MW}, C_{MA}, and C_{MD} are the concentrations of the metal in the mixed solvation shell of the organic phase: in extractant, water, active solvent, and diluent, respectively, at equilibrium.

$$\sum_{1}^{n} P_{\mathrm{Mi}} = 1 \qquad (10.7)$$

where P_{Mi} is the generalized partitioning factor of the metal ion in homogenous media.[26] Partitioning factor P_{Mi} is

$$P_{\mathrm{Mi}} = \frac{k_{\mathrm{Mi}} C_{\mathrm{i}}}{\sum_{1}^{n} k_{\mathrm{i}} C_{\mathrm{i}}} \qquad (10.8)$$

$$C_{\mathrm{Mi}} = C_{\mathrm{Morg}} P_{\mathrm{Mi}} \qquad (10.9)$$

or for the system considered:

$$C_{\mathrm{ME}} = C_{\mathrm{Morg}} \frac{k_{\mathrm{ME}} C_{\mathrm{E}}}{k_{\mathrm{ME}} C_{\mathrm{E}} + k_{\mathrm{MW}} C_{\mathrm{Worg}} + k_{\mathrm{MA}} C_{\mathrm{A}} + k_{\mathrm{MD}} C_{\mathrm{D}}} \qquad (10.10)$$

$$C_{MW} = C_{Morg}\frac{k_{MW}C_{Worg}}{k_{ME}C_E + k_{MW}C_{Worg} + k_{MA}C_A + k_{MD}C_D} \quad (10.11)$$

$$C_{MA} = C_{Morg}\frac{k_{MA}C_A}{k_{ME}C_E + k_{MW}C_{Worg} + k_{MA}C_A + k_{MD}C_D} \quad (10.12)$$

$$C_{MD} = C_{Morg}\frac{k_{MD}C_D}{k_{ME}C_E + k_{MW}C_{Worg} + k_{MA}C_A + k_{MD}C_D} \quad (10.13)$$

For solvent A′, partially soluble in both aqueous and organic phases, the partitioning factor is

$$C_{MA'} = C_{Morg}\frac{k_{MA'}C_{A'org}}{k_{ME}C_E + k_{MW}C_{Worg} + k_{MA'}C_{A'org} + k_{MD}C_D} \quad (10.14)$$

where $C_{A'\,org} = C_A^0 * F_{A'}$; C_A^0 is the total concentration of the solvent A′; F'_A is its partition factor between the organic and aqueous phases, determined experimentally at the absence of metal in the system.

Metal ions are present in both the complexed and the solvated forms with extractant in the solvation shell. Therefore, partitioning factor P_{ME} can be divided into two:

$$P_{ME} = P_{MEc} + P_{MEs} \quad (10.15)$$

where P_{MEc} and P_{MEs} represent partitioning factors of the metal ions in the complexed and solvated forms, respectively. So, metal ion concentrations in the complexed, C_{MEc}, and solvated, C_{MEs}, forms in the solvation shell are determined by equations:

$$C_{MEc} = C_{Morg}\frac{k_{MEc}C_E}{k_{MEc}C_E + k_{MEs}C_E + k_{MW}C_{Worg} + k_{MA}C_A + k_{MD}C_D} \quad (10.16)$$

$$C_{MEs} = C_{Morg}\frac{k_{MEs}C_E}{k_{MEc}C_E + k_{MEs}C_E + k_{MW}C_{Worg} + k_{MA}C_A + k_{MD}C_D} \quad (10.17)$$

where k_{MEc} and k_{MEs} are affinity constants of the metal ion toward extractant in the complexed and solvating forms, respectively.

The actual value of the physicochemical property X (e.g., chemical shift, δ, in NMR, spectral shift, ν, or molar extinction coefficient, ε, in UV–Vis, changes in free energy, ΔG, etc.) measured in mixed solvents is considered to be the weighed sum of this same property measured in pure solvent constituents.[7,9,10,26–30] In a mixed system, the contribution of each solvent on the shift of this property X is additive:

$$X = P_{ME}X_{ME} + P_{MW}X_{MW} + P_{MA}X_{MA} + P_{MD}X_{MD} \quad (10.18)$$

3.2. Analysis of the M–E–W System

When the classical theories are experimentally tested, water in the initial solution of metal ion-in-water, M_W^0, is replaced by extractant E, to arrive, finally, to the solution of metal-in-pure E. The corresponding variation of the physicochemical property of M constitutes a substitution profile, the linearization of which makes it possible to determine the thermodynamic and spectroscopic properties of M–E–W interactions and distributions. As a rule, these theories do not distinguish quantitatively between complexation and solvation in the M–E–W interactions. Also, they do not predict quantitatively the influence of different solvents and adducts (enhancers, modifiers, synergistic agents, and mixed extractants).

Validity of the presented approach, verified experimentally, gives a key for quantitative analysis and preliminary prediction of suitable extraction systems for different metals' separation.

Let us consider an M–E–W extraction system in which metal ion, M^{z+}, N_M-molecules – extractant (E), N_E-molecules – water (W), N_W-molecules participate in a single reaction act. Here, z is the oxidation state of metal ion.

For simplicity, let us introduce some preliminary assumptions:

1. One metal ion, M^{z+}, participates in the formation of the single solvation shell, $N_M = 1$.
2. Acidic organic extractant is monobasic, E = HL, where L is the dissociated form (ligand) of the extractant molecule, and completely insoluble in the aqueous phase.
3. The diluent D is inert enough and does not participate in solvation.
4. All components of the system are monomers in a pure, initial state.
5. The volume change on mixing of the solvents is neglected.
6. Water in the organic phase after equilibration, separation, and centrifugation is present as bonded only in the solvation shell, i.e., any solubility of water in the bulk organic solvents' mixture is excluded.

Considering three-component M–E–W system:

$$P_{ME} + P_{MW} = 1 \qquad (10.19)$$

$$C_{W\mathrm{org}} = \frac{1 - C_E \nu_E}{\nu_W} \qquad (10.20)$$

where ν_E and ν_W are partial molar volumes (dm^3/mol) of extractant and water.

Using Eqn (10.18), we obtain, after some algebra, the linear equation for the measured property X:

$$\frac{X_{MW} - X}{C_E} = \frac{k_{ME}}{k_{MW}} \nu_W (X_{MW} - X_{ME}) - \left[\frac{k_{ME}}{k_{MW}} \nu_W - \nu_E\right] (X_{MW} - X) \qquad (10.21)$$

and the saturation factor Z at $C_E > C_M^0$:

$$Z = \frac{C_{Morg}}{C_M^0 - C_{Morg}} = \frac{X_{MW} - X}{X_{MW} - X_{ME}} = \frac{k_{ME}\nu_W C_E}{k_{MW} + (k_{ME}\nu_W - k_{MW}\nu_E)C_E}, \quad 0 < Z < 1,0 \tag{10.22}$$

or, using Eqns (10.19) and (10.20):

$$Z = \frac{C_E \nu_E}{1 - \left(\nu_W - \frac{k_{MW}}{k_{ME}}\nu_E\right)C_{Worg}} = \frac{P_{ME}}{P_{ME} + \frac{k_{MW}}{k_{ME}}\nu_E C_{Worg}} \tag{10.23}$$

and

$$\frac{1}{Z} = 1 + \frac{k_{MW}C_{Worg}}{k_{ME}P_{ME}\nu_E} = 1 + \frac{k_{ME}C_{Worg}}{k_{MW}C_E} \tag{10.24}$$

From the slope of Eqn (10.21) plots $(X_{MW} - X)$ versus $(X_{MW} - X)/C_E$ one can obtain the affinity constant ratio k_{ME}/k_{MW}; the intercepts give the hypothetical differences between the property, measured in pure extractant and in pure water. Thus, the agreement between the direct experimental determination of the measured property (e.g., molar extinction coefficients in UV–Vis) in pure solvents and their graphically obtained values of $X_{MW} - X_{ME}$ may be examined.

It can be seen that only affinity constant ratios can be obtained but not individual k's. This is due to the fact that only relative values can be measured in solution because of the ubiquitous nature of molecular interactions. Affinity constant ratio's value of unity suggests the same values of solvation effects of the metal ion with extractant in the organic phase and with water in the aqueous phase; a large value of k_{ME}/k_{MW} means strong or very strong solvation–complexation effects with extractant. A positive sign of a slope means that the solvation of the reactant molecules is stronger than the product molecules. The negative slope shows that the solvation of product molecules is stronger than the reactant molecules. The solvation of the proton can be neglected since its solvation is constant for solvents used.

Using different measured properties X, we obtain relations, derived from Eqn (10.21):

For NMR, where $X = \delta$ (ppm) is chemical shift:[26]

$$\frac{\delta_{MW} - \delta}{C_E} = \frac{k_{ME}\nu_{Worg}}{k_{MW}}(\delta_{MW} - \delta_{ME}) - \left[\frac{k_{ME}}{k_{MW}}\nu_{Worg} - \nu_E\right](\delta_{MW} - \delta) \tag{10.25}$$

For UV–Vis, where $X = \nu$ (nm) is spectral shift:[27]

$$\frac{\nu_{MW} - \nu}{C_E} = \frac{k_{ME}\nu_{Worg}}{k_{MW}}(\nu_{MW} - \nu_{ME}) - \left[\frac{k_{ME}}{k_{MW}}\nu_{Worg} - \nu_E\right](\nu_{MW} - \nu) \quad (10.26)$$

For UV–Vis, where $X = \varepsilon$ is molar extinction coefficient:[26]

$$\frac{\varepsilon_{MW} - \varepsilon}{C_E} = \frac{k_{ME}\nu_W}{k_{MW}}(\varepsilon_{MW} - \varepsilon_{ME}) - \left[\frac{k_{ME}}{k_{MW}}\nu_W - \nu_E\right](\varepsilon_{MW} - \varepsilon) \quad (10.27)$$

$\frac{A}{C_{Morg}} = \varepsilon$, $\frac{A_{ME}}{C_{Morg}} = \varepsilon_{ME}$, $\frac{A_{MW}}{C_{Morg}} = \varepsilon_{MW}$ and A is measured absorbency.

For standard molar Gibbs free energy (or enthalpy) change,[27–29] where $X = \Delta G$ (at constant temperature and pressure):

$$\Delta G = -n\,RT\,\ln\left(\frac{k_{ME}}{k_{MW}} \times \frac{C_{ME}\nu_E}{C_{Worg}\nu_W}\right) \text{ or } \Delta G = -n\,RT\,\ln\left(\frac{k_{ME}}{k_{MW}} \times \frac{N_E}{N_W}\right) \quad (10.28)$$

At $N_M = 1$ and definite solvation (or coordination) number, $n = N_E + N_W$ for monodentate ligand/extractant:

$$\Delta G = -n\,RT\,\ln\left(\frac{k_{ME}}{k_{MW}} \times \frac{N_E}{n - N_E}\right) \quad (10.29)$$

According to the theory, partitioning factors in this case are

$$P_{ME} = N_E/n + 1 \quad (10.30)$$

and

$$P = N/n + 1 \text{ or } P_{MW} = (n - N_E)/n + 1 \quad (10.31)$$

N_E represents the number of molecules of extractant E = HL, bonded both as the charged anion L^- (complexed bonds) and as solvated (solvation, or coordination bonds) neutral molecules HL. The metal central ion is coordinated by z charged ligands L^-, by $N_E - z$ neutral HL molecules, and by N_W water molecules, filling metal ion coordination sites up to saturation.

According to Eqn (10.15) partitioning factors P_{MEc} and P_{MEs} may be determined separately:

$$P_{MEc} = z/n + 1 \quad (10.32)$$

$$P_{MEs} = (N_E - z)/n + 1 \quad (10.33)$$

Some experimental techniques are able to detect separately the complexed (charge transfer bonds) and the solvated (coordinate bonds) forms or to distinguish between them (e.g., UV–Vis spectrometry for copper), while others may monitor only the average complexed and solvated forms together (e.g., NMR spectroscopy, potentiometric titration). For example, for Cu–DEHPA–water system[11,22–24,40,41,44] measuring UV–Vis spectral shift we can determine k_{MEc}/k_{MW} and k_{MEs}/k_{MW} by Eqn (10.25).

Measuring absorbency A at a given wavelength we can distinguish between the complexed and the solvated forms:

For the complexed form at the absorbencies, A_c

$$A_c = \varepsilon_{MEc} C_{MEc} \tag{10.34}$$

For the solvated form at the absorbencies, A_s

$$A_s = \varepsilon_{MEs} C_{MEs} \tag{10.35}$$

Introducing ε^f_{MEc} (or ε^f_{MEs}) as a hypothetical molar extinction coefficient of the complexing (or solvating) interaction:

$$\varepsilon^f_{MEc}(\text{or } \varepsilon^f_{MEs}) = \varepsilon_{MEc}(\text{or } \varepsilon_{MEs}) \frac{k_{MEc}(\text{or } k_{MEs})}{k_{ME}} \tag{10.36}$$

we obtain:

$$\frac{C_{Morg}}{A_c(\text{or } A_s)} = \left[1 - \frac{\nu_E k_{MW}}{\nu_W k_{ME}}\right] \frac{1}{\varepsilon^f_{MEc}(\text{or } \varepsilon^f_{MEs})} + \frac{k_{MW}}{k_{ME} \nu_W C_E \varepsilon^f_{MEc}(\text{or } \varepsilon^f_{MEs})} \tag{10.37}$$

or

$$\frac{C_{Morg} C_E}{A_c(\text{or } A_s)} = \frac{k_{MW}}{k_{ME} \nu_W \varepsilon^f_{MEc}(\text{or } \varepsilon^f_{MEs})} + \left[1 - \frac{\nu_E k_{MW}}{\nu_W k_{ME}}\right] \frac{C_E}{\varepsilon^f_{MEc}(\text{or } \varepsilon^f_{MEs})} \tag{10.38}$$

3.3. Analysis of More Complicated Extraction Systems

The simplest M–E–W extraction system was analyzed above. The real extraction systems are much more complicated, in which two or more metal ions (M_1, M_2, M_3) have to be selectively separated, and/or mixture of extractants (E_1, E_2) and/or different active solvents as adducts (A_1, A_2) are used, for synergistic effect, etc. In the following chapter these systems will be analyzed in details with quantifying equations developed and predictions made on the basis of the presented theory. Here, it only has to be stressed that analysis is based on the experimentally determined affinity constant ratios and solvation (coordination) numbers of the aggregates, formed at different concentrations of the solute (metals) and solvents (extractants, adducts, etc.). For example, the contribution of every component in the complicated M–E–A–W system, where A is any active Solvent or adduct, may be evaluated at the determined affinity constant ratios k_{ME}/k_{MW}, k_{ME}/k_{MA}, k_{MA}/k_{MW}, and

k_{EA}/k_{EW}. These data may be realized by conventional equilibrium extraction experiment series with the simple three-component systems M–E–W, M–E–A, M–A–W, and E–A–W. Equations similar to Eqns (10.28)–(10.41) of the M–E–W system can be developed.

3.4. Independence and Transferability of Affinity Constant Values

The reasonable question is arising: What is the advantage of the presented theory if so many experiments have to be done?

The basic idea of the competitive preferential solvation concept is the independence and transferability of affinity constant values from one system to the other.

The theory of partitioning in the homogenous media implies that the components act independently in their interaction with a given solute (metal). In other words, the values of k_{ME} are independent on the values of k_{MW} or k_{MA} and are transferable from one system to the other. Once determined and tabulated, the data may be used in any system, containing these components. As a result, predictions of the influence of different solutes, extractants, adducts on the effectiveness, and selectivity of the extraction process are possible once some of their parameters are ascertained.

Transferability can be proved by different independent experiments with three-component systems, proving relations:

$$\frac{k_{M_1E}}{k_{M_2E}} = \frac{k_{(M_1E)_1}}{k_{(M_1W)_1}} \times \frac{k_{(M_2W)_2}}{k_{(M_2E)_2}} = \frac{k_{(M_1E)_1}}{k_{(M_1A)_1}} \times \frac{k_{(M_2A)_2}}{k_{(M_2E)_2}} \qquad (10.39)$$

$$\frac{k_{ME_1}}{k_{ME_2}} = \frac{k_{(ME_1)_1}}{k_{(MW)_1}} \times \frac{k_{(MW)_2}}{k_{(ME_2)_2}} = \frac{k_{(ME_1)_1}}{k_{(MA)_1}} \times \frac{k_{(MA)_2}}{k_{(ME_2)_2}} \qquad (10.40)$$

$$\frac{k_{MA_1}}{k_{MA_2}} = \frac{k_{(MA_1)_1}}{k_{(MW)_1}} \times \frac{k_{(MW)_2}}{k_{(MA_2)_2}} = \frac{k_{(MA_1)_1}}{k_{(ME)_1}} \times \frac{k_{(ME)_2}}{k_{(MA_2)_2}} \qquad (10.41)$$

where the systems (all in the same inert diluent) experimentally verified are: M_1–E–W, M_1–E–A, and M_2–E–W, M_2–E–A, respectively, for Eqn (10.39); M–E_1–W, M–E_1–A, and M–E_2–W, M–E_2–A, respectively, for Eqn (10.40); M–A_1–E, M–A_1–W, and M–A_2–E, M–A_2–W, respectively, for Eqn (10.41).

Concentrations of extraction components in the organic phase should be known or determined in all experiments. Water concentration in the organic phase should be determined in order to check the differences of the water concentration in the solvation shell in Regions I–III (see Fig. 10.1).The same should be done for adduct A′ that is partially soluble in water. The solubility of A′ in the organic phase is determined in the system E–A–W (in the absence of metal).

The independence and transferability of affinity constants was experimentally proved by Purnell with coworkers[45,46] for about 180 organic systems, using chromatographic techniques. Nagy and coworkers[7,26,47] have proved it for many organic systems by NMR, UV–Vis, potentiometric titration, and kinetic measurement techniques.

Nevertheless, the transferability of affinity constants has to be checked for different extraction systems of interest.

3.5. Comparison of Classical and CCST Mathematical Descriptions

As mentioned above, the most transition metals, post-transition, and some main group elements have well-defined solvation, or coordination numbers in solution:[4,8,13,27–30] the metal ion is coordinated by a definite number, n, of charged ligands and neutral molecules with a well-defined geometry. However, in the mixture of solvents (i.e., extractant and water) the competition between constituents for coordination site takes place according to their affinity constants and concentrations; thus, the contribution of each of them in a definite coordination number cannot be well defined. Many other cations, such as alkali metal cations and most organic cations, form diffusive, ill-defined solvation shells.

If solvates of definite geometry and coordination number, n, are formed, their stability constants can generally be estimated, i.e., the equilibrium constants for the replacement of some solvent (in our case, water molecules, W) in the solvate M(W)Z+ by another solvent (extractant molecules, E) to form the solvated ion M(E)Z+.[8] This replacement reaction is generally studied in mixed solvents, but the values may be extrapolated to apply to the pure solvents. Grunwald,[48] Covington,[27] Cox,[49] and Marcus[28,50] derived equations relating the equilibrium constants for solvent replacement to standard molar Gibbs free energy of ion M transfer from a solvent (W) to its mixture with another solvent (W + E). According to Cox:[49]

$$\Delta G_{\mathrm{t}}^{\infty}(\mathrm{M, W} -> \mathrm{E}) = -n\,RT \ln \overline{K}_{\mathrm{ME}} \tag{10.42}$$

where the $\overline{K}_{\mathrm{ME}}$ is the average equilibrium constant for W→E solvents replacement and n is the solvation (coordination) number.

According to Scatchard–Deranleau equation[51–53] for UV–Vis (ε) (if all metal species obey Beer's law) or NMR (ν) measurements:

$$\frac{\varepsilon_{\mathrm{ME}} - \varepsilon}{C_{\mathrm{E}}} = \overline{K}_{\mathrm{ME}}(\varepsilon_{\mathrm{ME}} - \varepsilon_{\mathrm{MW}}) - \overline{K}_{\mathrm{ME}}(\varepsilon_{\mathrm{ME}} - \varepsilon) \tag{10.43}$$

or

$$\frac{\varepsilon_{\mathrm{ME}} - \varepsilon}{\varepsilon_{\mathrm{ME}} - \varepsilon_{\mathrm{MW}}} = \frac{C_{\mathrm{ME}}}{C_{\mathrm{M}}^{0}} = \frac{\overline{K}_{\mathrm{ME}} C_{\mathrm{E}}}{1 + \overline{K}_{\mathrm{ME}} C_{\mathrm{E}}} \quad \text{at } 0 < Z < 1 \tag{10.44}$$

Comparing Eqns (10.42)–(10.44) with Eqns (10.28) and (10.29) of the presented theory we obtain

$$\overline{K}_{\mathrm{ME}} = \frac{k_{\mathrm{ME}}}{k_{\mathrm{MW}}} \times \frac{N_{\mathrm{E}}}{n - N_{\mathrm{E}}} \tag{10.45}$$

$$\overline{K}_{\mathrm{MEc}} = \frac{k_{\mathrm{MEc}}}{k_{\mathrm{MW}}} \times \frac{z}{n - N_{\mathrm{E}}} \tag{10.46}$$

$$\overline{K}_{\mathrm{MEs}} = \frac{k_{\mathrm{MEs}}}{k_{\mathrm{MW}}} \times \frac{N_{\mathrm{E}} - z}{n - N_{\mathrm{E}}} \tag{10.47}$$

where

$$\overline{K}_{\mathrm{ME}} = \overline{K}_{\mathrm{MEc}} \times \overline{K}_{\mathrm{MEs}} \tag{10.48}$$

One can see that the classical equilibrium constant parameter $\overline{K}_{\mathrm{ME}}$ represents, in fact, an affinity constant ratio, $k_{\mathrm{ME}}/k_{\mathrm{MW}}$, when the active solvent (i.e., water in organic phase) is included in the thermodynamic treatment.

4. EXPERIMENTAL VERIFICATION OF THE CCST APPROACH

4.1. Experimental Techniques Used with the CCST

The correctness of the relation (Eqns (10.45)–(10.47)) between the equilibrium constant parameter $\overline{K}_{\mathrm{ME}}$ of the classical theories and affinity constant ratios, $k_{\mathrm{ME}}/k_{\mathrm{MW}}$, of the presented theory, obtained by the measurements of the same property: chemical shifts, extinction coefficients, potentiometric titrations, etc, is one of the main tasks at the experimental verification of the presented model.

Any measured physicochemical property of the metal ion in pure solvents (extractant, water, and adduct) is measured directly. The molar extinction coefficient (or chemical shift, or free energy change) of the metal in pure solvents (here, water), $\varepsilon_{\mathrm{MW}}$, and the variation of the apparent molar extinction coefficient, ε, are measured as the water molecules are stripped off stepwise from the solvation shell of the metal ion and replaced by molecules of the extractant E, throughout the whole concentration range, until reaching the corresponding value in pure extractant, $\varepsilon_{\mathrm{ME}}$. The affinity constant ratios are determined by the slope of the plot $(\varepsilon_{\mathrm{MW}} - \varepsilon)$ versus $(\varepsilon_{\mathrm{MW}} - \varepsilon)/C_{\mathrm{E}}$. The intercept of the plot gives the hypothetical molar extinction coefficient (or chemical shift in NMR measurements) of the metal in pure water or in pure extractant (using Eqns (10.23)–(10.26)). Thus, the agreement between the direct experimental determination of the chemical shifts in pure solvents and their graphically obtained values at $\varepsilon_{\mathrm{MW}} - \varepsilon = 0$ or $\varepsilon_{\mathrm{ME}} - \varepsilon = 0$ can be compared. The changes in the linearities of the plots (intersection points), or the different slopes in the loading curve (see Fig. 10.1) over the concentration range, mean that the different charge-transfer (chemical complexation mechanisms), and consequently the different values of k_{MEc}, or the different solvation mechanisms (hydrogen bonds change to Van der Waals) and consequently different values of k_{MEs}, take place at different metal concentration. It is evident that in this case we will obtain different values of the hypothetical chemical shifts in pure water. It means that we have different compositions of the solvation shell, depending on solute concentration. We have to use one more experimentally measured property, such as UV–Vis or quantitative IR measurements, to determine either k_{MEc} or k_{MEs}, or both. This analysis is useful when studying polyvalent metals or polybasic extractants, and especially when studying mixtures of extractants.

For the M–E–W system, the determination of the affinity constant ratios, k_{ME}/k_{MW}, may be realized by conventional equilibrium extraction experiment series. However, water concentration in the organic phase should be determined in all experiments, in order to check the differences of the water concentration in the solvation shell in Regions I–III (see Chapter 6, Fig. 6.1).

Molar volumes (l/mol or dm^3/mol) of solvents and composite solutions are determined by density measurement.

4.2. Titanium(IV) Extraction by DEHPA

For example, reinvestigated, according to the CCST, data of titanium(IV) extraction by DEHPA[2,6] are presented in Tables 10.1 and 10.2 and Fig. 10.7. The results demonstrate correctness of the above suggestions. Table 10.1 shows some properties of the initial components of Ti–DEHPA–water extraction systems. Table 10.2 presents the values of affinity constant ratios, averaged equilibrium constants, and intercepts, calculated, using the presented theory equations, for different regions of the extraction isotherm. Fig. 10.7 shows a good agreement between experimentally obtained (points) and calculated (lines) data.

Returning to our analysis of titanium(IV) extraction with DEHPA in the Section 2.2, the reaction (1) may be suggested between Ti species and extractant in the Regions 2a and 2b of Fig. 10.7. Here, acid–base act is initiated by solvation of the dissolved compounds. Four anionic ligands, two OH− and two DEHPA−, are bonded (with asymmetric bonds) to titanium central atom. Two water molecules occupy titanium(IV) ion coordination sites up to saturation in the first (inner) coordination sphere. Two neutral, DEHPA molecules are bonded by hydrogen bonding to the water dipoles in the second (outer) coordination sphere.[2,6,11] So, in this colorless nucleus aggregate we have four (two of them with $DEHP^-$ ligands) strong, charge transfer bonds and two DEHPA weak, hydrogen bonds to one atom of titanium.

As mentioned above, this aggregate is not stable and slowly (during several weeks of aging) loses its water molecules and partly DEHPA, forming yellow complex with bidentate bonds of titanium atom with oxygens of DEHP phosphoryl group (see Eqn (10.2)).

TABLE 10.2 Affinity Constant Ratios and Averaged Equilibrium Constants of Titanium(IV) Extraction by DEHPA (at Benzene as a Reference Diluent), Calculated for Different Regions of the Extraction Isotherm Using the Equations of the CCST and Experimental Data from Ref. 2,6,11

Regions in the extraction isotherm	Acidity of initial aqueous phase, M/dm^3	Affinity constant ratios, k_{ME}/k_{MW}	Averaged equilibrium constants $\overline{K}_{ME}$	Hypothetical Z^a(interc.) in pure E at $Q=0$	RSQ	Suggested aggregate formed at coordination number, $N=6$
Formation of nuclei aggregates, Regions 1, 2	0.1	36.6	38.7	0.008	0.982	Linear aggregate, formed by bridging with titanium molecule
	2.0^b	0.034	0.036	0.023	0.993	
	7.6	13.5	14.0	0.05	0.966	
Formation of linear or cyclic aggregates, Region 3	0.1	0.37	0.45	0.22	0.988	Linear aggregates, formed by bridging with titanium or water molecule
	2.0	0.30	0.34	−0.073	0.999	Linear or cyclic aggregate, formed by bridging with molecules of DEHPA and/or titanium
	7.6	0.11	0.13	0.31	0.991	Cyclic aggregate, formed by bridging with molecules of DEHPA or HCl

[a]Hypothetical saturation factor, Z, calculated from the intercept of the graphically obtained hypothetical molar extinction coefficient ε^f_{ME} of metal in pure extractant, E.
[b]Distribution of titanium(IV) into organic phase at these conditions may be explained partly by its interaction with MEHPA, very strong complexing chelating agent, always present in DEHPA at some level (In the DEHPA from Sigma about 3%; for details see Ref. 6).

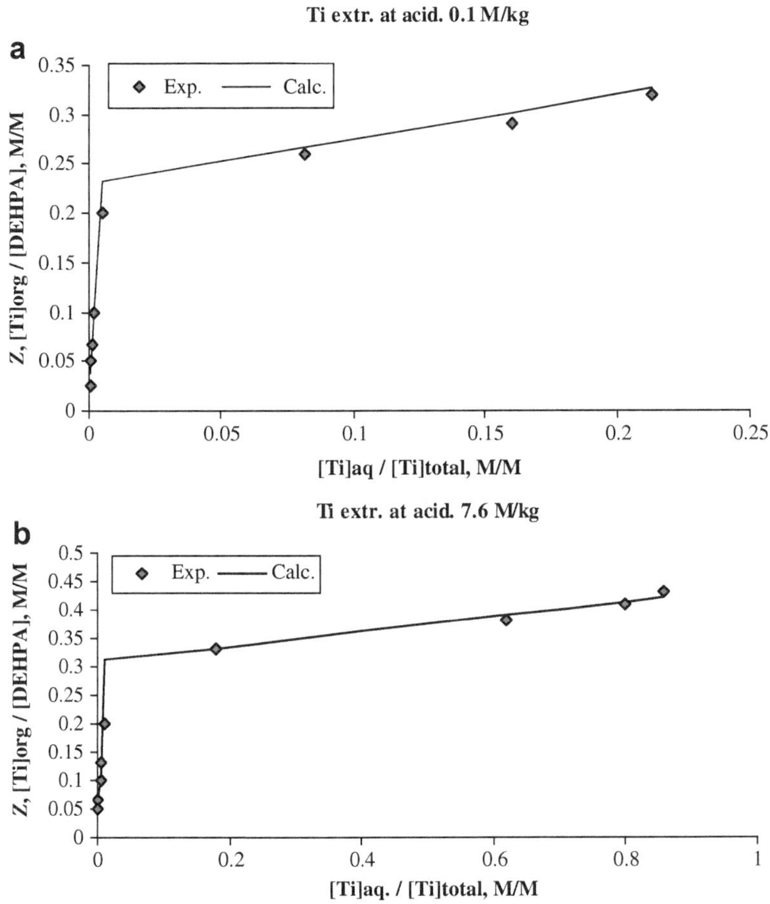

FIGURE 10.7 Comparison of experimentally obtained (points) and calculated (lines) data of extraction of titanium(IV) chlorides by DEHPA in benzene from the aqueous solutions at initial acidities (HCl): (a) 0.1 mol/kg, (b) 7.6 mol/kg.

Analogically, the structure of the nucleus aggregate formation at extraction of titanium from the strongly acidic aqueous phase (2b: 6.0–8.0 M HCl) is described by Eqn (10.3). This aggregate is not stable, but dehydration and desolvation with bidentate bonds formation is fast and is completed within short time of the phase equilibration contact[6,11] (see Eqn (10.4)).

As organic phase loaded by titanium with higher concentrations formation of linear or cyclic aggregates are clearly seen. At extraction of titanium from the low acidity (see Fig. 10.7a) aqueous phase titanium ion (or molecule) serves as a bridging component, interacting with DEHPA molecules of two (or three) nuclei aggregates. The slope or the value of the affinity constant ratio $k_{TiE}/k_{TiW} = 0.37$.

At extraction from the high acidity aqueous phase (see Fig. 10.7b) the value of affinity constant ratio ($k_{TiE}/k_{TiW} = 0.11$) is much less. In this case bridging of nuclei aggregates is suggested mainly through the HCl molecules with formation of stable cyclic aggregates. This mechanism may be compared with one, considered at extraction of strong mineral acids by tertiary amines (for details see Chapter 9).

Visible extraction of titanium by DEHPA from the aqueous phases of medium acidities begins only at relatively high titanium concentrations (see Fig. 10.3). The small affinity constant ratio, $k_{TiDEHPA}/k_{TiW} = 0.3$, permits to suggest that extraction here is driven mainly through the solvation, typical for the Region 3.

4.3. Divalent Cations Extraction

At extraction of divalent transition metals (cobalt, nickel, and copper) with acidic organophosphorus extractants, e.g., DEHPA,[7,22,23] nucleus metal-extractant aggregate is tetraligated intermediate species of type $M(DEHP)_2 \cdot 2H_2O \cdot 2DEHPA$ with two H-bonded neutral DEHPA molecules, bonded to the metal central atom in the first coordination sphere, and two water molecules in the second coordination sphere. Experiments showed that dehydration kinetics with copper was very fast and the authors determined $Cu(DEHP)_2 \cdot 2DEHPA$ species after extraction experiment.[7,22] For Ni dehydration kinetics is slow, so the authors determined $Ni(DEHP)_2 \cdot 2H_2O \cdot 2DEHPA$ species. At extraction of copper with Cyanex C302 complexation mechanism and intermediate transformation kinetics was similar to nickel with DEHPA.[24]

The authors proved the formation of linear, cyclic aggregates and reversed micelles using some different experimental and analytical techniques (see above). The linear and subsequently three-dimensional aggregation was observed with other organophosphorus extractants, namely, phosphonic and phosphinic acid esters.[41]

Cross-linking aggregation and cluster-like three-dimensional aggregate formation are also suggested.[2]

5. SUMMARIZING REMARKS

The readers can see that the presented model overcomes some limitations of "chemical modeling approach",[12–14] which is typical in the recent developments of the classical theories of extraction. The authors consider the system as a static, where one of the mechanisms (ion exchange, coordination, or hydrogen bonding) is a dominating one, which is not changed at changing concentrations of extracted solute. They have developed the mathematical models in which hypothetical complexes of a specific stoichiometry or their mixtures are formed. This approach (chemical modeling) is a useful tool to describe the data quantitatively, if complexation is strong (as in the Region 2 of general extraction isotherm).

In the presented theory, different regions are analyzed separately and the mathematical descriptions are developed for different interactions between the components of the extraction system. Interchanges in the linearity of Eqn (10.20), or the different slopes in the loading curve over the metal concentration range mean that the affinity constant ratios and/or concentrations of the components in the aggregate (or coordination complex, or solvation shell) are different. It means, also, that the different charge-transfer complexation mechanisms, and consequently the different values of k_{MEc}, or the different solvation mechanisms and consequently different values of k_{MEs}, take place at different metal concentrations. So, we have to use one more experimentally measured property, such as UV–Vis or quantitative IR measurements, to determine either k_{MEc} or k_{MEs}. Regions 1–4 in the distribution curve (see Chapter 6, Fig. 6.1) with different values of the slopes testify different compositions of the aggregates (solvation shells) formed and different influences of the constituents of extraction system on the interaction mechanisms over the metal concentration range.

This theory introduces an active solvent (including water) as a quantitative parameter, which is participating and influencing the formation of different compositions of the aggregates (solvation shells) at changing its concentration in the organic phase.

The CCST approach is an attempt to describe the mechanisms of solvent extraction with quantitative evaluation of different compounds, formed in the organic phase at different concentrations of the solute in the aqueous phase. Of course, this theory has many simplifications and limitations (see above and the following section). But we believe that it may be a starting point for the quantification of solvent extraction theories with an aim to predict suitable extraction systems for the separation of different metals.

REFERENCES

1. Kislik V, Eyal A. Acidity dependence of Ti(IV) extraction: a critical analysis. *Solvent Extr Ion Exch* 1993;**11**(2):258–72.
2. Kislik V, Eyal A. Mechanisms of titanium(IV) and some other transition metals extraction by organophosphorus acids. *Proceedings of ISEC'93* 1993;**2**:1353–60.
3. Tsuboi I, Kasai S, Yamamoto T, Komasawa I, Kunugita E. Recovery of rare metals from coal fly ash. *Proceedings of ISEC'90*. Amsterdam: Elsevier; 1992. Part B, p. 1199–204.
4. Wilkinson G, Gillard RD, MaccLeverty JA, editors. Comprehensive coordination chemistry. John Wiley & Sons, 1987, vol. 2, p. 3.
5. Douglas B, McDaniel D, Alexander J, editors. *Concepts and models of inorganic chemistry*. 3rd ed. New York: John Wiley & Sons; 1994. p. 309–49. 379–557.
6. Kislik V, Eyal A. Extraction of titanium(IV) by mixtures of mono- and d-(2-ethylhexyl) phosphoric acids esters. *Solvent Extr Ion Exch* 1993;**11**(2):285–310.
7. Szpakowska M, Nagy OB. Application of the preferential solvation theory to coordinative solute-solvent interactions. *J Chem Soc Faraday Trans1* 1989;**85**(9):2891–9.

8. Marcus Y. *Ion solvation*. London: Wiley-Interscience; 1985 (a) p. 44–6, (b) p. 184, (c) p. 194–216, (d) p. 218–234.
9. Kislik V. Application of preferential solvation model for interpretation of the mechanisms of solvent extraction. In: Proceedings of the XVth international symposium on physico-chemical methods of mixtures separation, "ARS SEPARATORIA 2000", Poland, 2000, June 14–17, p. 34–40.
10. Kislik V. Application of preferential solvation concept for interpretation of the mechanisms of solvent extraction of acids by amines. *Sep Sci Technol* 2001;**36**(13):2899–926.
11. Kislik V, Eyal A. Unpublished experimental data, implemented for the private company, 1996–1999.
12. Eyal AM, Canari R. pH Dependence of carboxylic and mineral acid extraction by amine-based extractants: effects of pK_a, amine basicity, and diluent properties. *Ind Eng Chem Res* 1995;**34**(5):1789–98.
13. Tamada J, King C. *Extraction of Carboxylic Acids by Amine Extractants. Ph.D. Thesis.* Lawrence Berkeley Laboratory, University of California, Applied Science Division; 1989.
14. Yang S, White S, Hsu S. Extraction of carboxylic acids with Tertiary and Quaternary amines: effect of pH. *Eng Chem Res* 1991;**30**:1335.
15. Grinstead RR. Base strengths of amine in liquid-liquid extraction systems. *Proceedings of ISEC'66, Solvent Extraction Chemistry.* Amsterdam: North-Holland; 1967. p. 426–32.
16. Mrnka M, Kyrsova V, Celeda J. Mechanism of iron separation by extraction with tri-n-octylamine from chloride solutions. *Proceedings of ISEC'74* 1975;**1**:49–62.
17. Lo TC, Hanson C, Baird MHI, editors. *Handbook of solvent extraction*. New York: Wiley; 1983.
18. Periole X, Allouche D, Ramirez-Solis A, Ortega-Blake I, Daudey JP, Sanejound YH. New effective potential extraction method for the interaction between cations and water. *J Phys Chem B* 1998;**102**:8579–87.
19. Wardel JM, King CJ. Solvent equilibria for extraction of carboxylic acids from water. *J Chem Eng Data* 1979;**23**(2):144–8.
20. Ermer O, Eling A. Molecular recognition among alcohols and amines: super-tetrahedral crystal architectures of linear diphenol-diamine complexes and aminophenpls. *J Chem Soc Perkin Trans 2* 1994;**3**:925–44.
21. Loehlin JH, Etter MC, Gendreau C, Cervasio E. Hydrogen-bond patterns in several 2:1 amine-phenol cocrystals. *Chem Mater* 1994;**6**:1218–21.
22. Neuman RD, Nai-Fu-Zhou, Wu J, Jons MA, Gaonkar AG. General model for aggregation of metal-extractant complexes in acidic organophosphorus solvent extraction systems. *Sep Sci Technol* 1990;**25**(13–15):1655–74.
23. Neuman RD, Jons MA, Zhou NF. Photo-correlation spectroscopy applied to hydrometallurgical solvent-extraction systems. *Colloid Surf* 1990;**46**(1):45–61.
24. Kislik V, Eyal A. Heavy metals removal from wastewaters of phosphoric acid production: a comparison of hybrid liquid membrane (HLM) and aqueous hybrid liquid membrane (AHLM) technologies. Proceedings of the Conference on Membranes in Drinking and Industrial Water Production, Paris, France, October 3–6, 2000, vol. 1, p. 503–14.
25. Yu ZJ, Ibrahim TH, Neuman RD. Aggregation behavior of cobalt(II), nickel(II), and copper(II) bis(2-ethylhexyl) phosphate complexes in n-heptane. *Solv Extr Ion Exch* 1998;**16**(6):1437–63.
26. Nagy OB, Muanda H, Nagy JB. Competitive preferential solvation theory of weak molecular interactions. *J Chem Soc Faraday Trans* 1978;**1**(74):2210–29.

27. Covington AK, Newman KE. Thermodynamics of preferential solvation of electrolytes in binary solvent mixtures. *Adv Chem Ser* 1976;**155**:153–96.
28. Marcus Y. Preferential solvation of ions in mixed solvents. *Rev Anal Chem* 1980;**5**:267–82.
29. Taft RW, Abboud JLM, Kamlet MJ, Abraham MN. Linear solvation energy relations. *J Solut Chem* 1985;**14**(3):153–86.
30. Marcus Y. *The properties of solvents*. London: Wiley-Interscience; 1998.
31. Luder WF, Zuffanti S. *The electronic theory of acids and bases*. New York: Wiley; 1946.
32. Pearson RG, Vogelsong DC. Developments in crystal field theory. *J Am Chem Soc* 1958;**80**:1038.
33. Pearson RG. Hard and soft acids and bases. *J Am Chem Soc* 1963;**85**:3.
34. Devis MM. *Acid-base behavior in aprotic organic solvents*. Washington, D.C: National Bureau of Standards, Monograph 105; 1968.
35. Usanowitsch M. Theory of acids and bases. *J Allg Chem (USSR)* 1939;**9**:182.
36. Dwyer A, Mellor D, editors. *Chelating agents and metal chelates*. Academic Press; 1964. p. 10–93.
37. Komasawa I, Otake T, Hattori I. The effects of diluent in the liquid-liquid extraction of copper and nickel using 2-hydroxy-5-nonylbenzophenone oxime. *J Chem Eng Japan* 1983;**16**(3):210–18.
38. Hogfeldt E, Fredlund F, Odberg L, Merenge G. On aggregation of the complex between nitric acid and trilaurylamine dissolved in octane and m-xylene. *Acta Chem Scand* 1973;**18**(5):1781–94.
39. Marcus Y, Kertes AS. *Ion exchange and solvent extraction of metal complexes*. London: Wiley-Interscience; 1969.
40. Evers A, Hancock R. A study of aggregation in the organic phase of solvent extraction systems and its influence on the extraction process. *Report Natl Inst Metall, Repub S Afr* 1977;34.
41. Gaonkar AG, Garver TM, Neuman RD. H-1-NMR spectroscopic investigation of reversed micellization in metal organophosphorus surfactant system. *Colloid Surf* 1988; **30**(3–4):265–73.
42. Fumio Toda, Hyoda S, Okada K, Hirtsu R. Isolation of anhydrous hydrasine as stable inclusion complexes with hydroquinone and p-methoxyphenol and their solide state reaction with esters which gives pure hydrolyzides. *J Chem Soc Chem Commun* 1995;1531–2.
43. Kislik V. Reversible proton transfer and hydrogen-bond models in competitive complexation/salvation theory of solvent extraction. In: Proceedings of International Conference: Hydrogen bonding and Proton Transfer, Berlin, 2002, p. 324–27.
44. Kislik V, Eyal A. Hybrid liquid membrane (HLM) and supported liquid membrane (SLM) based transport of titanium (IV). *J Mem Sci* 1996;**111**:273–81.
45. Purnell JH, Vargas de Andrade JM. Solution and complexing studies. I. Gas-liquid chromatographic investigation of supposed complexing systems. II. Comparison and correlation of nuclear magnetic resonance and gas-liquid chromatografic data. *J Am Chem Soc* 1975;**97**:3585–90.
46. Laub RJ, Purnell JH. Solution and complexing Studies. III. Further evidence for a microscopic partitioning theory of solution. *J Am Chem Soc* 1976;**98**(1):30–5.
47. Parbhoo B, Nagy O. Application of the competitive preferential solvation theory to ion-molecule interactions. *J Chem Soc Faraday Trans* 1986;**82**(1):1789–93.
48. Grunwald E, Baughman G, Kohnstam G. The solvation of electrolytes in dioxan-water mixtures, as deduced from the effect of solvent change on the standard partial molar free energy. *J Am Chem Soc* 1960;**82**:5801–11.

49. Cox BG, Parker AJ, Waghorne WE. Coordination and ionic solvation. *J Phys Chem* 1974;**78**:1731–40.
50. Marcus Y. A quasi-lattice quasi-chemical theory of preferential solvation of ions in mixed solvents. *Austr J Chem* 1983;**36**:1719–31.
51. Scatchard G. The attraction of proteins for small molecules and ions. *Ann NY Acad Sci* 1949;**51**:660.
52. Deranleau DA. Theory of the measurement of weak molecular complexes. I. General considerations. *J Am Chem Soc* 1969;**91**:4044–9.
53. Deranleau DA. Theory of the measurement of weak molecular complexes. II. Consequence of multiple equilibria. *J Am Chem Soc* 1969;**91**:4050–62.

Chapter 11

Final Remarks on the Competitive Complexation/ Solvation Theory of Solvent Extraction and its Application

The main objective of this book is to present a novel, competitive complexation/solvation (CCS) theoretical approach to solvent extraction separation processes. This subject was described in Part II, Chapters 6–10, of this book. In this chapter some final remarks are presented in comparison with conventional (classical) principles approach.

CCS theory was developed for interpretation and quantification of interactions in solvent extraction systems. According to the theory the molecules of an extracting mixture compete for solute (hydrated or solvated metal ions, acids, etc.) by the extent of their affinity to association and concentration. The solute is partitioned among the solvent components (i.e., extractant, synergistic agent, diluent, and water) forming, as a first step, H-bonded solvation shells or reversible reactive intermediates (adducts). The following step is proton/electron transfer with ion-exchange or ion-pair structures' formation. As a consequence, a given measured property in mixed solvents can be calculated from values, measured in single solvent component systems. The theory establishes the correlation between complexation and solvation. At low concentrations of the solute in the organic phase the solute–extractant complex with its solvation shell forms a geometric structure, which is denoted as a single nucleus aggregate. Competition between strong (complexation) and weak (solvation) interactions causes the shape of the distribution isotherm between two limiting shapes: pure complexation and pure solvation.

At increasing loading of organic phase different possible stages of extraction behavior and interacting mechanisms are discussed. At higher concentrations of the solute in the organic phase nuclei aggregates interact ("Self-assembly," see supramolecular aggregates formation in Chapter 12) and form linear or cyclic aggregates. At high solute concentration the linear or cyclic aggregates eventually undergo a structural reorganization to form

Solvent Extraction. DOI: 10.1016/B978-0-444-53778-2.10011-1

micelle-like or cross-linking cluster-like three-dimensional association structures (as a rule, forming a so-called third phase).

Active solvents (extractant, solute, synergistic agent, etc.) are considered amphoteric and may behave as acids (electron acceptors) or bases (electron donors), depending on the nature of their functional groups and composition of the organic phase and on the nature of solute species and composition of the aqueous phase. Acidic extractant interacts as a conjugate acid with hydrated solute ions. The same acidic extractant behaves as a conjugate base to solute ions, solvated (coordinated) by strong acid molecules.

The main advantage of the presented approach lies in the mathematical description that provides a key for quantitative analysis and preliminary prediction of suitable extraction systems for different solutes separation. More parameters such as water and active solvent concentrations in the organic phase are introduced to the quantitative consideration of the extraction systems in every region of the general extraction isotherm.

The theory overcomes some limitations of the conventional stoichiometric ion-exchange models. It presents the novel theoretical insights, and explains some problems in a clear-cut manner, which required ad hoc arguments using conventional theories.

The presented CCS modeling approach constitutes a general and useful framework for interpretation of ion–molecular interaction data in the solvent extraction systems.

Affinity constant ratios are easily acceptable through the independent measurements conducted with simple three-component systems. Only relative values of affinity constant ratios can be measured in solution because of the ubiquitous nature of molecular interactions. Experimental determination of the affinity constant ratios and solvation shell compositions, quantitative, or even semiquantitative spectrometric determination of the solute in the complexed and solvated forms permits to set up exact form of the aggregates formed and predicts the behavior of the extraction system of interest and influence of its different components. Affinity constant ratio's value of unity suggests the same values of solvation effects of the solute with extractant in the organic phase and with water in the aqueous phase; a large value of k_{SE}/k_{SW} means strong complexation–solvation effects with extractant. A positive sign of a slope means that solvation of the reactant molecules is stronger than that of the product molecules. A negative slope shows that solvation of the product molecule is stronger than that of the reactant. Composition and behavior predictions of the extraction system of interest may be analyzed.

Interchanges in the linearity or the different slopes in the loading curve over the solute concentration range mean that the affinity constant ratios and/or concentrations of the components in the aggregate are different. It also means that the different charge-transfer complexation mechanisms, and consequently the different values of k_{SEc}, or the different solvation mechanisms, and consequently different values of k_{SEs}, take place at different solute concentrations. Therefore, we have to use an additional experimentally measured property, such as UV–VIS or quantitative IR measurements, to determine k_{SEc}/k_{SEs}. Separate mathematical descriptions are developed for different regions of the extraction isotherm.

The theory of partitioning in the homogenous media implies that the components act independently in their interaction with a given solute. In other words, the values of k_{SE} are independent of the values of k_{SW} or k_{SA}, and are transferable from one system to the other. Influence of different solvent adducts to extractant on the effectiveness and selectivity of the extraction process is predictable once some solvent parameters are ascertained.

The presented theory predicts the synergistic, enhancing effect of the weaker solute in the mixture of solutes on the distribution of the stronger solute to the organic phase, or the weaker extractant in the mixture of extractants on the distribution of the solute to the organic phase. Classical theories do not indicate the direct knowledge of this effect.

Relationship between equilibrium constant, K_{Si}, of the conventional theories and affinity constant ratio, k_{Si}/k_{Sj}, of the CCS theory, obtained by the measurements of the same property, permits to use available equilibrium constant data for calculations in the presented theory.

Cosolvent independence, established experimentally, allows to compare directly the solvating-complexing power (competition order) of various extractants, synergistic agents, and diluents at a fixed other components of the system. Once determined and tabulated, the data may be used in any system, containing these components.

The readers can see now that the presented model overcomes some limitations of conventional "chemical modeling approach", which is typical in the recent developments of the classical theories for extraction of metals. The authors consider the system as a static one, where one of the mechanisms (ion exchange, coordination, or H-bonding) is a dominating one, which is not changed at changing concentrations of extracted solute. They have developed the mathematical models in which hypothetical complexes of a specific stoichiometry or their mixtures are formed. This approach (chemical modeling) is a useful tool to describe the data quantitatively, if complexation is strong.

In the presented theory different regions are analyzed separately and the mathematical descriptions are developed for different interactions between the components of the extraction system. Interchanges in the linearity of equation or the different slopes in the loading curve over the metal concentration range mean that the affinity constant ratios and/or concentrations of the components in the aggregate (or coordination complex or solvation shell) are different. It also means that the different charge-transfer complexation mechanisms, and consequently the different values of k_{MEc}, or the different solvation mechanisms and consequently different values of k_{MEs}, take place at different metal concentrations. So, we have to use one more experimentally measured property, such as UV–VIS or quantitative IR measurements, to determine either k_{MEc} or

k_{MEs}. Regions 1–4 in the general distribution curve (see Fig. 6.1 in Chapter 6) with different values of the slopes testify different compositions of the aggregates (solvation shells) formed and different influences of the constituents of extraction system on the interaction mechanisms over the metal concentration range. This approach is an attempt to describe the mechanisms of solvent extraction with quantitative evaluation of different compounds, formed in the organic phase at different concentrations of the solute in the aqueous phase.

Although conventional theories of solution (Chapter 1) and formation of extractable complexes (Chapters 2 and 3) now are well advanced, predictions of distribution ratios are mainly done by comparison with known similar systems (Chapters 4 and 5). Solvatochromic parameters, solubility parameters, and donor numbers, as discussed in the first part of this book, are so far mainly empirical factors. Continuous efforts are made to predict such numbers. It is likely that with novel theory greater ranges of conditions for more systems will successively encompass. Still much has to be done. In the future, the theory may allow the assignment of exact numbers to the solvatochromic parameters, thus also permitting theoretical predictions of distribution constants for known as well as for hypothetical (not yet synthesized) extractants.

Of course, this theory has many simplifications and limitations (see above). But I believe that it may be a starting point for the quantification of solvent extraction theories with the aim to predict suitable extraction systems for different Solutes' separation.

Part III

Modern and Future Trends in Fundamentals and Applications of Solvent Extraction

1. INTRODUCTION

Conventional solvent extraction, a technology for selective removal and recovery of solutes employs an organic solvent and an aqueous solution as the two immiscible phases. The ability to utilize a number of different diluents, extractants, and aqueous phases makes solvent extraction a powerful separation method possessing a number of favorable characteristics. These include rapid extraction kinetics for many separations, the adaptability of the method to a wide number of solutes, the feasibility of back-extraction or stripping of the solute, and the possibility of recycling the solvent and/or diluent. Furthermore, liquid–liquid extraction is capable of large volume throughput, is amenable to large-scale separations, and can be engineered for high selectivity and efficiency by the use of multistage contactors.

The suitability of using solvent extraction for a given separation is determined by thermodynamic and kinetic considerations. The main thermodynamic parameter is the solute distribution ratio, D_M, between the organic and the aqueous phase. The magnitude of D_M and selectivity determines the feasibility of the separation as an industrial process. Another consideration affecting the design of extraction processes is the extraction rate, kinetics of the process, as it determines the residence time of the phases in the contactor, and consequently its size.

In spite of the numerous advantages, there are several drawbacks to traditional solvent extraction. Even with today's environmental standards, a number of extraction systems still utilize toxic and flammable organic diluents. When the diluent is coupled with a highly selective extractant, the cost of the solvent system can become very expensive. Partitioning of a polar or charged solute from an aqueous phase into an organic medium requires dehydration, the extent of which depends on the organic solvent. In some separation schemes this is trivial, whereas in others, particularly for the separation of metal ions, it can be very complex.

The principle of solvent extraction - the distribution of chemical species between two immiscible liquid phases - has been applied to many areas of chemistry. A typical one is liquid partition chromatography, where the principle of solvent extraction provides the most efficient separation process available to organic chemistry today; its huge application has become a field (and an industry!) of its own. The design of ion selective electrodes is another application of the solvent extraction principle; it has also become an independent field. Both these applications are only briefly touched upon in this book on analytical applications. Nevertheless, fundamental research on solvent extraction will provide further important input into these two applications.

What are the future trends of solvent extraction?

In the following chapters (12-14) I am trying to indicate some areas in this field in which important advances in the future are expected. It may be divided on three main subjects: 1) developments of the theory: chemistry, thermodynamics, kinetics; 2) development of materials: extractants, solvents, diluents; and 3) development of processes and techniques.

Chapter 12

Modern and Future Trends in Fundamentals of Solvent Extraction

Chapter Outline

1. INTRODUCTION

Various instruments of theoretical chemistry have been widely used to describe separate steps of solvent extraction. Because of the complexity of solvent extraction systems, there is still no unified theory and no successful approach aimed at merging the extraction steps. The challenging problem for researchers dealing with solvent extraction, in particular with thermodynamic calculations,

Solvent Extraction. DOI: 10.1016/B978-0-444-53778-2.10012-3

is to evaluate correctly the solvent effects by the use of the most accurate explicit solvation models and quantum mechanical (QM) calculations. However, such calculations on extremely large sets consisting of hundreds or even thousands of molecules—necessary to model all aspects of the extraction systems—are still impossible due to many limitations.

Chemical quantum mechanics are now promising to contribute to prediction of distribution ratios, particularly when used in interactive computer modeling of chemical structures and reactions. This will provide better understanding of solute–solvent interactions and the theory of solubility, which is the foundation for prediction of distribution ratios. Progress is likely to speed up as efficient computing programs with large data bases become more easily available on the market at more attainable prices. Such interactive computer research will be able to explain poor extractability due to steric hindrance, hydration, the synergistic effect of specific additives, etc., and, as a consequence, give clues to better extraction reagents and conditions.

2. NOVEL MOLECULAR MODELS THEORETICAL APPROACH

The methods applied to the description of complex formation, hydration, and solvation are characterized by different degrees of approximation made when solving the equations. But even the introduction of quantum mechanics (QM), molecular mechanics (MM), and molecular Dynamics (MD) methods does not allow one to make effective calculations. This is because it is common in this approach to neglect interactions for pairs of atoms separated by distances greater than a selected cutoff value. The range of interactions in the solution is thus limited by the size of the system studied. This arbitrarily assumed division of the liquid phase into molecules that interact with one another and molecules that are neglected in the calculations and whose contribution is estimated by corrections makes the calculations possible. Nevertheless, in spite of this simplification, such calculations are still very difficult, time-consuming, and expensive; moreover, the results obtained are often insufficiently accurate, inconsistent with the experiment, and therefore inconclusive.

The development of other methods with less computing requirements and free of such limitations seems to be necessary in order to better understand the processes that take place in the liquid phase, understand their molecular mechanism, predict the influence of various factors on their equilibria and kinetics, and evaluate correctly the thermodynamic functions for the whole process and its particular steps.

The recent theoretical approach based on the information theory (IT) in studying aqueous solutions and hydration phenomena shows such a direction. IT is a part of the system based on a probabilistic way of thinking about communication.[1] It consists of quantitative description of the information by defining entropy as a function of probability.

The probability of cavity formation in bulk water, able to accommodate a solute molecule by exclusion of a given number of solvent molecules, was inferred from easily available information about the solvent, such as the density of bulk water and the oxygen–oxygen radial distribution function.[2]

MacKay's textbook[1] offers not only a comprehensive coverage of the IT, but also probabilistic data modeling and the mathematical theory of neural networks (NN). The wide range of applications of NN also comprises optimization issues. The information-theoretic capabilities of some NN algorithms are examined and NN are motivated as statistical models.

The applications of NN to solvent extraction suffer from an essential limitation in that they do not apply to processes of quantum nature; therefore, they are not able to describe complexes in extraction systems on the microscopic level. In fact, the networks can describe only the pure state of simplest quantum systems, without superposition of states. NN that indirectly take into account quantum effects have already been applied to chemical problems. For example, the combination of QM molecular electrostatic potential surfaces with NN makes it possible to predict the bonding energy for bioactive molecules with enzyme targets. Computational NN were employed to identify the QM features of the inhibitory molecules that contribute to bonding. This approach generates the relationships between the QM structure of the inhibitory molecule and the strength of bonding.

Interactions on the microscopic level can be treated by a new kind of NN—quantum neural networks (QNN).[3] Early publications were concerned with problems different from those discussed in this book. It may be expected that the QNN approach will also be helpful in describing intermolecular interactions accompanying the processes of extraction. The methods developed from either NN or IT undoubtedly reduce the requirements that must be applied.

Unfortunately, the application of these methods to the description of processes that take place in solutions, such as solvation, still requires the rejection of certain interactions between molecules that build the systems. This results from the tremendous number of interactions of the order of 2 N, where N denotes the number of molecules in the system. Optimization of such large systems consisting of mutually interacting elements requires special numerical procedures. It seems that application of approximate algorithms bring us closer to the understanding of the processes that take place in solutions, in spite of some disadvantages.

As far as the author is aware, the competitive complexation/solvation theory (CCST)—described in Part II of this book—has been used in the novel QM approach to the explicit solvation models, in particular to solvation of metal complexes. The author suggests that this direction is expected to be important in the future developments of theoretical principles of solvent extraction processes.

3. LOW ENERGY REACTIONS MEASUREMENT

Distribution ratio measurements offer a method to investigate low energy reactions in solutions, such as weak solute–solute and solute–solvent interactions, especially in the organic phase. A 10% change in the distribution ratio, which can be measured easily and accurately in the distribution range of 0.01–100 (most commonly used), corresponds to an energy change of only 0.2 kJ/mol. This technique has only been slightly exploited for this purpose. It is particularly noteworthy to find how few thermodynamic studies have been made of the distribution or extraction constants, as enthalpy and entropy values give a good indication of the driving force of the extraction and also may indicate the structure of the molecular species in the organic solvent.

Separation of metals by solvent extraction is usually based on the various complexing properties of the metals. Separation systems may be chosen on the basis of complex formation constants obtained from the literature. However, the literature often shows different values for "same systems" causing considerable concern for process design chemists. There is an obvious need for an objective presentation of the uncertainty in the published equilibrium constants—however conditional they may be.

The driving force of the transport of salts, proteins, etc., through the cell membrane from the nucleus to the body fluids, and vice versa, is a complicated biochemical process. As far as is known, this field has not been explored by traditional solution chemists, although a detailed analysis of these transfer processes indicates many similarities with solvent extraction processes (equilibrium as well as kinetics). It is possible that studies of such simpler model systems could contribute to the understanding of the more complicated biochemical processes.

Fundamental solvent extraction research will continue to contribute to the development of further selective analytical techniques.

Solvent extraction deals with the transport of chemical substances from one phase into another, the chemical kinetics of this process, and the final equilibrium distribution of the substances between the two phases. Such transport and distribution processes are the motors that make life in biological systems possible. Fundamental studies of such "solvent extraction" processes contribute to the better understanding of all processes in nature. Here, only the lack of imagination stands in the way of important new scientific discoveries.

4. EXPECTED ADVANCES IN KINETICS STUDIES

All chemical reactions occur at a certain rate. For aqueous systems, in which no redox reactions occur (e.g., simple complex formation), equilibrium is often attained rapidly, which is also true for adduct reactions in the organic phase, all at normal temperatures. However, the transfer of a solute from an aqueous solution to an organic solvent occurs via a phase boundary; such reactions depend on several parameters and may be quite slow. Therefore, the kinetics of solvent extraction, which is of paramount importance to all industrial applications, is largely determined by the interfacial chemistry. To speed up phase transfer, the engineer tries to maximize the interfacial surface by, e.g., violent stirring, producing billions of small droplets. Although the theory is well advanced, the interface is very complicated and its properties are difficult to investigate, particularly at the molecular level. Few applicable techniques are available, and the results are often difficult to interpret. The coalescence of the droplets (to produce two clean phases that are easy to separate) presents another interfacial surface problem, and so-called phase reversals may further complicate the separation. This is particularly important in the development of new types of contactor. Much progress is required here and as such presents a challenge to future scientists and engineers.

The largest research effort in extraction kinetics is likely to be in the development of solvent extraction-related techniques, such as various versions of liquid chromatography, liquid membranes, etc. These techniques require a detailed knowledge of the kinetics of the system to predict the degree of separation. It is a practical fact that most industrial solvent extractions are carried out under nonequilibrium conditions—however close the approach may be; e.g., centrifugal contactor-separators rarely operate at distribution equilibrium. An interesting possibility is to expand this into extractions further from equilibrium, if the kinetics of the desired and nondesired products is different. Such operations offer a real technological challenge.

5. SUPRAMOLECULAR (SUPRAS) THEORETICAL APPROACH

Supramolecular solvents (SUPRASs) is a recent term[4] to refer to nano-structured liquids generated from amphiphiles through a sequential, self-assembly process occurring on two scales: molecular and nano. This process gives, at first, three-dimensional aggregates that coacervate, in a second stage, producing water-immiscible liquids made up of large supramolecular aggregates dispersed in a continuous phase, generally water. Supramolecular solvents are well known to the analytical community and have been used for long years in extraction processes under different names (e.g., cloud point technique[5–7] to refer to the temperature at which nonionic surfactants start to undergo phase separation in aqueous solutions, thus becoming clouding, or coacervates[8,9] to refer to the phenomenon through which liquid–liquid phase separation is produced, i.e., coacervation). The use of the term "supramolecular solvents" is here encouraged to place greater emphasis on their solvent character, to differentiate them from molecular and ionic solvents, to consider the noncovalent interactions through which molecules are held together in the solvent, and to take into account the self-assembly processes through which they are formed.

5.1. Supramolecular Solvent Formation

The formation of supramolecular compounds (SUPRAS) from a homogeneous solution containing amphiphiles occurs through two sequential, self-assembly processes, which respectively occur at the molecular and nanometer level (see Fig. 12.1). First, amphiphiles give three-dimensional aggregates, mainly aqueous/reverse micelles or vesicles, above a critical aggregation concentration (CAC) and then, these nanostructures self-assemble in bigger aggregates that separate from the bulk of the colloidal solution as a new liquid phase (i.e., the supramolecular solvent) through a phenomenon known as coacervation.[10]

SUPRASs are amphiphile-rich and immiscible with the solvent (usually water) from which they originate, despite the solvent being a major component of SUPRASs and constituting the continuous phase in which the supramolecular assemblies disperse.[11]

The solution in equilibrium with the SUPRAS contains monomers of amphiphiles at the CAC. Below, basic aspects of the phenomena involved in SUPRAS formation, namely, self-assembly and liquid–liquid phase separation are briefly discussed.

5.1.1. Principles of Self-assembly

Self-assembly is the process by which isolated components organize autonomously and spontaneously into ordered and/or functional structures.[12] It occurs when components interact with one another through a balance of attractive and repulsive interactions. These interactions are generally weak and noncovalent

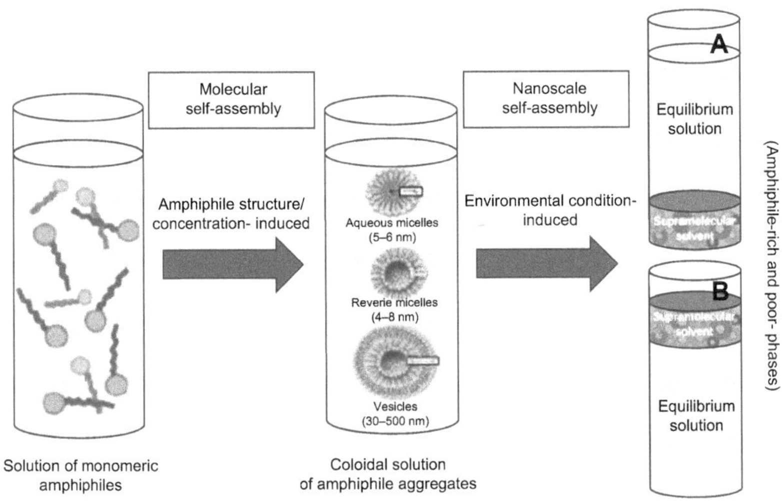

FIGURE 12.1 Self-assembly processes in supramolecular solvent formation.

(Coulomb and hydrophobic interactions, and hydrogen bonds) but relatively weak covalent bonds (coordination bonds) are increasingly recognized appropriate for self-assembly.[13]

Components that assemble can have sizes from the molecular to the macroscopic scales provided the appropriate conditions are met. Although much of the work in self-assembly has focused on molecular components, many of the most interesting applications of self-assembling processes can be found at larger sizes (nanometers to micrometers). In fact, self-assembly is the preferred way for nature to build its animate creations on various length scales and is a ubiquitous process in chemistry and material science. Today, self-assembly has emerged as a discrete field of study and as a synthetic strategy.[14,15]

Directing self-assembling systems is possible by proper design of the components, the environment, and the driving force.[16] The inherent structure of components determines the complexity of the final self-assembled system, so, tailoring it provides the first means of controlling the process. Likewise, the components have to be embedded in a proper environment (e.g., a solution) for the binding force to act. Tailoring of the environment or dynamically changing it provides an important means of controlling the self-assembling system. The environment can be manipulated to change the conformation of components or to alter binding forces in the system. If conditions are changed, aggregates will be reformed and indistinguishable from the original, so they behave as adaptive material.[17]

On the other hand, components must be mobile for self-assembly to occur. In solution, thermal noise is the driving force for molecular self-assembly since it provides the major part of the motion required for bringing molecules into contact. In nanoscale, mesoscopic, and macroscopic self-assembly systems, Brownian motion rapidly becomes irrelevant, and gravity and friction become important. The proper design of external agitation is fundamental to arrive at a high yield process.[12]

Reversibility or adjustability is a key factor for generating ordered structures in self-assembly. Thus, the association of components either must be reversible or must allow them to adjust their positions within the aggregate once it has formed. The strength of the bonds between the components, therefore, must be comparable to the forces tending to disrupt them.[12]

5.1.2. The First Self-assembly Process (Self-assembly of Amphiphilic Molecules)

Aggregation of amphiphilic molecules in a variety of nanostructures is the first self-assembly process in supramolecular solvent formation. Amphiphiles spontaneously associate, as a CAC is achieved, to minimize unfavorable solvophobic interactions. At CAC, it becomes energetically favorable for the amphiphiles to interact with one another. Thus, colloidal self-assembled structures arise from a delicate interplay between solute–solvent and solute–solute interactions.[18] The morphology of the aggregate depends on the relative size of the head group and hydrocarbon chain of the amphiphile and can be predicted by using the so-called packing factor (g) introduced by Israelachvili et al.[19]

$$g = \frac{V}{a_0 l_c}$$

where V is the volume of the hydrophobic chain, a_0 is the mean cross-sectional area of the head group in the aggregate, and l_c is the length of the fully extended chain (see Fig. 12.2).

V and l_c are empirical quantities, but a_0 is less accessible, although trends in a_0 are easily identified. The parameter g depends on the molecular geometry of the

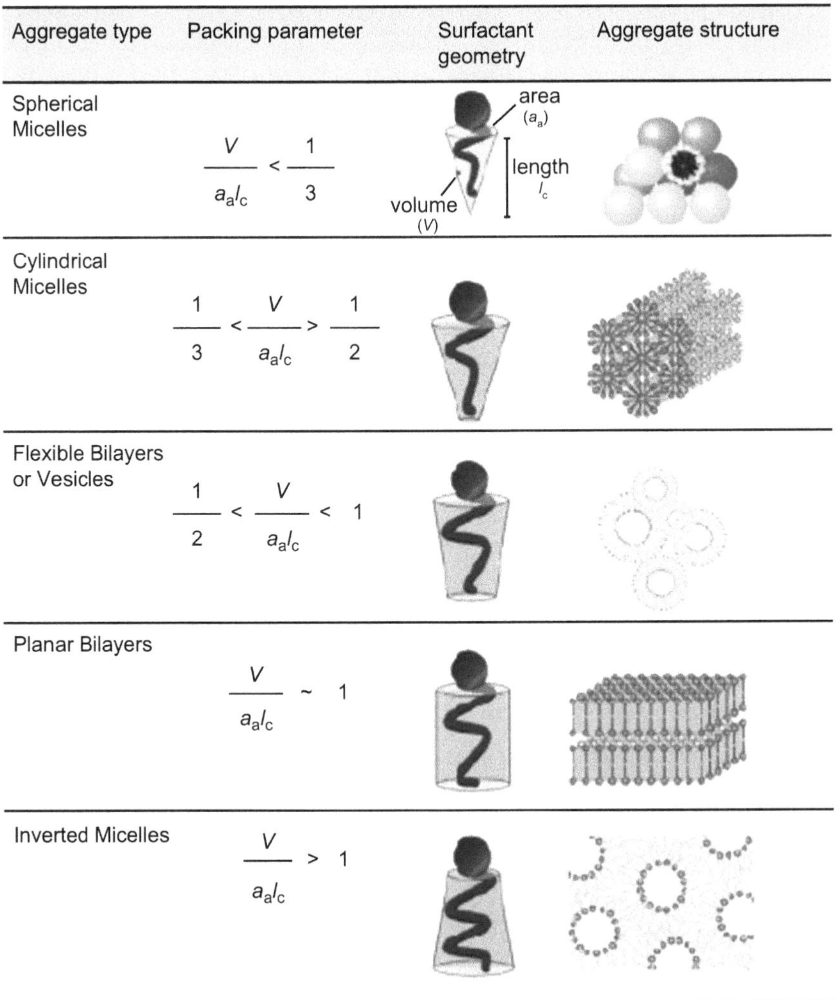

FIGURE 12.2 Structure of supramolecular aggregates according to the packing factor of amphiphiles.

surfactant molecules, such as the number of hydrocarbon chains and carbon atoms, the degree of chain saturation, and the size and charge of the polar head group. In addition, the effects of solution conditions, including ionic strength, pH, surfactant concentration, and temperature, are included implicitly in V, a_0, and l_c.

Figure 12.2 depicts the structures formed by amphiphiles in a given solution according to the g value. The size of an aggregate is affected by a number of factors, but can be considered in the abstract to depend primarily on the hydrocarbon chain length and the width of the head group in terms of intermolecular forces. Small micelles composed of low numbers of amphiphiles will be formed as the hydrocarbon chain is short and the head group is wide.

Aggregation of amphiphiles is a start–stop process—adding more surfactant results in the formation of more aggregates of the same size. For most typical surfactants, the stop process emanates from head group–head group repulsion.[20]

5.1.3. Second Self-assemble: From Nanometer-Sized Amphiphilic Aggregates to Micrometer-Sized Liquid Phases Separation

Aggregate growth from the previously formed, nanometer-sized, amphiphilic aggregates is the second self-assembly process in supramolecular solvent formation. This process cannot exist without the preceding one being in place. Aggregate growth in this step will proceed until a separate, amphiphilic-rich, liquid phase is produced. The phenomenon of liquid–liquid phase separation, typically occurring in colloidal solutions, is named coacervation.

Aggregate growth in this step involves reducing the head group–head group repulsions that stopped aggregation in the first self-assembly process. How to accomplish this goal depends on the particular system. The free energy cost of bringing the polar heads together is much smaller when the head group is uncharged than when it is charged.[16]

Aggregate growth in ionic systems can be promoted by adding a cosurfactant with a small head group—an electrolyte or an amphiphilic counterion—as well as by a pH change. In nonionic systems, one very effective way to promote aggregate growth is to lower the number of solvent molecules available for solvation, which can be achieved by modifying the temperature or by adding a poor solvent for the aggregate to be coacervated.[20]

Adding a cosurfactant like a long-chain alcohol is a general strategy to promote growth of ionic aggregates, but this method is difficult to implement since we can easily reduce the head group repulsion too much and cause a transition to a lamellar phase. Effective screening of long-range electrostatic interactions can also be achieved by adding high concentrations of electrolyte. In this case, the longer the apolar chain, the less salt is needed to induce growth. The larger the aggregate radius, the smaller is the change in curvature for a sphere-to-rod transition. For aqueous mixtures of oppositely charged surfactants, the growth of aggregates spontaneously occur when the attraction between the two types of amphiphiles is stronger than that required to form stable micelles in solution but weaker than that leading to precipitation. In order to promote coagulation of micelles against precipitation, the addition of salt (to weaken the electrostatic attraction between the oppositely charged surfactants due to electrostatic screening) is a common strategy. With this aim, an appropriate amount of NaCl (>0.2 mol/kg) was added to the catanionic system composed of a gemini cationic surfactant and sodium dodecyl sulfate in excess of the latter.[21] The coacervates formed showed globular isotropic vesicular phases or mixed vesicular and lamellar microstructures.

A more specific way to reduce ionic head group repulsion is to add an amphiphilic counterion. Thus, sodium salicylate induces growth in micelles of hexadecyl tetramethyl ammonium bromide. The aromatic ring of the salicylate enters the micelle and results in extensive aggregate growth even at low concentrations. The solution contains long thread-like aggregates that behave much like a polymer. In the growth process, adding a monomer simply adds to the length of a rod. On the other hand, the addition of tetrabutylammonium bromide to sodium dodecyl sulfate causes the formation of clusters of small ellipsoidal micelles that result from the interactions between the butyl chains belonging to different micelles.[22]

Fixing the pH of the solution below the pK_a of the amphiphilic ionic group has been a successful strategy to coacervate alkylsulfates, sulfonates, and sulfoccinates, the major problem being the high hydrochloric acid required (2–3 M) to produce phase separation.[8]

For nonionic aggregates formed by surfactants of the CnEm type (*m* referring to the number of oxyethylene groups), an increase in temperature primarily leads to the removal of some water from the palisade layer of ethylene oxide units and this has two effects. First, the area per head group is reduced, which causes micellar growth, and second, the head group regions of two micelles overlap, thus providing a mechanism for an attractive micelle–micelle interaction that ultimately triggers phase separation. Studies made through different techniques have revealed that micelles do grow in size as the cloud point is approached, but the extent of the growth depends on the ratio of *n* to *m*.[23] For surfactants with a moderate *n* to *m* ratio such as $C_{12}E_5$ the increase of the temperature induces a spherical-to-rodlike micelle transition.[24]

In contrast, no sphere-to-rod transition has been observed for $C_{12}E_8$.[25,26] Increasing the number of oxyethylene groups at a constant alkyl chain length decreases the packing parameter and the spherical shape becomes the preferred one.[27] In these cases, micelle growth is not the inherent cause of the liquid–liquid phase separation.

Instead, it is the result of an attractive micelle–micelle interaction. For nonionic micelles made up of zwitterionic head groups, a liquid–liquid phase separation can be observed as the temperature is lowered. Due to the strongly polar character of zwitterionic head groups, we should expect electrostatic interactions to play a major role for both intra- and inter-aggregate interactions. However, these interactions are of short range and their effects are qualitatively different from those of ionic surfactants. Sponge morphologies, i.e., a bicontinuous structure with surfactant bilayer films connected over macroscopic distances, have been observed for coacervates made up of zwitterionic gemini surfactants.[28,29] Water fills the pores of the continuous structure formed by the surfactant.

Partial removal of solvent molecules available for solvation of nonionic micelles can also be achieved by adding a poor solvent for micelles that is miscible with the solvation solvent. Competition for the solvation solvent will desolvate surfactant polar groups, thus causing micellar growth and phase separation. An example is the phase separation of reverse micelles of alkyl-carboxylic acid in tetrahydrofuran (THF), promoted by the addition of water.[30]

As a general rule, the environmental conditions required for phase separation depend on the hydrocarbon chain length of the amphiphile. Thus, for a homologous series of polyoxyethylene nonionic surfactants, the temperature required for phase separation increases as the number of carbon atoms in the hydrocarbon chain decreases.[31] Different models for estimating the cloud point of nonionic surfactants on the basis of their structures have been reported.[32–34] In general, cloud points for aqueous mixtures of two nonionic surfactants are intermediate between those of the two, whereas cloud points for aqueous mixtures of nonionic and ionic surfactants are higher than those of pure nonionic ones.

Likewise, the water content required for liquid phase separation of reverse micelles of alkyl carboxylic acids depends on the hydrocarbon chain length of the amphiphile. The shorter the apolar chain the larger is the water amount to induce growth.[30]

Several novel separation techniques, based on the supramolecular solvents application will be presented in Chapter 14. Supramolecular chemistry in general and different specific separation systems, thermodynamics, stability, and kinetics are expected to be perspective directions for future development.

REFERENCES

1. MacKay DJC. Information theory, inference and learning algorithms. Draft 3.1415, www.inference.phy.cam.ac.uk/mackay/itprnn/book.html; January 2003.
2. Hummer G, Garde S, Garcia AE, Pratt LR. New perspectives on hydrophobic effects. *Chem Phys* 2000;**258**(2–3):349–70.
3. Purushothaman G, Karayiannis NB. *IEEE Trans Neural Networks* 1997;**8**:679.
4. Ballesteros-Gymez A, Rubio S, Pйrez-Bendito DA. A potential of supramolecular solvents for the extraction of contaminants in liquid foods. *J Chromatogr* 2009;**1216**:530–9.
5. Hinze WL. Cloud point extraction using NDTT reagent for preconcentration. *Ann Chim (Rome)* 1987;**77**:167–208.
6. Pramauro E, Pelizzetti E. *Surfactants in analytical chemistry. Applications of organized amphiphilic media*. New York: Elsevier; 1996.
7. Carabias R, Marthnez E, Rodrhguez-Gonzalo B, Moreno-Cordero JL, Pйrez- Pavyn C, Garcha-Pinto E, Laespada EF. A. *J Chromatogr* 2000;**902**:251–65.
8. Casero I, Sicilia D, Rubio S, Pйrez-Bendito D. An acid-induced phase cloud point separation approach using anionic surfactants for the extraction and preconcentration of organic compounds. *Anal Chem* 1999;**71**:4519–26.
9. Cantero M, Rubio S, Pйrez-Bendito D. A. *J Chromatogr* 2004;**1046**:147–53.
10. Compendium of Chemical Terminology, IUPAC Recommendations, 2nd ed. compiled by AD. McNaught and A. Wilkinson, Blackwell, 2004;31:613.

11. Menger FM, Sykes BM. Anatomy of a coacervate. *Langmuir* 1998;**14**:4131–7.
12. Whitesides GM, Boncheva M. Supramolecular chemistry and self-assembly special feature: beyond molecules: self-assembly of mesoscopic and macroscopic components. *Proc Natl Acad Sci USA* 2002;**99**:4769–74.
13. Olenyuk B, Whiteford JA, Fenchtenkotter A, Stang PJ. Self-assembly of nanoscale cuboctahedra by coordination chemistry. *Nature* 1999;**398**:796–9.
14. Philip D, Stoddart JF. Self-assembly in natural and unnatural systems. *Angew Chem Int Ed Engl* 1996;**35**:1155–96.
15. Lehn JM, Ball P. In: Hall N, editor. *The new chemistry*. Cambridge, UK: Cambridge University Press; 2000. p. 300–51.
16. Pelesco JA. *Self assembly. The Science of things that put themselves together*. New York: Chapman & Hall/CRC, Taylor & Francis Group; 2007. pp. 67–80, Chapter 4.
17. Lehn JM. Supramolecular chemistry: concepts and perspectives. *Proc Natl Acad Sci USA* 2002;**99**:4763–8.
18. Steed JW, Turner DR, Wallace KJ. *Core concepts in supramolecular chemistry and nanochemistry*. Chichester: John Wiley & Sons; 2007. p. 1–27, Chapter 1.
19. Israelachvili JN, Mitchell DJ, Ninham BW. Surfactant diffusion: new results and interpretations. *J Chem Soc Faraday Trans* 1976;**72**:1525–68.
20. Evans DF, Wennerstrцm H. *The colloidal domain. Where physics, chemistry, biology, and technology meet*. New York: Wiley-VCH; 1999. p. 197–198.
21. Hao LS, Nan YQ. A. *Colloids Surf* 2008;**325**:186–93.
22. Zana R, Schmidt J, Talmon Y. *Langmuir* 2005;**21**:11628–36.
23. Glatter O, Fritz G, Lindner H, Brunner-Popela J, Mittelbach R, Strey R, Egelhaaf SU. Self-assembly of medium-chain alkyl monoglucosides in ammonium sulfate solutions with poly(ethylene glycol). *Langmuir* 2000;**16**:8692–701.
24. Bernheim-Groswasser A, Wachtel E, Talmon Y. Micellar growth, network formation, and criticality in aqueous solutions of the nonionic surfactant C12E5. *Langmuir* 2000;**16**:4131–40.
25. Kato T, Anzai S, Seimiya T. Light scattering from semidilute solutions of nonionic surfactants (C12E5 and C12E8) and the scaling law. *J Phys Chem* 1990;**94**:7255–9.
26. Danino D, Talmon Y, Zana R. Cryogenic temperature transmission electron microscopy in the study of surfactant systems. *J Colloid Interface Sci* 1997;**186**:170–9.
27. Ezrahi S, Tuval E, Aserin A. Properties, main applications and perspectives of worm micelles. *Adv Colloid Interface Sci* 2006;128–30, 77–102.
28. Menger FM, Peresypkin AV, Caran KL, Apkarian RP. A sponge morphology in an elementary coacervate. *Langmuir* 2000;**16**:9113–6.
29. Menger FM, Peresypkin A. A Combinatorially-derived structural phase diagram for 42 zwitterionic geminis. *J Am Chem Soc* 2001;**123**:5614–5.
30. Ruiz FJ, Rubio S, Pйrez-Bendito D. Determination of bisphenols A and F and their diglycidyl ethers in wastewater and river water by coacervative extraction and liquid chromatography-fluorimetry. *Anal Chem* 2007;**79**:7473–84.
31. Taechangam P, Scamehorn JF, Osuwan S, Rirksomboon T. *Colloids Surface Physicochem Eng Aspects* 2009;**347**:200–9.
32. Schott H. A linear relation between the cloud point and the number of oxyethylene units of water-soluble nonionic surfactants valid for the entire range of ethoxylation. *J Colloid Interface Sci* 2003;**260**:219–24.
33. Materna K, Szymanowski J. Separation of phenols from aqueous micellar solutions by cloud point extraction. *J Colloid Interface Sci* 2002;**255**:195–201.
34. Ren Y, Liu H, Yao X, Liu M, Hu Z, Fan B. The accurate QSPR models for the prediction of nonionic surfactant cloud point. *J Colloid Interface Sci* 2006;**302**:669–72.

Chapter 13

Advances in Development of Solvents for Liquid–Liquid Extraction

Chapter Outline

Solvent Extraction. DOI: 10.1016/B978-0-444-53778-2.10013-5

1. INTRODUCTION

Chances of new extractant molecules being synthesized, developed, and commercialized are not good due to the following:

1. Today's commercially available extractants are good, and a better—less costly to use—molecule is not likely to conservative technologists.
2. The cost of developing and commercializing a new molecule is high for a variety of reasons, not the least of which is registration of any new molecule and the accompanying toxicity testing.
3. There are not a lot of solutes that could be recovered in sufficient amounts to justify the expense of a new molecule.
4. Over the years, many molecules have been made and investigated by reagent suppliers, but the results have not been made public. Thus, there are fewer new molecules to investigate than the public realizes.

As a result, new reagents will likely come from combinations of existing reagents and new solvent extraction applications will come from fitting flow sheets to the properties of existing reagents.

Since the beginning of the green chemistry movement 15 years ago, the search for alternatives to classical procedures for solute extraction and separation has been one of the major challenges that scientists have faced. It has led to substantial research effort worldwide within different emerging topics including replacement of toxic reagents and/or organic solvents by innocuous ones. The pollution prevention requires a new way to diminish the environmental side effects based on the reduction or elimination of wastes and pollutant chemicals at the source. It promotes changes in development processes and also in laboratory practices concerning chemicals and other raw materials. The pollution prevention initiative in solvent extraction results in the modification of techniques in order to reduce or avoid the use of toxic reagents and products, and specially the solvents used for solutes of interest solution and extraction. This is the main trend in the improvement of existing solvent extraction technologies and the development of new processes.

1.1. Strategies for Solvent Replacement

At the beginning of the "Green Era," clean or environment-friendly methods were adopted by removing contaminants from the waste streams, using filters, mineralization of organic compounds, and precipitation and passivation of metals. They were considered the most appropriate and easy-to-apply procedures for "greening" leaching and extraction methods, and they are known as *end-of-the-pipe* technologies. Later, the direction that the scientists were following to adopt greener procedures changed completely. The "end-of-the-pipe" concept was gradually replaced by the "greening at the origin." As the name indicates, it was based on the elimination of the contamination sources. While applying this concept to extraction technologies, it is evident

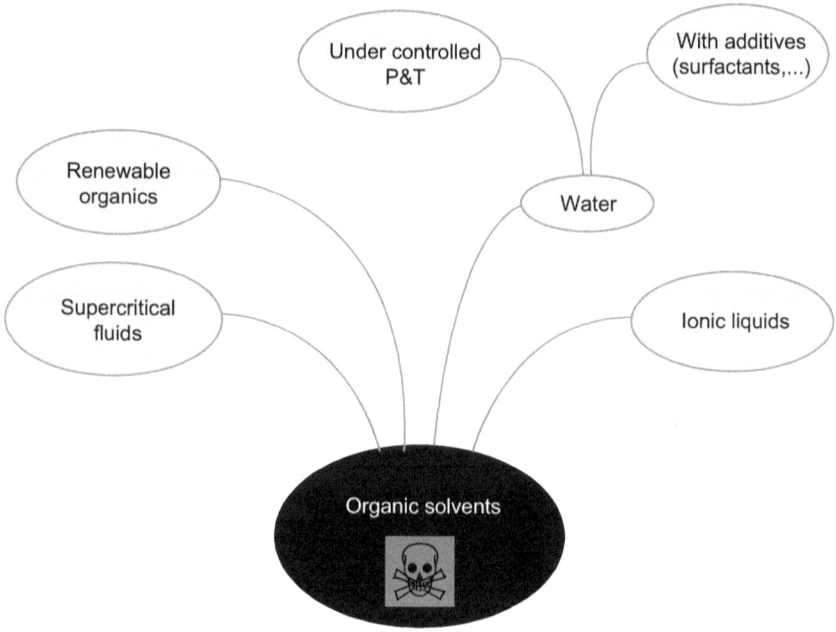

FIGURE 13.1 Greener alternatives to the organic toxic solvents in solvent extraction.

that organic solvents are an important source of contamination, which should be avoided and, in this sense, four directions toward green solvents have been developed:

1. Substitution of hazardous solvents by environment-friendly ones[1]
2. Use of "biosolvents"
3. Substitution of organic solvents with supercritical fluids[2]
4. Substitution of VOCs with ionic liquids (ILs)[3]

First of all, the use of water under special temperature and pressure conditions has to be considered, followed by the use of supercritical fluids as alternative solvents, the use of renewable organic or ILs, and finally the use of aqueous solutions of amphiphilic compounds or supramolecules, as depicted in Fig. 13.1.

2. ORGANIC SOLVENTS

Organic solvents are widely used in separation technologies, using solvent extraction, sonication, microwave-assisted extraction, and pressurized fluid extraction. The choice of solvent is largely dependent on recommended standard methods and experience, manufacturers' guidelines, or such as those compiled and promulgated by the U.S. Environmental Protection Agency

(US EPA) or the American Society for Testing and Materials (ASTM) and European Committee for Standardization, Environmental Management.

From the Environmental, Health, and Safety (EHS)[4] assessment results obtained for 26 commonly used pure organic solvents (see Fig. 13.2), overall high scores were obtained for formaldehyde, dioxane, formic acid, acetonitrile, and acetic acid. Formaldehyde had high scores for acute and chronic toxicity, irritation, and air hazard—dioxane had a high persistency, while both acetic and formic acids showed high scores for irritation. On the other hand, low overall scores were obtained for methyl acetate, ethanol, and methanol, which particularly pose low environmental hazards and relatively low health hazards.

The use of tetrahydrofuran, butylacetate, cyclohexanone, and 1-propanol is not recommended from a life-cycle perspective because these solvents cause high environmental impacts during petrochemical production. In addition, formic acid, ethyl acetate, acetonitrile, dioxane, 1-butanol, and dimethylformamide are solvents of significantly high environmental impact. At the other end, hexane, heptane, and diethyl ether are environmentally favorable solvents. With regard to the alkanes, these results are due to the energy recovery by incineration in combination with relatively low environmental impacts of their production.

From Table 13.1 it can be seen that commonly used solvents in solvent extraction have been included in the list of hazardous air pollutants published in 2002 by the US EPA.[5] Today, acetonitrile is by far the preferred organic solvent used in liquid chromatography (LC) because of its physical properties. A recent downturn in plastic production has resulted in severe shortages and escalating costs of acetonitrile, which is produced as a by-product of acrylonitrile.[6] In January 2009, the price of acetonitrile was recorded at US$ 100 per liter with disposal costs about double this price because of its toxic properties. Aqueous waste streams containing acetonitrile are typically disposed of as chemical waste.[7] Moreover, acetonitrile is ranked by the EPA as a hazardous solvent, and waste has to be incinerated, releasing NO_2, partly responsible for acid rain. Taking into consideration all the problems related to acetonitrile, the LC community is considering greener replacements for acetonitrile.[8] Moreover, a common additive used in LC is trifluoroacetic acid (TFA), which is cytotoxic, corrosive, and persistent in the environment.

Carbon tetrachloride, chloroform, and dichloromethane were the solvents that had been widely used in spectroscopy for their transparency in this spectral range and for extraction, but they are not used now owing to their toxicity and carcinogenic and ozone-depleting effects. Since then, considerable research effort is directed toward finding alternatives to chlorinated solvents.

The same problem is with volatile organic compounds (VOCs). Solvents commonly used in solvent extraction are often VOCs, and over the previous decades, they were associated with a series of direct and indirect environmental and health damages. Among the direct effects of VOCs, the following should be noted: their toxicity and carcinogenic effects, which depend on the nature of the

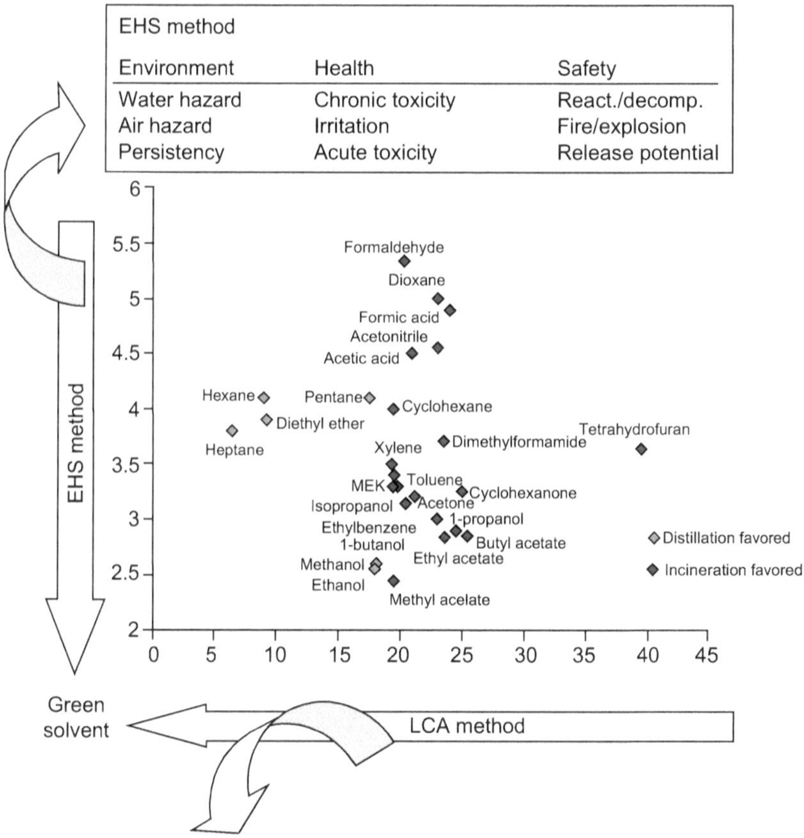

LCA method	Solvent production CED MJ (eq kg^{-1})	Solvent distillation CED MJ (eq kg^{-1})	Solvent incineration CED MJ (eq kg^{-1})
Acetonitrile	88.5	−79.6	**−29.7**
Methanol	40.7	**−21.7**	−22.2
Ethanol	50.1	**−31.2**	−31.7
Hexane	64.4	**−46.7**	−55.2
Tetrahydrofuran	270.8	−230.7	**−37.5**

FIGURE 13.2 The environmental assessment of solvents using a combination of EHS and LCA criteria. (The values given in bold represent the most appropriate solvent recovery method as a function of the solvent.)

compound, the exposure and duration, their flammability and related fire hazards, and the possibility of peroxide formation, especially in the case of ethers. The most important indirect environmental problems related to VOCs are as follows: (i) their ozone-depletion properties, especially in the case of chlorofluorocarbons, now phased out; (ii) the global warming potential due to the implication of VOCs in the production of photochemical smog; and (iii) the

TABLE 13.1 Common Organic Solvents Present in the List of Hazardous Air Pollutants Published in 2002 by the US EPA

Chemical name	Effects on humans and environment
Acetonitrile	Acid rain
Benzene	Recognized human carcinogen (Group 1)
Ethyl benzene	Photochemical smog formation
Toluene	Skin irritation
Xylene	Skin irritation
Phenol	Moderately toxic
Cresol	Moderately toxic
Carbon disulfide	Coronary heart disease
Carbon tetrachloride, chloroform, dichloromethane, trichloroethylene, tetrachloroethylene, chloroethane	Ozone-depleting agents/toxic to the liver, heart and kidneys/reasonably anticipated to be a human carcinogen (Group 2B)
Bromoform, dibromoethane, bromomethane	Recognized human carcinogen (Group 1)
Hexane	Nerve cells degeneration
Methanol	Produces formaldehyde, which causes headache, insomnia, gastrointestinal problems, and blindness
Methyl ethyl ketone	Potentiates toxicity of haloalkanes and *n*-hexane
Methyl isobutyl ketone	Potentiates toxicity of haloalkanes and *n*-hexane
N-butyl ketone	Nerve-cell degeneration
Methyl tert-butyl ether	Possible human carcinogen (Group 2B)
Diethanolamine	Negative effects on liver, kidney, and blood
Formaldehyde	Allergic contact dermatitis/reasonably anticipated to be a human carcinogen
Triethylamine	Reversible corneal edema
1,3-Butadiene	Recognized human carcinogen (Group 1)

environmental persistence. For the aforementioned reasons, the replacement of commonly used solvents by green ones is absolutely mandatory.

3. WATER AS A SOLVENT

It is clear that from an environmental point of view the best alternative for organic solvents is water: it is cheap, safe, nontoxic, noninflammable, and recyclable. The main problem with water as a solvent is that while it dissolves polar compounds, most of the commonly analyzed organic compounds are hydrophobic with poor or extremely poor water solubility.

Subcritical water, or pressurized hot water, provides liquid water under pressure at temperatures between the usual boiling point (100 °C) and the critical temperature (374 °C). The polarity of superheated water significantly decreases on increasing temperature and pressure, and the properties of superheated water seem to be those of an organic solvent, such as methanol or ethanol. Over the superheated temperature range, the extensive hydrogen bonds break down, changing the properties more than usually expected by increasing temperature alone. Solubility of organic solutes increases by several orders of magnitude and the water itself can act as a solvent, including extractions and separations.

3.1. Subcritical Hot Water as a Solvent for Extraction

The polarity of water decreases markedly at temperatures between 80 °C and 250 °C and high pressure, being able to replace organic modifiers and providing environment-friendly green methods.[9] The use of liquid water at temperatures above its boiling point in order to improve the solubility of organic compounds is not a new topic. For many years, it has been used as a cleaning agent to enhance the extraction of oil shale,[10] sulfur from ore bodies,[11] and as a vapor steam that is commonly used in hydrodistillation for the isolation of essential oils from plant materials.[12] The recent interest of chemistry in superheated or subcritical water extraction (SWE), or pressurized hot water extraction (PHWE) methods make it an interesting and environment-friendly alternative to organic solvents.

"Sub-critical water extraction (SWE)" is a commonly used term. The terms "pressurized hot water extraction (PHWE)", "superheated water extraction (SWE)", "extraction with water at elevated temperatures and pressures", "high-temperature water extraction", and "extraction using hot compressed water" have also been applied. To my opinion, one term has to be chosen and that is SWE. At high temperatures (above boiling point: 100–374 °C), steam needs to be pressurized for efficient transfer through the sample.

The SWE is emerging as a powerful alternative for the extraction of solid samples.[13] It has been used to extract pollutants with a wide range of polarities from environmental samples[14,15] and for the extraction of pesticides[16] and polycyclic aromatic hydrocarbons[17] from soils. The use of SWE as a very promising alternative to conventional and supercritical CO_2 extraction methods for isolation of essential oils was proposed by Basile et al.[18] Since this date, the technique has shown its applicability in the field of essences[19–22] as compared with conventional techniques like steam distillation[23,24] and solvent extraction,[25] which have some well-known disadvantages, namely, low extraction efficiency, long extraction time, and large amounts of toxic solvent waste. It also has the advantage of being selective.

The major advantage of SWE is that it considerably reduces the use of organic solvents when compared to classical extraction methods, which cause considerable concern as the use of large amounts of volatile organic solvents causes severe environmental side effects (ozone hole and global warming problems).

However, the main drawbacks of SWE are that the extract is a dilute aqueous solution and possible problems of precipitation or readsorption into the matrix can occur on cooling water. These problems can be avoided using different approaches based on liquid–liquid extraction (LLE) or solid-phase extraction (SPE). In LLE, the aqueous solution is collected and the solute is extracted in a small volume of organic solvent;[26] in SPE,[27] solid-phase microextraction (SPME),[28] or stir bar extraction (SBE),[29] the water solution passes through a filter retaining the solutes that are eluted with a small volume of an organic solvent.

Water has been used in analytical techniques to extract solutes from solid matrices in both static and dynamic modes. The main areas of application of SWE are the analysis of environmental matrices, such as the extraction of polychlorinated biphenyls,[30] polycyclic aromatic hydrocarbons,[31] and pesticides[32] from soils and plants, and volatile compounds from plant materials.[18]

Moreover, several strategies have been successfully developed to improve extraction efficiencies, such as pH control by the addition of a buffer,[33] solvent-assisted SWE using water–ethanol mixtures,[34] and SWE using micellar agents such as sodium dodecyl sulfate (SDS).[35]

4. RENEWABLE WATER-BASED SOLVENTS

There is a continuing demand for selective extractants that can be developed, e.g., by tailor-making organic molecules with preorganized metal-binding sites. The development of such extractants may be designed on computer-based models. Environmental considerations require that the extractants and diluents used are either nontoxic, nonvolatile and/or recovered within the process. The increasing combination of extraction and distillation, as used in biotechnology, places a whole host of new demands on the diluents employed. In the extraction of biologically active compounds, care must be taken to avoid the loss of activity that often occurs by contact with organic diluents. Thus, a series of systems have been developed specifically with these compounds in mind.

The first of these uses mixtures of aqueous solutions containing polymers and inorganic salts that will separate into two phases that are predominately water (two-phase separation systems).

A second system uses supercritical conditions in which the original two-phase system is transformed into one phase under special temperature–pressure conditions. Also, the active organic compound can be shielded from the organic diluent by encapsulation within the aqueous center of a micelle of surface-active compounds. All these systems are currently an active area for research as is discussed below.

4.1. Aqueous Two-Phase Systems as Extractants

Under certain conditions, polymer incompatibility in aqueous solutions can lead to the formation of two phases with high water content. With such a system, it is possible to separate sensitive biological molecules, such as proteins, without denaturation, which would be the case for an ordinary aqueous–organic solvent system.

Intracellular enzymes and other microbial proteins are of increasing interest. Processes for their purification may be divided into five stages:

1. Disruption of the cells
2. Removal of cell debris
3. Concentration and enrichment
4. High-resolution purification
5. Concentration and finishing.

An overall goal of these steps is to obtain a high yield while retaining the biological activity of the proteins. The required purity of the protein is determined by its end use. Enzymes that are to be used as technical catalysts require a lower purity than if they are to be used for analytic purposes or as pharmaceuticals. Utilization of aqueous two-phase systems as a separation method for these biological materials was extensively studied. The ability of different water-soluble polymers to form aqueous two-phase systems was investigated the same as ability of biological macromolecules and cell particles to partition in these phase systems.[36]

The polyethylene glycol (PEG)–dextran–water system is still the most used and best-studied aqueous polymer two-phase system. A phase diagram for a typical two-phase system is shown in Fig. 13.3 for the PEG–dextran system. Both polymers are separately miscible with water in all proportions. As the polymer concentration increases, phase separation occurs, with the formation of an upper phase, rich in PEG, and a lower phase, rich in dextran, each phase consisting of more than 80% water. Within the two phase regions, any mixture of the three components splits into two phases, with compositions dictated by the intersections of the tie-line passing through the mixture point with the binodal. The formation of two aqueous phases can be exploited in the recovery of proteins using LLE techniques.

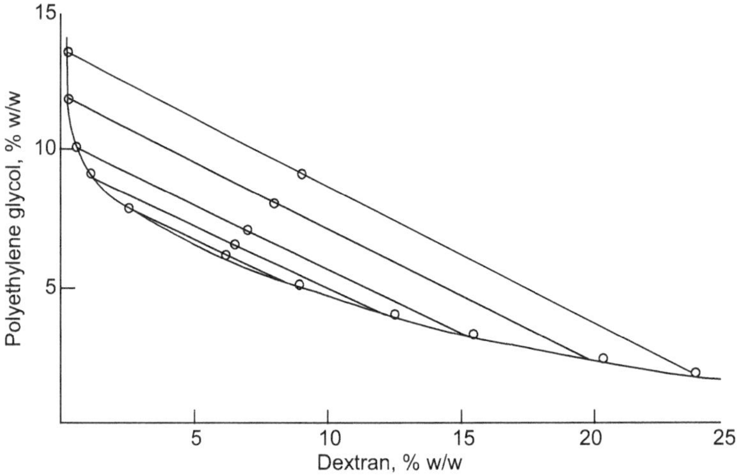

	Total system polyethylene			Bottom phase polyethylene			Top phase polyethylene		
System	Dextran	Glycol	H_2O	Dextran	Glycol	H_2O	Dextran	Glycol	H_2O
A	6.14	6.09	87.77	8.91	4.99	86.10	2.52	7.82	89.66
B	6.50	6.50	87.00	12.48	3.93	83.59	1.00	9.09	89.91
C	7.00	7.00	86.00	15.50	3.25	81.25	0.44	10.07	89.49
D	8.00	8.00	84.00	20.34	2.28	77.38	0.15	11.80	88.05
E	9.00	9.00	82.00	23.81	1.90	74.29	0.13	13.46	86.41

FIGURE 13.3 Phase diagram and phase compositions of the dextran–polyethylene glycol (PEG) system D 48-PEG 4000 at 293 K. All values are % w/w.

Many factors contribute to the distribution of a protein between the two phases. Smaller solutes, such as amino acids, partition almost equally between the two phases, whereas larger proteins are more unevenly distributed. This effect becomes more pronounced as protein size increases. Increasing the polymer molecular weight in one phase decreases partitioning of the protein to that phase. The variation in surface properties between different proteins can be exploited to improve selectivity and yield. The use of more hydrophobic polymer systems, such as fatty acid esters of PEG added to the PEG phase, favors the distribution of more hydrophobic proteins to this phase.

In Fig. 13.4, partition coefficients for several proteins in a dextran–PEG system are given.[37] Cells and cell particles generally partition almost entirely to one of the two phases or to the interface and one phase. Thus, it is possible to partition proteins away from the cell debris and, thereby, reduce the need for downstream processing.

The phase behavior of the polymers is also dependent on the type and concentration of salt present. Many times, a sufficiently high concentration of salt in a single polymer solution can induce phase separation to form one salt-rich

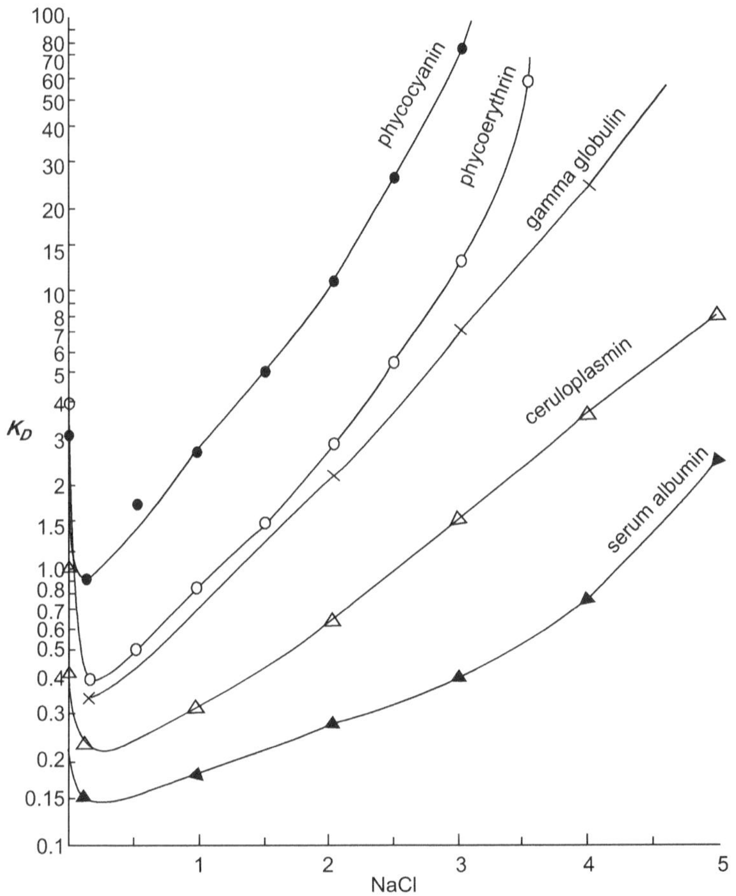

FIGURE 13.4 Distribution coefficients of several proteins in the dextran–polyethylene glycol (PEG) system in 0.0005 M KH_2PO_4, and with increasing concentration of NaCl.

and one polymer-rich phase. Sodium and potassium phosphates are commonly used salts. The efficiency of protein partitioning in aqueous two-phase systems can be considerably enhanced by the introduction of specific ligands into the phase system. The technique is called affinity partitioning, which is analogous to the well-known affinity chromatography concept widely used in analytical and preparative protein separations. In this technique, a small fraction of one of the phase-forming polymers is substituted by the same polymer with the specific ligand covalently attached to it. Since ligands interact strongly with specific proteins, the effect on the partition coefficient can be very dramatic, and there are examples when distribution coefficients have been changed by a factor of 10,000. The ligands that are used for affinity partition are mostly different dyes linked to a PEG molecule. Recent developments in aqueous two-phase partitioning have been published.[38]

Biotechnological processes, in general, are run at low concentrations of reactants and products. One reason for this is that most biocatalysts are sensitive to product inhibition. The reaction may stop when only a small fraction of the substrate has been converted into product. In such cases, there is a strong need for continuous extractive procedures during which the product is continuously removed. The integration of bioconversion and extraction is called extractive bioconversion. The large differences in surface tension and dielectric constants between water and common organic solvents cause enzymes to unfold when they are exposed to the interface between the two solvents. This problem can be overcome through extractive bioconversions in aqueous two-phase systems. A collection of cutting-edge methods intended to provide practical guidelines for the area of separations in two-phase systems has been reviewed.[39]

4.2. Supercritical Fluid Extractants

When a fluid is forced to a pressure and temperature above its critical point (see Fig. 13.5) it becomes a supercritical fluid. A supercritical fluid can be defined as a substance at a temperature and pressure above its critical point. Under these conditions, various properties of the fluid are placed between those of a gas and a liquid. Although the density of a supercritical fluid is similar to a liquid and its

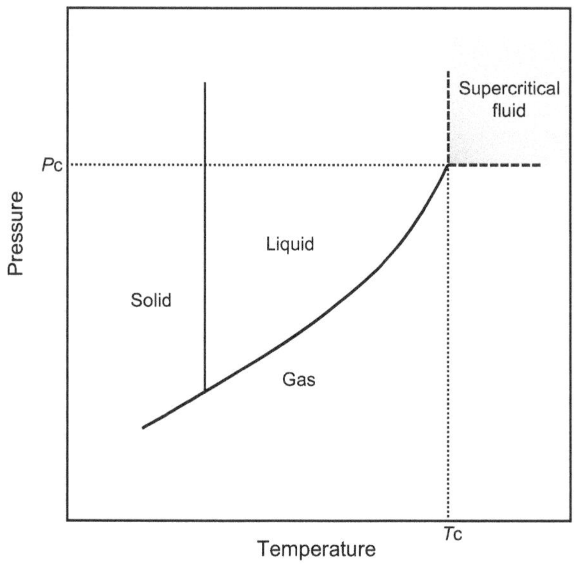

FIGURE 13.5 Phase diagram for a pure compound.

TABLE 13.2 Range Values of Several Physicochemical Properties of Gases, Liquids and Supercritical Fluids

State of fluid	Density (ρ, g/cm^3)	Diffusivity (D_{AB}, cm^2/s)	Viscosity (μ, g s/cm)
Gas $p=1$ atm; $T=21$ °C	10^{-3}	10^{-1}	10^{-4}
Liquid $p=1$ atm; $T=15$–30 °C	1	$<10^{-5}$	10^{-2}
Supercritical $p=p_c$; $T=T_c$	0.3–0.8	10^{-3}–10^{-4}	10^{-4}–10^{-3}

viscosity is similar to a gas, its diffusivity is intermediate between the two states, as can be seen in Table 13.2. Thus, the supercritical state of a fluid has been defined as a state in which liquid and gas are indistinguishable from each other, or as a state in which the fluid is compressible (i.e., similar behavior to a gas) even though possessing a density similar to a liquid and, therefore, similar solvating power.

Figure 13.6 illustrates the solubility of naphthalene in supercritical ethene as a function of pressure at 298 and 318 K and of temperature. The solubility of

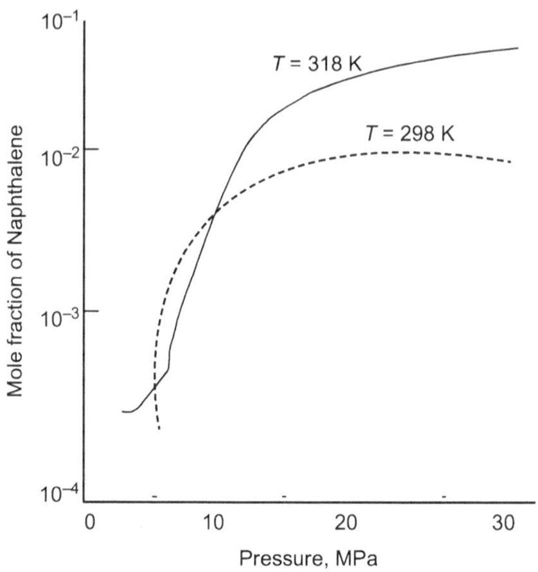

FIGURE 13.6 Solubility of naphthalene in ethene at 298 and 318 K.

naphthalene increases dramatically when passing the critical point of ethene ($T_c = 282$ K, $P_c = 5$ MPa). The increasing solubility is due to the rapid increase of the density of ethene after passing the critical point. It can be seen that the temperature dependence of solubility is different in different pressure areas. At high pressure, an increase in temperature is followed by an increase in solubility, whereas at lower pressures the opposite effect occurs.

In practice, because of the physicochemical characteristics of the carbon dioxide, more than 90% of all applications involving supercritical fluids in extraction techniques use carbon dioxide as solvent.[40] CO_2 is nontoxic, nonflammable, and available at low cost with a high degree of purity. In the supercritical state, CO_2 has a polarity comparable to liquid pentane or cyclohexane and is, therefore, best suited for lipophilic compounds—the lack of polarity for the extraction of polar analytes being its main drawback.[41] In Table 13.3, some basic properties of supercritical CO_2 are presented.

TABLE 13.3 Physical Properties of Some Supercritical Fluids

Solvents	Critical point	Density[a] (g L^{-1})	Solubility in water[b] (g L^{-1})
Carbon dioxide	73 atm and 31 °C	850	1.45
Propane	42 atm and 97 °C	582	0.07
Butane	38 atm and 152 °C	600	0.06
Pentane	33 atm and 197 °C	626	0.04
Hexane	30 atm and 234 °C	655	0.013
Dimethylether	54 atm and 127 °C	271	71
Nitrous oxide	72 atm and 36 °C	1223	1.5
Sulfurhexafluoride	38 atm and 45 °C	1880	0.045
Tetrafluoroethylene[c]	39 atm and 33 C	–	0.19
Tetrafluoromethane	37 atm and −45 °C	1603	0.015
Trifluoromethane	48 atm and 26 °C	1431	3.12
Perfluorobutane	24 atm and 113 °C	1594	–
Octafluoropropane	27 atm and 72 °C	1601	–

[a]*Density of the liquid form (at boiling point) except for dimethylether, which is for the critical point and for sulfur hexafluoride at triple point.*
[b]*Solubility in water at 25 °C and 1 atm.*
[c]*Reasonably anticipated to be a human carcinogen.*

As for the solvents, there is a wide range of compounds that can be used as supercritical fluids. Some examples of other compressed fluids can be seen in Table 13.3. These are propane, which has the potential to be a highly effective solvent for vegetable oil extraction,[42] and dimethyl ether, which is an excellent solvent for phospholipids[43] (see Table 13.3). Other compounds examined as supercritical fluid extractants (SFE) solvents are as follows: (1) hydrocarbons, such as hexane, pentane, and butane;[44] (2) nitrous oxide,[45] which is considered to be better suited for polar compounds because of its permanent dipole moment—unfortunately, it can cause violent explosions in combination with samples having high organic content; (3) sulfur hexafluoride, which is a nonpolar molecule that has been demonstrated to selectively extract aliphatic hydrocarbons up to around C-24 from a mixture containing both aliphatic and aromatic hydrocarbons; and (4) fluorinated hydrocarbons,[46] which have shown to increase the extraction efficiency compared to CO_2. Table 13.3 lists the physical properties of some supercritical fluids used in chemistry.

Due to their low viscosity and relatively high diffusivity, supercritical fluids have better transport properties than liquids, can diffuse easily through solid materials, and can therefore give faster extraction yields. One of the main characteristics of a supercritical fluid is the possibility of modifying the density of the fluid by changing its pressure and/or its temperature. Since density is directly related to solubility by altering the extraction pressure, the solvent strength of the fluid can be modified.

Supercritical CO_2 is a good solvent for lipophilic compounds, and is less effective in the extraction of polar solutes. Therefore, for the isolation of amphiphilic compounds, CO_2 has to be combined with a polar cosolvent to considerably increase the solubility of those compounds in supercritical CO_2. Ethanol and methanol are the most commonly used cosolvents.

Carbon dioxide is a gas at room temperature, so once the extraction is completed, and the system decompressed, a substantial elimination of CO_2 is achieved without residues, yielding a solvent-free extract. On an industrial scale, when carbon dioxide consumption is high, the operation can be controlled to recycle it. However, supercritical CO_2, because of its low polarity, is less effective in extracting more polar compounds from natural matrices. To overcome this problem, modifiers (cosolvents) are commonly used. Modifiers are highly polar compounds that, added in small amounts, can produce substantial changes of the solvent properties of neat supercritical CO_2. Another drawback of SFE is the high investment costs as compared to traditional atmospheric pressure extraction techniques.

5. IONIC LIQUIDS AS SOLVENTS FOR EXTRACTION

ILs are organic salts with melting point lower than 100 °C. They are called *room-temperature ILs* if the salt is liquid at room temperature. They are typically composed of a poorly coordinating, bulky organic cation, and an organic or inorganic anion. ILs attract attention of the chemists at the synthesis of air- and water-stable ILs in the beginning of 1990s.[47] ILs have numerous applications in chemistry, where they have been termed as a "green alternative." It is well known that they have low vapor pressures and are rarely flammable or explosive, which implies lower environmental effects than VOCs. ILs have negligible vapor pressure as well as good solubility for inorganic and organic compounds, thus being useful in LLE, liquid phase microextraction (LPME), and solid phase microextraction (SPME).[48]

5.1. Properties of ILs

ILs have many interesting and unique properties, among which is an exceedingly low vapor pressure. At normal process operating conditions, ILs essentially do not evaporate. Thus, there has been considerable interest in the use of ILs as industrial solvents to replace VOCs and thus eliminate a source of air pollution, as well as hazards due to inhalation and explosion.[49–52] Also, ILs are stable as liquids over a very wide temperature range (e.g., −70–400 °C).[53] Both of these properties (negligible vapor pressure and large liquidus range) will facilitate the recovery and reuse of ILs in the context of LLE, and thus provide economic benefits (e.g., extremely low solvent loss and make-up requirement).

The hydrophobic character of some ILs allows them to extract hydrophobic compounds. However, cationic compounds can be successfully extracted from hydrophobic ILs using appropriate chelators or ligands that could form complexes to increase the hydrophobicity of the metal species.[54] Parameters affecting the extraction efficiency of metal ions are as follows:

1. The structure of the IL and side chain length, which change its hydrophobicity and improve the partition coefficients[55].
2. The ligand used should optimize selectivity.
3. The pH of the system.[56]

By changing the anion, cation, and/or cation substituents, an essentially endless variety of ILs can be created. Thus, it may be possible to "design" or "tune" ILs that preferentially extract desired compounds from water. The unique properties of ILs allow them to be conveniently adopted as extraction solvents in analytical chemistry. Moreover, the nonvolatility of ILs makes them potentially useful for long-time headspace LPME of volatile analytes for obtaining high enrichment factor. [C8MIM][PF6] was used for the LPME of PAHs from water samples using the single-drop microextraction (SDME).[57] Compared to 1-octanol, a larger drop volume of IL can be suspended during longer extraction times with high enrichment factors. Formaldehyde extraction from mushrooms by LPME using ILs has been successfully developed.[58] SPME with a disposable IL has also been successfully developed.[59] The method was applied to the extraction of benzene, toluene, ethylbenzene, and xylenes in paints. The SPME fiber was coated with IL before every extraction, the analytes were extracted and desorbed on the injection port of GC instrument, and the IL coating was washed with solvents.

5.1.1. Toxicology Considerations

The possibilities offered by ILs, with several million structures available by the combination of different cations and anions, also represent an unusual problem related to the definition or identification of the toxicity and environmental side effects of those compounds. ILs are new molecules still not fully accepted in the wider community: it is difficult for conservative scientists to throw away concepts grown from the fertile ground of molecular solvents over thousands of years. However, there are now many laboratories worldwide (and the growth in China is spectacular) that work with ILs. Of course, deep and thorough understanding of potential health hazards and environmental impact of ILs is essential.

Unfortunately, it has been demonstrated that the first generation of ILs, based on imidazolium or pyridinium cations, turned out to be as toxic as conventional solvents, and sometimes even worse.[60] Development of simple toxicity tests to enable rapid and inexpensive identification of the best IL structures is the task for elimination of the most toxic ones. The remaining candidate structures can be investigated using either minimum inhibitory concentrations (MICs) and minimum biocidal concentrations (MBCs) tests or growth rate measurements and EC_{50} (medium effective concentration) calculations, selecting only the most appropriate candidates for application.

Some researchers[61] found that imidazolium, pyridinium, phosphonium, and ammonium cations could not be described as having low freshwater toxicity with EC_{50} values below 100 mg/L. The substituted alkyl chain length on the cation played a significant role by influencing the toxicity, i.e., ILs with eight atoms of carbon C8 were found to be more toxic than those with six and four, regardless of the type of cation.

The future of ILs seems really bright in separation techniques due to their large structural diversity.[62] Because some ILs are toxic, they cannot be considered as general green alternative solvents while some are benign and technologists need detailed investigations to accelerate the introduction of new nontoxic ILs in the developed new separation techniques.

6. AMPHIPHILIC COMPOUNDS (SUPRAMOLECULES) AS SOLVENTS IN SOLVENT EXTRACTION

Supramolecular solvents have a unique array of physicochemical properties that render them very attractive to replace organic solvents in solvent extraction. Main intrinsic properties of these solvents include: use of self-assembly-based synthetic procedures that are within everyone's reach; ubiquity of amphiphiles in nature and synthetic chemistry, which makes them easily accessible; tunability of solvent properties by varying the hydrophobic or polar group of the amphiphile; presence of different polarity regions in the supramolecular aggregates that provide excellent solvation properties for a variety of organic and inorganic compounds; multiligand ability through the multiple polar groups present in a supramolecular aggregate, which is an ideal platform for amplification of solute binding; and nonvolatility and nonflammability of supramolecular aggregates, which permit the implementation of safer processes.

The use of amphiphilic compounds, such as surfactants, cyclodextrins, or crown ethers, in aqueous solutions provides the so-called organized media, creating special characteristic points in the aqueous solution with different dielectric properties and with increased solvent characteristics. Surfactants are the less expensive and the most versatile amphiphilic compounds.

Since Watanabe et al.[63] firstly reported the cloud point technique, SUPRASs made up of nonionic surfactant-based micellar aggregates have been extensively applied to the extraction of hydrophobic organic compounds and metals, mainly from environmental and biological liquids. Excellent published reviews compile the phase-separation behavior of nonionic surfactants as a function of the temperature, the type of applications developed, and the advantages/limitations of separations based on the cloud point phenomenon.[64,65] As a result, parameters affecting extraction efficiencies and pre-concentration factors achieved by this technique are well known, and that facilitates its implementation in new processes. In the last decade, progress in this area has focused on the development of SUPRASs based on ionic surfactants, other aggregates than aqueous micelles (namely, reversed micelles and vesicles), and other stimuli than temperature (e.g., pH, electrolytes, or solvents) as the driving force phase separation. As a result, the scope of SUPRASs in solvent extractions has been greatly extended.

This review mainly covers progress related to SUPRAS-based extraction reported over the last decade. Advances in SUPRAS formation, mainly derived from recent developments in supramolecular chemistry, are outlined with the aim of better understanding the mechanisms of solvent production and are presented in the previous chapter. This knowledge is of primary importance for the synthesis of new solvents. The scarce studies reported so far on the structure of SUPRASs are presented and their physicochemical properties, mechanisms of interaction with solutes, extraction formats, compatibility with separation and detection techniques, and capability for the extraction of organic contaminants from a variety of matrices are critically discussed. A great deal of research is currently being conducted related to the extraction of metals by SUPRASs.[66]

It is clearly established nowadays that supramolecular solvents constitute a promising and powerful alternative to organic solvents in solvent extraction. The major asset of SUPRASs is their high capability of solubilization for a variety of solutes, which renders them ideal for multiresidue analysis in the environmental and agrifood fields—an outstanding feature hardly explored so far. Besides, they are ideally suited for developing generic sample treatments adaptable to the extraction of one or various solutes in very different types of samples with minor modifications.

Significant developments concerning both theoretical and practical aspects related to SUPRASs have been reported over the last decade. Self-assembly is today a discrete field of study and it seems to offer one of the most general strategies now available for generating nanostructures of interest in materials science, condensed matter science, and engineering. The wide expertise gained in self-assembly-based synthetic processes is a promising basis to develop supramolecular solvents with targeted properties that provide selective extraction in specific applications.

As pointed above, great effort has been made in recent years by researchers for targeting the extraction of polar solutes with the synthesis of new SUPRASs that provide a variety of interactions (ionic, cation, and hydrogen bonding). In this way, the scope of the SUPRASs has been greatly expanded. Extraction of contaminants from solids such as food, soil, sludge, sediment, etc. has received special attention in the last decade. The results obtained in this sense prove that SUPRASs offer advantages in terms of extraction efficiency and rapidity and are ideally suited for miniaturization of the extraction process.

SUPRASs have been successfully applied to the extraction of organic compounds, including PAHs, pesticides, surfactants, bioactive compounds, dyes, endocrine disruptors, etc. High concentration of amphiphiles has been found for SUPRASs made up of carboxylic acids, thus permitting to get the highest concentration factors reported so far (e.g., 769 in[67,68]).

There are still important efforts to be made in order to consolidate the use of SUPRAS. Some of the topics that deserve attention are the design and synthesis of supramolecular solvents with specific functions and tasks that extend their scope in extraction, the validation of methods including supramolecular-based extractions, and the development of new strategies for automation and compatibility with separation and detection techniques in analytical chemistry. The application of SUPRASs to the extraction of solutes from solid samples is another item hardly explored and research in this area is worthy of further attention.

6.1. Surfactants

Surfactants are the less expensive and the most versatile amphiphilic compounds.[69] As indicated in Fig. 13.7, a surfactant molecule can be described as a polar head with an unpolar tail, and different media can be prepared using different concentrations of surfactant molecules. Surfactant monomers can associate forming dimers or trimers, in general, called *premicellar aggregates* because the natural tendency of surfactants in aqueous solutions is to form an interphase between air and water or between the recipient walls and water to act as a surface-active linker, which can reduce the surface tension of the bulk phase.

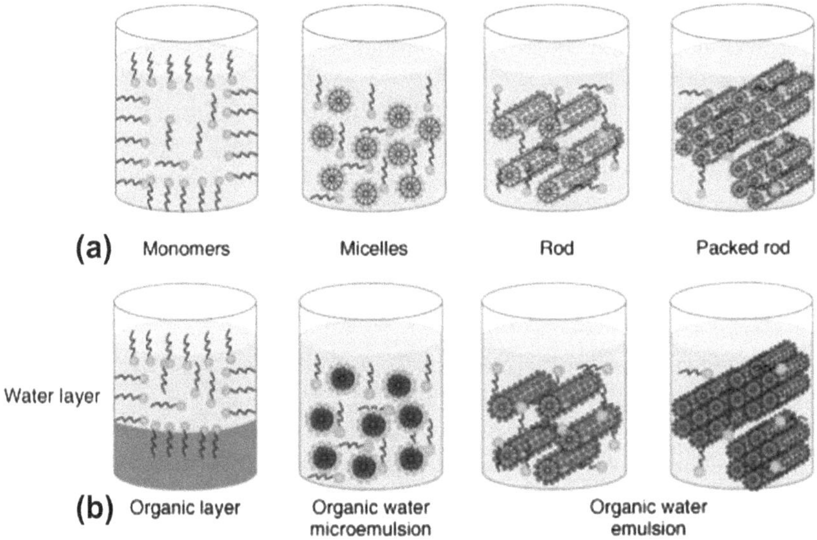

FIGURE 13.7 Different media created in water by the dissolution of increasing concentrations of (a) surfactant and (b) surfactant plus organic phase.

This particular behavior is very important in detergents, but as a solvent, aqueous solutions of surfactants with specific properties are of interest when the monomers interact to form supramolecular aggregates. As it can be seen in the normal micelles, the polar heads tend to be oriented inside the bulk phase and interact with the water dipoles and the dissolved ionic compounds. However, the unpolar tails tend to be arranged in a manner so that there is a minimum interaction with the dipoles and, thus, they create the micelle structures with an unpolar core and a polar surface. A special characteristic of these structures is that they are in dynamic equilibrium with the monomers that are formed and thus there is a continuous exchange between the monomers inside and outside the micelles.

The normal shape of micelles in aqueous media is rode like, but as a function of the surfactant concentration, these structures can move to hexagonal phases. In the presence of organic solvents, micelles can create microemulsions with the organic phase, thus providing a totally different medium inside and outside. The surfactant-created ordered medium has been extensively used in separation methods and in atomic and molecular spectroscopy because of their improvement of extraction coefficients, fluorometric and phosphorimetric quantum yields, and especially because of the solubility and mobility of unpolar compounds in water.

6.2. Toxicity of Surfactants

Although toxicity of surfactants is much lower than that of organic solvents, release of surfactants to the environment results in serious health and environmental problems related to the effectiveness of surfactants in solubilizing water-insoluble or poorly soluble compounds. Therefore, several investigations have been performed for reducing surfactant concentrations in the streams[70] and they should be applied for the online decontamination of wastes containing surfactants. The most commonly used surfactants are the anionic SDS, cationic cetyltrimethylammonium bromide (CTAB), and nonionic Brij-35. SDS is classified as "readily biodegradable" with "low bioaccumulation" with an acute oral toxicity LD_{50} of 0.8–1.10 g/kg in rats. The primary target of SDS toxicity is due to lipid peroxidation, increased glutathione production, and changes in carbon metabolism.[71] On the other hand, CTAB has low toxicity with an LD_{50} of 410 mg/kg in rats.

7. EXTRACTION BY MICROEMULSION

The term "microemulsion" is applied in a wide sense to different types of liquid–liquid systems. In this chapter, it refers to a liquid–liquid dispersion of droplets in the size range of about 10–200 nm that is both thermodynamically stable and optically isotropic. Thus, despite being two-phase systems, microemulsions look like single phases to the naked eye. There are two types of microemulsions: oil in water (O/W) and water in oil (W/O). The simplest system consists of oil, water, and an amphiphilic component that aggregates in either phase, or in both, entrapping the other phase to form the dispersion. The aggregates formed in the aqueous phase, called micelles, have their molecules orientated with their hydrophobic tails pointing to the interior of the aggregate and their hydrophilic head toward the continuous aqueous phase, whereas the aggregates in the organic phase, called reverse micelles, have the opposite orientation, as shown in Fig. 13.8. The microemulsion droplets are therefore the cores of either micelles or reverse micelles, stabilized by a surfactant layer. If more components besides the surfactant are present in the system, they may also be incorporated into the micelles or reverse micelles; these are then called mixed micelles or reverse micelles.

The ternary equilibrium diagram in Fig. 13.8 illustrates the effect of component concentration on the structure and the number of phases in a system consisting of an organic solvent, water, and a surfactant soluble in both. In the regions where the concentration of either water or organic solvent tends to zero, the system consists of one phase containing reverse micelles or micelles, respectively, with a negligible content of the third component in their cores. As the concentration of this component increases, more of its molecules transfer to the core of the micelles, forming either an O/W or a W/O microemulsion. In the multiphase region, the system has two or more phases, depending on the components of the system and the characteristics of the amphiphilic molecule.

Winsor[72] classified the phase equilibria of microemulsions into four types, now called Winsor I–IV microemulsions, illustrated in Fig. 13.9. Types I and II

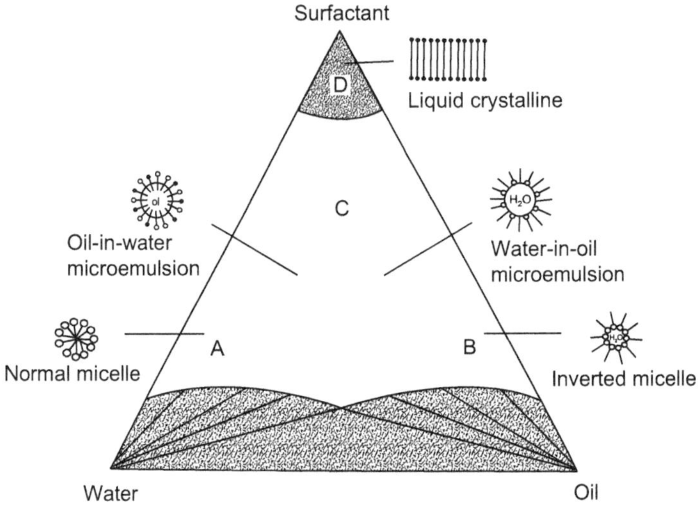

FIGURE 13.8 Schematic ternary-phase diagram of an oil–water–surfactant microemulsion system consisting of various associated microstructures. (a) normal micelles or O/W microemulsions; (b) reverse micelles or W/O microemulsions; (c) concentrated microemulsion domain; and (d) liquid-crystal or gel phase. Shaded areas represent multiphase regions.

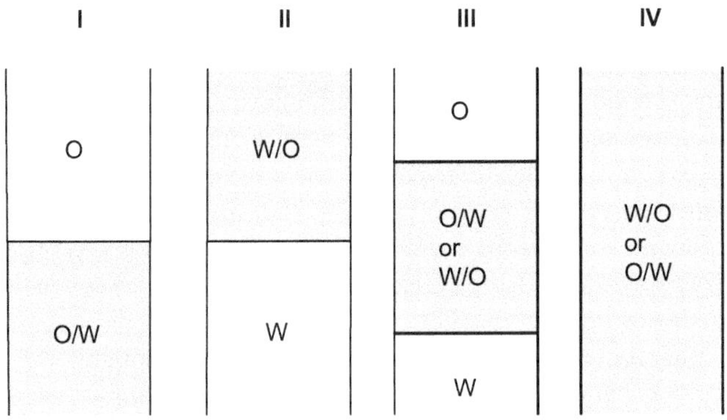

FIGURE 13.9 Types of Winsor microemulsions.

are two-phase systems where a surfactant-rich phase, the microemulsion, is in equilibrium with an excess organic or aqueous phase, respectively. Type III is a three-phase system in which a W/O or an O/W microemulsion is in equilibrium with an excess of both the aqueous and the organic phases. Finally, type IV is a single isotropic phase. In many cases, the properties of the system components require the presence of a surfactant and a cosurfactant in the organic phase in order to achieve the formation of reverse micelles; one example is the mixture of SDS and pentanol.

7.1. Critical Micelle Concentration

Critical micelle concentration (CMC) is the surfactant concentration below which the formation of reverse micelles does not occur, while the number of surfactant molecules per micelle is referred to as the aggregation number, n. The CMC is obtained through physical measurements, and varies from 0.1–1.0 mmol/dm^3 in water or the nonpolar solvents.

7.2. Water Solubilization Capacity

The amount of water solubilized in a reverse micelle solution is commonly referred to as W_0, the molar ratio of water to surfactant, and this is also a good qualitative indicator of micelle size. This is an extremely important parameter since it will determine the number of surfactant molecules per micelle and is the main factor affecting micelle size. For an (AOT)/isooctane/H_2O system, the maximum W_0 is around 60,[72] and above this value the transparent reverse micelle solution becomes a turbid emulsion, and phase separation may occur. The effect of salt type and concentration on water solubilization is important. Cations with a smaller hydration size, but the same ionic charge, result in less solubilization than cations with a large hydration size.[73] Micelle size depends on the salt type and concentration, solvent, surfactant type and concentration, and also temperature.

7.3. Mechanisms of Solute Extraction with Microemulsions

The Winsor II microemulsion is the configuration that has attracted most attention in solvent extraction from aqueous feeds, as it does not affect the structure of the aqueous phase. The organic extracting phase, on the other hand, is now a W/O microemulsion instead of a single phase. The main reason for the interest in W/O microemulsions is that the presence of the aqueous microphase in the extracting phase may enhance the extraction of hydrophilic solutes by solubilizing them in the reverse micellar cores. However, this is not always the case and it seems to vary with the characteristics of the system and the type of solute. Furthermore, in many instances the mechanism of extraction enhancement is not simply solubilization into the reverse micellar cores. Four solubilization sites are possible in a reverse micelle, as illustrated in Fig. 13.10.[74] An important point is that the term “solubilization” does not apply only to solute transfer into the reverse micelle cores, but also to insertion into the micellar boundary region called the palisade. The problem faced by researchers is that the exact location of the solute in the microemulsion phase is difficult to determine with most of the available analytical tools, and thus, it has to be inferred.

Some insights can be obtained from the mechanisms of extraction in two-phase systems. As in conventional solvent extraction, the mechanism of transfer of the solute is either physical or chemical. In conventional solvent

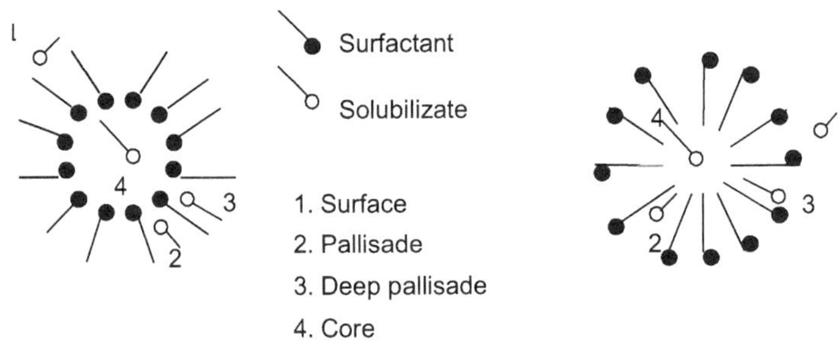

FIGURE 13.10 Possible solubilizate locations in a micelle. *(Source: From Ref. 19)*

extraction, physical transfer is used for species that prefer the organic phase, i.e., their distribution coefficient D allows the use of conventional solvent extraction. In some cases of low solubility in the organic phase, microemulsions have proved to enhance extraction. An important example in this category is the extraction of biological molecules.

8. RECOVERY OF REAGENTS

The recovery of reagents provides a satisfactory way to cut down on the side effects of extraction techniques, as it is an important step towards achieving zero emissions in research. Wastes are not discarded directly into the environment but are treated off-line. However, this practice increases the running costs of production and creates a problem through the accumulation of toxic residues.

A green alternative to waste storage is, therefore, the online recovery of solvents and toxic or expensive reagents. One way to reduce consumption of reagents is using solid reagents—another is to recover solvents online. A useful way of recovering reagents online is to incorporate a cation exchange minicolumn after a flow-through cell. In this way, the main reagent can be regenerated online and heavy-metal ions can be removed from the waste, thus reducing both reagent consumption and waste generation.[75]

In determination of propyphenazone and caffeine in pharmaceuticals, incorporating a distillation unit after the measurement cell provides online recycling of $CHCl_3$, which is used as carrier and solvent, thus reducing both cost and side effects of waste.[76] The system developed permits a dramatic reduction of reagent consumption, and makes sampling and cleaning of the measurement cell easy and fast.

8.1. Online Decontamination of Wastes

As described above, special emphasis should be placed on assessing the environmental impact of new techniques in addition to traditional goals (e.g., accuracy, precision, and sensitivity). Basically, online treatment of waste involves adding a decontamination step after basic flowsheet to obtain clean waste.

To detoxify wastes online, the following processes are suggested:

1. Thermal degradation
2. Oxidative detoxification
3. Photodegradation
4. Biodegradation[77]

Some of these strategies have yet to be applied to specific methods.

A detoxification step based on in-line TiO_2-catalyzed photodegradation reduces the toxicity of wastes. It has been applied for the determination of formetanate in waters[78] using their reaction with *p*-aminophenol (PAP). After the measurement step, the analytical waste was merged with a TiO_2 (anatase)-catalyst slurry and then detoxified by online UV irradiation. After that, the catalyst can be recovered through flocculation and reused.

Metal ions are nondegradable pollutants, so it is impossible to decontaminate metal wastes. However, it is possible to passivate wastes containing heavy metals and to reduce greatly the volume of waste (e.g., in the determination of Hg in milk by AFS).[79] The analytical waste was merged with NaOH and that solution was mixed with a solution of Fe(III), which precipitates $Fe(OH)_3$ and co-precipitates Hg, thus contributing to deactivation of traces of Hg and other heavy metals present in samples and standards, reducing the toxic wastes from several liters to less than 1 g of an hydroxide mixture with a high content of Fe.

9. FUTURE TRENDS IN THE DEVELOPMENT OF NEW SOLVENTS

The future will bring further increase in concern over the environmental impact of chemical operations. The liquid effluents must be rendered harmless to the environment. The main efforts correspond to the improvement of solvent characteristics of water under appropriate temperature and pressure conditions or the use of water-miscible cosolvents or additives, such as surfactants. However, other alternatives such as development of new solvents with better separation properties, replacement of petrol-derived organic solvents by renewable ones such as alcohols, or the use of supercritical fluids and ILs can offer cheaper, highly innovative, and unexplored environment-friendly options.

To evaluate the different green alternatives to the commonly employed organic solvents, the strengths, weaknesses, opportunities, and threats (SWOT) analysis[80] may be applied. Figure 13.11 presents the SWOT analysis of current strengths and weaknesses of superheated water, renewable organics, supercritical fluids, ILs, and surfactant solutions together with the

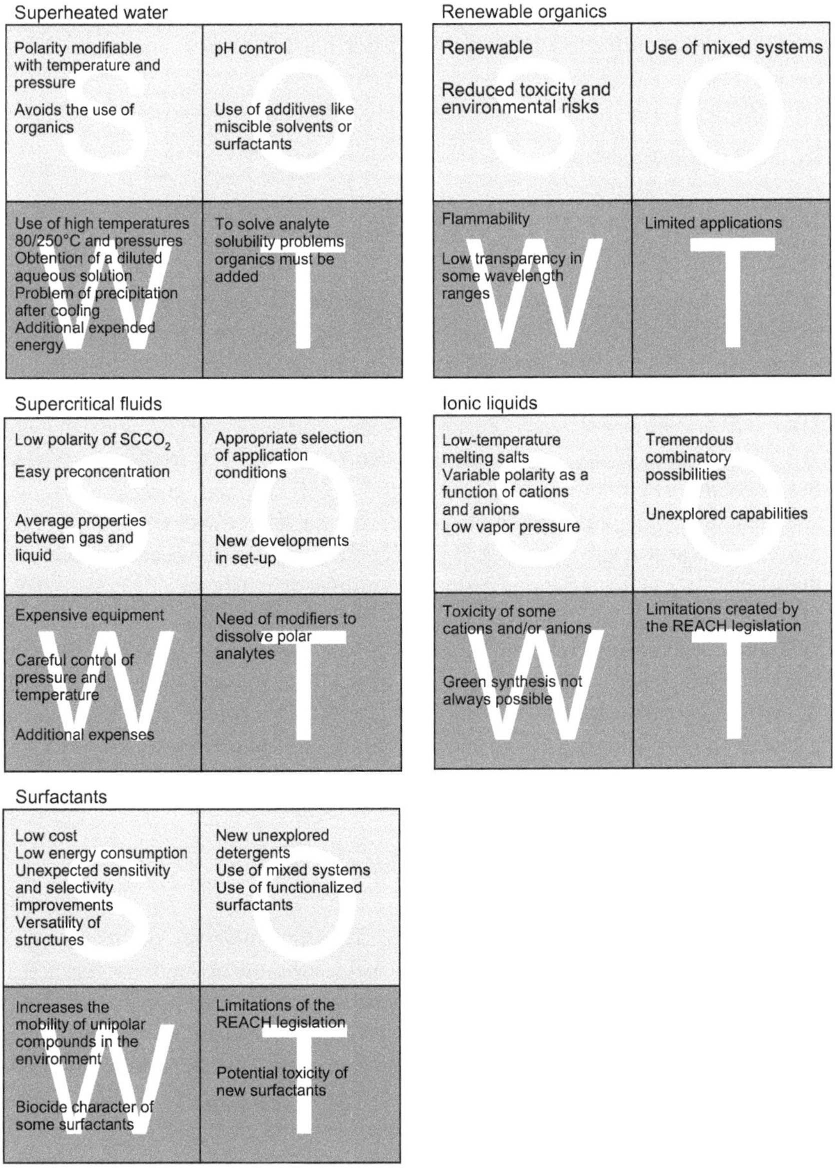

FIGURE 13.11 SWOT analysis of the different green alternatives available for the replacement of toxic organic solvents in analytical chemistry.

considerations of potential future strengths (opportunities) and potential future weaknesses (threats), considering simultaneously both the green parameters of available methods and their main partition features. Nowadays, the possibilities of using alcohols and supercritical fluids seem well known, but there are tremendous unexplored possibilities involving the use of ILs and surfactant solutions.

10. CONCLUDING REMARKS

The future will bring further increase in concern over the environmental impact of chemical operations. This requires removal of the hazardous substances. For many of the dilute waste solutions, solvent extraction has proved to be an effective process. This is even more truer for recycling of mixed metals from various industries. Nevertheless, the increasing amounts of wastes from human activities require much more to be done in this field.

In principle, solvent extraction is an environment-friendly process with no air or water pollution, provided the plant flows are properly designed. It could, therefore, replace many of the present polluting processes. A particular problem, however, is the solvent extraction effluents, which may contain biochemically active substances posing "new" hazards to the environment. These can be handled by various solid sorbents, which can then be incinerated, but the advantage of the solvent extraction process may be lost. There is, therefore, a demand for biodegradable and environmentally benign solvent phases. In the future, additional attention is required to this field.

The principle of solvent extraction—the distribution of chemical species between two immiscible liquid phases—has been applied to many areas of chemistry. A typical one is liquid partition chromatography, where the principle of solvent extraction provides the most efficient separation process available to organic chemistry today—its huge application has become a field (and an industry!) of its own. The design of ion-selective electrodes is another application of the solvent extraction principle—it also has become an independent field. Both these applications are only briefly touched upon in Chapter 5 on analytical applications. Nevertheless, fundamental research on solvent extraction will provide further important inputs into these two applications. It is likely that fundamental solvent extraction research will continue to contribute to the development of further selective analytical techniques.

The driving force of the transport of salts, proteins, etc., through the cell membrane from the nucleus to the body fluids, and vice versa, is a complicated biochemical process[81]. As far as is known, this field has not been explored by traditional solution chemists, although a detailed analysis of these transfer processes indicates many similarities with solvent extraction processes (equilibrium as well as kinetics). It is possible that studies of such simpler model systems could contribute to the understanding of the more complicated biochemical processes.

Solvent extraction deals with the transport of chemical substances from one phase into another, the chemical kinetics of this process, and the final equilibrium distribution of the substances between the two phases. Such transport and distribution processes are the motors that make life in biological systems possible. Fundamental studies of such "solvent extraction" processes contribute to the better understanding of all processes in nature. Here, only the lack of imagination stands in the way of important new scientific discoveries.

REFERENCES

1. Gani R, Jimenez-Gonzalez C, Kate A, Crafts PA, Jones M, Powell L, Atherton JH, Cordiner JL. A modern approach to solvent selection. *Chem Eng* 2006;**1**:30–41.
2. Henry MC, Yonker CR. Supercritical fluid chromatography, pressurized liquid extraction, and supercritical fluid extraction. *Anal Chem* 2006;**78**:3909.
3. Zhao H, Xia S, Ma P. Use of ionic liquids as "green" solvents for extractions. *J Chem Technol Biotechnol* 2005;**80**:1089–96.
4. http://www.epa.gov/ttn/atw/188polls.html (2010).
5. European Committee for Standardisation. *Environmental Management — Life Cycle Assessment — Principles and Framework*. EN ISO 14040. Brussels, Belgium: European Committee for Standardisation; 1997.
6. Tullo A. A solvent dries up. *Chem Eng News* 2008;**86**:27.
7. Agilent Technologies. *Technical Note, Reducing HPLC Solvent Consumption*. Publication Number 5990-3472EN, http://www.chem.agilent.com/Library/technicaloverviews/Public/5990-3472EN.pdf; 2009.
8. Welch CJ, Brkovic T, Schafer W, Gong X. Performance to burn? Re-evaluating the choice of acetonitrile as the platform solvent for analytical HPLC. *Green Chem* 2009;**11**:1232–8.
9. Smith RM. Extractions with superheated water. *J Chromatogr A* 2002;**975**:31–46.
10. Ogunsola OM, Berkowitz N. Extraction of oil shales with sub-critical and near-critical water. *Fuel Proc Technol* 1995;**45**:95–107.
11. Williams DD, Lee MK, Crawford JE, Tyree PO. Analysis of convective heat transfer in deformed and stratified aquifers associated with frasch thermal mining. *Ground Water* 1999;**37**:517–22.
12. Cassel E, Vargas RMF, Martinez N, Lorenzo D, Dellacassa E. Steam distillation modeling for essential oil extraction process. *E Ind Crop Prod* 2009;**29**:171–6.
13. Luque de Castro MD, Jiménez CMM. Potential of water for continuous automated sample-leaching. *Trends Anal Chem* 1998;**17**:441–7.
14. Hawthorne SB, Yang Y, Miller DJ. Extraction of organic pollutants from environmental solids with sub- and supercritical water. *Anal Chem* 1994;**66**:2912–20.
15. Yang Y, Bowadt S, Hawthorne SB, Miller DJ. Subcritical water extraction of polychlorinated biphenyls from soil and sediment. *Anal Chem* 1995;**67**:4571–6.
16. Jiménez CMM, Manclús JJ, Montoya A, Luque de Castro MD. Sub- and supercritical-fluid extraction of trichloropyridinol from soil prior to immunoassay. *J Chromatogr* 1997;**785**:329–36.
17. Kipp S, Peyrer H, Kleiboehmer W. Coupling superheated water extraction with enzyme immunoassay for an efficient and fast PAH Screening in soil. *Talanta* 1998;**46**:385–93.
18. Basile A, Jiménez CMM, Clifford AA. Extraction of rosemary by superheated water. *J Agri Food Chem* 1998;**46**:5205–9.
19. Fernández PV, Jiménez CMM, Luque de Castro MD. An approach to the static-dynamic subcritical water extraction of laurel essential oil: comparison with conventional techniques. *Analyst* 2000;**125**:481–5.

20. Gámiz-Gracia L, Luque de Castro MD. Continuous subcritical water extraction of medicinal plant essential oil: comparison with conventional techniques. *Talanta* 2000;**51**:1179–85.
21. Jiménez CMM, Ubera JL, Luque de Castro MD. Comparison of continuous sub-critical water extraction and hydrodistillation of marjoram essential oil. *J Chromatogr A* 1999;**855**:625–32.
22. Jiménez CMM, Luque de Castro MD. Isolation of eucalyptus essential oil for GC-MS analysis by extraction with subcritical water. *Chromatographia* 1999;**50**:578–82.
23. Chalchat JC, Garry RP, Michet A. Chemical composition of essential oil of *Calendula officinalis* L. (pot marigold). *Flavour Fragr J* 1991;**6**:189–92.
24. De-Pooter HL, Aboutabl EA, El-Shabrawy AO. Chemical composition and antimicrobial activity of essential oil of lead, stem and rhizome of *Alpinia speciosa* (J.C. Wendl.) K. Schum. grown in Egypt. *Flavour Fragr J* 1995;**10**:63–7.
25. Verma RK, Uniyal GC, Gupta MM. High-performance liquid chromatography of poppy straw. *Indian J Pharmaceut Sci* 1990;**52**:276–8.
26. Ayala RS, Luque de Castro MD. Continuous subcritical water extraction as a useful tool for isolation of edible essential oils. *Food Chem* 2001;**75**:109–13.
27. Di Corcia A, Caracciolo AB, Crescenzi C, Guiliano G, Murtas S, Samperi R. Subcritical water extraction followed by liquid chromatography mass spectrometry for determining terbuthylazine and its metabolites in aged and incubated soils. *Environ Sci Technol* 1999;**33**:3271–7.
28. Daimon H, Pawliszyn J. High temperature water extraction combined with solid phase microextraction. *Anal Commun* 1996;**33**:421–4.
29. Wennrich L, Popp P, Breuste J. Determination of organochlorine pesticides and chlorobenzenes in fruit and vegetables using subcritical water extraction combined with sorptive enrichment and CGC-MS. *Chromatographia* 2001;**53**:S380–6.
30. Hawthorne SB, Grabanski CB, Hageman KJ, Miller DJ. Simple method for estimating polychlorinated biphenyl concentrations on soils and sediments using subcritical water extraction coupled with solid-phase microextraction. *J Chromatogr A* 1998;**814**:151–60.
31. Hyotylainen T, Andersson T, Hartonen K, Kuosmanen K, Riekkola ML. Pressurized hot water extraction coupled on line with LC-GC: determination of polyaromatic hydrocarbons in sediment. *Anal Chem* 2000;**72**:3070–6.
32. Crescenzi C, D'Ascenzo G, Di Corcia A, Nazzari M, Marchese S, Samperi R. Multiresidue herbicide analysis in soil: subcritical water extraction with an on-line sorbent trap. *Anal Chem* 1999;**71**:2157–63.
33. Crescenzi C, Di Corcia A, Nazzari M, Samperi R. Hot phosphate-buffered water extraction coupled on line with liquid chromatography/mass spectrometry for analyzing contaminants in soil. *Anal Chem* 2000;**72**:3050–5.
34. Curren MSS, King JW. Ethanol-modified subcritical water extraction combined with solid-phase microextraction for determining atrazine in beef kidney. *J Agric Food Chem* 2001;**49**:2175–80.
35. Fernandez-Perez V, Luque de Castro MD. Micelle formation for improvement of continuous subcritical water extraction of polycyclic aromatic hydrocarbons in soil prior to high-performance liquid chromatography-fluorescence detection. *J Chromatogr A* 2000;**902**:357–67.
36. Albertsson PA. *Partition of cell particles and macromolecules*. 2nd ed. New York: John Wiley & Sons; 1971.
37. Raghavarao KSMS, Guinn MR, Todd P. Recent developments in aqueous two phase extraction in bioprocessing. *Sep Purif Methods* 1988;**27**(1):1.
38. Hatti-Kaul R. *Aqueous two-phase systems: methods and protocols*. Totowa, New Jersey: Humana Press; 2000.

39. Hustedt H, Kroner KH, Kula MR. Extractive enzyme recovery. *Proceedings of European Congress on Biotechnology* 1984;**1**:597.
40. Pourmortazavi SM, Hajimirsadeghi SS. Supercritical fluid extraction in plant essential and volatile oil analysis. *J Chromatogr A* 2007;**1163**:2–24.
41. Wang SM, Ling YC, Giang YS. Forensic applications of supercritical fluid extraction and chromatography. *Forensic Sci J* 2003;**2**:5–18.
42. Acosta GM, Smith Jr RL, Walsh JE, Boni KA. Beef shank fat solubility in supercritical carbon dioxide-propane mixtures and in liquid propane. *J Food Sci* 1995;**60**:983–7.
43. Tallon S, Catchpole OJ, Grey JB, Fenton K, Fletcher K, Fletcher AJ. Extraction of lipids from aqueous protein-rich streams using near-critical dimethylether. *Chem Eng Technol* 2007;**30**:501–10.
44. Favati F, King JW, Mazzanti M. Supercritical carbon dioxide extraction of evening primrose oil. *J Am Oil Chem Soc* 1991;**68**:422–7.
45. Alexandrou N, Lawrence MJ, Pawliszyn J. Cleanup of complex organic mixtures using supercritical fluids and selective adsorbents. *Anal Chem* 1992;**64**:301–11.
46. Levy JM, Storozynsky E, Ravey RM. The use of alternative fluids in online supercritical fluid extraction-capillary gas-chromatography. *J High Resolut Chromatogr* 1991;**14**:661–6.
47. Wilkes JS, Zaworotko MJ. Air and water stable 1-ethyl-3-methylimidazolium based ionic liquids. *J Chem Soc Chem Commun* 1992;**13**:965–7.
48. Chapeaux A, Simoni LD, Ronan T, Stadtherr MA, Brennecke JF. Extraction of alcohols from water with 1-hexyl-3-methylimidazolium bis(trifluoromethyl sulfonyl)imide. *Green Chem* 2008;**10**:1301–6.
49. Fadeev AG, Meagher MM. Opportunities for ionic liquids in recovery of biofuels. *Chem Commun* 2001;**3**:295–6.
50. Brennecke JF, Maginn EJ. Ionic liquids: innovative fluids for chemical processing. *AIChE J* 2001;**47**:2384–9.
51. Wasserscheid P, Keim W. Ionic liquids—new solutions for transition metal catalysis. *Angew Chem Int Ed* 2000;**39**:3773–89.
52. Welton T. Room-temperature ionic liquids. Solvents for synthesis and catalysis. *Chem Rev* 1999;**99**:2071–83.
53. Fredlake CP, Crosthwaite JM, Hert DG, Aki SNVK, Brennecke JF. Thermophysical properties of imidazolium-based ionic liquids. *J Chem Eng Data* 2004;**49**:954.
54. Visser AE, Swatloski RP, Griffin ST, Hartman DH, Rogers RD. Liquid/liquid Extraction of metal ions in room temperature ionic liquids. *Sep Sci Technol* 2001;**36**:785–804.
55. Chun S, Dzyuba SV, Bartsch RA. Influence of structural variation in room-temperature ionic liquids on the selectivity and efficiency of competitive alkali metal salt extraction by a crown ether. *Anal Chem* 2001;**73**:3737–41.
56. Wei GT, Yang Z, Chen CJ. Room temperature ionic liquid as a novel medium for liquid/liquid extraction of metal ions. *Anal Chim Acta* 2003;**488**:183–92.
57. Liu JF, Jiang GB, Chi YG, Cai YQ, Hu JT, Zhou QX. Use of ionic liquids for liquid-phase microextraction of polycyclic aromatic hydrocarbons. *Anal Chem* 2003;**75**:5870–6.
58. Liu JF, Peng JF, Chi YG, Jiang GB. Determination of formaldehyde in shiitake mushroom by ionic liquid-based liquid-phase microextraction coupled with liquid chromatography. *Talanta* 2005;**65**:705–9.
59. Liu JF, Li N, Jiang GB, Liu JM, Jonsson JA, Wen MJ. Disposable ionic liquid coating for headspace solid-phase microextraction of benzene, toluene, ethylbenzene and xylenes in paints followed by gas chromatography-flame ionization detection. *J Chromatogr A* 2005;**1066**:27–32.

60. Kulacki KJ, Lamberti GA. Toxicity of imidazolium ionic liquids to freshwater algae. *Green Chem* 2008;**10**:104–10.
61. Zhao D, Liao Y, Zhang Z. Toxicity of ionic liquids. *Clean* 2007;**35**:42–8.
62. http://www.rsc.org/Publishing/ChemScience/Volume/2010/04/searching_for_greener.asp accessed year: 2010.
63. Watanabe H, Tanaka H. A nonionic surfactant as a new solvent. *Talanta* 1978;**25**:585–9.
64. Scamehorn JF, Harwell JH, editors. *Surfactant-based separation processes*. New York: Marcel Dekker; 1989. p. 139–54.
65. Hinze WL, Pramauro E. Solubilization in micellar systems. *Crit Rev Anal Chem* 1993;**24**:133–77.
66. Butler OT, Cook JM, Davidson CM, Harrington CF, Miles DL. Atomic spectrometry update. Environmental analysis. *J Anal At Spectrom* 2009;**24**:131–77.
67. Ballesteros-Gymez A, Ruiz FJ, Rubio S, Pйrez-Bendito D. Determination of bisphenols A and F and their diglycidyl ethers in wastewater and river water by coacervative extraction and liquid chromatography-fluorimetry. *Anal Chim Acta* 2007;**603**:51–9.
68. Ruiz FJ, Rubio S, Pйrez-Bendito D. Study of the influence of water matrix components on admicellar sorbents. *J Chromatogr A* 2007;**1163**:269–76.
69. de la Guardia M. Espectrometria molecular de medios micelares. Molecular spectrometry of micellar media. In: Valiente M, Hoffmann H, Ródenas E, editors. *Propiedades, Caracterización y Aplicaciones de los Sistemas Coloidales. Properties, Characterization and Applications of the Colloidal Systems.* Madrid, Spain: University of Alcala, Universidad de Alcalá; 1994.
70. Gupta S, Pal A, Ghosh PK, Bandyopadhyay M. Performance of waste activated carbon as a low-cost adsorbent for the removal of anionic surfactant from aquatic environment. *J Environ Sci Health A* 2003;**38**:381–97.
71. Nickerson KW, Aspedon A. Detergent-shock response in bacteria. *Mol Microbiol* 1992;**6**:957–61.
72. Winsor PA. *Solvent properties of amphiphilic compounds*. London: Butterworth Scientific; 1954.
73. Zulauf M, Eicke HF. Inverted micelles and microemulsions in correlation spectroscopy. *J Phys Chem* 1979;**83**(4):480.
74. Leodidis EB, Hatton TA. Specific ion effects in electrical double-layers – Selective solubilization of cations in aerosol-OT reversed micelles. In: Pileni MP, editor. *The structure and reactivity in reversed micelles*. Amsterdam: Elsevier; 1989. p. 270.
75. Zenki M, Minamisawa K, Yokoyama T. Clean analytical methodology for the determination of lead with Arsenazo III by cyclic flow-injection analysis. *Talanta* 2005;**68**:281.
76. Bouhsain Z, Garrigues S, de la Guardia M. Clean method for the simultaneous determination of propyphenazone and caffeine in pharmaceuticals by flow injection fourier transform infrared spectrometry. *Analyst* 1997;**122**:441.
77. de la Guardia M. An integrated approach of analytical chemistry. *J Braz Chem Soc* 1999;**10**:429.
78. Escuriola MJ, Morales-Rubio A, de la Guardia M. A clean analytical method for the spectrophotometric determination of formetanate. *Anal Chim Acta* 1999;**390**:147.
79. Cava-Montesinos P, Ródenas-Torralba E, Morales-Rubio A, Cervera ML, de la Guardia M. Cold vapour atomic fluorescence determination of mercury in … sampling using multicommutation. *Anal Chim Acta* 2004;**506**:145.
80. Deetlefs M, Seddon KR. Assessing the greenness of some typical laboratory ionic liquid preparations. *Green Chem* 2010;**12**:17–30.
81. Osseo-Asare K, Kenney ME. Aggregation, reversed micelles, and microemulsions in liquid-liquid extraction. *Proceedings of ISEC'80*, Liege, 1980, vol. 1, paper no. 80, p. 121.

Chapter 14

Recent Advances in Solvent Extraction Processes and Techniques

Chapter Outline

Solvent Extraction. DOI: 10.1016/B978-0-444-53778-2.10014-7

1. INTRODUCTION

Conventional solvent extraction is a well-established technology for the separation of solutes from relatively concentrated feeds such as those found in the industrial production of chemicals and of metals by hydrometallurgy. Dilute streams, on the other hand, pose a challenge. Treatment of these streams using conventional liquid–liquid extraction requires a very large value of distribution ratio, otherwise the organic phase volume would become unacceptably high from environmental and safety considerations.

The novel solvent extraction technologies developed in the last decades try to address these limitations. Their potential to improve the performance of conventional solvent extraction is analyzed in this chapter according to modern trends in the principles and in solvent development.

Preparation of treated solutions continues to play a basic role and a lot of effort has been devoted to the development of faster, safer, and more environment-friendly techniques for leaching, extraction, and extract cleanup in the last two decades. Reduction of organic solvent consumption has been the primary driving force in the development of these techniques. Some strategies including adsorption of solutes on solids (e.g., solid-phase extraction, SPE,[1] matrix solid-phase dispersion, MSPD[2]), miniaturization (e.g., headspace solid-phase microextraction, HS-SPME,[3] stir bar sorptive extraction, SBSE[4]), and enhanced solvent-extraction (e.g., pressurized liquid extraction, PLE,[5] microwave-assisted extraction, MAE,[6] etc.) currently play an important role in sample handling in analytical labs. Attention has also been paid to the use of alternative solvents, mainly supercritical fluids,[7] and more recently ionic liquids (ILs).[8] Many of these novel techniques are developed firstly by analytical chemists for preparation of samples and determination needs, but modern and future trends in extraction-based technologies use and develop them.

Some novel technologies that have been developed in the last decades will be discussed in the following sections.

2. EQUIPMENT DESIGN AND SCALE-UP CONSIDERATIONS

Laboratory solvent extraction studies are often carried out with test tubes, a process that is cheaper but more laborious, time-consuming, and usually yields more scattered data than using semicontinuously stirred baffled beakers or Lewis-type cells. Highly precise distribution data can be obtained efficiently either "mechanically" by continuous flow centrifugal separators or "physically" by using special phase separating filter devices—some of these methods are described. Although these techniques are going to be improved, new, more advanced methods are likely to appear in the near future.

Specific reagents with exotic structures are costly to produce, which requires the experimentalist to use as small amounts as possible of the reagents. This leads to a desire to shift from "milli-experiments" (e.g., mixing 10 mL of each phase) to micro experiments (mixing 50 μL of each phase), which is now taking place, to even smaller sizes (volumes and amounts)—the *nano experiments*. Progress in this field requires a high degree of ingenuity.

Although single-stage laboratory techniques provide the first step toward multistage industrial processes, such process development usually requires small-scale multistage and pilot-plant scale equipment. A large number of excellent designs are available, and we consider further fundamental improvements.

The industrial application of solvent extraction is a mature technique, and it is now possible to move from laboratory experiments on a new extraction system to full industrial practice with little technological risk. There is a sufficient variety of large-scale equipments available to cope with most problems encountered in this application, although much of the equipment remains rather massive. Attempts to miniaturize, for instance, by using centrifugal forces to mix and separate phases, still have to be developed further.

Many industrial processes begin with a leaching step, yielding a slurry that must be clarified before solvent extraction. The solid–liquid separation is a costly step. The solvent extraction of unclarified liquids (solvent-in-pulp) has been proposed to eliminate solid–liquid separation. The increased revenue and reduced energy cost make this an attractive process, but many problems remain unsolved: loss of solutes and extractants to the solid phase, optimization of equipment design, effluent disposal, etc.

An essential step in industrial solvent extraction is the regeneration of the extractant. This can be done in many ways, e.g., by distillation, evaporation, or stripping (back-extraction). While distillation and evaporation do not discriminate between solutes (the diluent is simply removed by heating), stripping, by careful choice of strip solution and conditions, can be made highly selective. Alternatively, all the solutes can be stripped and then subjected to a selective extraction by changing the extractant. The possibilities are many, and it may be worthwhile to explore new paths.

Membrane extraction is a relatively new technique for solvent extraction, in which a solute is transferred from one aqueous phase to another through a membrane holding an extractant dissolved in a diluent. This ingenious scheme has only been slightly explored, though it offers great potential for the future, e.g., for wastewater cleaning.

The step from laboratory experiments via pilot plants to industrial scale requires serious consideration of all the points here; practical experience is invaluable in order to avoid mistakes and excess costs.

3. SUPERCRITICAL FLUID EXTRACTION

Solvent extraction processes usually run at ambient pressures and temperatures. If higher pressures are applied, it is mostly because a higher extraction temperature is required when equilibrium or mass transfer conditions are more favorable at an elevated temperature. Distillation is based on the differences in vapor pressures of the components to be separated, whereas solvent extraction utilizes the differences in intermolecular interactions in the liquid phase. Supercritical fluid extraction (SFE) is a separation technique that, to a certain extent, unites the principles of distillation and solvent extraction.

Because of its different physicochemical properties, SFE provides several operational advantages over traditional extraction methods.[9] Due to their low viscosity and relatively high diffusivity, supercritical fluids have better transport properties than liquids, can diffuse easily through solid materials, and can therefore give faster extraction yields. One of the main characteristics of a supercritical fluid is the possibility of modifying the density of the fluid by changing its pressure and/or its temperature. Since density is directly related to solubility,[10] by altering the extraction pressure, the solvent strength of the fluid can be modified. Other advantages—compared to other extraction techniques—are the use of solvents generally recognized as safe, the higher efficiency of the extraction process (in terms of increasing yields and lower extraction times), and the possibility of direct coupling with analytical chromatographic techniques such as gas chromatography (GC) or supercritical fluid chromatography (SFC).

An increase in temperature at constant pressure, on one hand, leads to a decrease in solvent density, which would lower the solubility. On the other hand, an increase in temperature results in an increase in vapor pressure. At high pressures, the density dependence on temperature is small compared with the effect of vapor pressure, which results in an increased solubility. At lower pressures, the density effect dominates when increasing the temperatures, resulting in a decrease in solubility. Figure 14.1 is characteristic for most mixtures in which the mutual solubility of the components is limited. From

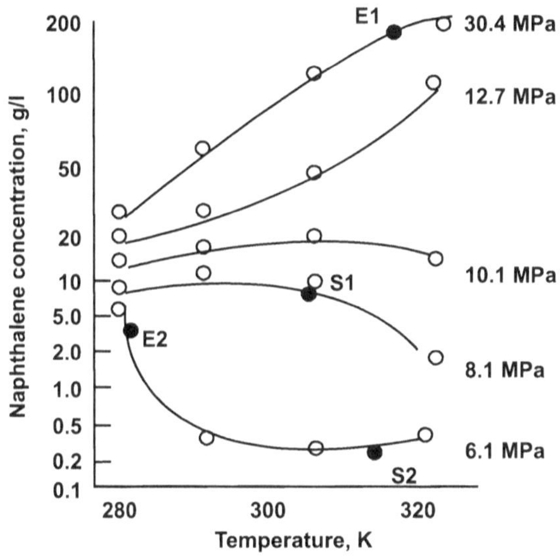

FIGURE 14.1 Solubility of naphthalene in supercritical ethene as function of temperature at different pressures.

the figure it is possible to derive simplified, general principles for extraction processes in the supercritical area, presented schematically in Fig. 14.2.

Figure 14.2a illustrates the principle for extraction and solvent recovery through a change in pressure. In Fig. 14.1, the point E1 represents the state in the extractor (30 MPa). The extract phase is passed through an expansion valve at which the pressure is reduced to 8 MPa (point S1 in Fig. 14.1). Naphthalene now precipitates in the separator. The solvent is recompressed and returned to the extractor.

In the scheme of Fig. 14.2b, naphthalene is extracted at 6 MPa and 285 K (point E2 in Fig. 14.1). The extract phase is passed through a heat exchanger in which the temperature is raised to 315 K (point S2 in Fig. 14.1). Also, in this case, naphthalene precipitates in the separator. The solvent passes through another heat exchanger to lower the temperature and is then returned to the extractor.

The extracted substances can also be recovered through adsorption (e.g., on activated carbon), which is shown in Fig. 14.2c. The potential to vary the density of the solvent without passing any phase borders make the supercritical area very interesting from the separation point of view. Through a gradual decrease in pressure, it is possible, e.g., to fractionate a mixture of substances.

There is a wide range of compounds that can be used as supercritical fluids (see Table 14.1 where critical properties of several solvents used in SFE are given). The attention is particularly focused on supercritical carbon dioxide solvent since it is nontoxic, environmentally acceptable, cheap, and has a low critical temperature (31.3 °C) and a moderate critical pressure (72.9 atm). The carbon dioxide is a rather nonpolar solvent, which dissolves mainly nonpolar

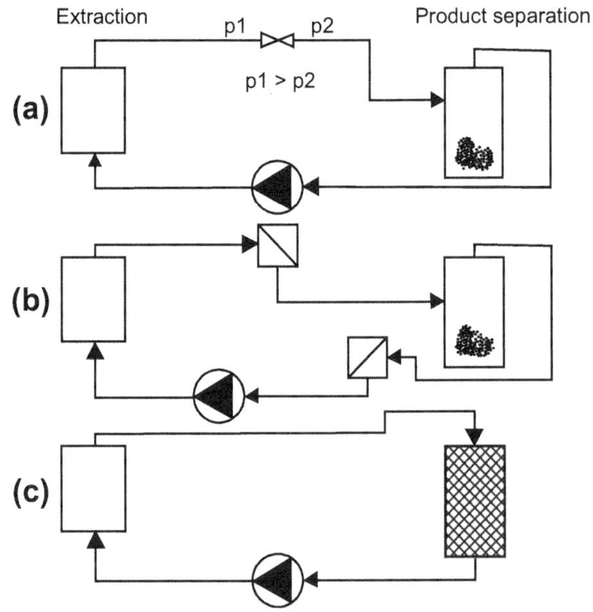

FIGURE 14.2 (a) Extraction and separation through reducing the pressure; (b) extraction and separation through a change in temperature; and (c) extraction and separation through adsorption.

TABLE 14.1 Critical Properties of Several Solvents Used in Supercritical Fluid Extraction (SFE)

	Critical property			
Solvent	Temperature (°C)	Pressure (atm)	Density (g/mL)	Solubility parameter δ_{SFC} (cal$^{-1/2}$ cm$^{-3/2}$)
Ethene	10.1	50.5	0.200	5.8
Water	101.1	217.6	0.322	13.5
Methanol	−34.4	79.9	0.272	8.9
Carbon dioxide	31.2	72.9	0.470	7.5
Ethane	32.4	48.2	0.200	5.8
Nitrous oxide	36.7	71.7	0.460	7.2
Sulfur hexafluoride	45.8	37.7	0.730	5.5
n-Butane	−139.9	36.0	0.221	5.2
n-Pentane	−76.5	33.3	0.237	5.1

solutes. Solubilities of selected compounds in liquid carbon dioxide are given in Table 14.2. The solubility and selectivity can be altered by adding small amounts of polar solvents, called entrainers (e.g., water or ethanol).

Carbon dioxide is a gas at room temperature, so once the extraction is completed, and the system decompressed, a substantial elimination of CO_2 is achieved without residues, yielding a solvent-free extract. On an industrial scale, when carbon dioxide consumption is high, the operation can be controlled to recycle it. However, supercritical CO_2, because of its low polarity (as can be seen in Table 14.1, where solubility parameter, *d*, is shown) that provides a measurement of the solvent polarity,[11] is less effective in extracting more polar compounds from natural matrices. To overcome this problem, modifiers (also called cosolvents or entrainers) are commonly used. Modifiers are highly polar compounds that, added in small amounts, can produce substantial changes of the solvent properties of neat supercritical CO_2.[12]

TABLE 14.2 Solubilities of Different Substances in Liquid Carbon Dioxide

Completely miscible	Fraction dissolving (% wt)		Insoluble
Stannous chloride	Water	0.1	Urea
Benzine	Iodine	0.2	Glycine
Benzene	Naphthalene	2.0	Phenylacetic acid
Pyridine	Aniline	3.0	Oxalic acid
Acetic acid	*o*-Nitroanisole	2.0	Succinic acid
Caprylic acid	Oleic acid	2.0	Malic acid
Ethyl lactate	Lactic acid	0.5	Tartaric acid
Amyl acetate	Butyl stearate	3.0	Citric acid
Glyceryl triacetate	Ethyl anthranilate	6.0	Ascorbic acid
Ethanol	Glyceryl monoacetate	1.0	Dextrose
Hexanol	Glycerol	0.05	Sucrose
Benzaldehyde	*n*-Decanol	1.0	–
Camphor	–	–	–
Limonene	–	–	–
Thiophene	–	–	–

3.1. Applications of SFE

Numerous vegetable matrices have been used as natural sources for compressed fluid extraction. Legumes, spices, aromatic plants, and even fruit beverages, such as natural orange juice have been processed to obtain natural antioxidant compounds. Today, there are more than 60 SFE extraction plants operating around the world. Much of the interest in SFE has been focused on using carbon dioxide to extract different natural products from solid and liquid materials. Examples of large application are processes in food, cosmetics, pharmaceutical industry, decaffeinating coffee beans, and oil and hop extraction. In the area of extracting solutes from aqueous solutions, many systems have been screened in feasibility tests that have used carbon dioxide as a solvent. SFE carbon dioxide extraction of metal ions from aqueous solutions is intensively studied.[13] A partial list of the solutes includes ethanol, acetic acid, dioxane, acetone, and ethylene glycol. The reason for these efforts has been potential low energy costs compared with distillation and the environmental advantages of using carbon dioxide.

According to the specific requirements, the design of a SFE system can be relatively simple or highly complex. It is possible to differentiate between analytical instruments and pilot- or industrial-scale systems. The analytical systems are utilized in sample preparation prior to, e.g., a chromatographic analysis. There are several configurations, depending on their degree of automation.

The pilot systems are used to extract grams of compounds when working on a pilot scale or kilograms on an industrial scale. Basically, this system on a pilot-scale plant consists (see Fig. 14.3) of a solvent pump, that delivers the fluid throughout the system, a modifier pump if necessary, an extraction cell or extraction column, according to the system configuration (i.e., for solids or liquids, respectively), and one or more separators (also called fractionation cells) in which the extract is collected and the solvent depressurized. Likewise, the extraction cell or column and the separators are commonly equipped with independent control of temperature and pressure, in such a way that fractionation of the extracted compounds can be carried out by a stepwise depressurization. Therefore, different compounds can be obtained within each separator, depending on their differential solubility in the supercritical fluid. In addition, it is possible to install a refrigerated system, especially designed to trap the most volatile compounds, as well as a recycling system to recycle the fluid employed.

As has been mentioned, the most important difference between pilot plants to process solid or a liquid sample is in the use of an extraction cell or an extraction column. Solid processing is always done in batch in a discontinuous or semicontinuous process, while liquid processing is usually carried out under countercurrent conditions in a continuous mode. In liquid-sample extractions, the supercritical fluid (usually CO_2) is moving upwards while the sample feed,

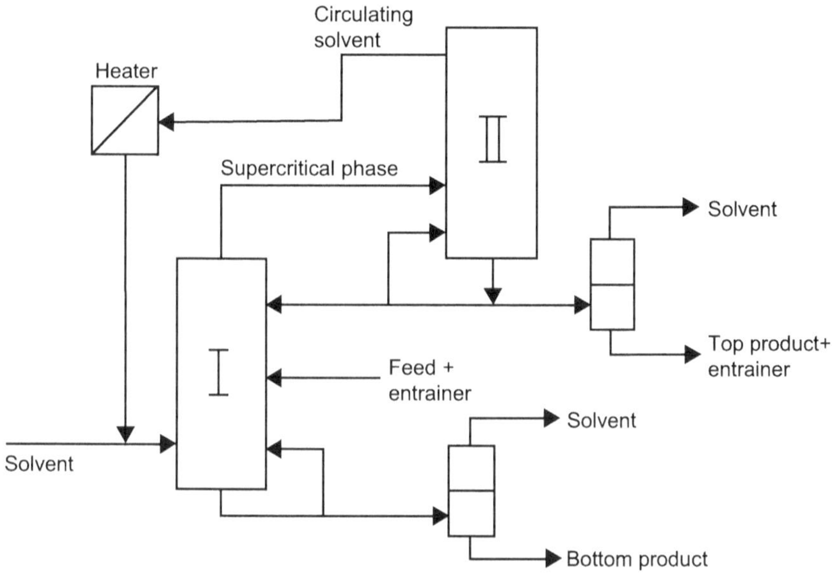

FIGURE 14.3 Process for separation of a nonvolatile component from a mixture.

introduced into the system from the top or center of the column, is moving downwards by gravity.

Other systems that have been investigated are the separation of biocides from edible oils and fractionation of different components in vegetable oils. Examples of the application of SFE are presented in the literature references.[14–17]

3.1.1. Separation of Monoglycerides from Triglycerides

Figure 14.3 illustrates a simplified process scheme of separation of a nonvolatile compound from a mixture, using an entrainer. In Fig. 14.4, the quasi-ternary system, carbon dioxide–acetone–glycerides, at 13 MPa is shown.[17,18] Acetone increases the solubility of glycerides in the carbon dioxide phase. The entrainer has the function of increasing the solubility of the solute and enhancing the selectivity. The process consists of two columns. In the first column, the nonvolatile monoglycerides are separated from the mixture, and in the second column, they are separated from the solvent. The mixture to be separated is fed, together with the entrainer, to the middle of the first column. Here, the solvent—carbon dioxide and acetone—is supercritical to provide high solubility of the monoglycerides. The supercritical phase leaves the top of column I and goes to the lower part of column II. In column II, the binary solvent entrainer is subcritical and in the bottom of this column, the monoglyceride leaves, together with the entrainer. Part of it is returned as reflux to column I, whereas the rest goes to distillation for the separation of acetone. With a bottom temperature of 353 K and a pressure of 13.5 MPa in column I and 383 K at the

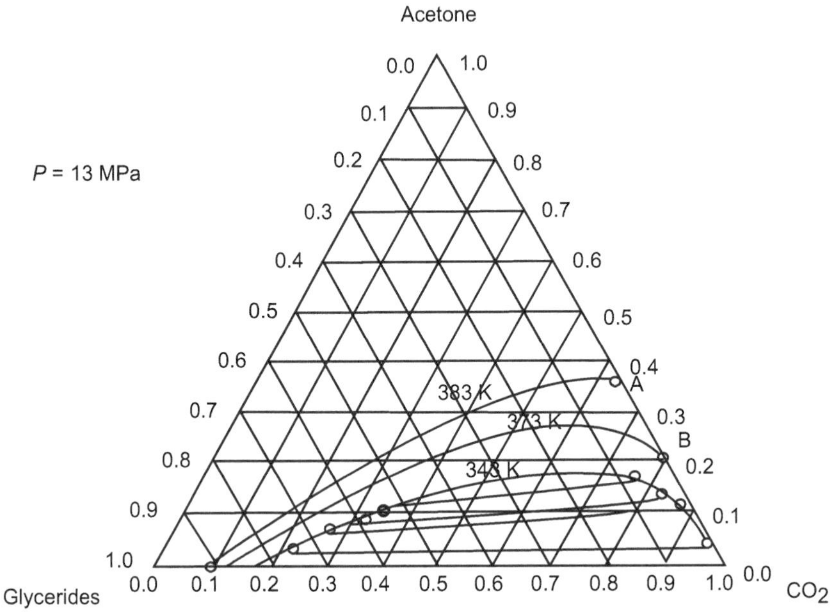

FIGURE 14.4 Phase equilibrium for the quasi-ternary system CO_2–acetone–glycerides at 13 MPa and 343, 373, and 383 K.

top of column II, 95% of the monoglycerides could be separated selectively from the mixture.

SFE with solid feed stocks are industrially carried out batchwise because of lack of equipment for feeding solid materials to a pressurized extractor. With liquid feed solutions, however, it is possible to work in a manner analogous to traditional solvent extraction. Pressurized columns can be of the packed-bed type or agitated by magnetic stirrers. Because of the efforts of pilot plant tests, much of the scale-up work has to be carried out in laboratory extractors. From solubility measurements, it is possible to determine parameters in thermodynamic models[19] (e.g., equations of state), which can be used for the simulation of large-scale applications.

3.2. Supercritical Fluid Chromatography

Supercritical fluids used as mobile phases in chromatography should act as substance carriers like the mobile phases in GC and also should dissolve these substances like mobile phases in LC. This chromatographic technique, called *supercritical fluid chromatography* (*SFC*), was discovered by Klesper et al.[20] and acts as a bridge between GC and LC. However, the technique became popular with the introduction of capillary columns. The main elements of those open tubular columns SFC instruments were the capillary columns, GC-like oven, pump, restrictor, and flame ionization detector.[21] Open tubular SFC was primarily applied in the petrochemical industry due to the lipophilic character of supercritical CO_2. However, the technique had severe limitations because of its poor reproducibility and limited application range in other sectors.[22]

Packed column SFC, with all its components similar to those of LC instruments including injectors, packed columns, UV detectors, electronically controlled back pressures, and organic modifiers to increase the solvating tendency and to decrease the retention time of polar solutes, confirms the future of the technique.[23] The main reasons for the last renaissance of the technique are as follows: the elimination of the technical problems related to back-pressure regulation, consistent flow rates, automatically controlled modifier addition, sample injection, automation and new stationary phase incorporation, the increased demand for environment-friendly processes that eliminate or substantially reduce the use of organic solvents,[24] and the hyphenation of packed column SFC to MS.[25] Nowadays, it is widely accepted that SFC is a kind of normal-phase LC, without many of the problems associated with normal-phase LC.

The major emphasis in SFC today concerns the use of packed columns, although several studies using open tubular columns continue to appear each year. Packed column SFC applications include chiral separations,[26] mass spectrometric detection of pharmaceutically relevant compounds,[27] and natural product applications with preparative separations.[28]

In conclusion, SFC constitutes a green alternative, which uses CO_2 as "mobile" phase and sometimes different proportions of green organic modifiers, such as ethanol or methanol, to reduce the organic solvent consumption in analytical chemistry.

3.3. Advantages and Disadvantages of Supercritical Extraction

The main disadvantage of SFE is the elevated pressure, requiring more expensive process equipment. The critical pressures, however, are below the pressures used in many high-pressure processes in the petrochemical industry today. Table 14.1 gives critical data for the most common solvents used in high-pressure extraction. It illustrates the favorable mass transport properties that can be achieved in the supercritical area owing to a low viscosity and a high diffusivity, compared with the liquid phase. The separation properties in SFE are dependent on the choice of solvents, as well as on the solutes.

4. SUBCRITICAL WATER EXTRACTION

Subcritical water extraction (SWE), i.e., extraction (critical point of water, 22.4 MPa and 374 °C) using hot water under pressure, has recently emerged as a useful tool to replace the traditional extraction methods. SWE is an environment-friendly technique that can provide higher extraction yields from solid samples.[29] SWE is carried out using hot water (from 100 to 374 °C,) under high pressure (usually from 10 to 60 bar) to maintain water in the liquid state.

4.1. Extraction Technique

The most important factor to consider in this type of extraction procedure is the variability of the dielectric constant with temperature. Water at room temperature is a very polar solvent, with a dielectric constant close to 80. However, this value can be significantly decreased to values close to 27 when water is heated up to 250 °C (see Fig. 14.5) while maintaining its liquid state by applying the appropriate pressure. This dielectric constant value is similar to that of ethanol.[30]

At high temperatures (above boiling point: 100–374 °C), steam needs to be pressurized for efficient transfer through the material.

The experimental device required for SWE is quite simple (see Fig. 14.6). Basically, the instrumentation consists of a water reservoir coupled to a high-pressure pump to introduce the solvent into the system, an oven, where the extraction cell is placed and extraction takes place, and a restrictor or valve to maintain the pressure. Extracts are collected in a vial placed at the end of the extraction system. In addition, the system can be equipped with a coolant device for rapid cooling of the resultant extract.

Although this technique has been mainly used as a batch process, studies of continuous techniques and the online coupling of a SWE system to an HPLC equipment via a solid phase trapping have been reported.[31]

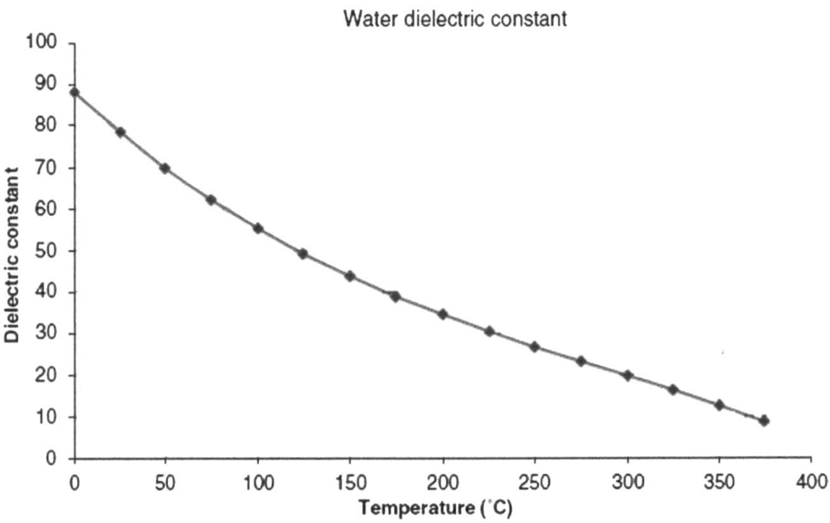

FIGURE 14.5 Graphical representation of dielectric constant of water versus temperature.

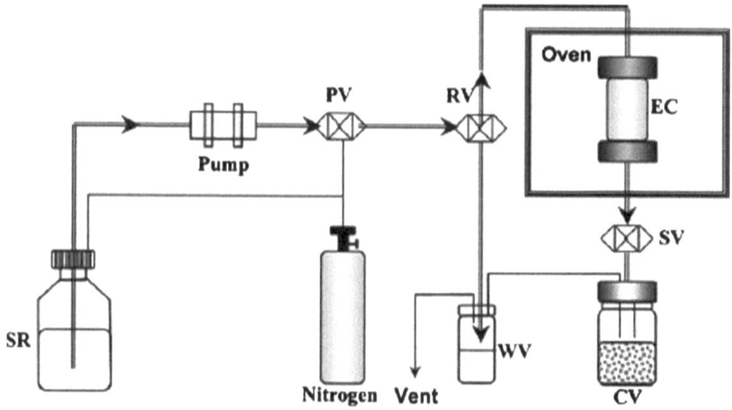

FIGURE 14.6 Diagram of a subcritical water extraction (SWE) system. SR, solvent reservoir; PV, purge valve; RV, pressure relief valve; EC, extraction cell; SV, static valve; CV, collector vial; WV, waste vial.

4.2. Extraction from Plants Using SWE

The SWE is emerging as a powerful alternative for the extraction of solid samples.[32] It has been used to extract pollutants with a wide range of polarities from environmental samples[33,34] and for the extraction of pesticides[35] and polycyclic aromatic hydrocarbons[36] from soils. The use of SWE as a very promising alternative to conventional and supercritical CO_2 extraction methods for isolation of essential oils was proposed by Basile et al.[37] Investigations showed that the SWE-proposed method is quicker, cheaper, and more effective than hydrodistillation. Since this date, the technique has shown its applicability in the field of essences[38–44] as compared with conventional techniques like steam distillation[45,46] and solvent extraction,[47] which have some well-known disadvantages: low extraction efficiency, long extraction time, and large amounts of toxic solvent waste. It has also the advantage of being selective.

SWE has been widely used to extract different compounds from several vegetable matrices. Likewise, one of the most deeply studied materials with SWE has been rosemary (*Rosmarinus officinalis* L.). De-Pooter et al.[48] studied the extraction of antioxidant compounds of rosemary by SWE over a wide range of temperatures. Several temperatures, from 25 °C to 200 °C, were tested to study the extraction selectivity toward antioxidant compounds. There was a clear effect of water temperature on the extraction yield, which increased at higher extraction temperatures. The authors verified that the most polar compound (i.e., rosmanol) was the main compound extracted at low temperatures (25 °C). When the extraction was performed at 200 °C, a decrease in the capability of water to dissolve the most polar compounds was observed, while a high concentration of other compounds, such as carnosic acid, was obtained. Antioxidant extraction by

SWE could be obtained comparable to those achieved using supercritical carbon dioxide.

In general, the use of SWE provides a number of advantages over traditional extraction techniques (i.e., hydrodistillation, organic solvents, and solid–liquid extraction). These are, mainly, rapid extraction and the use of low working temperatures, thus avoiding losses and degradation of volatile and thermolabile compounds. It has also the advantage of being selective, because by means of kinetic experiments, it is possible to manipulate the extract composition under given working conditions. More advantages are higher quality of the extracts (mostly for essential oils), lower costs of the extracting agent, and an environmentally compatible technique.

The extraction of essential oil from *Thymbra spicata*[49] was studied. The influences of several factors, such as temperature (100, 125, 150 and 175 °C), pressure (20, 60 and 90 bar), and flow rate (1, 2, and 3 mL/min) were studied. It was shown that the best extraction yields (3.7%) were obtained at 150 °C and 60 bar, using a flow rate of 2 mL/min for 30 min. The essential oils of *T. spicata* were found to inhibit mycelial growth of several fungi species.

4.3. Pressurized Liquid Extraction (Also Known as Accelerated Liquid Extraction, ALE)

PLE is a technique similar to SWE—the difference is only in solvents used. PLE is based on the use of organic solvents or diluted organic acids, such as acetic and formic acids, at a high pressure and/or high temperature without reaching the critical point. For example, the PLE system tested for pretreatment of biological material for trace element leaching and extraction is presented below.[50]

This system consists of a stainless steel extraction cell in which the sample is placed and where the programmed parameters (temperature, pressure, static extraction time, and extraction steps) are kept at their specified values by electronically controlled heaters and pumps. Once the extraction cell is loaded with the sample, solvent is added and the cell is heated and pressurized to the set values. After that, a static extraction period follows when solutes are released from the sample to the solvent. When the static step is concluded, the valve is opened and the solvent is transferred to the collection vial. The solvent residue is purged from the sample to the collection vial using a suitable gas. By pressurizing the sample cell (from 4 to 20 MPa), it is possible to keep the solvent in a liquid phase even at a relatively high extraction temperatures (up to 200 °C). This improves efficiency over classic extractions by shortening extraction times and reducing solvent volumes.

PLE procedures are considered to be relatively straightforward because the number of parameters to be optimized is rather limited,[51] thereby reducing the time devoted to the development of the extraction procedure. The application of PLE to assist trace metals acid leaching could offer several advantages when compared to mechanical shaking, ultrasound (water bath or probe), or microwave-assisted acid leaching methods.

The following are the advantages of the technique:

1. Less corrosive acids, such as weak organic acids, can be used instead of strong concentrated mineral acids, commonly needed for leaching. In addition, low volumes of such acids are required, which implies low toxic wastes.
2. Short leaching or extraction times are enough to complete the extraction. This last fact increases the sample throughput.
3. High degree of automation is offered by PLE and no additional filtration step is required after acid leaching assisted PLE.
4. Since diluted organic acids can be used to leach metals, chemical species integrity is guaranteed and PLE can be used for organometallic speciation studies.[52]

The feasibility of PLE to extract or leach total elements may be fully exploited in ores leaching and extraction avoiding concentrated mineral acids application.

Under PLE conditions, diluted weak carboxylic acids can attack the matrix, and major and trace elements can be efficiently leached. Variables affecting the carboxylic acid, PLE process may be tested using multivariate approaches in order to find out compromise acid leaching and extraction conditions.

4.4. Superheated Water Chromatography

In reversed-phase liquid chromatography (RPLC), water is used to reduce the elution strength of the organic modifiers, such as methanol, acetonitrile, and tetrahydrofuran. However, the same enhanced solvation strength described in the previous section also enables superheated water to be used as a mobile phase in LC. This mode of separation has been named *superheated water chromatography (SHWC)*, *pressurized water chromatography*, or *subcritical water chromatography*.

The description of separation using hot water as the mobile phase[53–56] provides an introduction to the technique, and indicates the enormous potential of the method for greening analytical chemistry. The instrumentation required is essentially the same as that used for conventional LC, with the inclusion of a high-temperature oven and a method of controlling the column back pressure (see Fig. 14.7). Another advantage of SHWC is that water can be used with a wide range of detectors. It is compatible with UV and fluorescence, refractive index, electrochemical, evaporative light scattering, and mass spectroscopy (MS) detection.[57] Moreover, it is also compatible with flame ionization detection,[58] being useful for the detection of analytes lacking chromophores, such as aliphatic alcohols, amino acids, and carbohydrates.

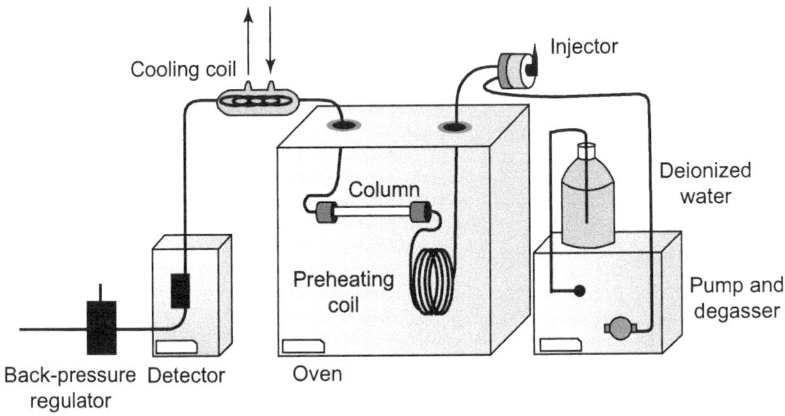

FIGURE 14.7 Schematic setup of superheated water chromatography.

However, the limited availability of commercial systems seems to be the main reason for lack of widespread adoption of the method.

The selectivity changes on heating water are not equivalent to the effect of adding an organic modifier, such as methanol, acetonitrile, or propan-2-ol.[59] However, high-temperature water could provide a complementary selectivity and was most suited for polar analytes. A 3.5 °C rise in water temperature corresponds to a 1% increase in methanol and a 5–8 °C rise corresponded to a 1% increase in acetonitrile,[60] using a series of aromatic analytes.

SHWC has been successfully applied in the pharmaceutical sector where the polarity of analytes is often compatible with an aqueous eluent. For instance, anticancer drugs have been separated on a polystyrene–divinylbenzene column with buffered superheated water as the mobile phase. The temperature range studied was from ambient temperature up to 160 °C, and the pH of the water was adjusted to 11.5 and 3.5 with phosphate buffer. Total elution time was less than 13 min.[61] The separation of alkyl aryl ketones and barbiturates using water rich and superheated water at high temperature (100–200 °C) as the eluent has been successfully performed.[62]

SHWC may be investigated in the environmental and food industries to offer some advantages.

5. AQUEOUS TWO-PHASE SYSTEMS

Under certain conditions, polymer incompatibility in aqueous solutions can lead to the formation of two phases with high water content. With such a system, it is possible to separate sensitive biological molecules, such as proteins, without denaturation, which would be the case for an ordinary aqueous–organic solvent system. Principles and properties of the system were described in Chapter 13.

Biotechnological processes, in general, are run at low concentrations of reactants and products. One reason for this is that most biocatalysts are sensitive to product inhibition. The reaction may stop when only a small fraction of the substrate has been converted into product. In such cases, there is a strong need for continuous extractive procedures during which the product is continuously removed. The integration of bioconversion and extraction is called extractive bioconversion. The large differences in surface tension and dielectric constants between water and common organic solvents cause enzymes to unfold when they are exposed to the interface between the two solvents. This problem can be overcomed through extractive bioconversions in aqueous two-phase systems. A collection of cutting-edge methods intended to provide practical guidelines for the area of separations in two-phase systems has been reviewed.[63]

5.1. Application of the Process

An important issue to be addressed in industrial applications of two-phase polymer systems for enzyme recovery is the economic or product quality requirement that the phase-forming polymers and salts be recycled. In some rare cases, the polymer is accepted in the product (e.g., in a process for purification of enzymes for detergent purposes). This process utilizes the system PEG–Na_2SO_4–water, and more than 95% of the enzyme is extracted in a two-stage countercurrent process. No recycling of PEG is necessary, since the polymer is a part of the product.

Usually, recycling of phase constituents is necessary, and several methods can be applied. Enzymes can be recovered from salt solutions by ultrafiltration, which can also be used to separate PEG from lower-molecular-weight products. If the product is charged, an ion exchange chromatographic step could be sufficient for isolation. Several methods to precipitate the polymer are also available.

Because of the small density difference between the two aqueous phases, the phase separation properties of the system are poor. Solid-free systems can be separated in a few hours or overnight in simple cylindrical vessels. Centrifugal separators are usually necessary for phase separation when cell debris is present because of the relatively high viscosity of the bottom phase. The performance of centrifugal separators is greatly reduced by a high solid concentration in the top phase. Thus, when extracting proteins from disintegrated cells, conditions should be such that the protein is partitioned to the top phase and the cell material to the bottom phase.

6. EXTRACTION PROCESSES WITH IONIC LIQUIDS

ILs are a new group of extractants of great interest studied as potential "green solvents."[64–68] Practically, zero vapor pressure of IL and temperature stability make them attractive solvents in many applications. ILs are gaining wide recognition as novel solvents in chemistry, although they are not new, e.g., $[EtNH_3]^+ NO_3^-$, was discovered in 1914.[65] They are also known as room-temperature (RT) ILs (not all ILs are RTILs), nonaqueous ILs, molten salts, liquid organic salts, or fused salts. The first example of a new IL, which is currently receiving much attention as a novel medium for homogeneous catalysis, is ethyl methylimidazolium tetrafluoroborate (reported in 1992).[66] The application of ILs in analytical chemistry, especially to separate analytes, is worthwhile because these solvents have some unique properties (e.g., negligible vapor pressure, good thermal stability, tunable viscosity, and miscibility with water and organic solvents, as well as good extractability for various organic compounds and metal ions[64]). Because they are nonvolatile and nonflammable, they have been proposed as environment-friendly solvents for "clean processes" and "green chemistry,"[67,68] and they are good substitutes for traditional organic solvents that are volatile and flammable.[69] A higher viscosity at room temperature could be their less favorable property.[67]

6.1. ILs in Separation Techniques

Extraction with ILs as extractants of metal ions, such as lanthanides,[70] radioactive metals, such as actinides,[71] are of particular industrial interest. Moreover, the extraction of Na^+, Cs^+, and Sr^{2+};[72] Cd^{2+}, Co^{2+}, Ni^{2+}, and Fe^{3+};[73] and Hg^{2+} and Cd^{2+} from wastewater using task-specific ILs[74] has been successfully performed. ILs were also applied to extract various organic compounds including substituted benzene derivatives,[75,76] carbohydrates,[77] biofuels,[78] and erythromycin A[79] from aqueous solutions.

Experiments with solvent extraction of organic acids by ILs showed promising results[80–83] in extraction of organic acids, especially of lactic acid.[80] In extraction of butyric acid, lactic acid, and phenol, fairly higher distribution coefficients were found for solvents with tested developmental ionic liquid IL-A compared to the solvents containing tertiary amines. The value of the distribution coefficient of lactic acid is up to 30 at lower acid concentrations, which is promising. IL-A acts as an extractant forming undissociated lactic acid/IL-A complexes having ratios of 1:1 and 2:1 (lactate anions are not extracted).[83]

In the pertraction of LA through Supported liquid membrane, SLM, the value of the overall mass-transfer coefficient increases with decreasing concentration of lactic acid in the aqueous phase what correlates with increasing value of its distribution coefficient.[81] Increased concentration of the carrier IL-A did not change the value of the mass-transfer coefficient in pertraction of lactic acid contrary to the increased value of the distribution coefficient. This may indicate that the slower kinetics of the interfacial reaction in decomposition of the complex plays a role or higher viscosity of membrane is responsible for this.

Separation of taurine (2-aminoethanesulfonic acid) and sodium sulfate by leaching a solid mixture by ILs is another example of its application.[84] Dialkylimidazolium chloride IL as leaching agent and organic solvent as precipitating agent (lower alcohols, like ethanol, are effective) were developed. Selective separation of taurine from a solid mixture containing a large amount of sodium sulfate could be realized with 67–98.5% yield in a single separation step.

Transport of amines and neutral organic substances through liquid membranes with IL was described in references.[85,86] Pertraction of organic acids through liquid membranes facilitated by enzymatic reactions on L–L interfaces using ILs as a liquid membrane was also studied.[87,88] Tailoring of ILs to achieve good partitioning of target solutes and acceptable viscosity of IL may lead to interesting results.

Formaldehyde extraction from mushrooms by LPME using ILs has been successfully developed.[89] Moreover, the methodology has been applied to the screening of chlorinated anilines,[90] dichlorodiphenyltrichloroethane and its metabolites,[91] chlorobenzenes,[92] phenols,[93] and chlorophenols[94] in water samples. However, this SDME procedure is incompatible with GC due to the nonvolatility of the IL.

6.2. Ionic Liquids in Analytical Separations

Currently, scientists are interested in the applications of ILs in separation and analysis.[95] Reports on ILs used as running electrolytes in capillary electrophoresis (CE)[96] appeared frequently. Moreover, ILs possess many favorable properties, such as nonvolatility, nonflammability, good solubility of many compounds, high viscosity, and polarity, which make them unique stationary phases in GC;[97] however, both the topics are out of the scope of this chapter and are not discussed in detail. On the other hand, ILs cannot be used themselves as mobile phase due to their high viscosity being only useful as additives in mobile phases.[98] For detailed information on the use of ILs in LC, readers can refer to the work of Marsza and Kaliszan.[99]

7. LIQUID MEMBRANE AND NONDISPERSIVE, MEMBRANE-BASED SOLVENT EXTRACTION TECHNIQUES

Membrane separation is a relatively new and fast-growing field of supramolecular chemistry. Liquid membrane process is a three-phase system consisting of two phases of a similar nature but different compositions (aqueous–aqueous, organic–organic, gas–gas) separated by a third phase of a different nature and as insoluble as possible into the other two. The middle phase is nominated as liquid membrane.

There are three main configurations in liquid membrane separation techniques that have been more widely investigated due to their potential industrial application:

1. Supported liquid membrane (SLM)
2. Emulsion liquid membrane (ELM), also called the surfactant liquid membrane
3. Bulk liquid membrane (BLM)

In the ELM configuration, the liquid membrane is formed by dispersing into the feed an emulsion of the stripping phase in an organic phase containing an emulsifying agent.

In the SLM process, the liquid membrane phase impregnates into microporous solid support placed between the two bulk phases. The liquid membrane is stabilized by capillary forces or through some kind of bonding. Two types of support configurations are used: hollow fiber or flat sheet membrane modules.

BLM consist of three bulk phases—two phases of a similar nature but different composition and between them a third, immiscible, phase. The phases may be separated by hollow fiber or flat sheet membrane supports, or may be without support. Many different BLM techniques are published lately, but all of them are based on membrane-based nondispersive selective solvent extraction coupled to permselective diffusion of solute–extractant complexes and selective stripping of the solute in one continuous dynamic process.

Nondispersive solvent extraction is one of the configurations of the BLM.

The term "nondispersive solvent extraction" arises from the fact that instead of producing a drop dispersion of one phase in the other, the phases are contacted through porous membrane.

There are two different arrangements for this process.[100] One uses two modules—one for extraction and the other for stripping—making it formally closer to conventional solvent extraction (see Fig. 14.8). The other configuration uses one three-liquid phase module in which all the three phases flow through. In the hollow fiber module, the liquid membrane phase is in the shell, and the feed and the stripping phase passes through the lumen of different fibers in the module. The solution to be extracted is fed to the lumen side of a microporous hollow fiber membrane module, with extracting organic solvent being fed to the shell side. The membrane is selected to be preferentially wetted by one of the phases (usually the organic phase), and the pressure adjusted to prevent capillary penetration of the aqueous phase into the membrane. In this manner, the solvent–water interface is stabilized at the membrane–aqueous solution boundary, and extraction of the desired component into the organic phase takes place by transmembrane diffusion and convection. The solute-enriched organic phase is then delivered to a second membrane unit where solute is transferred into an aqueous solution in which it has elevated solubility—the solute-depleted organic is then recycled to the first unit.

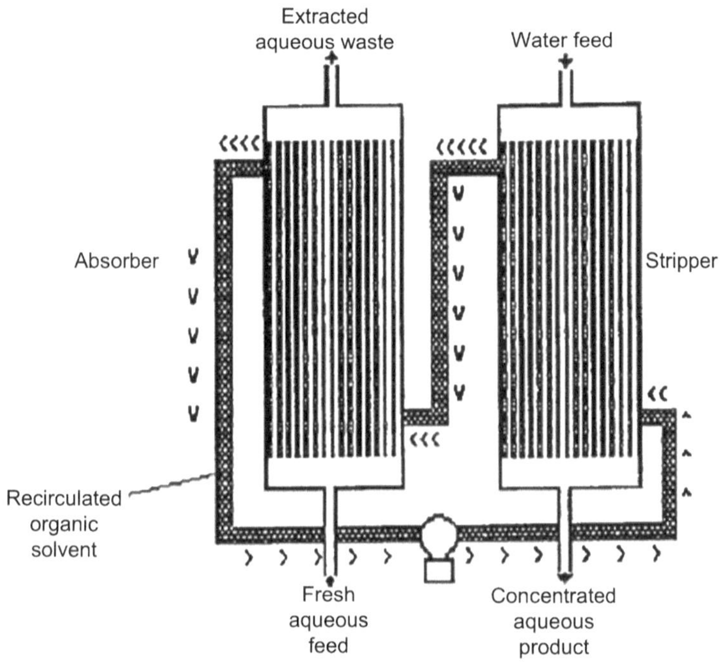

FIGURE 14.8 Membrane solvent extraction. *(Source: From Ref. 100 with permission)*

This technique has important advantages over conventional solvent extraction—it substantially eliminates emulsification, and the need for phase separation equipment such as centrifugal separators; it provides a stable, large interfacial area for interphase mass transfer; and it prevents contamination of the organic phase and extracted product by particulate impurities in the feed. Indeed, if hollow fiber membranes of suitably large lumen diameter are used, it is possible to process whole fermentation broths containing high concentrations of suspended biomass. This latter feature makes possible the process of "continuous extractive fermentation," in which the product-bearing whole broth is fed to the extraction unit, and the product-depleted broth (containing viable cells) is returned to the fermenter.

A number of commercial applications of this system have been already successfully realized. A bubble-free membrane-based carbonation line, using Liqui-Cel equipment, is in operation by Pepsi in West Virginia since 1993.[101] The systems are also used in beer production—the CO_2 removal stage is followed by nondispersive nitrogenation to obtain a dense foam head.[102]

This technology clearly has potential for use in organophilic solute recovery or removal from chemical process streams, and industrial or municipal wastewaters. Another important field of application of Bulk organic hybrid liquid membrane, BOHLM is the production of ultrapure water for semiconductor manufacturing.[103,104]

To my opinion, inclusion of this process to the solvent extraction techniques is questionable, but more and more publications in the periodic and reports on the ISEC are included this process to solvent extraction. The last is an equilibrium-based separation process. So, the membrane-based or nondispersive solvent extraction process has to be equilibrium based also. "Liquid membrane separation is a rate process and the separation occurs due to a chemical potential gradient, not by equilibrium between phases."[105] According to these definitions, many authors who refer to their works as membrane-based or nondispersive solvent extraction processes are not correct. The theory of membrane-based solvent extraction suggests[106] that overall mass transfer of treated solute consists of several steps: diffusion of the solute through the aqueous layer from the bulk source aqueous solution to the phases' interface (nonequilibrium process), interaction of the solute with extractant and formation of the solute–extractant complex (as a rule, the process is rapid and reaches equilibrium), diffusion of the solute–extractant complex through the membrane support itself (nonequilibrium process), and diffusion of the solute–extractant complex through the organic layer to the bulk organic solution (nonequilibrium process). Here, we can see that only one of the components may be equilibrium based. So, we can conclude that all systems may be equilibrium based in two cases: (1) if the kinetics of solute–extractant interaction is a rate-controlling process, which is not true in the majority of separations published in the literature, or (2) if the overall mass transfer of the solute from the bulk source solution to the bulk membrane solution comes to equilibrium.

The authors of very many works on so-called membrane-based or nondispersive solvent extraction could not prove that the process reaches equilibrium. Therefore, we cannot confirm the processes, published in these works, as membrane-based solvent extraction, but can confirm them as liquid membrane processes. Liquid membrane separations are dynamic nonequilibrium processes, in which only local equilibrium at immiscible phases' interface may be suggested. The works, in which the equilibrium state is not proved, or cannot be reached (e.g., the process where the feed—membrane—strip flows are arranged in one continuously operating module) are referenced as hybrid liquid membrane or hollow fiber liquid membrane processes.

The reader interested in details of the liquid membrane separation principles and applications is referred to the book published last year.[106]

8. SUPRAMOLECULAR-BASED EXTRACTION TECHNIQUES

In the last decades, several novel separation techniques were developed on the basis of supramolecular extractants. Below the reader will find some examples of these techniques, which have many perspectives in the future trends of separation technologies.

8.1. Nano- and Microtechnological Extraction

As it was described in the previous chapter, nanometer-sized amphiphilic aggregates, formed by self-assembly process, continue to grow by adding surfactant, and this is the second self-assembly process in supramolecular solvent formation. This process cannot exist without the preceding one being in place. Aggregate growth in this step will proceed until a separate, amphiphilic-rich, liquid phase is produced. The phenomenon of liquid–liquid phase separation, typically occurring in colloidal solutions, is named coacervation. These are micrometer-sized aggregates or reversed micelles.

Molecules making up the ordered structures in supramolecular solvents contain a hydrophilic and a hydrophobic part. As a result, these structures have regions of different polarity that provide a variety of interactions for solutes. The type of interaction may be tuned varying the hydrophobic part or the polar group of the amphiphile and, in theory, one may design the most appropriate aggregate for a specific application because amphiphiles are ubiquitous in nature and synthetic chemistry.

The nature of polar groups in the ordered structures determines the type of polar compounds that can be extracted by supramolecular extractants. So far, the more frequent polar groups used in analytical applications include polyethylene oxides, carboxylic acids, sulfates, sulfonates, carboxylates, and ammonium and pyridinium ions. Binding interactions involved during extraction of polar compounds primarily include ionic, hydrogen bonding, π-cation, and π–π. Hydrogen bonding is an extremely effective retention mechanism for polar compounds. On the other hand, when the surfactant contains a benzene ring, electrophilic interactions are provided through delocalized electrons in π-orbitals. These delocalized electrons interact with conjugated groups such as aromatic rings or double/triple bonds. A major property of supramolecular compounds is their capability to extract amphiphilic compounds (e.g., surfactants, drugs, pesticides, etc.) through the formation of mixed aggregates with the amphiphiles making up the ordered aggregates. Both hydrophobic and polar interactions govern the formation of mixed aggregates. These nanometer- and micrometer-sized aggregates or micelles are novel extractants for many separation techniques.

For example, experimental setup of microextraction technique in analytical chemistry is shown in Fig. 14.9. In liquid phase microextraction (LPME), the solvent is placed in a hollow fiber.[107] Analytes are transferred from the donor phase (usually aqueous) to a receptor phase through an organic phase immobilized in pores of the hollow fiber. In two-phase mode, the solvent in the pores of the fiber is the same as that present in the fiber. In three-phase mode, the solvent placed in the pores of fiber differs from that inside the fiber. The three-phase mode is applied for extracting polar analytes, whereas the two-phase mode is applied for extracting nonpolar and semi-polar analytes. The amount of solvent in the fiber is about 10–25 L.[108] The advantage of LPME over SDME is

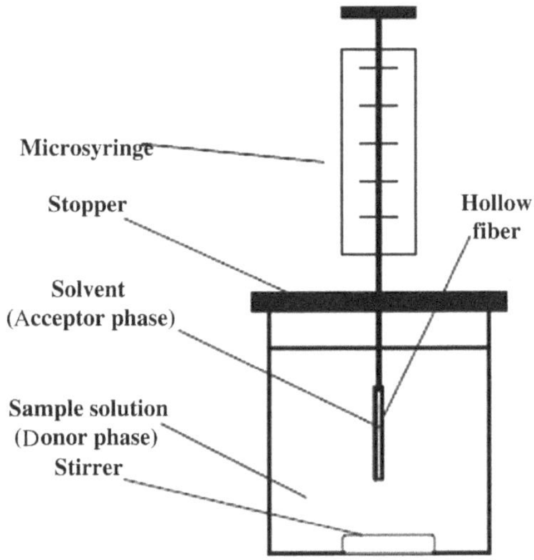

FIGURE 14.9 Main components of systems liquid-phase microextraction (LPME).

the presence of the fiber that supports organic solvent, so the process of dissolution or evaporation of the solvent is decelerated.

8.2. Surfactant-Based Extraction

Alternative extraction approaches, based on the use of surfactant solutions, have been proposed to reduce the environmental impact of analytical methodologies. Aqueous solutions of some surfactants are used in micellar extraction (ME) and cloud point extraction (CPE). The first technique is based on the fact that the micellar aggregates are able to solubilize different compounds and have a size that prevents them from crossing certain ultrafiltration membranes.[109] The CPE technique is based on the cloud point effect that is exhibited by the aqueous micellar solution of some surfactants when they are heated or cooled above or below certain temperature.[110] For details, see Chapter 5.

CPE, also known as *micelle-mediated extraction*, has been widely used for the extraction and preconcentration of trace metal analysis,[111] organic acids,[112] and hydrophobic proteins.[113] However, recently, CPE has been applied for the extraction of organic molecules other than biomolecules. CPE has been successfully exploited for the extraction and preconcentration of PAHs as a preliminary step to their determination by liquid chromatography (LC) using nonionic surfactants[114] and anionic surfactants, such as SDS, sodium dodecylbenzene sulfonic acid (SDBSA), and sodium dodecanesulfonic acid (SDSA).[115] Several nonionic surfactants were successfully used in the preconcentration of PCBs[116] and dibenzofurans (PCDFs)[117] from seawater. The presence of organophosphorus pesticides, such as methyl and ethyl parathion, paraoxon, and fenitrothion, has been determined in river water samples by using CPE with the nonionic surfactant Triton X-114 prior to their separation by LC.[118] The use of the nonionic surfactant poly(oxyethylene glycol) mono-ooctyl ether led to good extraction efficiency even for polar species, such as chlorinated phenols and anilines, with recoveries of 88–99%.[119]

8.3. Solvent Extraction with Microemulsions

Principles of microemulsion formation and solute extraction were presented in Chapter 13. Below, the reader will find some examples of microemulsion application in solvent extraction of metal ions and biological compounds. The influence of extraction parameters are discussed shortly.

8.3.1. Metal Ion Extraction with Microemulsion

Comparative to conventional extraction, in this case, the organic phase is replaced by a W/O microemulsion containing the reactant. There are two possible ways of forming a W/O microemulsion in the solvent phase:

1. The extractant forms reverse micelles in the organic phase leading to the formation of a microemulsion when this phase is contacted with the aqueous one (in which case, although perhaps unknown, the organic phase cannot be anything but a microemulsion).
2. The extractant does not form reverse micelles under the conditions of the process, in which case a surfactant, and sometimes also a cosurfactant, must be added to the organic phase in order to produce a reverse micellar phase. In this case, the reverse micelles are usually mixed, i.e., they include in the micellar shell the reactant and the additives.

Only case (2) can provide a comparison between conventional and micellar extraction. The few comparisons reported in the literature on the metal extraction performance of microemulsions containing an extractant with that of the extractant on its own are, at first sight, contradictory. In some cases, microemulsions produce both synergism and extraction rate enhancement with respect to the single surfactant, whereas in others they substantially reduce the metal distribution coefficient and the extraction rate, or leave them unchanged. An interesting example is found in the extraction with di(2-ethylhexyl)phosphoric acid (DEHPA). DEHPA does not form microemulsions in aliphatic solvents at pH 4 or below; however, it forms microemulsions on addition of a surfactant and a cosurfactant, e.g., sodium dodecylbenzene sulfonate and *n*-butanol. Improvements in the extraction of trivalent and tetravalent metals with respect to the conventional system with a DEHPA microemulsion were reported.[120] At extraction of Al(III) using the same microemulsion enhancements were also obtained[121] but not at the extraction of Zn(II), which was significantly reduced with respect to the conventional DEHPA system. The authors[121] attempted to explain this phenomenon of the microemulsion by thc different interfacial behavior of their complexes—the aluminum complex as more hydrophilic have a greater desorption energy than the zinc complex, making the interface its preferred location, whereas the zinc complex is more soluble in the organic phase. This suggestion remains to be confirmed. A similar behavior of microemulsion in the extraction of Bi(III) and Zn(II) was observed[122]—an increase in the extraction of Bi(III), but not of zinc.

Microemulsion extraction is a complex process with three phases: two interfaces and a usually unknown phase morphology. The prediction of its performance is quite difficult, particularly when dealing with solutes of diverse properties. The literature indicates that it does not always improve metal extraction, and in some cases, it may even hinder it due to the effect of the emulsifying additives. The author does not see much perspective in the future development in this direction.

8.3.2. Extraction of Biological Molecules

Separation mechanism and parameters that affect the distribution coefficient and kinetics of biological molecules such as proteins and amino acids extraction by microemulsion have been studied more extensively than the extraction of inorganic solutes. This is mainly due to the variety of size and structure of these molecules and, furthermore, to the fact that their characteristics may be adversely affected by contact with solvents and surfactants.

The distribution of biomolecules between the aqueous and the reverse micelle phases depends on system parameters such as pH, ionic strength, salt type, solvent structure and type, temperature, surfactant and cosurfactant concentration, and some others. Furthermore, protein size and hydrophobicity are also important in determining its partitioning behavior in the reverse micelle phase.

The pH of a solution affects the solubilization characteristics of a protein by modifying the charge distribution over the protein surface. At pH values below its isoelectric point (pI), or point of zero net charge, a protein acquires a net positive charge, while above its pI the protein will be negatively charged. Thus, if electrostatic interactions are the dominant factor, solubilization should be possible only with anionic surfactants at pH values less than the pI of the protein. The opposite effect would be anticipated in the case of cationic surfactants.

The ionic strength of the aqueous solution in contact with a reverse micelle phase affects protein partitioning in a number of ways.[123,124] The first is through modification of electrostatic interactions between the protein surface and the surfactant head groups by modification of the electrical double layers adjacent to both the charged inner micelle wall and the protein surface. The second effect is to "salt out" the protein from the micelle phase because of the increased propensity of the ionic species to migrate to the micelle water pool, reduce the size of the reverse micelles, and thus displace the protein.

An increase in surfactant concentration results in an increase in the number of micelles rather than any substantial change in size, and this enhances the capacity of the reverse micelle phase to solubilize proteins. Increasing the surfactant concentration leads to the increasing of protein solubilization in the reverse micelle phase.[125] At low "minimal" concentrations (6–20 mmol/dm^3 AOT), reverse micelles could be highly selective in separating very similar proteins from fermentation broths.[126] Recovery in activity of up to 95% could be achieved.

The temperature markedly affected the transfer of a-chymotrypsin in a chloroform–trioctylmethylammonium chloride (TOMAC) system.[127] By increasing the temperature from 25 to 40 °C, about 50% higher transfer yield was realized. No appreciable transfer of glucagon took place at room temperature, whereas at 37 °C transfer was possible.

Back-extraction of proteins encapsulated in AOT reverse micelles was evaluated by adding a counterionic surfactant, either TOMAC or DTAB, to the reverse micelles.[128] This novel backward transfer method gave higher backward extraction yields compared to the conventional method. The back-extraction process with TOMAC was found to be 100 times faster than back-extraction with the conventional method, and as much as three times faster than forward extraction. The 1:1 complexes of AOT and TOMAC in the solvent phase could be efficiently removed using adsorption onto montmorillonite so that the organic solvent could be reused.

8.3.3. Potential Applications of Micellar Extraction

The series of binary and ternary protein mixtures were studied using reverse micelle extraction.[126,129] Cytochrome c and lysozyme were quantitatively separated from a ternary mixture of these proteins with ribonuclease A. Separation of a mixture of ribonuclease A and concanavalin A showed no interaction between the proteins.

Recovery of an extracellular alkaline protease from a whole fermentation broth using microemulsion extraction was demonstrated.[130] Purification factors as high as 6 and yields of 56% were achieved in a three-stage cascade. The combination of a cascade with a higher aqueous/organic ratio, and the use of true cross-flow designs, shows promise for purification without dilution.

Application of the reverse micelle technique by extracting an a-amylase broth of *Bacillus licheniformis* using a CTAB/isooctane/5% octanol reverse micelle system showed that in a two-step extraction the protein concentration was reduced by a factor of 10 with a purification factor of 8.9, and a maximum yield of 89% a-amylase activity.[131]

Extraction as high as 98% of lysozyme from egg white was achieved[132] and a variety of demulsifiers, added to the mixture, enhanced yields substantially.

Several investigations of extraction of intracellular enzymes with reverse micelles were published recently.[133–135] CTAB in octane with hexanol as cosurfactant liberated enzyme rapidly into the water from whole cells reported to be able to lyse quickly and accommodate the pool of surfactant aggregates.[133,134] In another case, a periplasmic enzyme, cytochrome c553, was extracted from the periplasmic fraction using reverse micelles.[135] The purity achieved in one separation step was very close to that achieved with extensive column chromatography. These results show that reverse micelles can be used for the extraction of intracellular proteins.

8.3.4. Scale-up Considerations

Extractors developed for conventional liquid–liquid extraction are, in principle, also suitable for this application. Spray columns,[136] Graesser contactor (raining bucket),[137,138] were successfully tested and integrated system of separation and back-extraction was developed.[139] To scale up the process of biomolecular compounds extraction using reverse micelles to an industrial-sized unit, still much has to be done.

8.4. Micellar Liquid Chromatography

Micellar liquid chromatography (MLC) is an RPLC, in which the mobile phase consists of an aqueous solution of surfactant above its critical micellar concentration (CMC).[140] From a green analytical chemistry point of view, the idea of replacing organic solvents by surfactants as mobile phases in LC is highly appreciated. MLC is based on the solubility changes provided by the variation of the surfactant concentration above the CMC, which implies the increase of the concentration of amphiphilic aggregates or micelles in the solution, remaining constant the number of monomers of surfactant in the mobile phase. A detailed description of the retention mechanisms in MLC can be found in the work.[141] The advantages of MLC are the elimination of organic solvents and the fast way offered to perform gradient elutions with a convenient phase equilibrium.

Since its introduction, because of two main problems, MLC was less used when compared to conventional RPLC. The first is the excessive retention observed for hydrophobic compounds due to the weak eluting power of micellar mobile phases[142] and the second is the reduced efficiency due to the resistance to analyte mass transfer and/or increased flow anisotropy.[143] In MLC routine applications, the addition of a small amount of an organic solvent to the micellar solution is needed to achieve retention in practical time windows and to improve peak efficiency and resolution.[144] In 1983, the addition of a short-chain alcohol, 1-propanol or butanol, to the micellar eluent was proposed to enhance the efficiency of the separations. Propanol is the most commonly used alcohol additive, followed by pentanol, butanol, and acetonitrile.[145] From an environmental point of view, it is clear that a high percentage of organic solvent is not desirable. Moreover, using a high percentage of organic modifier would diminish or annul the role of the micelle and the surfactant molecules adsorbed in the stationary phase can be completely removed. The concentration of organic solvent that preserves the integrity of micelles is approximately 15% for propanol and acetonitrile, 10% for butanol, and 6% for pentanol.[146]

MLC has proved to be a useful technique in the control of diverse groups of drugs in pharmaceutical preparations, such as antihistamines and phenethylamines,[147] diuretics,[148] and benzodiazepines[149] using mobile phases that contain SDS and an alcohol, such as propanol, butanol, or pentanol. Moreover, SDS and an organic modifier have been used as mobile phases to determine antioxidants in foods, such as for the quantification of phenolic antioxidants in oils,[150] synthetic antioxidants in meat,[151] sulfonamides in milk,[152] and amines in wine.[153]

9. SOLID–LIQUID EXTRACTION SYSTEMS

Most techniques described in this and next sections of this chapter are developed and used in analytical applications, but the author considers that they are very perspective for development of commercial technologies in pharmaceutical, food, and environmental industries.

9.1. Solid-Phase Extraction

SPE is currently the most frequently employed technique for isolating or enriching solutes from water solutions. The isolation process comprises partition of organic compounds dissolved in water between a solid sorbent and an aqueous phase, due to which solutes are selectively trapped on this sorbent.[154] The trapping process is governed by the following mechanisms:

1. Adsorption of the solutes dissolved in water, on the surface of the sorbent (solid phase), mainly due to Van der Waals forces, hydrogen bonds, hydrophobic effects, π–electron interactions, and cation- and anion-exchange processes
2. Possible dissolution of a solute in some organic solid sorbents
3. Partition of solutes between sample and liquid sorbent (liquid stationary phase), in the form of thin film, coated on or chemically bonded to a solid support. SPE techniques can be applied in several modes, which are presented in Table 14.3.[155]

9.2. Solvent-Impregnated Resins

SPE columns are one of the most commonly applied methods of cleanup. Various column packings are now commercially available and include silica, charcoal, Florisil, C8, C18 reversed phase, and aluminum oxide.[156,157] These surfaces are chemically modified to allow the adsorption of the solute of interest or the impurities. In summary, the feed solution is applied to a preconditioned SPE column, the solute is retained on the column, which is washed to remove impurities, then eluted by means of a solvent wash. In the majority of

TABLE 14.3 Solid-Phase Extraction (SPE) Modes Used in Analyte Isolation

Extraction columns	The process of isolation and enrichment of analytes is realized during the flow of a water sample (due to gravitation or applied pressure) through the column packed with the sorbent (most often modified silica gel). In the case of samples containing suspensions, mechanical clogging of the pores of the adsorbing substance is relatively frequent, and the extraction time is rather long.
Empore disks (ED)	The sorbent is embedded in the porous Teflon film, which enables the sample to pass at high flow rate. This technique is suitable for isolating analytes from large volume liquid environmental samples and those containing significant amounts of suspended matter.
Speed disks (SD)	Modified form of Empore disk with the layer of the sorbent bed being placed between two glass-fiber filters, which considerably shortens the extraction time due to the possibility of the sample passing through at a significantly greater flow rate.
Sorbent adding to aqueous sample	Suitable sorbent is added to the water sample, and, after extraction, it is removed by decantation or filtration.

cases, the eluate is evaporated to dryness and reconstituted in a particular solvent or buffer depending on the analytical technique being used. Comprehensive review[156] of a variety of SPE column packings is available. The disadvantage of using such columns for the trichothecenes is that these toxins often differ greatly with respect to polarity and solubility, so recovery for some may be compromised. With this in mind, commercial companies have sought to overcome this problem (e.g., Varian (now Agilent Technologies, Inc., USA)), which produced a Bond Elut Mycotoxin cartridge designed for the simultaneous cleanup of 12 type A and B trichothecenes in addition to zearalenone—another important Fusarium mycotoxin food contaminant. This will significantly reduce the cost of technique in addition to making it more rapid and straightforward.

Another well-established method for the preparation of trichothecenes is the use of MycoSep columns (Romer Lab, Inc.).[157] Three types of columns are available, depending on the types of samples being analyzed and also the amount of resultant extract required for the analytical procedure, including:

1. MycoSep/MultiSep 227 Trich^{+}, applicable to complex food and feed matrices

2. MycoSep/MultiSep 225 Trich, for grain and simple sample matrices
3. MycoSep 113 Trich, employed for analytical procedures requiring small amounts of extracts

These columns comprise different adsorbents (e.g., charcoal and ion-exchange resins). A plastic tube is packed with the desired adsorbent between filter discs and contains a rubber flange at the lower end of the tube, which contains a porous frit and a one-way valve. Immunoaffinity columns (IACs), ion-exchange columns, and the more traditional liquid–liquid partitioning.[158] Antibodies are attached to an inert support and will specifically bind the solute of interest while allowing interfering components to pass through the column. As with other SPE formats, preconditioning, rinsing, and elution of the toxin are required and, in this instance, the extract must be an aqueous solution containing little or no organic solvent that will have a detrimental effect on the antibody and subsequently the antibody–antigen binding event. A major advantage of this approach is the specific interaction between the antibody and the analyte in question, but, as with all antibody-based methods, nonspecific interactions, due to cross-reactivities with other trichothecenes, may impair the results achieved.

9.3. Solid-Phase Microextraction

Solid-phase microextraction (SPME) is a relatively new technique, introduced in the early 1990s by Pawliszyn and Lord.[159] It can be easily automated[160] and has been increasingly used in sample preparation for chromatographic analysis, with a rapid increase in applications since the first fibers became commercially available in 1993.[161] The most common applications, limited by the properties of commercially available fibers, are in the field of organic compound determination (i.e., volatile organic compounds, alkylbenzenes, polycyclic aromatic hydrocarbons, and pesticides as target analytes). The technique is based on partitioning of an analyte between a sample and a fused-silica fiber coated with stationary phase, which can be a liquid polymer or a solid polymer dispersed in the liquid polymer. The analyte is then desorbed from the fiber directly into a suitable separation and detection system.[162]

10. ASSISTED SOLVENT EXTRACION

10.1. Microwave-Assisted Extraction

Microwave-assisted extraction (MAE) is one of the milestones in the development of sample-preparation strategies.[163] Microwaves, initially used in the food and agricultural industries for conditioning food products, have begun to be used for sample digestion.[164] They are used in the solvent extraction of organic analytes from solid samples. Enhancement of extraction is based on absorption of microwave energy by molecules of chemical compounds. MAE can be performed in two ways:

1. Pressurized MAE, which employs a microwave transparent, closed extraction vessel, and a solvent of high dielectric constant. Such solvents absorb microwave radiation, and thus are heated to a temperature exceeding the solvent boiling points under standard pressure conditions. However, boiling does not occur because the vessel is pressurized.
2. Atmospheric MAE system in open vessels—also called focused microwaves (FMW). This technique employs solvents with low dielectric constants. Such solvents are essentially microwave-transparent, so they absorb very little energy, and extraction can, therefore, be performed in open vessels. The temperature of the sample increases during extraction, because it usually contains water and other components with high dielectric constants. Because extraction conditions are milder, this mode of operation can be used to extract thermolabile analytes (e.g., polycyclic aromatic hydrocarbons, polychlorinated biphenyls, organochlorine pesticides, and alkanes).

The main advantages of MAE from the point of view of green chemistry are significant reduction in the amount of solvent used (25–50 mL), which reduces waste generation, shortens extraction time (from 1 h to 3 min), and reduces the amount of the sample required and correspondingly reduces energy input and cost.

10.2. Ultrasound-Assisted Extraction

Ultrasound-assisted extraction (UAE) is one of the techniques promising to speed up and to simplify sample treatment.[165] Ultrasonic energy, when imparted to solutions, causes acoustic cavitation (i.e., bubble formation and subsequent implosion).[166] The collapse of bubbles created by the sonication of solutions generates extremely high local temperature and pressure gradients, which may be regarded as localized "hot spots" that can be employed for extracting analytes from solid matrices, so that species may then be measured by instrumental techniques (e.g., atomic spectrometry). This technique is regarded as environment friendly because not only is the amount of solvent required decreased (100–300 mL) and the extraction time (3 min–1 h) shortened, but also ultrasound research is finding new applications in environmental remediation.

11. EXTRACTION OF GAS IN ANALYTICAL APPLICATIONS

Gas extraction is used in static or dynamic headspace modes. The static headspace (gas-phase extraction) technique includes transfer of the part headspace or gas phase above the sample in a condensed phase, by means of

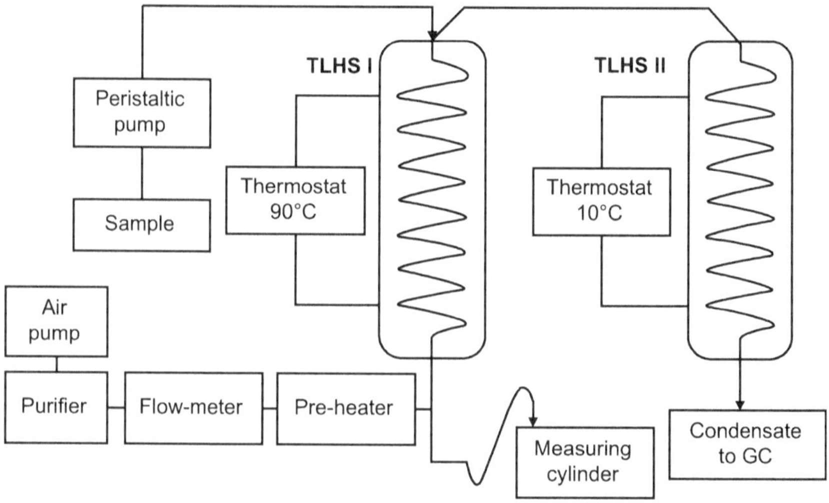

FIGURE 14.10 The main components of a thin-layer headspace (TLHS) device system.

gastight syringe or by transfer line into a GC machine. Dynamic headspace is the collection of the analytes in a trap from flowing gas phase above or through the sample, and then the trapped analytes by desorption into GC for their determination.

Headspace techniques are completely solventless, since inert gas is used as an extracting agent. The other advantage is that only volatiles are injected into the GC instrument. This is very important since nonvolatile compounds, if present, could decrease column quality. A new column requires material and energy inputs; by applying gas extraction, the problem is avoided.[167]

Another gas-extraction technique is thin-layer headspace (TLHS), which involves only water as a solvent, so is considered a "green technique." Figure 14.10 shows a setup for TLHS with autogenic generation of liquid solvent. The aqueous sample is pumped with peristaltic pump to TLHS "column I." In heated "column I," the aqueous stream flowing down in a form of thin layer contacts a countercurrent stream of heated clean air. Volatile analytes are transferred to the headspace together with some water vapor and then to "column II," where they dissolve in condensing water and exit the column as aqueous concentrate. Nonvolatile compounds remain in the sample waste, so TLHS is suitable for determination of volatiles in samples heavily contaminated with nonvolatile organics, inorganics, and particulate matter.

REFERENCES

1. Hennion MC. Solid-phase extraction method development, sorbents, and coupling with liquid chromatography. *J Chromatogr A* 1999;**856**:3–54.

2. Gilbert-Lypez B, García-Reyes JF, Molina-Díaz A. Sample treatment and determination of pesticide residues in fatty vegetable matrices: A review. *Talanta* 2009;**79**:109–28.
3. Aulakh JS, Malik AK, Kaur V, Schmitt-Kopplin P. A review on solid phase microextraction-high performance liquid chromatography (SPME–HPLC) analysis of pesticides. *Crit Rev Anal Chem* 2005;**35**:71–85.
4. S6nchez-Rojas F, Bosch-Ojeda C, Cano-Pavyn JM. 40 years of chromatographia: reviews of current techniques and developments. *Chromatographia* 2009;**69**:S79–94.
5. Rodriguez E, Navarro F, Moreno-Bondi MC, Marazuela MD. Multiresidue determination of fluoroquinolone antimicrobials in baby foods by liquid chromatography. *J Chromatogr A* 2010;**1217**:605–13.
6. Du G, Zhao HY, Zhang QW, Li GH, Yang FQ, Wang Y, Li YC, Wang YT. A rapid method for simultaneous determination of 14 phenolic compounds in Radix Puerariae using microwave-assisted extraction and ultra high performance liquid chromatography coupled with diode array detection and time-of-flight mass spectrometry. *J Chromatogr A* 2010;**1217**:705–14.
7. Hauthal WH. Chemical synthesis using supercritical fluids. *Chemosphere* 2001;**43**:123–35.
8. Buszewski B, Studzinska S. A review of ionic liquids in chromatographic and electromigration techniques. *Chromatographia* 2008;**68**:1–10.
9. Anklam E, Berg H, Mathiasson L, Sharman M, Ulberth F. Supercritical fluid extraction (SFE) in food analysis: a review. *Food Additiv Contamin* 1998;**15**:729–50.
10. Del Valle JM, Aguilera JM. Review: high pressure CO_2 extraction. Fundamentals and applications in the food industry. *Food Sci Technol Int* 1999;**5**:1–24.
11. Ravento M, Duarte S, Alarco R. Application and possibilities of supercritical CO_2 extraction in food processing industry: an overview. *Food Sci Technol Int* 2002;**8**:269–84.
12. Hildebrand JH, Scott RL. *The solubility of nonelectrolytes*. New York: Reinhold; 1958.
13. Valcárcel M, Tena MT. Applications of supercritical fluid extraction in food analysis. *Fresenius J Anal Chem* 1997;**358**:561–73.
14. Erkey C. Supercritical carbon dioxide extraction of metals from aqueous solutions: a review. *J Supercrit Fluids* 2000;**17**:259–87.
15. Bungert B, Sadowski G, Arlt W. Separations and material processing in solutions with dense gases. *Ind Eng Chem Res* 1998;**37**(8):3208.
16. Teja AS, Eckert CA. Commentary on supercritical fluids: research and applications. *Ind Eng Chem Res* 2000;**39**(12):4442.
17. Tonthubthimthong P, Douglas Peter L, Douglas S, Luewisutthichat W, Teppaitoon W, Pengsopa L. Extraction of nimbin from neem seeds using supercritical CO_2 and a supercritical CO_2-methanol mixture. *J Supercrit Fluids* 2004;**30**(3):287–301.
18. Fernández-Ronco MP, Cismondi M, Gracia I, De Lucas A, Rodríguez JF. High-pressure phase equilibria of binary and ternary mixtures of carbon dioxide, triglycerides and free fatty acids: measurement and modeling with the GC-EOS. *Fluid Phase Equilibr* 2010;**295**(1):1–8.
19. Perakis C, Loulis V, Voutsas E, Magoulas K. Supercritical CO_2 extraction of dittany oil: experiments and modelling. *J Supercrit Fluids* 2010;**55**:573–8.
20. Klesper E, Corwin AH, Turner DA. High pressure gas chromatrography above critical temperatures. *J Org Chem* 1962;**27**:700–1.
21. Griffiths PR. Contemporary SFC — accomplishments and limitations. *Anal Chem* 1988;**60**:593A–7A.
22. Harris CM. The SFC comeback — Pharmaceuticals give supercritical fluid chromatography a fighting chance. *Anal Chem* 2002;**74**:87A–91A.

23. Taylor LT. Supercritical fluid chromatography for the 21st century. *J Supercrit Fluids* 2009;**47**:566–73.
24. Welch CW, Leonard WR, DaSilva JO, Baba M, Albaneze-Walker J, Henderson DW, Laing B, Mathre DJ. Preparative chiral SFC as a green technology for rapid access to enantiopurity in pharmaceutical process research. *LC–GC* 2005;**23**:16–29.
25. Bolanos B, Greig M, Ventura M, Farrell W, Aurigemma CM, Li H, Quenzer TL, Tivel K, Bylund JMR, Tran P, Pham C, Phillipson D. SFC/MS in drug discovery at Pfizer, La Jolla. *Int J Mass Spectrom* 2004;**238**:85–97.
26. Toribio L, del Nozal MJ, Bernal JL, Alonso C, Jimenez JJ. Enantiomeric resolution of bifonazole by supercritical fluid chromatography. *J Sep Sci* 2006;**29**:1373–8.
27. Pinkston JD, Wen D, Morand KL, Tirey DA, Stanton DT. Comparison of LC/MS and SFC/MS for screening of a large and diverse library of pharmaceutically relevant compounds. *Anal Chem* 2006;**78**:7467–72.
28. Ramirez P, Santoyo S, Garcia-Risco MR, Senorans FJ, Ibanez E, Reglero G. Use of specially designed columns for antioxidants and antimicrobials enrichment by preparative supercritical fluid chromatography. *J Chromatogr A* 2007;**1143**:234–42.
29. Luque de Castro MD, Jiménez-Carmona MM, Fernández-Pérez V. Towards more rational techniques for the isolation of valuable essential oils from plants. *Trends Anal Chem* 1999;**19**:708–16.
30. Miller DJ, Hawthorne SB. Solubility of liquid organic flavor and fragrance compounds in subcritical (hot/liquid) water from 298 to 473 K. *J Chem Eng Data* 2000;**45**:315–8.
31. Li B, Yang Y, Gan Y, Eaton CD, He P, Jones AD. On-line coupling of subcritical water extraction with high-performance liquid chromatography via solid-phase trapping. *J Chromatogr A* 2000;**873**:175–84.
32. Luque de Castro MD, Jiménez CMM. Potential of water for continuous automated sample-leaching. *Trends Anal Chem* 1998;**17**:441–7.
33. Hawthorne SB, Yang Y, Miller DJ. Extraction of organic pollutants from environmental solids with sub- and supercritical water. *Anal Chem* 1994;**66**:2912–20.
34. Yang Y, Bowadt S, Hawthorne SB, Miller DJ. Subcritical water extraction of polychlorinated biphenyls from soil and sediment. *Anal Chem* 1995;**67**:4571–6.
35. Jiménez CMM, Manclús JJ, Montoya A, Luque de Castro MD. Sub- and supercritical-fluid extraction of trichloropyridinol from soil prior to immunoassay. *J Chromatogr* 1997;**785**: 329–36.
36. Kipp S, Peyrer H, Kleiboehmer W. Coupling superheated water extraction with enzyme immunoassay for an efficient and fast PAH Screening in soil. *Talanta* 1998;**46**:385–93.
37. Basile A, Jiménez CMM, Clifford AA. Extraction of rosemary by superheated water. *J Agric Food Chem* 1998;**46**:5205–9.
38. Fernández PV, Jiménez CMM, Luque de Castro MD. An approach to the static-dynamic subcritical water extraction of laurel essential oil: comparison with conventional techniques. *Analyst* 2000;**125**:481–5.
39. Gámiz-Gracia L, Luque de Castro MD. Continuous subcritical water extraction of medicinal plant essential oil: comparison with conventional techniques. *Talanta* 2000;**51**:1179–85.
40. Jiménez CMM, Ubera JL, Luque de Castro MD. Comparison of continuous sub-critical water extraction and hydrodistillation of marjoram essential oil. *J Chromatogr A* 1999;**855**:625–32.
41. Jiménez CMM, Luque de Castro MD. Isolation of eucalyptus essential oil for GC-MS analysis by extraction with subcritical water. *Chromatographia* 1999;**50**:578–82.

42. Chalchat JC, Garry RP, Michet A. Chemical composition of essential oil of *Calendula officinalis* L. (pot marigold). *Flavour Frag J* 1991;**6**:189–92.
43. De-Pooter HL, Aboutabl EA, El-Shabrawy AO. Chemical composition and antimicrobial activity of essential oil of lead, stem and rhizome of *Alpinia speciosa* (J.C. Wendl.) K. Schum. grown in Egypt. *Flavour Frag J* 1995;**10**:63–7.
44. Soto Ayala R, Luque de Castro MD. Continuous subcritical water extraction as a useful tool for isolation of edible essential oils. *Food Chem* 2001;**75**:109–13.
45. Verma RK, Uniyal GC, Gupta MM. High-performance liquid chromatography of poppy straw. *Indian J Pharm Sci* 1990;**52**:276–8.
46. Ibañez E, Kubatova A, Señorans FJ, Cavero S, Reglero G, Hawthorne SB. Subcritical water extraction of antioxidant compounds from rosemary plants. *J Agric Food Chem* 2003;**51**:375–82.
47. Kubatova A, Miller DJ, Hawthorne SB. Comparison of subcritical water and organic solvents for extracting kava lactones from kava root. *J Chromatogr A* 2001;**923**:187–94.
48. De-Pooter HL, Aboutabl EA, El-Shabrawy AO. Chemical composition and antimicrobial activity of essential oil of lead, stem and rhizome of *Alpinia speciosa* (J.C. Wendl.) K. Schum. grown in Egypt. *Flavour Frag J* 1995;**10**:63–7.
49. Ozel MZ, Gogus F, Lewis AC. Subcritical water extraction of essential oil from. *Thymbra spicata. Food Chem* 2003;**82**:381–6.
50. Moreda-Piñeiro J, Alonso-Rodríguez E, López-Mahía P, Muniategui-Lorenzo S, Fernández-Fernández E, Prada-Rodríguez D, Moreda-Piñeiro A, Bermejo-Barrera A, Bermejo-Barrera P. Pressurized liquid extraction as a novel sample pre-treatment for trace element leaching from biological material. *Anal Chim Acta* 2006;**572**:172–9.
51. Lou X, Janssen HG, Cramers CA. Parameters affecting the accelerated solvent extraction of polymeric samples. *Anal Chem* 1997;**69**:1598.
52. Alonso-Rodríguez E, Moreda- Piñeiro J, Moreda- Piñeiro A, López-Mahía P, Muniategui-Lorenzo S, Fernández-Fernández E, Bermejo-Barrera P, Prada-Rodríguez D. *Trends Anal Chem* 2006;**25**:511.
53. Guillemin CL, Millet JL, Dubois J. Thermal aqueous liquid chromatography — the TALC technique. *J High Resolut Chromatogr* 1981;**4**:280–6.
54. Smith RM, Burgess RJ, Chienthavorn O, Stuttard JR. Superheated water: a new look at chromatographic eluents for reversed-phase HPLC. *LC–GC N Am* 1999;**17**:938.
55. Coym JW, Dorsey JG. Superheated water chromatography: a brief review of an emerging technique. *Anal Lett* 2004;**37**:13–23.
56. Su Y, Jen JF, Zhang W. The Method development of subcritical water chromatography. *J Chin Chromatogr* 2005;**23**:238–42.
57. Smith RM. Superheated water: the ultimate green solvent for separation science. *Anal Bioanal Chem* 2006;**385**:419–21.
58. Ingelse BA, Janssen HG, Cramers CA. HPLC-FID with superheated water as the fluent: improved methods and instrumentation. *J High Resol Chromatogr* 1998;**21**: 613–6.
59. Pawlowski TM, Poole CF. Solvation characteristics of pressurized hot water and its use in chromatography. *Anal Commun* 1999;**26**:71–5.
60. Kondo T, Yang Y. Comparison of elution strength, column efficiency, and peak symmetry in subcritical water chromatography and traditional reversed-phase liquid chromatography. *Anal Chim Acta* 2003;**494**:157–66.
61. Teutenberg T, Lerch O, Gotze H-J, Zinn P. Separation of selected anticancer drugs using superheated water as the mobile phase. *Anal Chem* 2001;**73**:3896–9.

62. Sanagi MM, See HH. High temperature liquid chromatography on a poly(styrene-divinylbenzene) stationary phase. *J Liq Chromatogr* 2005;**28**:3065–76.
63. Hatti-Kaul R. *Aqueous two-phase systems: methods and protocols*. Totowa, NJ: Humana Press; 2000.
64. Wilkes JS, Zaworotko MJ. Air and water stable 1-ethyl-3-methylimidazolium based ionic liquids. *J Chem Soc Chem Commun* 1992;**13**:965–7.
65. Rogers RD, Seddon KR. *Ionic liquids in industrial applications to green chemistry*. Washington DC: ACS; 2002. p. 474.
66. Rogers RD, Seddon KR. *Ionic liquids as green solvents. Progress and prospects. ACS Symposium Series*. Washington DC: American Chemical Society; 2003. p. 599.
67. Rogers RD, Swatloski RP, Visser AE. Liquid–liquid extraction of organics in room-temperature ionic liquids: cationic effects. *Abstr Papers Am Chem Soc* 1999;**218**:22.
68. Rogers RD, Visser AE, Swatloski RP, Hartman DH. Room temperature ionic liquids as alternatives to volatile organic solvents in liquid/liquid extraction. *Abstr Papers Am Chem Soc* 2000;**219**:3.
69. Cornmell RJ, Winder CL, Tiddy GJT, Goodacre R, Stephens G. Accumulation of ionic liquids in escherichia coli cells. *Green Chem* 2008;**10**:836–41.
70. Nakashima K, Kubota F, Maruyama T, Goto M. Ionic liquids as a novel solvent for lanthanide extraction. *Anal Sci* 2003;**19**:1097–8.
71. Visser AE, Swatloski RP, Reichert WM, Mayton R, Sheff S, Wierzbicki A, Davis Jr JH, Rogers RD. Task-specific ionic liquids for the extraction of metal ions from aqueous solutions. *Chem Commun* 2001;**1**:135–6.
72. Visser AE, Swatloski RP, Reichert WM, Griffin ST, Rogers RD. Traditional extractants in nontraditional solvents: groups 1 and 2 extraction by crown ethers in room-temperature ionic liquids. *Ind Eng Chem Res* 2001;**39**:4596–604.
73. Visser AE, Swatloski RP, Griffin ST, Hartman DH, Rogers RD. Liquid/liquid extraction of metal ions in room temperature ionic liquids. *Sep Sci Technol* 2001;**36**:785–804.
74. Visser AE, Swatloski RP, Reichert WM, Mayton R, Sheff S, Wierzbicki A, Davies JH, Rogers RD. Task-specific ionic liquids incorporating novel cations for the coordination and extraction of Hg^{2+} and Cd^{2+}: synthesis, characterization, and extraction studies. *Environ Sci Technol* 2002;**36**:2523–9.
75. Liu Q, Janssen MHA, van Rantwijk F, Sheldon RA. Room-temperature ionic liquids that dissolve carbohydrates in high concentrations. *Green Chem* 2005;**7**:39–42.
76. Vidal L, Psillakis E, Domini CE, Grane N, Marken F, Canals A. An ionic liquid as a solvent for headspace single drop microextraction of chlorobenzenes from water samples. *Anal Chim Acta* 2007;**584**:189–95.
77. Fadeev AG, Meagher MM. Opportunities for ionic liquids in recovery of biofuels. *Chem Commun* 2001;**1**:295–6.
78. Cull SG, Holbery JD, Vargas-Mora V, Seddon KR, Lye GJ. Room-temperature ionic liquids as replacements for organic solvents in multiphase bioprocess operations. *Biotechnol Bioeng* 2000;**69**:227–33.
79. Banerjee T, Verma KK, Khanna A. Liquid–liquid equilibrium for ionic liquid systems using COSMO-RS: effect of cation and anion dissociation. *AIChE J* 2008;**54**:1874–85.
80. Huddleston JG, Willauer HD, Swatloski RP, Visser AE, Rogers RD. Room temperature ionic liquids as novel media for 'clean' liquid-liquid extraction. *Chem Commun* 1998;**1**:1765–6.
81. Marták J, Schlosser Š. New formulations of liquid membranes and solvents based on ionic liquids. In: *Proceedings of conference PERMEA 2003*, 2003, p. 1. Full text on CD ROM, Tatranské Matliare (SK), September 7–11.

82. Marták J, Schlosser Š. Screening of ionic liquids for application in solvent extraction and pertraction. In: *Proceedings of 31st conference of SSCHI*, 2004, p. 12. CD ROM with full papers, Tatranské Matliare (SK), May 25–28.
83. Marták J, Schlosser Š. Ionic liquids in pertraction and extraction of organic acids. In: *Proceedings of 19th Symposium Ars Separatoria*, Zloty Potok (PL), 2004, p. 106 (http://www.ars separatoria.chem.uni.torun.pl/MATERIALY4.htm).
84. Gu Y, Shi F, Yang H, Deng Y. Leaching separation of taurine and sodium sulfate solid mixture using ionic liquids. *Separat Purif Technol* 2004;**35**:153–9.
85. Fortunato R, Branco L, Afonso CAM, Reis MA, Crespo JG. Supported liquid membranes using ionic liquids: study of transport mechanism and stability. *Membr News (EMS)* 2002;**5**:35–8.
86. Branco LC, Crespo JG, Afonso CAM. Highly selective transport of organic compounds by using supported liquid membranes based on ionic liquids. *Angew Chem Int* 2002;**41**:2771.
87. Miyako E, Maruyama T, Kamiya N, Goto M. Enzyme-facilitated enantioselective transport of (S)-ibuprofen through a supported liquid membrane based on ionic liquids. *Chem Commun* 2003:2926–7.
88. Miyako E, Maruyama T, Kamiya N, Goto M. Use of ionic liquids in a lipase-facilitated supported liquid membrane. *Biotechnol Lett* 2003;**25**:805.
89. Wood N, Stephens G. Accelerating the discovery of biocompatible ionic liquids. *Phys Chem Chem Phys* 2010;**12**:1670–4.
90. Bekou E, Dionysiou D, Qian R-Y, Botsaris GD. Extraction of chlorophenols from water using room temperature ionic liquids. In: Rogers RD, Seddon KR, editors. *Ionic liquids as green solvents. Progress and prospects*. Washington DC: ACS; 2003. p. 544.
91. Liu JF, Jonsson JA, Jiang G. Application of ionic liquids in analytical chemistry. *Trends Anal Chem* 2005;**24**:20–7.
92. Liu JF, Peng JF, Chi YG, Jiang GB. Determination of formaldehyde in shiitake mushroom by ionic liquid-based liquid-phase microextraction coupled with liquid chromatography. *Talanta* 2005;**65**:705–9.
93. Peng JF, Liu JF, Jiang GB, Tai C, Huang MJ. Ionic liquid for high temperature headspace liquid-phase microextraction of chlorinated anilines in environmental water samples. *J Chromatogr A* 2005;**1072**:3–6.
94. Ye CL, Zhou QX, Wang XM. Headspace liquid-phase microextraction using ionic liquid as extractant for the preconcentration of dichlorodiphenyltrichloroethane and its metabolites at trace levels in water samples. *Anal Chim Acta* 2006;**572**:165–71.
95. Liu JF, Jonsson JA, Jiang GB. Application of ionic liquids in analytical chemistry. *Trend Anal Chem* 2005;**24**:20–7.
96. Ye CL, Zhou OX, Wang XM. Determination of phenols in environmental water samples by ionic liquid-based headspace liquid-phase microextraction coupled with high-performance liquid chromatography. *J Sep Sci* 2007;**30**:42–7.
97. Jiang TF, Gu YL, Liang B, Li JB, Shi YP, Ou QY. Dynamically coating the capillary with 1-alkyl-3-methylimidazolium-based ionic liquids for separation of basic proteins by capillary electrophoresis. *Anal Chim Acta* 2003;**479**:249–54.
98. Breitbach ZS, Armstrong DW. Characterization of phosphonium ionic liquids through a linear solvation energy relationship and their use as GLC stationary. *Anal Bioanal Chem* 2008;**390**:1605–17.
99. Marszałł MP, Kaliszan R. Application of ionic liquids in liquid chromatography. *Crit Rev Anal Chem* 2007;**37**:127–40.

100. Michaels AS. Membranes, membrane processes, and their applications: needs, unsolved problems, and challenges of the 1990's. *Desalination* 1990;**77**:5–34.
101. Chou KH, Bell LN. Caffeine content of prepackaged national-brand and private-label carbonated beverages. *J Food Sci* 2007;**72**(6):C337–42.
102. Seike S, Oishi K. Method of removing carbon dioxide using polymer electrolyte membrane, and apparatus and system therefore. Japan Patent JP 2003306788 A, 2003, 10/31.
103. Li Guan-Shan. Method for recycling wastewater from paper industry. Faming Zhuanli Shenqing Gongkai Shuomingshu, 2008, p. 9.
104. Drioli E, Criscuoli A, Curcio E. Integrated membrane operations for seawater desalination. *Desalination* 2002;**147**:77.
105. Bartsch RA, Way JD, editors. *Chemical separations with liquid membranes*. Washington DC: American Chemical Society; 1996. ACS Symposium Series 642.
106. Kislik VS, editor. *Liquid membranes. Principles and applications in chemical separations and wastewater treatment*. Amsterdam: Elsevier; 2010.
107. Lambropoulou DA, Albanis TA. *J Biochem Biophys Methods* 2007;**70**:195.
108. Rasmussen KE, Pedersen-Bjergaard S. *Trends Anal Chem* 2004;**23**:1.
109. Tani H, Kamidate T, Watanabe H. Micelle-mediated extraction. *J Chromatogr A* 1997;**780**:229–41.
110. Carabias-Martínez R, Rodríguez-Gonzalo E, Moreno-Cordero B, Perez-Pavon JL, García-Pinto C, Fernandez Laespada E. Surfactant cloud point extraction and preconcentration of organic compounds prior to chromatography and capillary electrophoresis. *J Chromatogr A* 2000;**902**:251–65.
111. Watanabe H, Tanaka H. Nonionic surfactant as a new solvent for liquid-liquid extraction of zinc(II) with 1-(2-pyridylazo)-2-naphtol. *Talanta* 1978;**25**:585–9.
112. Ballesteros-Gymez A, Ruiz FJ, Rubio S, Perez-Bendito D. *Anal Chim Acta* 2007;**603**:51–9.
113. Bordier C. Phase-separation of integral membrane-proteins in triton X-114 solution. *J Biol Chem* 1981;**256**:1604–7.
114. Boeckelen A, Niessner R. Combination of micellar extraction of polycyclic aromatic hydrocarbons from aqueous media with detection by synchronous fluorescence. *Fresenius J Anal Chem* 1993;**346**:435–40.
115. Casero I, Sicilia D, Rubio S, Perez-Bendito D. An acid-induced phase cloud point separation approach using anionic surfactants for the extraction and preconcentration of organic compounds. *Anal Chem* 1999;**71**:4519–26.
116. Eiguren-Fernandez A, Sosa-Ferrera Z, Santana-Rodriguez JJ. Cloudpoint extraction and preconcentration of polychlorinated biphenyls and their determination using high performance liquid chromatography. *Quim Anal* 1997;**16**:283–9.
117. Eiguren-Fernandez A, Sosa-Ferrera Z, Santana-Rodriguez JJ. Application of cloud-point methodology to the determination of polychlorinated dibenzofurans in sea water by high-performance liquid chromatography. *Analyst* 1999;**124**:487–91.
118. Garcia-Pinto C, Perez-Pavon JL, Moreno-Cordero B. Cloud point preconcentration and high performance liquid-chromtaographic determination of organophosphorous pesticides with dual electrochemical detection. *Anal Chem* 1995;**67**:2606–12.
119. Frankewich RP, Hinze WL. Evaluation and optimization of the factors affecting nonionic surfactant-mediated phase separations. *Anal Chem* 1994;**66**:944–54.
120. Bauer D, Cote G, Komornicki J, Mallet-Faux S. *Can Soc Chem Eng* 1989;**2**:425.
121. Brejza EV, Perez de Ortiz ES. Phenomena Affecting the Equilibrium of Al(III) and Zn(II) Extraction with Winsor II Microemulsions. *J Colloid Interface Sci* 2000;**227**:244–6.

122. Pepe EM, Otu EO. The synergistic Extraction of thorium(IV) and uranium(VI) by 2-ethylhexyl phenyl phosphonic acid and Micells of Dinonyl naphthalene sulphonic acid. *Solv Extr Ion Exch* 1996;**14**(2):247.
123. Leodidis EB, Hatton TA. Specific ion effects in electrical double-layers – Selective solubilization of cations in aerosol-OT reversed micelles. In: Pileni MP, editor. *The structure and reactivity in reversed micelles*. Amsterdam: Elsevier; 1989. p. 270.
124. Marcozzi G, Correa N, Luisi PL, Caselli M. Protein extraction by reverse micelles: a study of the factors affecting the forward and backward transfer of a-chymotrypsin and its activity. *Biotechnol Bioeng* 1991;**38**:1239.
125. Woll JM, Hatton TA. A simple phenomenological thermodynamic model for protein partitioning in reversed micellar systems. *Bioprocess Eng* 1989;**4**:193.
126. Jarudilokkul S, Poppenborg LH, Stuckey DC. Selective reverse micellar extraction of three proteins from filtered fermentation broth using response surface methodology separation science and technology. *Sep Sci Tech* 2000;**35**:503–17.
127. Luisi PL, Bonner FJ, Pellergrini A, Wiget P, Wolf R. Micellar solubilization of proteins in aprotic solvents and their spectroscopic characterization. *Helv Chim Acta* 1979;**62**:740.
128. Jarudilokkul S, Poppenborg LH, Stuckey DC. Backward extraction of reverse micellar encapsulated proteins using a counterionic surfactant. *Biotech Bioeng* 1999;**62**(5):593–601.
129. Goklen KE, Hatton TA. Liquid-liquid extraction of low molecular weight. In: *Proceedings of ISEC '86*, Munchen, 1986;**3**:587.
130. Rahaman RS, Lee JY, Cabral JMS, Hatton TA. Recovery of an extracellular alkaline protease from whole fermentation broth using reversed micelles. *Biotechnol Prog* 1988;**4**(4):218.
131. Krei GA, Hustedt H. Extraction of enzymes by reversed micelles. *Chem Eng Sci* 1992;**47**(1):99.
132. Jarudilokkul S, Paulsen E, Stuckey DC. The effect of demulsifiers on lysozyme extraction from hen egg white using reverse micelles. *Biosep* 2000;**9**:81–91.
133. Laane C, Dekker M. The effect of demulsifiers on lysozyme extraction from hen egg white using reverse micelles. In: *Proceedings of 6th International Symposium on Surfactants in Solution*. Mittal EL, editor. vol. 9, New York: Plenum Press; 1989, p. 9.
134. Giovenco S, Verheggen F, Laane C. Purification of intracellular enzymes from whole bacterial cells using reversed micelles. *Enzyme Microb Technol* 1987;**9**:470.
135. Jarudilokkul S, Poppenborg LH, Valetti F, Gilardi G, Stuckey DC. Separation and purification of periplasmic cytochrome *c*553 using reversed micelles. *Biotech Techniques* 1999;**13**:159–63.
136. Lye GY. Ph D Thesis, Reverse micellar mass-transfer processes: spray column extraction of lysozyme. University of Reading, 1993.
137. Jarudilokkul S, Paulsen E, Stuckey DC. Lysozyme extraction from egg white using reverse micelles in a graesser contactor: mass transfer characterization. *Biotech Bioeng* 2000;**69**:618–26.
138. Jarudilokkul S, Paulsen E, Stuckey DC. Extraction of monoclonal antibodies (IgG1) using anionic and anionic/nonionic reverse micelles. *Biotech Progress* 2000;**16**:1071–8.
139. Jarudilokkul S, Stuckey DC. Protein extraction from fermentation broths using reverse micelles: selectivity, activity recovery and contactor design. *Sep Sci Technol* 2001;**36**: 657–70.
140. Berthod A, García-Álvarez-Coque MC. Micellar liquid chromatography. In: Cazes J, editor. *Chromatographic science series*, vol. 83. New York: Marcel Dekker; 2000, pp. 276–83.
141. Ruiz-Ángel MJ, Carda-Broch S, Torres-Lapasió JR, García-Álvarez-Coque MC. Retention mechanisms in micellar liquid chromatography. *J Chromatogr A* 2009;**1216**:1798–814.

142. Khaledi MG. Micelles as separation media in high-performance liquid chromatography and high-performance capillary electrophoresis: overview and perspective. *J Chromatogr A* 1997;**780**:3–40.
143. Berthod A. Causes and remediation of reduced efficiency in micellar liquid chromatography. *J Chromatogr A* 1997;**780**:191–206.
144. Dorsey JG, DeEchegaray MT, Landy JS. Efficiency enhancement in micellar liquid chromatography. *Anal Chem* 1983;**55**:924–8.
145. Ruiz-Ángel MJ, García-Álvarez-Coque MC, Berthod A. New insights and recent developments in micellar liquid chromatography. *Sep Purif Rev* 2009;**38**:45–96.
146. Ruiz-Ángel MJ, Caballero RD, Simo-Alfonso EF, García-Álvarez-Coque MC. Micellar liquid chromatography: suitable technique for screening analysis. *J Chromatogr A* 2002;**947**:31–45.
147. Gil-Agusti M, Capella-Peiro ME, Monferrer-Pons LL, García-Álvarez-Coque MC, Esteve-Romero J. Chromatographic analysis of phenethylamine-antihistamine combinations using C-8, C-18 or cyano columns and micellar sodium dodecyl sulfate-pentanol mixtures. *Analyst* 2001;**126**:457–64.
148. Carda Broch S, Esteve Romero JS, García-Álvarez-Coque MC. Chromatographic determination of diuretics in urine samples using hybrid micellar mobile phases with fluorimetric detection. *Anal Chim Acta* 1998;**375**:143–54.
149. Gil-Agusti M, Carda-Broch SC, García-Álvarez-Coque M, Esteve Romero J. Use of micellar mobile phases for the chromatographic determination of clorazepate, diazepam, and diltiazem in pharmaceuticals. *J Chromatogr Sci* 2000;**38**:521–7.
150. Noguera-Orti JF, Villanueva-Camañas RM, Ramis-Ramos G. Determination of phenolic antioxidants in vegetal and animal fats without previous extraction by dilution with n-propanol and micellar liquid chromatography. *Anal Chim Acta* 1999;**402**:81–6.
151. Gil-Agusti M, Esteve-Romero J, Carda-Broch S. Anserine and carnosine determination in meat samples by pure micellar liquid chromatography. *J Chromatogr A* 2008;**1189**:444–50.
152. Raviolo MA, Rambla-Alegre M, Clausell-Tormos J, Capella-Peiro ME, Carda-Broch S, Esteve-Romero J. Determination of sulfonamides in milk after precolumn derivatisation by micellar liquid chromatography. *Anal Chim Acta* 2007;**593**:152–6.
153. Gil-Agustí M, Carda-Broch S, Monferrer-Pons L, Esteve-Romero J. Simultaneous determination of tyramine and tryptamine and their precursor amino acids by micellar liquid chromatography and pulsed amperometric detection in wines. *J Chromatogr A* 2007;**1156**:288–95.
154. Marczak M, Wolska L, Chrzanowski W, Namieśnik J. Microanalysis of volatile organic compounds (VOCs) in water samples – methods and instruments. *Microchim Acta* 2006;**155**:331.
155. Escher BI, Quayle P, Muller R, Schreiber U, Mueller JF. Passive sampling of herbicides combined with effect analysis in algae using a novel high-throughput phytotoxicity assay (Maxi-Imaging-PAM). *J Environ Monit* 2006;**8**:456.
156. Pavón JLP, Martín SH, Pinto CG, Cordero BM. Determining the Henry's Law constants of THMs in seawater by means of purge-and-trap gas chromatography (PT-GC): the influence of seawater as sample matrix. *Anal Chim Acta* 2008;**629**:6.
157. De Almeida Bezerra M, Arruda MAZ, Ferreira SLC. Cloud point extraction as a procedure of separation and pre-concentration for metal determination using spectro analytical techniques: a review. *Appl Spectrosc Rev* 2005;**40**:269.
158. Yoshimura K, Waki H, Ohashi S. Enhancement of sensitivity of ion-exchanger absorptiometry by using a thick ion-exchanger layer. *Talanta* 1986;**23**:449.

159. Lord H, Pawliszyn J. Evolution of solid-phase microextraction technology. *J Chromatogr A* 2000;**885**:153.
160. Blasco C, Fernández M, Picó Y, Font G. Solid-phase microextraction-liquid chromatography-mass spectrometry applied to the analysis of insecticides in honey. *J Chromatogr A* 2004;**1030**:77.
161. Zygmunt B, Jastrzebska A, Namieśnik J. Solid-phase microextraction a convenient tool for the determination of organic pollutants in environmental matrices. *Crit Rev Anal Chem* 2001;**31**:1.
162. Poppa P, Bauer C, Wennrich L. Determination of human serum albumin by chem-iluminescence immunoassay with luminol using a platinum-immobilized flow-cell. *Anal Chim Acta* 2001;**436**:1.
163. Hayes BL. Recent advances in microwave-assisted synthesis. *Aldrichim Acta* 2004;**37**:66.
164. Gallagher PA, Schegel CA, Parks A, Gamble BM, Wymer L, Creed JT. Preservation of As(III) and As(V) in drinking water supply samples from across the United States using EDTA and acetic acid as a means of minimizing iron-arsenic coprecipitation. *Environ Sci Technol* 2004;**38**:2919.
165. Ashley K, Andrews RN, Cavazos L, Demange M. Ultrasonic extraction as a sample preparation technique for elemental analysis by atomic spectrometry. *J Anal At Spectrom* 2001;**16**:1147.
166. Cintas P, Luche JL. Green chemistry: the sonochemical approach. *Green Chem* 1999;**4**:115.
167. Piunno PAE, Krull UJ. Trends in the development of nucleic acid biosensors for medical diagnostics. *Trends Anal Chem* 2009;**28**:946.

Appendix

1. FUNDAMENTAL CONSTANTS

e	electron (proton) charge, 1.6022 10^{-19} coulomb (C)
F	Faraday constant, 96,485 C mol^{-1}
NA	Avogadro number, 6.022 10^{23} mol^{-1}
kB	Boltzmann constant, 1.3807 10^{-23} joule/kelvin (JK^{-1})
R	molar gas constant, 8.314 $JK^{-1}mol^{-1}$ (0.08206 L atm $mol^{-1}K^{-1}$)
vM	molar volume of ideal gas, 0.02241 m^3 mol^{-1}

2. SI FUNDAMENTAL (A) AND DERIVED (B) UNITS USED IN THE BOOK

a)

Physical property	Unit symbol	Symbol name	Unit quantity
Length	*l*	meter	m
Mass	*m*	kilogram	kg
Time	*t*	second	s
Electric current	*I*	ampere	A
Temperature	*T*	Kelvin	K

b)

Physical property	Unit symbol	Unit name	Explanation
Physical mass (weight)	m(w)	gram, g;	mA is weight of pure substance A
Molar mass	M	atomic mass	molecular weigh units (amu)
Amount of substance	n	moles, mol;	nA = mAMA 1 mol = N_A molecules
Volume	V	liter, L, or dm^3	(SI-unit is m^3)
Molar volume	V/M	liter/mol or dm^3/mol	
Concentration	c or [ ]	molarity, M,	mol/L, or mol/dm^3
Density	ρ	g/cm^3 = 10^3 kg/m^3	(SI unit is kg/m^3)
Pressure	P	megapascal, MPa;	1 atm = 0.1013 MPa; 1 bar = 0.1 Mpa
Temperature	T	C°, degrees Celsius;	0 T°C = 273.15 K

3. ABBREVIATIONS FOR DIFFERENT PARAMETERS, CONSTANTS, COEFFICIENTS, COMMONLY USED IN BOOK

E_{ex}	Extraction equilibria (including adduct formation)
R%	Recovery factor
Q_M	Distribution ratio of M between nonaqueous and aqueous phase
D_M	Distribution ratio of M between organic solvent and aqueous phase (e.g., D_{Cu} for distribution of element copper)
Ki	stepwise formation constant for i-th complex (e.g., K_{Li})
K_{ex}	overall extraction constant
K_{DHL}	distribution constant of acid as a solute
K_a	acid dissociation (concentration) constant; $pK_a = -\log K_a$
K_{al}	first acid dissociation constant for multibasic acid
K_{adl}	first adduct formation constant (organic phase)
K_{Dx}	distribution constant for x (specified); for example, x = R, or = HA, for undissociated extractant (reagent), = C for neutral (e.g., metal-containing) complex
K_{sp}	solubility product constant
β	stability (formation) constant
β_n	overall complex formation, or stability, constant for formation of n-th complex
P	thermodynamic distribution constant
s_{AC}	solubility of AC in moles or grams per liter (mL^{-1} or gL^{-1})
K_{ass}	association constant for ion pair formation
K_{BS}	Bunsen coefficient for solubility of gas in a liquid
CMC	Critical micelle concentration
PF	Preconcentration factor
SC	Synergistic coefficient
b	solvation number
n	coordination number
NTU	number of transfer units
HTU	Height of the extraction column
HETS	height equivalent to a theoretical stage
MTF	Murphree mass transfer factor
CAC	Critical aggregation concentration
MIC	Minimum inhibitory concentrations
MBC	Minimum biocidal concentrations
EC_{50}	Medium effective concentration
O/W	Oil in water
W/O	Water in oil

4. ABBREVIATIONS FOR PROCESSES AND TECHNIQUES, DESCRIBED IN THE BOOK

Amex process	Recovery of uranium by amines
ASE	Accelerated solvent extraction
BLM	Bulk liquid membrane
BOHLM	Bulk organic hybrid liquid membrane
CE	Capillary electrophoresis
CPE	Cloud point extraction

CZE	Capillary zone electrophoresis
CURE	Clean use of reactor energy
Dapex process	Recovery of uranium by DEHPA
DIAMEX process	Diamide Extraction process
ELM	Emulsion liquid membrane
ISE	Ion-selective electrodes
ISFET	Ion-selective field effect transistors
IT	Information theory
HF	Hollow fiber
HS-SPME	Headspace solid-phase microextraction
LPME	Liquid-phase microextraction
LLPC	Liquid-liquid partition chromatography
MAE	Microwave-assisted extraction
MBSE	Membrane-based solvent extraction
MBSS	Membrane-based solvent stripping
MD	Molecular dynamics
ME	Micelle-mediated extraction
MIMS	Membrane introduction mass spectrometry
MHS	Multimembrane hybrid system (LM between two ion-exchange membranes)
MM	Molecular mechanics
MLC	Micellar liquid chromatography
MSPD	Matrix solid-phase dispersion
MZP	Zincex process
NN	Theory of neural networks
QM	Chemical quantum mechanics
QNN	Quantum neural networks
PHWE	Pressurized hot water extraction
PIM	Polymer inclusion membranes
PLE	Pressurized liquid extraction
PT	Pertraction
PUREX process	Plutonium–Uranium Extraction process (for spent nuclear fuel)
RPC	Reversed-phase chromatography
RPLC	Reversed-phase liquid chromatography
SANEX	trivalent actinide–lanthanide separation processes
SBE	Stir bar extraction
SBSE	Stir bar sorptive extraction
SDME	Single-drop microextraction
SFE	Supercritical fluid extraction
SFC	Supercritical fluid chromatography
SHWC	Superheated water chromatography
SLM	Supported liquid membrane
SPE	Solid-phase extraction
SPIN	solvent partition–incineration program
SPME	Solid-phase microextraction
SREX	Strontium recovery process
SWE	Subcritical water extraction
SWC	Subcritical water chromatography
Thorex process	Solvent extraction reprocessing of thorium fuels
TRUEX process	Separation and recovery of actinides (Am, Pu, Np from liquid nuclear wastes)
UAE	Ultrasound-assisted extraction

5. ABBREVIATIONS FOR SOLVENT EXTRACTION EQUIPMENT

CF HF	Cross-flow hollow fiber contactor
EC	Equilibrium cell for contacting two liquids
FSC	Flat sheet contactor
HF	Hollow fiber
IAC	Immunoaffinity columns
PF HF	Parallel flow hollow fiber contactor
RDC	Extractor
RFC	Rotating film contactor with BLM
TLHS	thin-layer headspace

6. DETECTION TECHNIQUES IN ANALYTICAL CHEMISTRY, CONNECTED WITH SOLVENT EXTRACTION PRETREATMENT

AAS	Atomic absorption spectrophotometry
AMS	accelerator mass spectrometry
CCFA	Completely continuous flow analysis
CPE	Cloud point extraction
CZE	Capillary zone electrophoresis
ESI-MS	electrospray ionization mass spectroscopy
EXAFS	Extended X-ray absorption fine-structure spectroscopy
FAAS	Furnace atomic absorption spectrophotometry
FIA	Flow injection analysis
GC	Gas chromatography
GC-MS	Gas chromatography - mass spectrometry
GDMS	Glow discharge mass spectrometry
GF-AAS	Graphite furnace atomic absorption spectrophotometry
HPLC	High-performance liquid chromatography
ICP	Inductively coupled plasma spectrophotometry
ICP-AES	Inductively coupled plasma with atomic emission spectrophotometry
ICP-MS	Inductively coupled plasma mass spectrometry
ISE	Ion-selective electrode
ISFET	Ion-selective field effect transistors
LLPC	Liquid–liquid partition chromatography
LS	Light-scattering
LSC	Liquid scintillation counters
MIMS	Membrane introduction mass spectrometry
NAA	Neutron Activation Analysis
NMR	Neutron magnetic resonance
RIMS	resonance ionization mass spectrometry
SFA	Segmented flow analysis
SERS	Surface-enhanced Raman spectroscopy
SIMS	secondary ion mass spectrometry
SPF	solid-phase fluorimetry
SPS	Solid-phase spectroscopy
TIMS	Thermal ionization mass spectrometry
TRLIF	Time-resolved laser-induced fluorescence spectroscopy
UFA	Unsegmented flow analysis
ZP	Zeta-potential

7. ABBREVIATIONS FOR INORGANIC COMPOUNDS

HLW	high-level radioactive wastes
LLW	low level radioactive wastes
LLR	long-lived radioactive elements
REE	Rare earth elements
SLR	Short-lived radioactive isotopes
TRU	transuranic wastes
TPE	Transplutonium elements

8. ABBREVIATIONS FOR ORGANIC COMPOUNDS

AA	acetylacetone; 2,4-pentanedione
ABC	acid–base-coupled extractants
ABS	Aqueous biphasic systems
ABEC	aqueous biphasic extraction chromatography
Adogen 364	mixtures from trioctyl to tridecylamines
AG1X2	2[2-(3-5-dibromopyridyl) azo]-5-dimethylaminobenzoic acid in ion exchange resin
ALAMINE 336	Mixture of trioctyl- and tridecylamine
ALIQUAT 336	Methylated (quaternized) amine 336
AOT	Sodium di(2-ethylhexyl)sulfosuccinate (anionic surfactant)
6-APA	6-Aminopenicillanic acid
BA	benzoylacetone
BDPDA	Bis(2,2′-dipyridyl-6-yl)diazene
BODMAC	benzyl octadecyldimethyl ammonium chloride
BPHA	*N*-benzoyl phenyl hydroxylamine
CHN	cyclohexanone
ChCoDiC	Chlorinated cobalt dicarbollide
CLX 50	Acorga
CMPO	octyl(phenyl)-N,N-di-isobutyl carbamoyl methyl phosphine oxide
C5-BTBP	6,6′-Bis-(5,6-dipentyl-[1,2,4]triazin-3-yl)-[2,2′]bipyridinyl
CTAB	Cationic cetyltrimethylammonium bromide
CYANEX 272	Organophosphinic acid
CYANEX 301	bis-(2,4,4-trimethylpentyl)dithiophosphinic acid
Cyanex 302	monothiophosphinic acid
Cyanex 471	phosphine sulfides
CYANEX 471X	trialkylphosphine sulfide
Cyanex 923	mixture of four types of trioctyl phosphine oxides (TOPO)
DBP	dibutyl phosphate
DBBP	dibutyl phosphonate
DBPA	dibutylphosphoric acid
DCTA	1,2-diaminocyclohexane tetraacetic acid
DEHPA	Di(2-ethylhexyl)phosphoric acid
DMDOMA	Dimethyl dioctyl methylphosphinic acid
DMDOHEMA	*N,N*′-dimethyl-*N,N*′-dioctylhexylethoxymalonamide
DMDOSA	Dimethyl dioctyl sulfonic acid
DMSO	Dimethylsulfoxide
DNNSA	Dinonylnaphtalenesulfonic acid
DOBA	N,N′ dialkylamides
DOiBA	iso form of DOBA

DOTPA	di-*o*-tolyl phosphoric acid
DTPA	diethylenetriamine pentaacetic acid
EDTA	ethylene diaminetetraacetic acid
FAA	hexafluoroacetylacetone
H2MBP	monobutylphosphoric acid
Henkel	1:1 mixture of an aldoxime and ketoxime
HIBA	a-hydroxyisobutyric acid
HITP	isopropyltropolone
HTTA	thenoyltrifluoroacetone
IL	Ionic liquids
IPT	Isopropyltropolone
Kelex 100 (Sherex)	8-quinolinol
La 2	Amberlite
LIX 34 (Henkel)	8-alkylarylsulfonamidoquinoline
LIX 54	β-Diketone
LIX63	mixture of aliphatic a-hydroxyoximes
LIX64	o-hydroxybenzophenone oxime
LIX 84I	2′-Hydroxy-5′-nonylacetophenone oxime (Ketoxime)
MBP	butyl phosphate
MEHPA	2-ethylhexyl phosphate
MIBK	methyl-isobutyl ketone
MIPK	methyl-isopropyl ketone
MPCA	5-Methyl-2-pyrazinecarboxylic acid
Neutr.	MIBK, MIPK, CHN, TTAc, MBP, TBP, TOPO, Cyanex 923, etc.
NTA	nitrilotriaminoacetic acid
Oxine	8-hydroxyquinoline
PAA	Phenylacetic acid
PAH	Polynuclear aromatic hydrocarbon
PAN	1-(2-pyridylazo)-2-naphthol
PAR	4-(2-pyridylazo)resorcinol
PC88A	2-ethylhexylphosphonic acid 2-ethylhexyl ester
PCB	Polychlorinated biphenyl
PCDF	Polychlorinated dibenzofuran
PEHPA	pyroethylhexyl phosphate
PEG	Poyethylene glycole
P50	ICI Acorga
PPG	Polypropylene glycole
PVP	Polyvinyl pyrrolidone
PONPE	Polyoxyethylene nonyl phenyl ether
PTFE	Polytetrafluoroethene
PVC	Porous polyvinyl chloride
PT5050	tridecanol
R	organic radical in general
RNH^{+3}	organic primary ammonium ion
$R2NH^{+2}$	organic secondary ammonium ion
$R3NH^{+}$	organic tertiary ammonium ion
R3NHCl	tri-n-octylamine hydrochloride

$R4N^+$	organic quaternary ammonium ion
SDVB	Styrene divinylbenzene
SDS	sodium dodecyl sulfate
SDBSA	Sodium dodecylbenzene sulfonic acid
SDSA	Sodium dodecanesulfonic acid
SHA	salicyl hydroxamic acid
Shellsol H	Aliphatic hydrocarbon mixture
Shellsol A	Aromatic hydrocarbon mixture
SME 529	Shell (methyl chain)
SUPRAS	Supramolecular solvents
TAA	triamyl amine
TAMA	trialkyl methyl amine
TBA	tributyl amine
TBeA	tribenzyl amine
TBOA	Tributyl oksalic acid
TBMA	Tributyl maleic acid
TBSA	Tributyl sulfonic acid
TBP	Tributylphosphate
TiBP	iso form of TBP
TBPO	tributyl phosphine oxide
TDA	tridodecyl amine
TEA	triethyl amine
TEDAHP	triethyl-diamide heptyl 18C6, DC18C6, DB18C6-crown ethers
TrEDAHP	tetraethyl-diamideheptyl
Thioxine	8-quinolinethiol
THF	Tetrahydrofuran
THM	Trihalomethane
THPO	trihexyl phosphine oxide
TFA	Trifluoroacetic acid
TLA	trilaurylamine; C12
TLMA * HNO3	trilaurylmethylammonium nitrate
TOA	Trioctyl amine
TODGA	*N,N,N′,N′*-tetraoctyldiglycolamide
TOMAC	Trioctylmethylammonium chloride
TOPO	Trioctyl phosphine oxide
Triton X-100	surfactant with coproporphyrin III
TTA	thenoyltrifluoroacetone
TVEX	New family of resins
Versatic 10	Versatic acid
VOC	Volatile organic compounds
XAD2	8-(benzene-sulfonamido)quinoline in Amberlite
XAD4	perylene dibutyrate in Amberlite
18C6, DC18C6, DB18C6	crown ethers
DtBuCH18C6	4,4(5),di-(t-butyl cyclohexano)-18-crown-6
PCDF	Polychlorinated dibenzofurans
Triton X-114	Poly(oxyethylene glycol) monooctyl ether

9. STRUCTURES OF SOME EXTRACTANTS

Dinonylnaphtylsulfonic acid

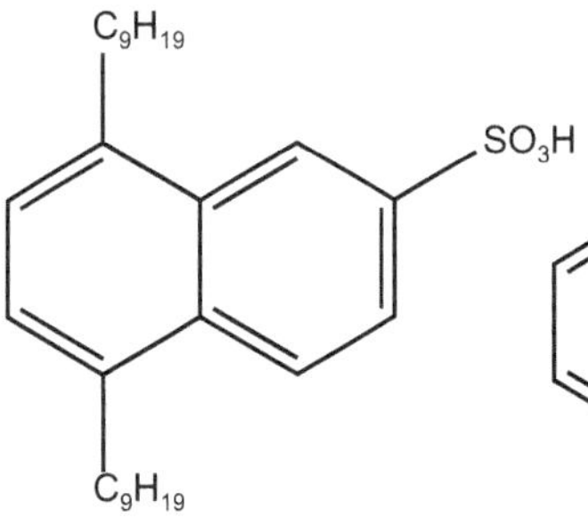

Benzoic acid

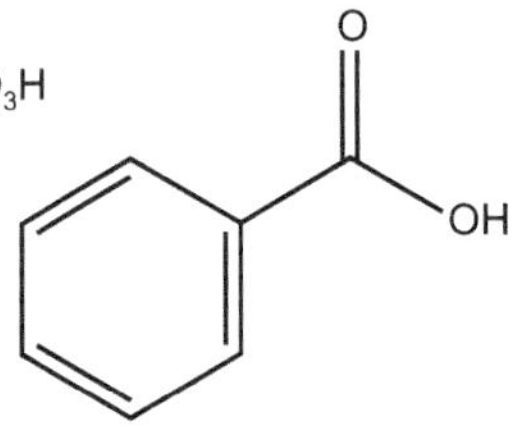

Salicylic acid

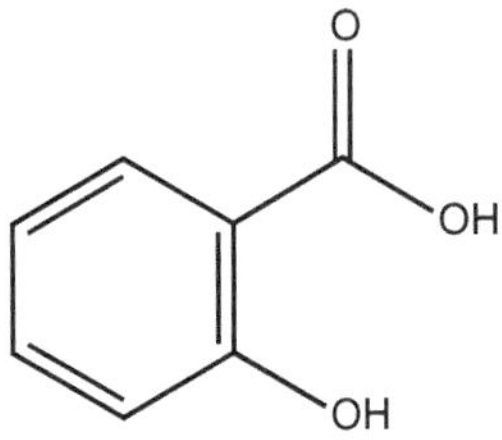

Isopropyltropolone

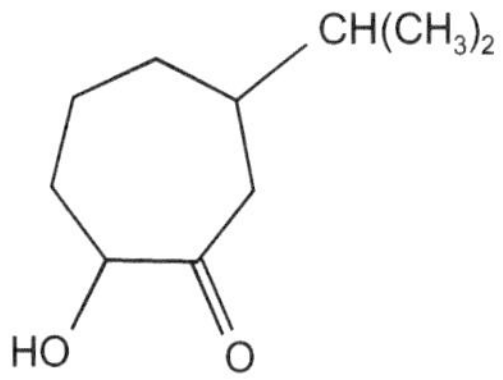

thenoyltrifluoroacetone, HTTA:

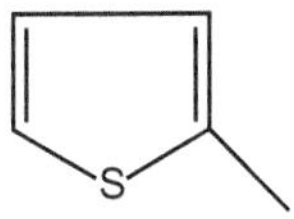

8-Hydroxoquinoline, oxine

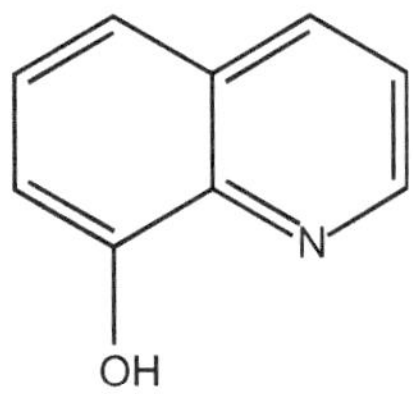

(a) b-Diketones: ketoform; **(b)** enol form;

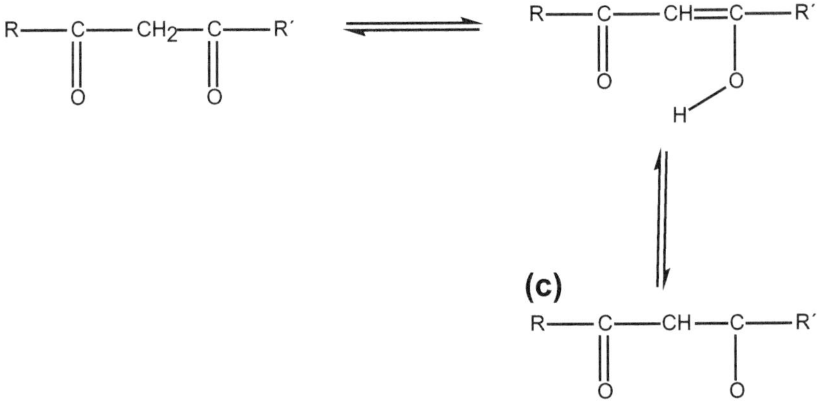

enolate ion

Oxime: R_1 and R_2 are longchained, branched aliphats or aromates

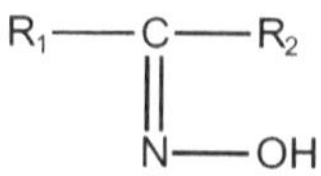

N-Phenylbenzo-hydroxamic acid

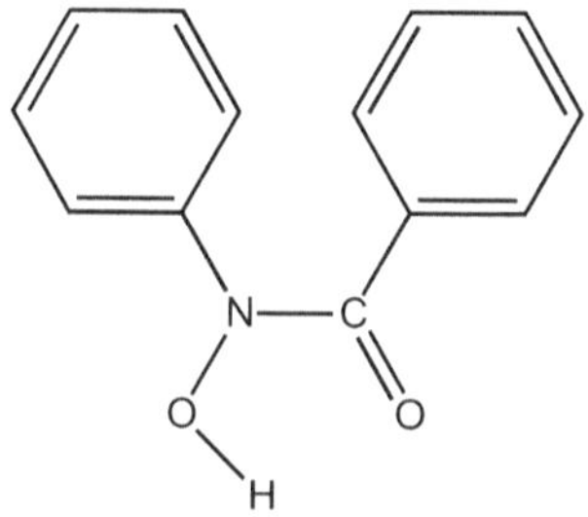

N-Nitrosophenylhydroxyl amine (ammonium salt is cupferron)

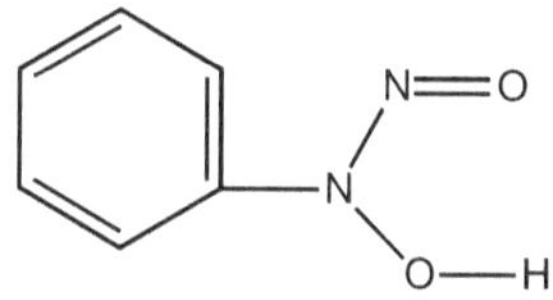

Carbamoylmethylene phosphonate

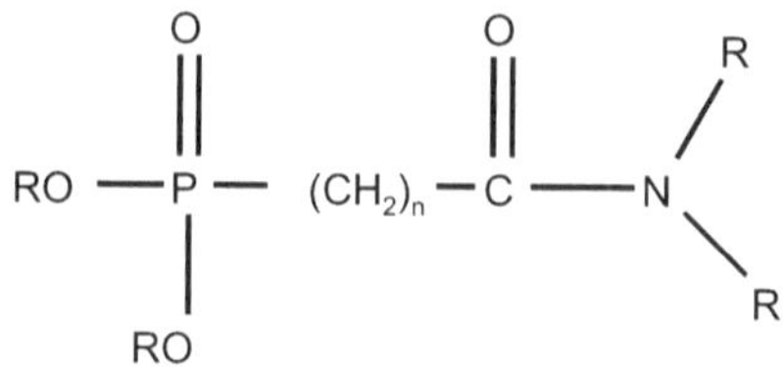

Dialkyl phosphonate

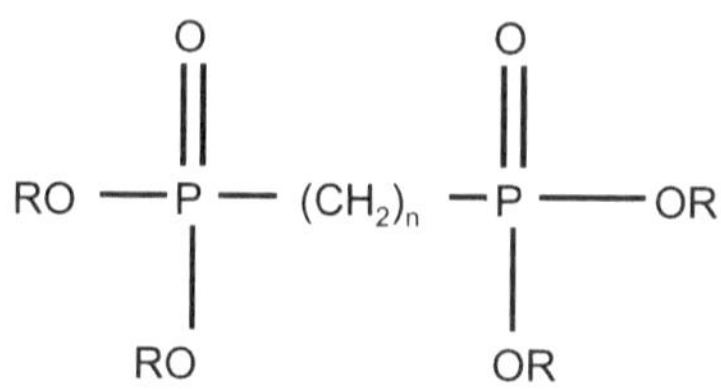

carbamoylphosphinate: R1 = R2 = alkyl, R = phosphinate

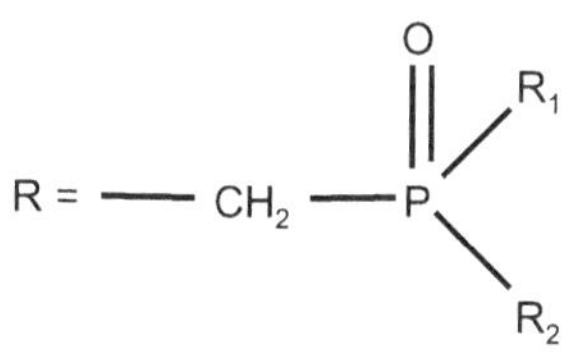

Amide R=R′=R″=alkyl

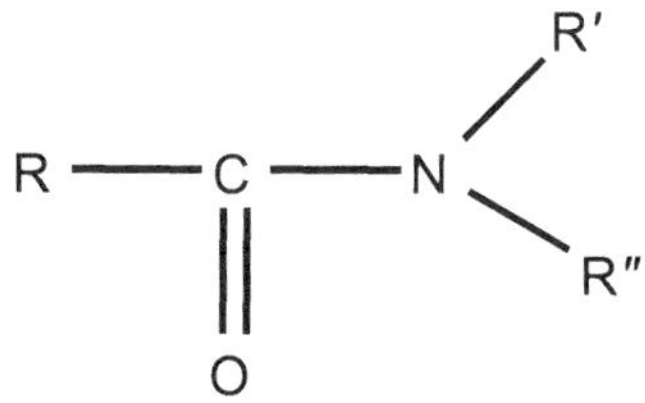

Di(2-ethylhexyl) phosphoric acid, HDEHP, or D2EHPA, or DEHPA

monomeric form;

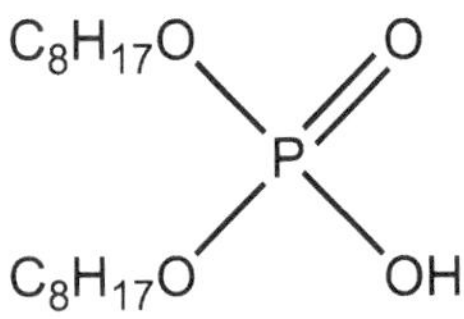

dimeric form;

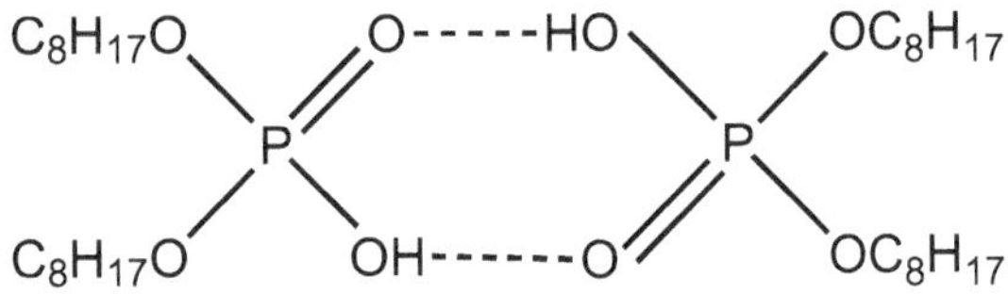

Trioctyl phosphine oxide, TOPO

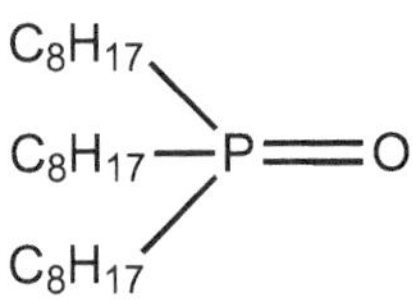

Dialkyl dithiophosphonic acid

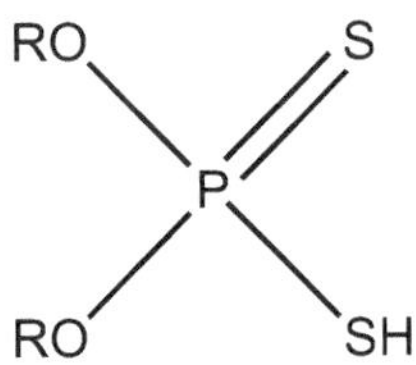

Tributylphosphate, TBP

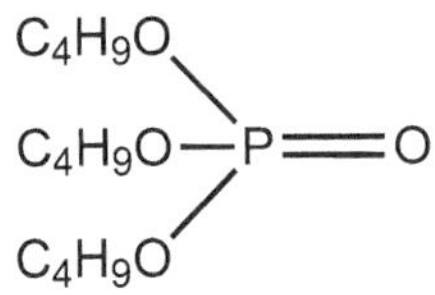

Methyl isobutyl ketone, MIBK, hexone

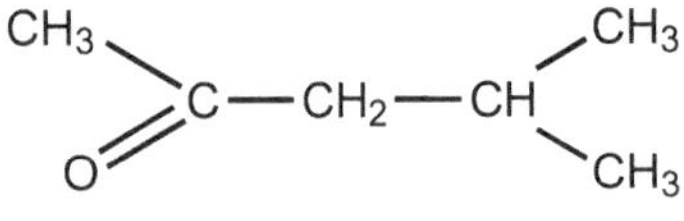

Diethyl dithiocarbamate

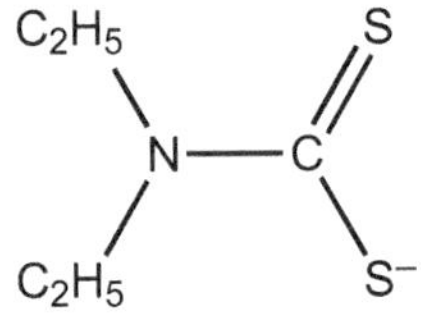

o-Phenanthroline

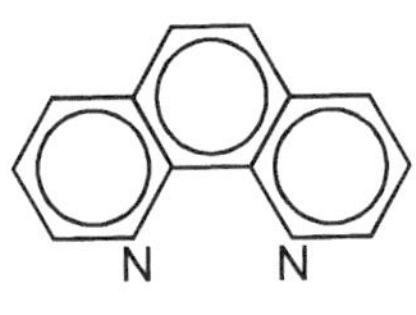

Crown ethers:

dicyclohexyl-18-crown-6, DC18-C6

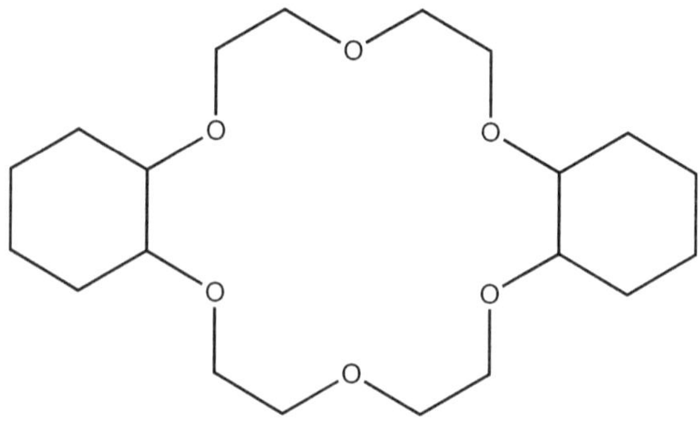

15-crown-5, 15-C5

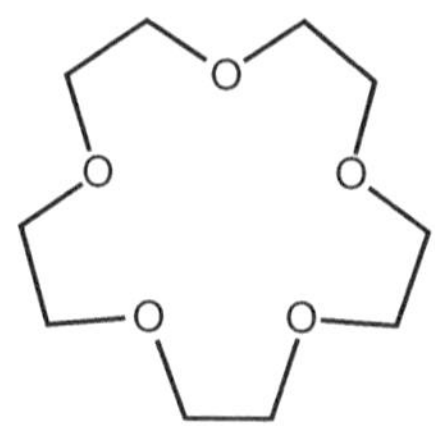

Bonding fixation places in $18C_6$ system

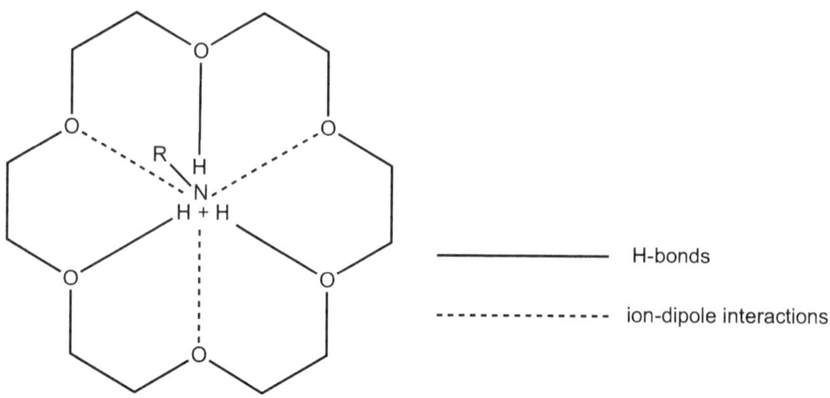

10. SOME EXAMPLES OF METAL COMPLEXES

HDEHP (DEHPA) – U(VI) complex:
$UO_2(DEHP)_2$*2HDEHP complex

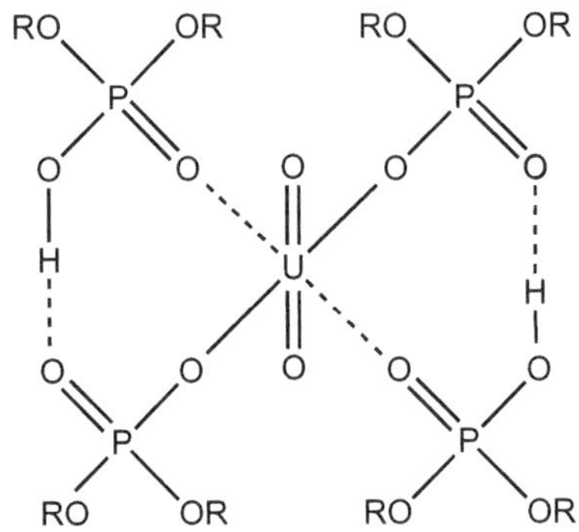

Bis-acetylacetonate copper(II) complex

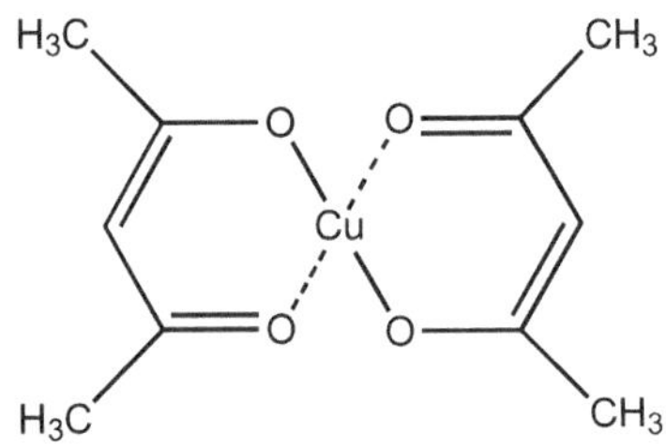

bis(5-methylsalicylaldoximato)copper(II) complex.

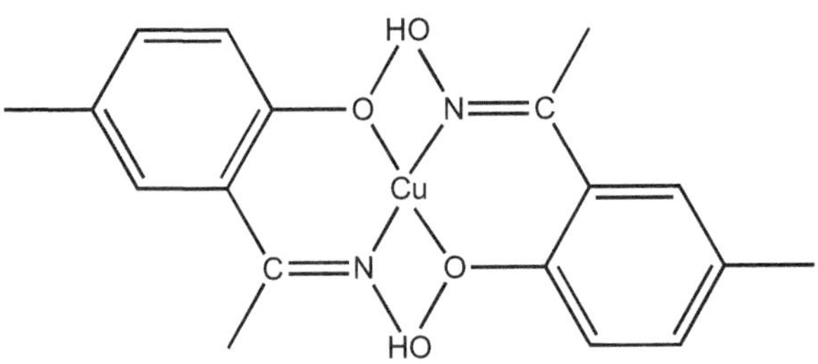

11. ABBREVIATIONS FOR ENVIRONMENTAL PROTECTION ORGANISATIONS AND PROGRAM MENTIONED IN BOOK

US EPA	US Environmental Protection Agency
ASTM	American Society for Testing and Materials
EuC SEM	European Committee for Standardization, Environmental Management
EHS	Environmental, Health, and Safety
SWOT	Strengths, weaknesses, opportunities, and threats analysis

REFERENCED SOURCES FOR APPENDIX DATA

1. Stary J. *The Solvent Extraction of Metal Chelates*. New York: Pergamon Press; 1964.
2. Zolotov Yu A. *Extraction of Chelate Compounds*. London: Ann Arbor- Humphrey Science Publishers; 1970.
3. Marcus Y, Kertes AS. *Ion Exchange and Solvent Extraction of Metal Complexes*. London, New York: Wiley – Inerscience; 1969.
4. Hoegfeldt E. *Stability Constants of Metal-Ion Complexes. Part A: Inorganic ligands; IUPAC Chemical Data Series No 22*. New York: Pergamon Press; 1982.
5. Freiser H, Nancollas GH. *Compendium of Analytical Nomenclature. Definitive Rules 1987; IUPAC*. Oxford: Blackwell Scientific Publications; 1987 (www.IUPAC.com/ publications).
6. IUPAC, *Quantities Units and Symbols in Physical Chemistry*. 2nd ed. Blackwell Science; 1993.
7. McNaught AD, Wilkinson A. *IUPAC Compendium of Chemical Terminology*. 2nd ed. Blackwell Science; 1997.

Index

A

B

L

M

N

O

P

Q

R

FSC
www.fsc.org
MIX
Papier | Fördert gute Waldnutzung
FSC® C083411

Kolibri 360 GmbH
Ferdinand-Jühlke-Straße 7
99095 Erfurt, Deutschland
produktsicherheit@kolibri360.de